DESIGN OF
EQUILIBRIUM
STAGE PROCESSES

BUILDING THE LITERATURE OF A PROFESSION

Fifteen prominent chemical engineers first met in New York more than 30 years ago to plan a continuing literature for their rapidly growing profession. From industry came such pioneer practitioners as Leo H. Baekeland, Arthur D. Little, Charles L. Reese, John V. N. Dorr, M. C. Whitaker, and R. S. McBride. From the universities came such eminent educators as William H. Walker, Alfred H. White, D. D. Jackson, J. H. James, Warren K. Lewis, and Harry A. Curtis. H. C. Parmelee, then editor of CHEMICAL & METALLURGICAL ENGINEERING, served as chairman and was joined subsequently by S. D. Kirkpatrick as consulting editor.

After several meetings, this first Editorial Advisory Board submitted its report to the McGraw-Hill Book Company in September, 1925. In it were detailed specifications for a correlated series of more than a dozen texts and reference books which have since become the McGraw-Hill Series in Chemical Engineering.

Since its origin the Editorial Advisory Board has been benefited by the guidance and continuing interest of such other distinguished chemical engineers as Manson Benedict, John R. Callaham, Arthur W. Hixson, H. Fraser Johnstone, Webster N. Jones, Paul D. V. Manning, Albert E. Marshall, Charles M. A. Stine, Edward R. Weidlein, and Walter G. Whitman. No small measure of credit is due not only to the pioneering members of the original board but also to those engineering educators and industrialists who have succeeded them in the task of building a permanent literature for the chemical engineering profession.

THE SERIES

BUFORD D. SMITH

Associate Professor of Chemical Engineering
Purdue University

DESIGN OF
EQUILIBRIUM
STAGE PROCESSES

CHAPTER 14
Tray Hydraulics: Bubble-cap Trays
by *William L. Bolles*, Monsanto Chemical Company

CHAPTER 15
Tray Hydraulics: Perforated Trays
by *James R. Fair*, Monsanto Chemical Company

McGRAW-HILL BOOK COMPANY, INC.
New York   San Francisco   Toronto   London

CHEMISTRY

DESIGN OF EQUILIBRIUM STAGE PROCESSES

PREFACE

This text was prepared primarily for the plant engineer. In the pursuit of his profession, he must deal with a great array of technical problems and is continually faced with the necessity to review the technical calculation methods available for the solution of the problem facing him at the moment. The review must be fast because the problems come fast. For review, the engineer must depend upon his own notes, upon calculation manuals prepared by his company, or upon a textbook in the field. It is hoped that this text will be a useful reference for those engineers who must deal with separation problems in which the equilibrium-stage concept can be utilized.

Although it is anticipated that the major use of this book will be by practicing engineers, a strong attempt was made to make it suitable also for classroom use. The notes from which the text was prepared have been used for some years in senior elective and graduate courses at Purdue University. The presentation of the material assumes the reader to have the equivalent of an undergraduate education in chemical engineering. No mathematics beyond the usual undergraduate courses is required. The calculus of finite differences is used in Chap. 8, and the application of matrices to the multicomponent separation problem is discussed briefly in Chap. 10, but previous background in these mathematical tools is not essential.

It is, unfortunately, impossible to satisfy both the practicing engineer and the engineering professor with one book. The practicing engineer will become impatient with a superfluity of background material, while the professor will be irritated by what he considers excessive detail in the applications. Whenever possible in the various chapters, an attempt was made to separate the "theoretical" from the "art and experience" material. Such a division is ridiculous from the over-all engineering viewpoint, but it does have the immediate advantage of concentrating the more quantitative material for use in the classroom, where the available time is always severely limited. Some of the material included is obviously not intended for classroom use. For example, most of the material in Chap. 4 makes very dull reading but the detailed examples should be useful to the engineer who makes flash calculations only infrequently. In fact, these examples were assembled in the first place several years ago because the author became tired of relearning the best calculation method every time such a problem arose. Many organiza-

vii

242

tions prepare "idiot charts" for such calculations, but the use of such aids is not very satisfying and can be dangerous if the user is not familiar with the calculation method upon which the charts are based. At least, the example problems illustrate clearly some practical applications of the three basic tools, namely, equilibrium relationships, material balances, and the enthalpy balance.

It would be difficult, if not impossible, to write a revolutionary book on a subject such as distillation which had its beginning in the Middle Ages. A. J. Liebman [*J. Chem. Educ.*, **33,** 166 (1956)] notes that distillation was so intimately associated with the birth of the science of chemistry that during the sixteenth century the word chemistry was taken to mean "the art of distilling." Man's progress in many of the sciences has been almost in lockstep with his progress in the art of separating materials into their pure forms. As a result the separation processes have been the subject of intensive investigation for centuries, and needless to say, a voluminous literature concerning them exists today. It is, therefore, difficult to say something about them which has not been said before, and such originality was not one of my primary aims. Nevertheless, the organization of the text does have one novel aspect. All the equilibrium-stage processes discussed—distillation, absorption, stripping, extraction, and washing—are treated as simple variations of the same basic process. Even more novel is the use of a common nomenclature. These innovations should materially reduce the time required for a student to become familiar with these chemical engineering operations. It is unreasonable to require a student to learn a new nomenclature and a new method of attack for each of the various processes.

The nomenclature used is essentially that recommended by Subcommittee 12 of the American Institute of Chemical Engineers in their report in *Chemical Engineering Progress,* June, 1956. The use of a common nomenclature for the various processes made it inevitable that many well-established conventions had to be discarded. For example, a great advantage in simplicity is gained in the multicomponent problems if the stages in all the processes are numbered from the same end. It was finally decided to number all the vapor-liquid processes from the bottom and the liquid-liquid processes from the extract end. In the majority of cases, the extract phase is the heavier phase and the extract end is the bottom end of the column. Numbering from the extract end conforms to the usual convention in extraction and stripping. A clear-cut convention does not exist in distillation, but it is believed that the majority of design engineers number from the bottom. The tradition in absorption has been to number from the top. The advantages of a common convention for all processes are particularly evident in Chap. 8. In the binary graphical solutions in Chaps. 5, 6, and 7 it is immaterial which way the stages are numbered. The most convenient way will vary with the problem specifications.

A considerable amount of space in the book is devoted to the derivation of equations. An equation is never written unless accompanied by a derivation or a description of how the derivation can be made. The reasons for this emphasis are twofold. First, if the book is to serve as a useful reference book for the practicing engineer, it must be easily understood by the reader without the expenditure of a large amount of time on his part in the development of background for the equations and methods. Second, inclusion of the derivations frees the teacher from the drudgery of presenting them in class and permits him to move more quickly to the application of the relationships.

Enough physical data are included in Appendix B to permit the solution of most example and homework problems. Occasionally, the student is required to consult some other source such as the steam tables or the compilation of thermodynamic properties by the American Petroleum Institute's Project 44. Inclusion of more data is impractical, since it would be impossible to include more than a small fraction of the total data available to the engineer in sources such as J. B. Maxwell's "Data Book on Hydrocarbons," the API Project 44 compilation, the "Engineering Data Book" compiled by the Natural Gasoline Supply Men's Association, and others.

Every author owes a debt of gratitude to associates, reviewers, and former teachers who contributed directly and indirectly to the preparation of a manuscript. Special thanks are extended to W. N. Lyster of the Humber Oil and Refining Company for some very direct aid with the material presented in Chap. 10. J. R. Friday, who was engaged in his Ph.D. research on the multicomponent multistage problem during the writing of the text, contributed significantly with his comments on several of the chapters and also furnished the improved method presented in Chap. 12 for the solution of the multicomponent extraction problem. I was particularly fortunate to obtain the contributions of William L. Bolles and James R. Fair of the Monsanto Chemical Company who wrote the material in Chaps. 14 and 15 on tray hydraulics. These chapters represent the latest word on the design of bubble-cap and perforated trays and therefore should be of great value to engineers who deal with these devices. Each of the reviewers left his mark on the book. Dr. Otto Redlich provided a very helpful review of Chaps. 1 and 2. Professor W. C. Edmister of Oklahoma State University commented on some of the early chapters of the book. Professor J. A. Gerster of the University of Delaware reviewed the entire book and provided many suggestions that helped strengthen the material. Professor Erwin Amick of Columbia University meticulously reviewed all but a few chapters and suggested several helpful changes.   Messrs. J. A. Davies and F. W. Winn provided a like service for Chaps. 14 and 15.

Perhaps the greatest aid in the preparation of a manuscript is a patient and helpful wife.   In this respect this author was indeed fortunate.   My wife's skillful work on the manuscript reduced its final preparation to almost a routine task and relieved me of a myriad of details.

BUFORD D. SMITH

CONTENTS

INTRODUCTION

The equilibrium-stage concept makes possible the adequate design of separation processes despite our inability to deal adequately with the complex heat- and mass-transfer operations that occur in an actual contact stage. A hypothetical process whose contact stages are all true equilibrium stages is created on paper to accomplish the separation desired in the actual plant process. The number of equilibrium stages required in the hypothetical process is related to the required number of actual contact stages by proportionality factors (stage efficiencies) which describe the extent to which the performance of an actual contact stage duplicates the performance of an equilibrium stage. Construction of the hypothetical equilibrium-stage process permits the design engineer to bypass temporarily the complex dynamics problem involved in the heat- and mass-transfer operations which occur in a contact stage. It is true that these transfer operations must be taken into account in the prediction of stage efficiencies, but it is much simpler to approximate a stage efficiency than solve the dynamic problem directly. From this viewpoint the equilibrium-stage concept represents a worthwhile simplification.

The concept of the equilibrium stage is based upon the presumption that the phases leaving the stage are in thermodynamic equilibrium. The condition of equilibrium between phases can be obtained if the streams entering the stage are mixed so thoroughly that the light phase contacts the heavy phase for a sufficient length of time to permit the completion of the required heat and mass transfer. Separation of the phases then produces exit streams which are in equilibrium. Equilibrium between two phases can be achieved by experimental devices and is approached closely by some plant devices. Since an equilibrium stage can be approximated in industrial equipment, it is natural to analyze such staged operations as distillation, absorption, and extraction in terms of ideal equilibrium stages.

Use of the equilibrium-stage concept separates the design of a staged process into three major parts. First, if equilibrium stages are to be assumed, a method must be available for the prediction of the equilibrium-phase compositions. Second, the number of equilibrium stages required to accomplish a specified separation (or the separation which will be accomplished in a given number of equilibrium stages) must be calculated.

1

Third, the number of equilibrium stages must be converted to an equivalent number of actual contact stages. Of these three parts, the second is by far the simplest. Yet it has never been solved in a completely satisfactory manner for the general multicomponent, multistage problem. The basic equilibrium, material-balance, and enthalpy-balance relationships are simple, but they can be combined in a seemingly endless variety of ways. Hardly a year passes without the publication of a new calculation procedure. Since the fascination of the problem is such that it regularly acquires a large number of new addicts each year, it is unlikely that the stream of publications will cease in the foreseeable future.

The first and third parts of the total design problem rival each other in complexity. The accurate prediction of equilibrium compositions in nonideal liquid mixtures requires an understanding far beyond that presently attained. Similarly, the precise relation of an equilibrium stage to an actual stage involves problems in momentum, heat, and mass transfer in a situation so complex that available methods of analysis are inadequate. In both instances empiricism and simplified mathematical models must be utilized to obtain practical answers.

This text attempts to deal with the three parts in the order in which they are listed. Chapters 1 and 2 discuss briefly the thermodynamics and correlation methods for ideal and nonideal vapor-liquid and liquid-liquid equilibrium mixtures. Chapters 3 through 13 present methods for the calculation of the number of stages required in the hypothetical equilibrium-stage process. Chapters 14, 15, and 16 deal with the methods whereby the hypothetical equilibrium stages are translated into actual plant devices. This order of presentation corresponds to the order in which the three parts must be handled in design.

CHAPTER 1

DISTRIBUTION COEFFICIENTS

The concept of the equilibrium stage is based upon the presumption that the phases leaving the stage are in thermodynamic equilibrium. The use of the concept in design calculations requires the prediction of the equilibrium relationships in terms of the stage temperature, pressure, and phase compositions. Therefore, any discussion of equilibrium-stage processes appropriately begins with the available methods for the prediction of equilibrium data.

The equilibrium relationship for any component in an equilibrium stage is conveniently defined in terms of the distribution coefficient K, where $K_i = y_i/x_i$. The significance of the distribution coefficient and its dependence upon the phase-rule variables of temperature, pressure, and phase concentrations were recognized long before the prediction of such coefficients became important in process design. The first attempts to predict K values for design purposes made use of Raoult's law. The vapors and liquids were assumed to be ideal, and the K values for vapor-liquid processes were obtained with

$$K_i = \frac{y_i}{x_i} = \frac{p_i^*}{p}$$

Such a simple relationship can apply only in special situations, whereas separation processes today operate over wide temperature and pressure ranges and separate components with widely different molecular structures. As the separation operations became more complex, it has been necessary to develop improved methods for the calculation of equilibrium data. Few subjects in chemical engineering have received attention equivalent to that devoted to the measurement, correlation, and prediction of K values. It will be the purpose of the first two chapters of this book to review briefly the major developments in this field. A complete review is beyond the scope of this book; indeed, an entire volume the size of this book would hardly suffice to contain an exhaustive review of the literature available on the subject. Enough material has been included to indicate to the student where he might most profitably begin his study and to indicate to the practicing engineer where he might find the most accurate K values for his purpose. If the reader has at hand a

3

satisfactory source of K values and is interested only in methods whereby they can be used to calculate a separation, he should omit these two chapters and pass on to later ones.

The calculation of K values depends upon the calculation of component fugacities in liquid and gas solutions. It is useful therefore to review the basic relationships between fugacity and pressure in gas and liquid solutions and show how these can be combined to provide the working equations for the calculation and correlation of K values in liquid-liquid and vapor-liquid systems. An orderly development of these relationships emphasizes the assumptions inherent in the major correlations used to predict K values in vapor-liquid systems.

1-1. Definition of Fugacity and Activity

The Gibbs free energy of a pure component is defined as

$$F = H - TS = E + pV - TS$$

or, in the differentiated form,

$$dF = dE + p\,dV + V\,dp - T\,dS - S\,dt$$

Combination of the first and second laws of thermodynamics provides for a reversible change:

$$dE = T\,dS - p\,dV$$

At constant temperature then,

$$dF_T = V\,dp$$

Hereafter, the subscript T on dF to denote the constant-temperature restriction will be omitted, but the reader must always keep this restriction in mind.

The use of the ideal gas relation

$$V^* = \frac{RT}{p}$$

to eliminate V provides the following relation between free energy and pressure for an ideal gas:

$$dF = RT\frac{dp}{p} = RT\,d\ln p \tag{1-1}$$

The relationship between free energy and total pressure is not valid for nonideal gases. The simple form of the equation can be retained for use with nonideal gases by defining (18)† a special property f and substituting it for pressure in Eq. (1-1).

$$dF = RT\,d\ln f \tag{1-2}$$

† Numbers in parentheses refer to references at end of chapter.

The special property f is called fugacity and is partially defined by Eq. (1-2).

If fugacity is to be of practical use, it must be related to some property of the gas subject to direct measurement. It can be related to pressure in the following manner: At constant temperature,

$$dF = V\,dp = RT\,d\ln f$$

Integrating at constant temperature between states 1 and 2 gives

$$F_2 - F_1 = \int_{p_1}^{p_2} V\,dp = RT\ln\frac{f_2}{f_1} \qquad (1\text{-}3)$$

For every set of lower and upper limits, the integral term in Eq. (1-3) has a definite value which depends upon the p-V relationships at constant temperature for the particular component under consideration. Corresponding to this integral there is an increment in F (described by $F_2 - F_1$) and an increment in f (described by $\ln f_2 - \ln f_1$) which are of such a size as to make the equalities hold in Eq. (1-3). In order to assign numerical values to F and f, it is necessary to select a reference state for each and arbitrarily assign a numerical value for each function in its reference state. The reference state chosen is arbitrary, and the only criterion is convenience. Many reference states have been used for free energy, but in the case of fugacity only one is widely used. Comparison of Eqs. (1-1) and (1-2) shows that for an ideal gas $f = p$. Since real gases approach ideality at low pressure, it is sufficient for practical reasons to assume that

$$\lim_{p\to 0}\frac{f}{p} = 1.0$$

and complete the definition of fugacity by setting $f = p$ at $p = 0$. This choice of reference state fixes the numerical values of fugacity for any given component. For any pressure p_2 at a given temperature there is a corresponding fugacity f_2 which can be calculated from experimental p-V-T data or a general equation of state (4, 8, 23). It would be possible, of course, to make $f_1 = p_1$ at some point on the pressure scale other than $p = 0$. A different numerical scale for fugacity would result, but since we always deal with the difference in fugacity between two states, the actual numerical values are basically unimportant. However, it is fortunate that only one reference state (zero pressure) is widely used for fugacity, thereby eliminating the confusion which would arise if several were common.

Now that the numerical scale for fugacity has been related to pressure and the numerical value of fugacity fixed at any given state of the component, the integration of Eq. (1-3) between any two limits can be considered. The fact that we normally deal with the ratio f_2/f_1 rather than individual values of fugacity makes it convenient to represent the ratio with a single symbol. The symbol a was chosen (18) and given the name activity. The activity is defined by

$$a = \frac{f_2}{f_1}$$

but its numerical value depends upon the limits of integration in Eq. (1-3) or in other words upon the choice of states 1 and 2. State 2 is whatever state of the system is being investigated. State 1, or the lower limit on the integration, can usually be chosen arbitrarily. The term *standard state* will be used for state 1, and the values of the thermodynamic functions in the standard state denoted by a superscript zero. The change in free energy in going from the standard state to state 2 (the state under consideration) is related to the corresponding change in fugacity and the activity by

$$F_2 - F^\circ = RT \ln \frac{f_2}{f^\circ} = RT \ln a$$

A clear distinction must be made between the terms standard state and reference state as they are used in this text in connection with fugacity. The term reference state refers to that part of the definition of fugacity which fixes numerical values of fugacity in relation to pressure. The term standard state is used solely in connection with activities and, later on, activity coefficients. The choice of standard state has no effect upon the numerical value of fugacity but does serve to establish numerical values of activity by fixing the lower limit on the integration of Eq. (1-3).

The choice of the standard state to which the numerical values of activity are referred is a matter of convenience. In chemical equilibrium and reaction kinetics, it is usually most convenient to choose the standard state for gases as unit fugacity ($f_1 = f^\circ = 1.0$) and the temperature of the system while solids and liquids are both referred to the pure solid or liquid component at the temperature and pressure of the system. In vapor-liquid phase-equilibrium work it is convenient to refer both the gas and liquid to the same standard state. In this text, the standard state always used will be the pure liquid component at the temperature and pressure of the system. This choice has one shortcoming in that the component may not exist as a liquid at the temperature and pressure of the system. If so, the standard state becomes a hypothetical state, but this causes no practical difficulties. This situation can be avoided (when the component is below its critical temperature) if the vapor pressure of the component rather than the system pressure is used as the standard-state pressure. The effect of this change usually has a negligible effect upon the numerical value of the activity as will be shown later.

The activity of a component in a mixture is related to its mole fraction by the definition of the activity coefficient. In a liquid phase, the activity coefficient γ for component i is defined by

$$\gamma_i = \frac{\bar{a}_i}{x_i}$$

Substitution of the definition of activity relates γ to fugacity as follows:

$$\gamma_i = \frac{\bar{a}_i}{x_i} = \frac{\bar{f}_i^L}{f_i^\circ x_i} = \frac{\bar{f}_i^L}{f^L_{i,p} x_i} \tag{1-4}$$

where $f^L_{i,p}$ is the standard-state fugacity (pure liquid component i at the temperature and pressure of the system). An expression for $f^L_{i,p}$ is developed later in Sec. 1-3.

An expression for the activity coefficient in the vapor phase which is analogous to Eq. (1-4) can be written.

$$\phi_i = \frac{\bar{a}_i^V}{y_i} = \frac{\bar{f}_i^V}{f_i^\circ y_i} = \frac{\bar{f}_i^V}{f^L_{i,p} y_i} \tag{1-5}$$

A different symbol ϕ is used above for the vapor-phase activity coefficient so that the symbol γ can be reserved strictly for the liquid.

Fugacity coefficients $\bar{f}_i^L/x_i$ and $\bar{f}_i^V/y_i$ can be defined in a manner analogous to the activity coefficients except that no reference need be made to a standard state. These ratios will be used later in the calculation of K values with an equation of state. Other fugacity coefficients have been in use for many years. In the calculation of fugacities at high pressures, it has been found convenient (23, 24) to use the ratio $\bar{f}_i^V/p y_i$. This coefficient is more analogous to the f/p ratio read from generalized charts for pure gases and to which the name fugacity coefficient long has been applied.

1-2. Fugacity of Gases

Before consideration of a vapor-liquid equilibrium system, the equations for the fugacity of a gas will be reviewed. These equations will relate fugacity to pressure at constant temperature in terms of the molar volumes of the gas. The molar volume of an ideal gas is, by definition, RT/p. In the following derivations this ratio is represented by V^*, which denotes the volume which 1 mole of the gas would occupy if it were ideal. The molar volume of the real gas is denoted by V, while $\bar{V}$ is used for the volume which the real gas occupies in a mixture of gases.

In general, the volume which 1 mole of a real gas will occupy at a given pressure and temperature will change slightly when the environment of the gas molecules is changed by mixing with another gas. The kinetic behavior of the gas is modified by the interaction of the individual molecules with their neighbors. They will, in general, interact differently with strange molecules than with molecules of their own kind. The change in kinetic behavior of gases due to mixing will be represented in this text by the difference $V - \bar{V}$. If $V = \bar{V}$ for each gas in a mixture, the effect of mixing is zero and the gas mixture is called an ideal gas solution. Or the volumes of the pure real gases are said to be additive. The $\bar{V}$ is often called the partial molar volume because it represents the change in the total volume of a gas mixture due to the addition of 1 mole of the pure real gas to a very large amount of the mixture. If the amount of mixture is large enough, the addition of 1 mole will not change the composition appreciably; that is, the "environment" presented by the mixture does not change during the addition. The increase in total volume then represents the molar volume of the added gas at the given mixture composition. Methods for the determination of partial molar volumes in gas mixtures are described in any standard chemical engineering thermodynamics text.

Pure Gases. The simplest way to derive the equation used to calculate the fugacity of a pure gas is to subtract from the left-hand side of

$$RT \, d \ln f = V \, dp$$

the expression $RT \, d \ln p$ and from the right-hand side the equivalent expression $V^* \, dp$ and then integrate the result from $p = 0$ to $p = p$ to give

$$\ln f = \ln p - \int_0^p \frac{V^* - V}{RT} \, dp \qquad (1\text{-}6)$$

The choice of reference state to make $f = p$ at $p = 0$ causes the two indeterminate terms obtained in the integration to cancel. The $V^* - V$ term in Eq. (1-6) is often termed the residual volume and represented by the symbol α. Experiment has shown that α does not in general go to zero at zero pressure (8).

If experimental data are not directly available for the evaluation of residual volumes but the second virial coefficient is known, the fugacity can be approximated by assuming that the p-V-T behavior of the gas can be represented by the first two terms of the virial equation of state,

$$pV = A + Bp = RT + Bp$$

or
$$V = \frac{RT}{p} + B$$

and replacing the $V^* - V$ term with $-B$.

A third method for the calculation of fugacity depends upon the use of compressibility factors from generalized charts based upon the law of corresponding states. Multiplication of the numerator and denominator of the integrand in Eq. (1-6) by p permits the equation to be rewritten as

$$\ln f = \ln p - \int_0^p \frac{1 - z}{p}\,dp$$

where $z = pV/RT =$ compressibility factor and $pV^*/RT = 1.0$ by definition. The ratio $(1 - z)/p$ approaches a finite value as p approaches zero.

Gas Mixtures. The basic relationship between fugacity and pressure holds for a gas in a mixture as well as for the pure gas.

$$RT\,d \ln \bar{f}_i = \bar{V}_i\,dp$$

The superscript bars are used to denote that the gaseous component i exists in a mixture of gases. In order to develop a more convenient expression for $\bar{f}_i$ it is convenient to define another residual volume $\bar{\alpha}_i$ where

$$\bar{\alpha}_i = \frac{RT}{p} - \bar{V}_i = V^* - \bar{V}_i$$

Substitution for $\bar{V}_i$ in the basic relationship between $\bar{f}_i$ and p gives

$$d \ln \frac{\bar{f}_i}{p} = - \frac{\bar{\alpha}_i}{RT}\,dp$$

In this text, the partial pressure p_i of component i will be defined as

$$p_i = y_i p$$

As p goes to zero, $\bar{f}_i = p_i = y_i p$, $\bar{f}_i/p = y_i$, and at a given temperature and composition

$$\int_{y_i}^{\bar{f}_i/p} d \ln \frac{\bar{f}_i}{p} = - \int_0^p \frac{\bar{\alpha}_i}{RT}\,dp$$

Integrating the left side and using

$$\bar{\alpha}_i = \alpha_i + V_i - \bar{V}_i$$

along with Eq. (1-6) written in terms of α_i to eliminate the $\bar{\alpha}_i$ and $\ln p$ terms give

$$\ln \bar{f}_i = \ln y_i f_i + \int_0^p \frac{\bar{V}_i - V_i}{RT}\,dp \qquad (1\text{-}7)$$

Equation (1-7) is the rigorous expression for the fugacity of gaseous component i in a mixture of gases at any given temperature and composition. If the gas mixture can be assumed to form ideal gas solutions (usually a good assumption for hydrocarbons below 200 psia), $\bar{V}_i = V_i$ and Eq. (1-7) reduces to the familiar Lewis-Randall rule (19) of ideal solutions:

$$\bar{f}_i = y_i f_i$$

If the gases can be assumed to be ideal (usually a good assumption for hydrocarbons below 30 psia), $f_i = p$ and $\bar{f}_i = y_i p$.

1-3. Vapor and Liquid in Equilibrium

Before the relationships for vapor-liquid equilibrium in multicomponent systems are considered, it is necessary to set down the equation for a pure liquid in equilibrium with its own vapor and from there proceed to the expression for the fugacity of a pure liquid at any pressure. The latter expression is necessary to describe the standard-state fugacity as specified in this text.

Pure Liquids. The fugacity of a component in the liquid state is identical with its fugacity in the vapor with which the liquid is in equilibrium. A pure component is in equilibrium with its vapor only at a pressure equal to the vapor pressure of the liquid. So for a pure liquid i at a total pressure equal to its vapor pressure

$$\ln f^L{}_{i,p_i{}^*} = \ln f^V{}_{i,p_i{}^*} = \ln p_i^* + \int_0^{p_i{}^*} \frac{V_i - V^*}{RT}\, dp \qquad (1\text{-}8)$$

where the p_i^* refers to the vapor pressure of pure i. The subscript p_i^* on a fugacity denotes the fugacity at a pressure equal to the vapor pressure.

If the liquid is under a pressure other than its vapor pressure, then a correction must be made for the effect of pressure on the fugacity of the liquid. From the basic definition of fugacity

$$\int_{p_i^*}^p d \ln f_i^L = \frac{1}{RT} \int_{p_i^*}^p V_i^L\, dp$$

$$\ln f^L{}_{i,p} = \ln f^L{}_{i,p_i{}^*} + \int_{p_i^*}^p \frac{V_i^L}{RT}\, dp$$

Substituting for $\ln f^L{}_{i,p_i{}^*}$ from (1-8) gives

$$\ln f^L{}_{i,p} = \ln p_i^* + \int_0^{p_i{}^*} \frac{V_i - V^*}{RT}\, dp + \int_{p_i^*}^p \frac{V_i^L}{RT}\, dp \qquad (1\text{-}9)$$

Mixtures. Besides the requirements that $t^V = t^L$ and $p^V = p^L$, it is also necessary at equilibrium that

$$\bar{f}_i^V = \bar{f}_i^L \qquad (1\text{-}10)$$

By definition for the liquid state

$$\bar{f}_i^L = \gamma_i x_i f_i^\circ \qquad (1\text{-}4)$$

This definition of γ can be used to replace $\bar{f}_i{}^L$, and Eq. (1-7) to replace $\bar{f}_i{}^V$.

$$y_i f_{i,p} \exp \left(\int_0^p \frac{\bar{V}_i - V_i}{RT} \, dp \right) = \gamma_i x_i f_i{}^\circ \tag{1-11}$$

The standard state chosen for a liquid component is the pure component at the temperature and pressure of the mixture. (Note that a standard state need not be specified for the vapor, since the activity of the vapor is not involved.) The $f_i{}^\circ$ can be replaced by the right side of Eq. (1-9), and equation (1-11) can be rewritten as

$$\ln y_i + \ln f_{i,p} + \int_0^p \frac{\bar{V}_i - V_i}{RT} \, dp = \ln \gamma_i + \ln x_i$$
$$+ \ln p_i^* + \int_0^{p_i*} \frac{V_i - V^*}{RT} \, dp + \int_{p_i*}^p \frac{V_i{}^L}{RT} \, dp \tag{1-12}$$

The third and fourth terms on the right-hand side can be replaced by $\ln f_{i,p_i*}$ [from Eq. (1-6)], and the equation written more simply as

$$\ln \gamma_i = \ln \frac{y_i f_{i,p}}{x_i f_{i,p_i*}} + \int_0^p \frac{\bar{V}_i - V_i}{RT} \, dp - \int_{p_i*}^p \frac{V_i{}^L}{RT} \, dp \tag{1-13}$$

where $f_{i,p}$ and f_{i,p_i*} refer to the gas fugacities of the pure component at the total pressure and the vapor pressure, respectively.

Equation (1-6) can be used to replace the fugacity terms, and Eq. (1-13) can be written in terms of pressure as

$$\ln \gamma_i = \ln \frac{y_i p}{x_i p_i^*} + \int_{p_i*}^p \frac{V_i - V^*}{RT} \, dp + \int_0^p \frac{\bar{V}_i - V_i}{RT} \, dp$$
$$- \int_{p_i*}^p \frac{V_i{}^L}{RT} \, dp \tag{1-14}$$

Equation (1-14) is often written with the second virial coefficient B_i substituted for $V_i - V^*$ to approximate the deviation from ideal gas behavior. This approximation plus the assumption of an ideal gas solution permits the equation to be written as

$$\ln \gamma_i = \ln \frac{y_i p}{x_i p_i^*} + \frac{B_i - V_i{}^L}{RT} (p - p_i^*) \tag{1-15}$$

where $V_i{}^L$ is assumed to be independent of pressure over the range involved.

It is of interest to note at this point the effect on Eqs. (1-13) and (1-14) if the standard-state pressure is taken as the vapor pressure of the component rather than the total system pressure. The standard-state fugacity $f_i{}^\circ$ is then f^L_{i,p_i*}, and Eq. (1-8) instead of (1-9) is used to replace $f_i{}^\circ$ in Eq. (1-11). The result is the elimination of the pressure correction on the fugacity of the pure liquid represented by the last term in Eqs. (1-13) and (1-14) and the $V_i{}^L$ term in (1-15).

Example 1-1. Consider a three-phase, two-component system with one vapor and two liquid phases at equilibrium. The two components are partially miscible in the liquid phase. The vapor forms an ideal solution. For the sake of clarity use the symbols shown on the diagram to designate the various phases. Show the relationships you would use to calculate the following items. Use the same standard states for liquid and gas phases. The effect of pressure on the fugacity of a liquid can be neglected.

(a) Activities in the liquid phases.
(b) Activities in the vapor phase.
(c) Liquid-phase activity coefficients.
(d) Total system pressure.
(e) Vapor-liquid distribution coefficients.
(f) Liquid-liquid distribution coefficients.

Solution. The pure liquid component at the temperature and pressure of the system will be chosen as the standard state for each component in both the vapor and liquid phases. At equilibrium,

$$\bar{f}^V{}_{i,p} = \bar{f}^Z{}_{i,p} = \bar{f}^L{}_{i,p}$$

Also, because the same standard state is used for both the vapor and liquids,

$$\bar{a}_i{}^V = \bar{a}_i{}^Z = \bar{a}_i{}^L$$

Since the vapor forms an ideal solution,

$$\bar{f}^V{}_{i,p} = y_i f^V{}_{i,p}$$

Finally, if the effect of pressure on the fugacity of a liquid is neglected, the following approximation can be used:

$$f^L{}_{i,p} \cong f^L{}_{i,p^*} = f^V{}_{i,p^*}$$

The above relationships are used repeatedly below.

(a)
$$\bar{a}_1{}^Z = \frac{\bar{f}^Z{}_{1,p}}{f_1{}^\circ} = \frac{\bar{f}^V{}_{1,p}}{f^L{}_{1,p}} \cong \frac{y_1 f^V{}_{1,p}}{f^V{}_{1,p^*}}$$

$$\bar{a}_1{}^L = \frac{\bar{f}^L{}_{1,p}}{f_1{}^\circ} = \frac{\bar{f}^V{}_{1,p}}{f^L{}_{1,p}} \cong \frac{y_1 f^V{}_{1,p}}{f^V{}_{1,p1^*}}$$

It can be seen that $\bar{a}_1{}^Z = \bar{a}_1{}^L$. Analogous expressions hold for component 2.

$$\bar{a}_2{}^Z = \bar{a}_2{}^L = \frac{\bar{f}^Z{}_{2,p}}{f_2{}^\circ} = \frac{\bar{f}^L{}_{2,p}}{f_2{}^\circ} = \frac{\bar{f}^V{}_{2,p}}{f^L{}_{2,p}} \cong \frac{y_2 f^V{}_{2,p}}{f^V{}_{2,p2^*}}$$

(b)
$$\bar{a}_1{}^V = \frac{\bar{f}^V{}_{1,p}}{f_1{}^\circ} = \frac{y_1 f^V{}_{1,p}}{f^L{}_{1,p}} \cong \frac{y_1 f^V{}_{1,p}}{f^V{}_{1,p1^*}}$$

The validity of the statement above that $\bar{a}_i{}^V = \bar{a}_i{}^Z = \bar{a}_i{}^L$ is shown by the expressions developed in (a) and (b). An analogous expression can be written for $\bar{a}_2{}^V$.

(c)
$$\gamma_1{}^Z = \frac{\bar{a}_1{}^Z}{z_1} \cong \frac{y_1 f^V{}_{1,p}}{z_1 f^V{}_{1,p1^*}}$$

$$\gamma_2{}^Z = \frac{\bar{a}_2{}^Z}{z_2} \cong \frac{y_2 f^V{}_{2,p}}{z_2 f^V{}_{2,p2^*}}$$

Similar expressions can be written for $\gamma_1{}^L$ and $\gamma_2{}^L$.

(d) The total system pressure is the sum of the individual partial pressures. Equation (1-14) can be solved for the partial pressure. The last two terms drop out because the vapors form ideal solutions and the effect of pressure on the fugacity of a liquid is neglected.

$$y_i p = \gamma_i x_i p_i^* \left[\exp \left(\int_{p_1*}^{p} \frac{V^* - V_i}{RT} dp \right) \right]^{-1}$$

The phase superscript is omitted on γ_i because the partial pressure of component i can be related to either liquid phase. Below both are related to the Z phase.

$$p = y_1 p + y_2 p$$

$$= \gamma_1^Z z_1 p_1^* \left[\exp \left(\int_{p_1*}^{p} \frac{V^* - V_1}{RT} dp \right) \right]^{-1}$$

$$+ \gamma_2^Z z_2 p_2^* \left[\exp \left(\int_{p_2*}^{p} \frac{V^* - V_2}{RT} dp \right) \right]^{-1}$$

(e) Equation (1-13) degenerates to give

$$K_1{}^Z = \frac{y_1}{z_1} = \frac{\gamma_1^Z f^V{}_{1,p_1*}}{f^V{}_{1,p}}$$

$$K_1{}^L = \frac{y_1}{x_1} = \frac{\gamma_1^L f^V{}_{1,p_1*}}{f^V{}_{1,p}}$$

Expressions for $K_2{}^Z$ and $K_2{}^L$ are obtained by replacing 1 with 2.

(f) From (c) it can be seen that

$$K_1 = \frac{z_1}{x_1} = \frac{K_1{}^L}{K_1{}^Z} = \frac{\gamma_1{}^L}{\gamma_1{}^Z}$$

Since both liquids are in equilibrium with the same vapor phases, all terms in the ratio of γ's cancel except the liquid mole fractions.

1-4. Distribution Coefficients When $\gamma = 1.0$

Those systems which form ideal liquid solutions at any pressure are, unfortunately, only a small minority in the total number of systems with which the design engineer must work. Usually only close-boiling homologs can safely be considered to form ideal liquid solutions and then only at relatively low pressures (<150 psia). The liquid state of matter is so poorly understood that nonideal behavior due to chemical dissimilarity or wide differences in molecular weight cannot be predicted in a quantitative manner over the entire composition range. As a consequence, distribution coefficients can be predicted with sufficient accuracy only for systems where the activity coefficients at low pressures can be assumed to be equal to unity. When dealing with systems which contain the "inert" gases, with gas and crude oil mixtures, or with systems containing chemically dissimilar components, the designer must rely on experimental data. Methods of correlating such nonideal equilibrium data are discussed in the next chapter. The remainder of this chapter will be devoted to methods for the calculation of K values in systems where the activity coefficients can safely be considered to be 1.0 at low pressures.

$p < $ **30 PSIA.** At low pressures, the vapor phase approaches ideal behavior and $\bar{V}_i = V_i = V^*$. The pressure at which this assumption can be made varies with the critical temperature and pressure of the system, but usually the assumption is valid below 30 psia. In any case, the "ideality" of the vapor-phase components can be quickly checked on a generalized compressibility or fugacity coefficient (f/p) chart.

If the assumption $\bar{V}_i = V_i = V^*$ can be made, and if the liquid activity coefficients can be assumed to be unity, Eq. (1-14) reduces to a combination of Dalton's and Raoult's laws.

$$y_i p = x_i p_i^* \quad \text{or} \quad K_i = \frac{y_i}{x_i} = \frac{p_i^*}{p} \tag{1-16}$$

The last term in Eq. (1-14) is negligible at the conditions where Eq. (1-16) is valid. It should be recognized that both p and p_i^* must be low if Eq. (1-16) is to be used. If the vapor pressure is high, the $V - V^*$ term may not be approximately zero over the range of integration in the first integral term in Eq. (1-14). Equation (1-16) is therefore generally restricted to mixtures of hydrocarbon or other homologs which boil over a relatively narrow temperature range.

$p < $ **150 PSIA.** Gas mixtures can be expected to form ideal solutions at pressures far above the point where they cease to behave as ideal gases; that is, $V_i = \bar{V}_i$ even though $V_i \neq V^*$. This assumption is generally valid up to pressures of 200 to 300 psia. If γ_i can be assumed to be 1, and if the effect of pressure on the fugacity of the liquid is neglected, Eq. (1-13) becomes

$$y_i f_{i,p} = x_i f_{i,p_i^*} \quad \text{or} \quad K_i = \frac{y_i}{x_i} = \frac{f_{i,p_i^*}}{f_{i,p}} \tag{1-17}$$

Equation (1-17) is the basis for the so-called MIT K values which appear in such sources as Perry (22) and Maxwell (20) for hydrocarbon systems.

Even though the assumption of ideal gas solutions is generally considered valid to at least 200 psia, distribution coefficients calculated from Eq. (1-17) begin to deviate from experimental data at pressures above 150 psia. Therefore, above 150 psia the use of the Lewis-Randall rule as written in Eq. (1-17) is not valid for the calculation of K values. The original calculation of K values by Souders, Selheimer, and Brown at the time when K was first defined in terms of the liquid and vapor fugacities included the effect of pressure on the liquid fugacity (27, 28).

$$K_i = \frac{y_i}{x_i} = \frac{f_{i,p_i^*}}{f_{i,p}} \exp\left(\int_{p_i^*}^{p} \frac{V_i^L}{RT} \, dp \right) \tag{1-18}$$

The inclusion of the exponential term improves the fit somewhat at higher pressures in some cases, but Eq. (1-18) is incapable of reproducing the experimental data at pressures far above 150 psia. The divergence at higher pressures of ideal-solution K values from experimental data is shown in Fig. 1-1.

$p > $ **150 PSIA.** At pressures where Raoult's law or the Lewis-Randall rule of ideal solutions is valid, the distribution coefficients for a given component are functions of temperature and pressure only. In nonideal solutions, they become functions of composition also. The calculation of

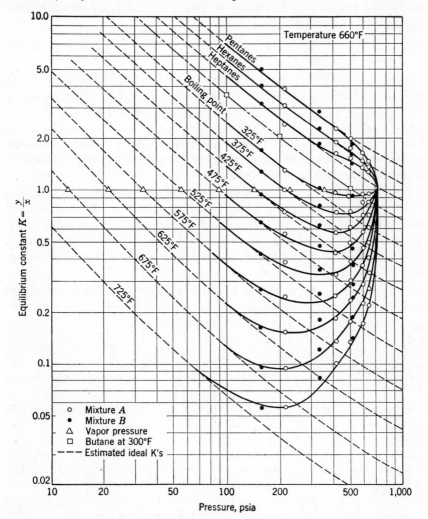

FIG. 1-1. Equilibrium-distribution coefficients for two complex mixtures plotted as a function of pressure at a constant temperature of 660°F. Note the deviation of the ideal-solution K values from the experimental curves. [*White and Brown, Ind. Eng. Chem.*, **34** 1162 (1942).]

K values at high pressures, therefore, depends upon the calculation of liquid and vapor fugacities in nonideal mixtures. These calculations require an equation of state which not only represents adequately the p-V-T data for the pure components but permits also the constants for the individual components to be combined to provide a set of equation constants which will adequately represent the p-V-T behavior of any desired mixture. Such an equation has been developed by Benedict, Webb, and Rubin, and they have presented the constants for the following paraffinic hydrocarbons (1, 2):

Methane	Propane	i-Pentane
Ethylene	i-Butane	n-Pentane
Ethane	i-Butylene	n-Hexane
Propylene	n-Butane	n-Heptane

The Benedict-Webb-Rubin equation contains eight constants which are functions of composition. It represents adequately the thermodynamic properties of mixtures of the light hydrocarbons listed above in both the liquid and vapor phases.

The calculation of K values by Benedict et al. was accomplished by the use of the equation

$$K_i = \frac{y_i}{x_i} = \frac{\bar{f}_i^L / x_i}{\bar{f}_i^V / y_i} \tag{1-19}$$

where $\bar{f}_i^L$ and $\bar{f}_i^V$ represent the fugacities in the equilibrium liquid and vapor phases at the system pressure and temperature. At equilibrium $\bar{f}_i^L$ must equal $\bar{f}_i^V$. The fugacity coefficients $\bar{f}_i^L / x_i$ and $\bar{f}_i^V / y_i$ were calculated by the use of the equation of state to evaluate the integral in the following equation, which can be written for both the vapor and liquid phases;

$$RT \ln \frac{\bar{f}_i^L}{x_i} = RT \ln (dRT) + \int_0^d \left[\left(\frac{\partial p V_t}{\partial n_i} \right)_{n,V,T} - RT \right] d(\ln d) \tag{1-20}$$

The symbol d is used in Eq. (1-20) as both the differential operator (outside the brackets) and as the molar density $d = \Sigma n_i / V_t$. The volume of the entire mixture is represented by V_t. Equation (1-20) can be obtained by combining Eqs. (1-6) and (1-7), changing the variable of integration from p to d, and replacing the partial molar volume $\bar{V}_i$ with $(\partial V_t / \partial n_i)_{n,V,T}$. Since the Benedict equation of state is good for both the liquid and vapor phases, it can be used to evaluate the integral in Eq. (1-20) for either the liquid or vapor state.

The Benedict-Webb-Rubin equation was used to calculate the Kellogg equilibrium charts (3, 15). To represent all possible mixtures of the 12 hydrocarbons on a reasonable number of charts, it was necessary to select a composition parameter. The molar average normal boiling point

$$\text{M.A.B.P.} = x_1\,(\text{B.P.})_1 + x_2\,(\text{B.P.})_2 + \cdots$$

with the normal boiling points in degrees Fahrenheit, was found to correlate the equation results with the least loss of accuracy. A set of 324 charts which represented 26 pressures between 14.7 and 3600 psia and covered a temperature range from -100 to $+400°F$ and a range of molar average boiling points from -255 to $+180°F$ was prepared.

The Benedict-Webb-Rubin equation is the best source of K values for the hydrocarbons listed. It has also been found to predict satisfactorily vapor-liquid equilibria in the nitrogen–carbon monoxide binary (26), where the two components have similar physical properties. The original equation does not work for such systems as the methane-nitrogen (29) or the carbon dioxide–propane (6) binaries. A change in some of the constants to extend the usefulness of the equation to the "inert" gases such as nitrogen and carbon dioxide has been indicated but such a change has not been generalized.

DePriester (7) replotted the Kellogg charts in the 14.7- to 1000-psia range. The molar average boiling point was retained as the composition variable. The number of charts required for this pressure range was reduced from 192 for the Kellogg correlation to 24. The inherent accuracy of the Kellogg correlation was increased in some regions by the availability of new experimental data. The accuracy with which the charts could be read was improved because of easier interpolation on the DePriester charts. Besides the 24 pressure-temperature-composition charts, DePriester presented nomographs (Fig. B-3a and b) which provide approximate K values as a function of pressure and temperature only. The nomographs are useful for the initial prediction of the liquid and vapor compositions which must be known before the pressure-temperature-composition charts can be used. The use of the nomographs reduces the trial-and-error effort inherent in the use of any K-value correlation which includes a composition parameter.

Both the Kellogg and DePriester charts are based on the calculation of the factors $\bar{f}_i^L/px_i$ and $\bar{f}_i^V/py_i$. In this manner the nonideal behavior of the liquid and gas phases is separated. The K values are then obtained by

$$K_i = \frac{y_i}{x_i} = \frac{\bar{f}_i^L/px_i}{\bar{f}_i^V/py_i} \tag{1-21}$$

Edmister and Ruby (9) modified Eq. (1-21) to

$$K_i = \frac{y_i}{x_i} = \frac{\hat{f}_i{}^L/p_i^*x_i}{\hat{f}_i{}^V/py_i} \frac{p_i^*}{p}$$

and have correlated $\hat{f}_i{}^L/p_i^*x_i$ and $\hat{f}_i{}^V/py_i$ as a function of the following reduced coordinates:

$$T_{ri} = \frac{\text{system temperature, °R}}{\text{critical temperature of } i\text{th component, °R}}$$

$$p_{ri} = \frac{\text{system pressure, psia}}{\text{critical pressure of } i\text{th component, psia}}$$

$$(\text{M.A.B.P})_{ri} = \frac{\text{M.A.B.P. of vapor or liquid mixture, °R}}{\text{atmospheric B.P. of } i\text{th component, °R}}$$

The Edmister-Ruby generalized correlations are based upon the fugacity values obtained by Benedict et al. but also permit estimates of K values for paraffin and olefin hydrocarbons other than the 12 covered in the Kellogg and DePriester charts. The generalized correlations can also be used to obtain values for partial enthalpy and volume.

The Kellogg, DePriester, and Edmister-Ruby charts are all based upon the Benedict-Webb-Rubin equation of state and therefore represent the equation-of-state approach to phase equilibria. Another major approach to the problem is the convergence-pressure correlation developed at the University of Michigan (13, 14, 30, 21) under the direction of G. G. Brown. The Michigan correlation is based on the concept of system convergence pressure and has been presented in chart form by the Fluor Corporation, Ltd., and by the Natural Gasoline Association of America. The Fluor and the NGAA charts (10) incorporate the Benedict et al., DePriester, and Edmister-Ruby equations and correlations in the component ranges where they are applicable.

The convergence pressure is defined (13, 21, 30) as the pressure at which the K values appear to converge to unity on an isothermal plot of log K vs log p (see Fig. 1-1). The concept was based upon the observation in high-pressure work (14, 30) that as the pressure was increased on a hydrocarbon system at a constant temperature, the K values of the various components converged to unity at some high pressure. In other words, the system could be made to exhibit the critical phenomenon (loss of identity between vapor and liquid) at any desired temperature. The pressure at which the critical phenomenon occurred was termed the convergence pressure. The convergence pressure was found to be a function of composition and was therefore suggested (30) as a parameter to represent the effect of composition at pressures above the range of the ideal-solution assumption. Several schemes have been proposed

(12, 17, 21, 25) for the prediction of convergence pressure as a function of composition. The method used in the Fluor and NGAA charts is that of Hadden (11). The Fluor and NGAA charts include a short-cut method for the initial estimate of composition (or convergence pressure) in order to reduce the trial-and-error effort. The charts are restricted to paraffinic systems and some of the "inerts" such as nitrogen, carbon dioxide, hydrogen sulfide, and hydrogen. They are not valid for systems containing appreciable amounts of aromatics. The effect of convergence pressure on the distribution coefficient of butane is shown in Fig. 1-2.

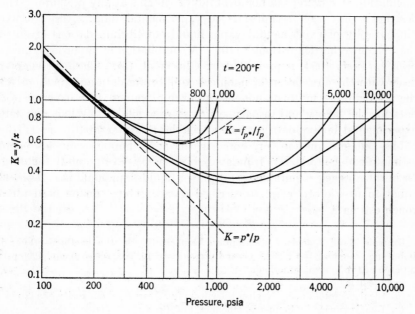

FIG. 1-2. Effect of convergence pressure on the distribution coefficients for normal butane.

Note the inaccuracy in the K values calculated from Raoult's law and the Lewis-Randall rule of ideal solutions.

Winn (31) has presented a nomograph for systems which have a convergence pressure of 5000 psia. Besides the paraffins up to n-heptane, the nomograph can be used to estimate K values for components such as carbon dioxide, hydrogen, water, hydrogen sulfide, methyl mercaptan, and methane in various solutions. Some aromatic solutions are included. Although the nomograph is constructed for a convergence pressure of 5000 psia, an accompanying grid allows its use for convergence pressures to 10,000 psia.

None of the correlations discussed in this section provide a generalized method for systems above 150 psia other than paraffinic ones. Below 150 psia, all aromatic or all naphthenic systems can be handled with the Lewis-Randall rule, $K_i = f_{i,p_i*}/f_{i,p}$, or by Raoult's law if the vapor behavior is ideal. No general correlation scheme is available for mixtures of paraffins with aromatics or with naphthenes where the liquid behaves nonideally at all pressures. Lenoir (16) has presented a correlation to predict the convergence pressures (or the critical locus) for nonideal binary systems, but no correlation has been developed for the general calculation of K values in nonideal liquid systems at any pressure. The designer's only safe recourse is to experimental data. An extensive bibliography of experimental vapor-liquid equilibrium data is given in the "Engineering Data Book" (10). A compilation of vapor-liquid data prior to June, 1954, is presented by Chu et al. (5). The next chapter deals with the correlation of nonideal equilibrium data and methods for the prediction of multicomponent equilibrium values when experimental data for the binary and ternary systems are available. When complete experimental binary data are not available, the integrated forms of the Gibbs-Duhem equation (presented in the next chapter) can be used to estimate vapor-liquid data from available azeotropic or liquid-liquid data. Or in the absence of suitable azeotropic or liquid-liquid data the empirical equation of Pierotti et al. (described in Sec. 11-1) can be used for a rough estimate.

Example 1-2. Calculate distribution coefficients for the methane-ethylene-isobutane system at 100°F (37.8°C) and pressures of 10, 100, 200, 500, and 1000 psia by each of the following methods:

(a) Ideal gases.
(b) Ideal gas solutions. (Neglect the effect of pressure on the fugacity of a liquid.)
(c) Equation used by Souders, Selheimer, and Brown, Eq. (1-18).
(d) DePriester charts (at 500 and 1000 psia only).
(e) Convergence-pressure charts (at 500 and 1000 psia only).

Plot the calculated values vs pressure on a log-log plot. M. Benedict, E. Solomon, and L. C. Rubin [*Ind. Eng. Chem.*, **37**, 55 (1945)] give the following experimental data at 37.78°C:

Component	500 psia			1000 psia		
	y	x	K	y	x	K
Methane	0.3355	0.069	4.86	0.657	0.284	2.31
Ethylene	0.4815	0.279	1.726	0.187	0.1705	1.098
Isobutane	0.1830	0.652	0.281	0.156	0.5455	0.286

Solution. The conditions of this problem are obviously outside the range of applicability of some methods. However, all five methods will be applied in so far as possible for comparison purposes. Physical data were obtained from Maxwell (20). The results of the example are plotted in Fig. 1-3.

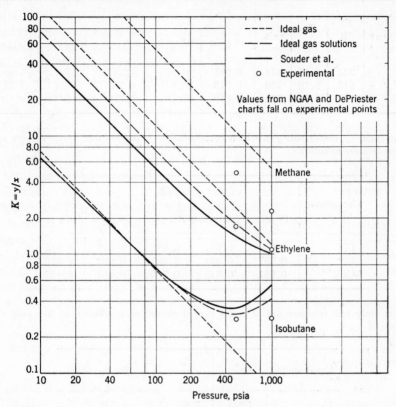

FIG. 1-3. Comparison of K values calculated in Example 1-3 with experimental values at 100°F reported by Benedict, Solomon, and Rubin [*Ind. Eng. Chem.*, **37** 55 (1945)].

(a) For ideal gases when γ is assumed to be unity, $K_i = y_i/x_i = p_i^*/p$. The vapor-pressure values for methane and ethylene below are extrapolated.

Component	p^*	K at 100°F				
		10 psia	100 psia	200 psia	500 psia	1000 psia
Methane	5290	529	52.9	26.5	10.6	5.29
Ethylene	1205	120	12.0	6.02	2.41	1.20
Isobutane	72.7	7.27	0.727	0.364	0.145	0.0727

(b) For ideal gas solutions when γ is assumed to be unity, $K_i = y_i/x_i = f_{i,p_i*}/f_{i,pi}$. The critical temperatures and pressures, the reduced temperatures, and the reduced pressures at the various total pressures are as follows:

Component	T_c, °R	T_R	p_c, psia	p_R				
				10 psia	100 psia	200 psia	500 psia	1000 psia
C_1	344	1.63	674	0.015	0.149	0.297	0.742	1.48
$C_2^=$	510	1.10	750	0.013	0.134	0.267	0.667	1.34
i-C_4	735	0.76	529	0.019	0.189	0.378	0.945	1.89

The fugacity coefficients f/p which correspond to these reduced temperatures and pressures are as follows:

Component	f/p				
	10 psia	100 psia	200 psia	500 psia	1000 psia
C_1	1.0	0.991	0.982	0.959	0.915
$C_2^=$	0.997	0.964	0.930	0.830	0.676
i-C_4	0.981	0.845	0.705	0.415	0.152

Since only one temperature is under consideration, each component has only one value of f_{i,p_i*}.

Component	p_i^*	p_R^*	$\left(\dfrac{f}{p}\right)^*$	f_{i,p_i*}
C_1	5290	7.85		
$C_2^=$	1205	1.61	0.612	738
i-C_4	72.7	0.137	0.885	64.4

The p_R^* for methane is off scale on the fugacity chart in Maxwell. The ideal-gas-solution K values for the other two components are shown below.

Component	K at 100°F				
	10 psia	100 psia	200 psia	500 psia	1000 psia
C_1					
$C_2^=$	74.0	7.65	3.97	1.78	1.09
i-C_4	6.56	0.762	0.456	0.31	0.423

(c) The Souder-Selheimer-Brown equation can be written as follows if the effect of pressure on the volume of the liquid is neglected:

$$K_i = \frac{y_i}{x_i} = \frac{f_{i,p_i^*}}{f_{i,p}} \exp \frac{V_i^L(p - p_i^*)}{RT}$$

Multiplication of the exponential term by the ideal-gas-solution K values from part (b) will give the desired answers for part (c). The V_i^L/RT term for each component is calculated below. The RT product is $(10.71)(560) = 6000$.

Component	Liquid specific gravity at 100°F	V_i^L, ft³/lb mole	$\dfrac{V_i^L}{RT}$
C_1	0.162†	1.58	2.64×10^{-4}
C_2^-	0.218†	2.06	3.43×10^{-4}
$i\text{-}C_4$	0.537	1.73	2.89×10^{-4}

† Values at critical point.

Pressure, psia	$p - p^*$		$\exp \dfrac{V_i^L(p - p^*)}{RT}$		K	
	C_2^-	$i\text{-}C_4$	C_2^-	$i\text{-}C_4$	C_2^-	$i\text{-}C_4$
10	−1195	−62.7	0.663	0.983	49.1	6.45
100	−1105	27.3	0.684	1.008	5.23	0.768
200	−1005	127.3	0.708	1.037	2.81	0.473
500	− 705	427.3	0.784	1.132	1.40	0.351
1000	− 205	927.3	0.932	1.308	1.02	0.553

It can be seen that the effect of pressure on the fugacity of the liquid is sizable when p^* varies widely from p.

(d) Values from the DePriester charts (7) will be obtained only at 500 and 1000 psia. The experimental compositions will be used to evaluate the molar average boiling points for the vapor and liquid phases. This, of course, eliminates the trial and error necessary if the phase compositions were not known beforehand. In practice, either the total feed composition (in flash calculations) or the composition of at least one phase (in stage-to-stage calculations) will be known. The use of ideal-solution K values will provide a good estimate of the phase compositions, which can then be used to get more accurate K values. The trial and error is continued until no further change in calculated phase compositions occurs.

In the tabulations below, B.P. refers to normal boiling point while A and B refer to the liquid and vapor factors read from the DePriester charts and used in the equation $K = AB$.

At 500 psia:

Component	B.P., °F	x	x(B.P.)	y	y(B.P.)	A	B	K
C_1	−258.9	0.069	−17.9	0.3355	− 86.9	4.85	0.97	4.71
$C_2^=$	−154.7	0.279	−43.2	0.4815	− 74.5	1.49	1.19	1.762
i-C_4	10.9	0.652	7.1	0.1830	2.0	0.153	2.0	0.306
			−54.0		−159.4			

At 1000 psia:

Component	B.P., °F	x	x(B.P.)	y	y(B.P.)	A	B	K
C_1	−258.9	0.284	−73.5	0.657	−170.1	2.25	1.039	2.34
$C_2^=$	−154.7	0.1705	−26.4	0.187	− 28.9	0.76	1.44	1.095
i-C_4	10.9	0.5455	6.0	0.156	1.7	0.095	3.06	0.291
			−93.9		−197.3			

(e) The convergence-pressure charts utilize a weight average t_c based on the liquid composition excluding the lightest component. The average t_c is then used with the figure on page 172 of the "Engineering Data Book" (10) to obtain the convergence pressure of the system.

Component	t_c, °F	500 psia		1000 psia	
		Wt. %	(Wt. %)(t_c)	Wt. %	(Wt. %)(t_c)
C_1					
$C_2^=$	50	0.171	8.5	0.131	6.5
i-C_4	275	0.829	228.0	0.869	239.0
		1.000	236.5	1.000	245.5

The convergence pressures at 100°F for the two pseudocritical temperatures above are both approximately 1500 psia. No charts are provided for a convergence pressure of 1500, so it is necessary to interpolate. K values at convergence pressures on both sides of 1500 were plotted, and the values at 1500 read from the curves. Interpolation is difficult in the region under consideration, and this plus the fact that no charts are provided for ethylene makes the convergence-pressure chart less useful for this particular system than it normally is. The values obtained for methane and isobutane are tabulated below. Both methane K's and the isobutane K at 500 psia are firm; that is, the interpolation curves from which they were read could be located with a high degree of certainty. The isobutane value at 1000 psia could not be determined as certainly.

Component	500 psia	1000 psia
C_1	4.8	2.28
$C_2^=$		
$i\text{-}C_4$	0.27	0.28

An attempt to obtain values for ethylene by interpolation between methane and ethane provided only very approximate values.

It can be seen from Fig. 1-3 that both the DePriester and the convergence-pressure charts check well with the experimental values. The ideal-gas-solution K values are better than the ideal gas K's. The pressure correction in the Souder-Selheimer-Brown equation is deleterious.

NOMENCLATURE

$a_i = f_i/f_i^\circ$ = activity of pure component i.

$\bar{a}_i = \bar{f}_i/f_i^\circ$ = activity of component i in solution.

A and B = first and second coefficients in the virial equation of state, $pV = A + Bp + \cdots$.

d = differential operator. Also used to denote the molar density, $d = \Sigma n_i/V_t$ in Eq. (1-20).

E = molar internal energy.

f_i = fugacity of pure component i. Superscripts L and V refer to liquid and vapor phases respectively. Subscripts p and p_i^* refer to pressure at which vapor fugacity is evaluated. If no subscript is used, p is inferred. If no superscript is used, V is inferred.

$\bar{f}_i$ = fugacity of component i in solution. Superscripts L and V refer to liquid and vapor solutions respectively.

f_i° = fugacity of component i in the standard state.

F = molar Gibbs free energy.

H = molar enthalpy.

$K_i = y_i/x_i$ = equilibrium distribution coefficient of component i.

n = number of moles in system.

p = total pressure

p_i^* = vapor pressure of component i.

p_i = partial pressure of component i.

R = gas law constant.

S = molar entropy.

T = absolute temperature.

V_i = molar volume of pure component i. Superscript L used to denote liquid phase.

$V^* = RT/p$ = molar volume of pure component if it were ideal gas.

$\bar{V}_i$ = partial molar volume of component i.

V_t = volume of total system.

x = concentration in heavy or extract phase.

y = concentration in light or raffinate phase.

$z = pV/RT$, compressibility factor for gases. Also used to denote concentrations in the second liquid phase in Example 1-1.

Greek Symbols

$\alpha = RT/p - V = V^* - V$ = residual volume of pure component.

$\bar{\alpha}_i = RT/p - \bar{V}_i = V^* - \bar{V}_i$ = residual volume of component i in solution.

γ_i = liquid phase activity coefficient of component i. See Eq. (1-4) for definition.

∂ = partial differential operator.

ϕ_i = vapor-phase activity coefficient of component i. See Eq. (1-5) for definition.

PROBLEMS

1-1. Starting with the definition of the activity coefficient $\gamma_i = \bar{a}_i/x_i$, develop the relationship for total pressure in terms of the activity coefficients for a multicomponent vapor-liquid equilibrium system containing C components at low pressure.

1-2. Consider a three-phase, two-component system with one vapor and two liquid phases at equilibrium. Assume the two components to be completely immiscible in the liquid state. The vapor forms an ideal solution. For the sake of clarity use V and y for the vapor phase, Z and z for the lighter liquid (component 1), and L and x for the heavier liquid (component 2). Show the relationships you would use to calculate the following items. Obtain numerical answers where possible. Use the pure liquids as the standard states for both the liquid and vapor phases.

V	y	Vapor
Z	z	Component 1
L	x	Component 2

(a) Activities in liquid phases.
(b) Activities in vapor phase.
(c) Liquid-phase activity coefficients of both components in both liquid phases.
(d) Total system pressure.
(e) Vapor-liquid distribution coefficients.
(f) Liquid-liquid distribution coefficients.

1-3. The activity coefficients for A and B as a function of composition at a temperature of 70°C are given below. Would you expect this system to form an azeotrope at 70°C? If so, what would be its composition and at what pressure would it boil?

Vapor pressure of A at 70°C is 713 mm Hg.

Vapor pressure of B at 70°C is 535 mm Hg.

x_A	A	B
0	3.0	1.0
0.1	2.04	1.03
0.2	1.62	1.075
0.3	1.38	1.125
0.4	1.25	1.18
0.5	1.15	1.25
0.6	1.10	1.33
0.7	1.07	1.42
0.8	1.03	1.55
0.9	1.02	1.73
1.0	1.0	2.0

1-4. What are the composition and temperature of the azeotrope in the benzene-methanol system at 3 atm pressure? Assume that the activity coefficients are independent of temperature and pressure over this range. The vapor pressures and activity coefficients are given below.

	Vapor pressures, mm Hg		Activity coefficients, 1 atm.		
°C	p_m^*	p_b^*	x_m	γ_m	γ_b
70	952.8	552.1	0.028	9.05	1.001
68	881.0	516.5	0.050	7.16	0.985
65	776.1	466.0	0.057	7.00	0.990
63	712.9	435.5	0.090	6.13	1.052
60	629.4	392.5	0.118	6.06	1.001
58	576.8	365.6	0.120	5.825	1.007
55	508.2	329.0	0.270	2.803	1.223
80.1		760.0	0.440	1.770	1.57
84.0	1520		0.586	1.39	2.00
103.8		1520.0	0.695	1.205	2.625
112.5	3800		0.817	1.051	3.935
			0.883	1.003	5.105
			0.902	1.000	5.470
			0.934	1.020	5.810
			0.945	1.000	6.075
			0.968	1.015	5.63
			0.988	1.010	8.32

1-5. The volumes of five propane-isopentane mixtures at 300°C and various pressures have been measured by W. E. Vaughan and F. C. Collins [*Ind. Eng. Chem.*, **34**, 885 (1942)]. Their data together with pure component volumes estimated from a generalized compressibility chart are shown below for 300°C.

Mole fraction pentane	Molar volumes, ft³/lb mole					
	5 atm	10 atm	25 atm	40 atm	60 atm	80 atm
0	149.4	74.0	28.8	17.5	11.3	8.28
0.101	149.6	74.10	28.77	17.41	11.17	8.103
0.206	149.6	74.18	28.77	17.43	11.17	8.095
0.412	149.6	73.80	28.19	16.81	10.60	7.531
0.607	150.0	74.18	28.19	16.72	10.37	7.342
0.899	149.9	74.18	27.41	15.85	9.50	6.552
1.0	149.1	73.8	26.5	14.9	8.64	5.912

(a) Calculate the vapor-phase-activity coefficients $\phi_i = \bar{a}_i/y_i$ for each component at each of the pressures for a 50 mole per cent pentane mixture at 300°C. Use the pure vapor at the t and p of the system as the standard state in each case.

(b) What portion of the free-energy change due to mixing at 80 atm is "excess" free energy? The excess free energy ΔF^E is defined as the difference between the free-energy change which actually occurs and the change which would have occurred if the materials mixed had formed an ideal solution.

$$\Delta F^E = \Delta F_{\text{nonideal}} - \Delta F_{\text{ideal}}$$

REFERENCES

1. Benedict, M., G. B. Webb, and L. C. Rubin: *Chem. Eng. Progr.*, **47**, 419 (1951).
2. *Ibid.:* **47**, 449 (1951).
3. Benedict, M., G. B. Webb, L. C. Rubin, and L. Friend: *Chem. Eng. Progr.*, **47**, 571, 609 (1951).
4. Bennett, C. O.: *Chem. Eng. Progr. Symposium Ser.*, **49**(7), 45 (1953).
5. Chu, J. C., S. L. Wang, S. L. Levy, and R. Paul: "Vapor-Liquid Equilibrium Data," J. W. Edwards, Publisher, Inc., Ann Arbor, Mich., 1956.
6. Cullen, E. J., and K. A. Kobe: *A.I.Ch.E. J.*, **1**, 452 (1955).
7. DePriester, C. L.: *Chem. Eng. Progr. Symposium Ser.*, **49**(7), 1 (1953).
8. Dodge, B. F.: "Chemical Engineering Thermodynamics," 1st ed., pp. 170, 233–237, McGraw-Hill Book Company, Inc., New York, 1944.
9. Edmister, W. C., and C. L. Ruby: *Chem. Eng. Progr.*, **51**, 95-F (1955).
10. "Engineering Data Book," 7th ed., p. 161, Natural Gasoline Supply Men's Association, Tulsa, Okla., 1957.
11. Hadden, S. T.: *Chem. Eng. Progr. Symposium Ser.*, **49**(7), 53 (1953).
12. Hadden, S. T.: *Chem. Eng. Progr.*, **44**, 37, 135 (1948).
13. Hanson, G. H., and G. G. Brown: *Ind. Eng. Chem.*, **37**, 821 (1945).
14. Katz, D. L., and G. G. Brown: *Ind. Eng. Chem.*, **25**, 1373 (1933).
15. The M. W. Kellogg Company: "Liquid-Vapor Equilibria in Mixtures of Light Hydrocarbons, MWK Equilibrium Constants, Polyco Data," 1950.
16. Lenoir, J. M.: *A.I.Ch.E. J.*, **4**, 263 (1958).
17. Lenoir, J. M., and G. A. White: *Petrol. Refiner*, **32**(10), 121 (1953); **32**(12), 115 (1953); **37**(3), 173 (1958).
18. Lewis, G. N.: *Proc. Am. Acad. Arts Sci.*, **37**, 49 (1901).
19. Lewis, G. N., and M. Randall: "Thermodynamics," 2d ed., McGraw-Hill Book Company, Inc., New York, 1961.
20. Maxwell, J. B.: "Data Book on Hydrocarbons," 1st ed., p. 46, D. Van Nostrand Company, Inc., Princeton, N.J., 1958.
21. Organick, E. I., and G. G. Brown: *Chem. Eng. Progr. Symposium Ser.*, **48**(2), 97 (1952).
22. Perry, J. H.: "Chemical Engineers' Handbook," 3d ed., pp. 568–572, McGraw-Hill Book Company, Inc., New York, 1950.
23. Prausnitz, J. M.: *A.I.Ch.E. J.*, **5**, 3 (1959).
24. Redlich, O., and J. N. S. Kwong: *Chem. Revs.*, **44**, 233 (1949).
25. Rzasa, M. J., E. D. Glass, and J. B. Opfell: *Chem. Eng. Progr. Symposium Ser.*, **48**(2), 28 (1952).
26. Schiller, F. C., and L. N. Canjar: *Chem. Eng. Progr. Symposium Ser.*, **49**(7), 67 (1953).
27. Selheimer, C. W., M. Souders, Jr., R. L. Smith, and G. G. Brown: *Ind. Eng. Chem.*, **24**, 515 (1932).
28. Souders, M., Jr., C. W. Selheimer, and G. G. Brown: *Ind. Eng. Chem.*, **24**, 517 (1932).
29. Stotler, H. H., and M. Benedict: *Chem. Eng. Progr. Symposium Ser.*, **49**(6), 25 (1953).
30. White, R. R., and G. G. Brown: *Ind. Eng. Chem.*, **34**, 1162 (1942).
31. Winn, F. W.: *Chem. Eng. Progr. Symposium Ser.*, **48**(2), 121 (1952).

CHAPTER 2

DISTRIBUTION COEFFICIENTS
Nonideal Liquid Systems

The ideal liquid solution is a very good approximation for close-boiling homologs but falls short of reality in other systems. The applicability of the assumption of ideal solutions to a given problem depends upon the degree of accuracy required in the calculations as well as the molecular structure of the components to be handled. Rough designs for preliminary estimates can be based upon the assumption of ideal liquid solutions even in systems where it is known to be widely in error. For the final design, however, it is usually important that the nonideality of the solutions be taken into account as accurately as possible. This means that in the majority of the systems with which the chemical engineer must deal, the activity coefficient cannot be taken as unity and the system cannot be handled by the correlations described in Sec. 1-4.

Any dissimilarity among components is sufficient to cause nonideal-liquid-solution behavior. The dissimilarity may take the form of a difference in molecular weight or a difference in molecular structure. The former effect is illustrated by the behavior of light hydrocarbons such as methane, ethylene, etc., in mixtures of heavier paraffins or crude oil. The following tabulation shows the dependence of K values for methane upon the molecular weight of the other component in the binary system.

Binary	t, °F	p, psia	Methane K
$C_1 - n\text{-}C_3$	104	588	3.64
$C_1 - i\text{-}C_4$	100	600	4.10
$C_1 - n\text{-}C_4$	100	600	4.40
$C_1 - n\text{-}C_5$	100	600	4.95
$C_1 - n\text{-}C_{10}$	100	600	5.65

A more extreme example of the effect of molecular weight is the immiscibility of propane with heavy crude residuum. The effect of differences in molecular structure is evident in the nonideal behavior of aromatic-paraffin systems even when the compounds involved have similar molecular weights. Figure 2-1 shows the magnitude of the

29

activity coefficients of hexane in aromatics with molecular weights similar to hexane. The nonideality in paraffin-naphthene mixtures is less marked but sufficient to influence the design of separation equipment. Even the relatively slight difference between propane and propylene is sufficient to affect markedly the number of stages required for their separation. If the system under consideration includes compounds which contain oxygen, nitrogen, or sulfur, the chances for intermolecular

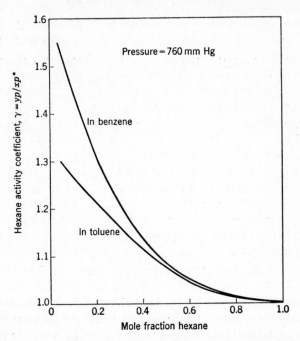

FIG. 2-1. Magnitude of deviations of hexane in aromatics with similar molecular weights. [*Data of H. S. Myers, Ind. Eng. Chem.*, **47**, 2215 (1955).]

interactions which cause nonideal behavior are greatly increased. Deviations of such systems are presented in Chap. 11, which discusses azeotropic and extractive distillation systems. Chemical dissimilarity plus the difference in molecular weight causes the "inerts" such as carbon dioxide, hydrogen, nitrogen, hydrogen sulfide, etc., to behave nonideally in mixtures with heavier components.

The nonideal behavior of chemically dissimilar molecules is often due to a particular type of intermolecular attraction called hydrogen bonding. Molecules which contain atoms such as oxygen, chlorine, fluorine, or nitrogen tend to be polar. A dipole is created because the electrons in the bonds between these atoms and hydrogen are not equally shared. The electrons tend to lie closer to the larger atoms. As a result the larger atoms become negatively charged with respect to the hydrogen, which becomes the positive end of the dipole. In a solution of polar substances, the molecules tend to arrange themselves so that the electron or charge deficiency of the hydrogen atoms is satisfied by an inter-molecular bond with a "donor" or negatively charged atom. These "hydrogen bonds" have energies of the order of several kilocalories per gram mole. The bonds may cause the formation of bimolecular com-plexes between like or unlike molecules, or they may form chainlike or three-dimensional aggregates between large numbers of molecules. The mixing of two components which results in the formation or destruction of hydrogen bonds is accompanied by relatively large heat effects, and the volatilities of the two components are altered as are all the thermo-dynamic properties. Ewell, Harrison, and Berg (6) have divided all liquids into five classes according to their tendency to form hydrogen bonds. This classification is useful for the qualitative prediction of nonideal behavior, and its use is discussed later in the chapter on azeo-tropic and extractive distillation. Bondi and Simkin (2) have discussed the effect of hydrogen bonding on the heats of vaporization. Prigogine (19) presents a good discussion of the hydrogen bond and of the spec-troscopic and thermodynamic properties of associated solutions.

Deviations from ideal behavior have been attributed (10, 22) to differences in internal pressures of the respective components. The name *internal pressure* has been applied to the left side of the thermo-dynamic equation

$$\left(\frac{\partial E}{\partial V}\right)_T = T\left(\frac{\partial p}{\partial T}\right)_V - p$$

The $T(\partial p/\partial T)_V$ term is called the thermal pressure and, for temperatures below the boiling point, is much greater than the external pressure p. The difference between the thermal pressure, which tends to expand the liquid, and the external pressure, which tends to compress it, is the internal pressure. The internal pressure represents the balance of the forces necessary to retain the material in the liquid and is therefore affected by any intermolecular interactions within the solution. The internal pressure can be approximated (10) for many liquids as shown in the following equation, where the internal pressure ($\partial E/\partial V)_T$ is repre-sented by p_i.

$$p_i \cong \frac{41.3\,\Delta E_v}{V} = \frac{41.3(\Delta H_v - RT)}{V}$$

where p_i = internal pressure, atm
ΔE_v = internal energy of vaporization, cal/g mole
ΔH_v = enthalpy of vaporization, cal/g mole
R = gas low constant, 1.987 cal/(g mole)(°K)
T = absolute temperature, °K
V = molar volume of liquid, cm³/g mole

Treybal (26) presents a table of calculated internal pressures for various substances and shows how differences in internal pressure can be used to indicate probable relative selectivities of two solvents for a solute.

The theories of hydrogen bonding and intermolecular interactions in general have not been developed to a point where the quantitative prediction of nonideal behavior from pure component properties is possible. Pierotti, Deal, and Derr (18) have presented a method for the prediction of activity coefficients at infinite dilution which is considerably more quantitative than the internal pressure approach or the classification of Ewell et al. in the prediction of the behavior of a mixture. Their method is based on an empirical equation which relates the infinite dilution activity coefficients to constants which are functions of the molecular structures of the solute and solvent. The equation is useful in the selection of solvents for such processes as extraction and azeotropic or extractive distillation, and its use will be more fully described in the chapter on azeotropic and extractive distillation.

Although the methods described above indicate the direction and probable extent of deviations from Raoult's law, they do not provide a complete set of accurate equilibrium data for a nonideal system. Such data must be obtained experimentally, and the deviation from ideal behavior correlated in terms of the activity coefficient γ. The activity coefficients are calculated from the experimental pressure-temperature-composition data by either Eq. (1-13) or (1-14). At low pressures and in the absence of any association in the gas phase, the activity coefficients are usually presented as

$$\gamma_i = \frac{y_i p}{x_i p_i^*}$$

It can be seen that this ratio measures the agreement with or departure from Raoult's law. The numerator is the partial pressure that actually exists in the vapor phase, while the denominator represents the partial pressure that would exist if Raoult's law held. If the activity coefficient is greater than unity, the deviation from Raoult's law is said to be positive since log γ will be a positive number. Conversely, if the activity coefficient is less than unity, log γ will be a negative number and the deviation from Raoult's law is said to be negative. Figures 2-2 and 2-3 show typical curves for systems with positive and negative deviations.

The experimental determination of thermodynamically consistent activity coefficients is a difficult task when only two components are involved. It becomes correspondingly more difficult as the number of components is increased. Because of the experimental inaccuracies involved, it is mandatory that the data be checked against the basic thermodynamic relationships for phase equilibria. The most useful of

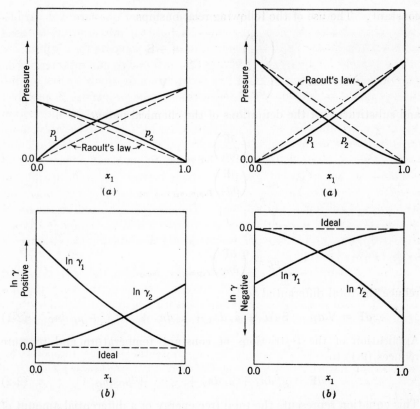

FIG. 2-2. Typical isothermal curves showing the direction of deviation from ideality
of (a) the partial pressures and (b) the
activity coefficients for a system with
positive deviations from Raoult's law.

FIG. 2-3. Typical isothermal curves showing the direction of deviation from ideality
of (a) the partial pressures and (b) the
activity coefficients for a system with
negative deviations from Raoult's law.

these relationships is the Gibbs-Duhem equation, and much of this
chapter will be devoted to its use in the correlation of phase-equilibrium
data.

2-1. Gibbs-Duhem Equation

The change in total free energy of a solution of C components can be
expressed as the total differential

$$ d\mathbf{F} = \frac{\partial \mathbf{F}}{\partial p} \, dp + \frac{\partial \mathbf{F}}{\partial T} \, dT + \frac{\partial \mathbf{F}}{\partial n_1} \, dn_1 + \frac{\partial \mathbf{F}}{\partial n_2} \, dn_2 + \cdots + \frac{\partial \mathbf{F}}{\partial n_C} \, dn_C $$

where each of the partials is taken holding all the other variables constant. The use of the following relationships

$$\left(\frac{\partial F}{\partial T}\right)_{p, n_1, \ldots, n_C} = -S$$

$$\left(\frac{\partial F}{\partial p}\right)_{T, n_1, \ldots, n_C} = V$$

and substitution of the definitions of the chemical potentials

$$\mu_1 = \left(\frac{\partial F}{\partial n_1}\right)_{T, p, n_2, \ldots, n_C}$$

$$\mu_2 = \left(\frac{\partial F}{\partial n_2}\right)_{T, p, n_1, n_3, \ldots, n_C}$$

$$\cdot$$
$$\cdot$$
$$\cdot$$

$$\mu_C = \left(\frac{\partial F}{\partial n_C}\right)_{T, p, n_1, \ldots, n_{C-1}}$$

reduce the total differential to

$$dF = V\,dp - S\,dT + \mu_1\,dn_1 + \mu_2\,dn_2 + \cdots + \mu_C\,dn_C \qquad (2\text{-}1)$$

Application of the restrictions of constant temperature and pressure reduces (2-1) to

$$dF = \mu_1\,dn_1 + \mu_2\,dn_2 + \cdots + \mu_C\,dn_C \qquad (2\text{-}2)$$

This equation represents the total free energy of a differential amount of solution formed when differential amounts $(dn_1, dn_2,$ etc.) of the various components with partial molar free energies $(\mu_1, \mu_2,$ etc.) are mixed together. A finite amount of solution formed by summing many differential amounts of the same composition, temperature, and pressure as the original material has a total free energy described by

$$F = \mu_1 n_1 + \mu_2 n_2 + \cdots + \mu_C n_C \qquad (2\text{-}3)$$

Differentiation of Eq. (2-3) without any restrictions provides an expression for the total differential of F which can be compared with Eq. (2-1) to show that

$$V\,dp - S\,dt = n_1\,d\mu_1 + n_2\,d\mu_2 + \cdots + n_C\,d\mu_C \qquad (2\text{-}4)$$

This is the general form of the Gibbs-Duhem equation and will be used later in the derivation of the constant-temperature and constant-pressure forms of the equation for binary systems. At the present time, the restrictions of constant temperature and pressure are applied to give

$$n_1\,d\mu_1 + n_2\,d\mu_2 + \cdots + n_C\,d\mu_C = 0$$

or, in terms of mole fractions,

$$x_1 \, d\mu_1 + x_2 \, d\mu_2 + \cdots + x_C \, d\mu_C = 0 \qquad (2\text{-}5)$$

Since

$$\mu_i = \frac{\partial \mathbf{F}}{\partial n_i} = \bar{F}_i$$

the chemical potential is related to the fugacity by

$$d\mu_i = d\bar{F}_i = RT \, d \ln \bar{f}_i \qquad (2\text{-}6)$$

Integration of (2-6) with the standard state as the lower limit gives

$$\mu_i - \mu_i{}^\circ = RT \ln \frac{\bar{f}_i}{f_i{}^\circ}$$

Substitution of $\bar{a}_i = \bar{f}_i / f_i{}^\circ$ gives the relationship between the chemical potential and the activity,

$$\mu_i - \mu_i{}^\circ = RT \ln \bar{a}_i$$

Or in differentiated form,

$$d\mu_1 = RT \, d \ln \bar{a}_i \qquad (2\text{-}7)$$

Equations (2-6) and (2-7) can be used to rewrite Eq. (2-5) in terms of $\bar{f}_i$ or $\bar{a}_i$ as follows:

$$x_1 \, d \ln \bar{f}_1 + x_2 \, d \ln \bar{f}_2 + \cdots + x_C \, d \ln \bar{f}_C = 0 \qquad (2\text{-}8)$$
$$x_1 \, d \ln \bar{a}_1 + x_2 \, d \ln \bar{a}_2 + \cdots + x_C \, d \ln \bar{a}_C = 0 \qquad (2\text{-}9)$$

Since $\bar{a}_1 = \gamma_i x_i$ for a liquid and $dx_1 + dx_2 + \cdots + dx_C = 0$, Eq. (2-9) can be written in terms of the liquid activity coefficients.

$$x_1 \, d \ln \gamma_1 + x_2 \, d \ln \gamma_2 + \cdots + x_C \, d \ln \gamma_C = 0 \qquad (2\text{-}10)$$

Equations (2-8), (2-9), and (2-10) can be applied rigorously only to data obtained at constant temperature and pressure. A binary system has only two degrees of freedom, and either temperature or pressure must be varied if equilibrium data are to be obtained over the entire composition range. The forms of the Gibbs-Duhem equation which are valid for isobaric and isothermal data will now be derived after the manner of Ibl and Dodge (12).

Isothermal Binary Form. The constant-temperature binary form of Eq. (2-4) when written for 1 mole of liquid solution is

$$V \, dp = x_1 \, d\mu_1 + x_2 \, d\mu_2$$

Substitution from (2-6) gives

$$\frac{V}{RT} \, dp = x_1 \, d \ln \bar{f}_1 + x_2 \, d \ln \bar{f}_2 \qquad (2\text{-}11)$$

Since $\bar{a}_i = \bar{f}_i/f_i^\circ$, Eq. (2-11) can be written in terms of activities as

$$\frac{V}{RT}\, dp = x_1\, d\ln \bar{a}_1 + x_2\, d\ln \bar{a}_2 + x_1\, d\ln f_1^\circ + x_2\, d\ln f_2^\circ$$

All the differentials are a function of x_1, so the equation can be written as

$$\frac{V}{RT}\frac{dp}{dx_1} = \frac{d\ln \bar{a}_1}{d\ln x_1} - \frac{d\ln \bar{a}_2}{d\ln x_2} + \frac{d\ln f_1^\circ}{d\ln x_1} - \frac{d\ln f_2^\circ}{d\ln x_2} \tag{2-12}$$

If the standard state is chosen as the pure liquid at the temperature and pressure of the system, the last two terms will not be zero as they were in the derivation of Eqs. (2-9) and (2-10) when both pressure and temperature were constant. In isothermal systems, the pressure of the system varies with composition and therefore f_1° and f_2° are functions of x also. The activity coefficient can be introduced by use of the definition $\bar{a}_i = \gamma_i x_i$. Making this substitution and noting that

$$\frac{d\ln x_1}{d\ln x_1} = \frac{d\ln x_2}{d\ln x_2} = 1.0$$

change Eq. (2-12) to

$$\frac{V}{RT}\frac{dp}{dx_1} = \frac{d\ln \gamma_1}{d\ln x_1} - \frac{d\ln \gamma_2}{d\ln x_2} + \frac{d\ln f_1^\circ}{d\ln x_1} - \frac{d\ln f_2^\circ}{d\ln x_2} \tag{2-13}$$

The definition of fugacity provides the relation

$$\frac{d\ln f_i^\circ}{dp} = \frac{V_i^\circ}{RT}$$

which together with the fact that each of the last two terms in (2-12) can be rewritten in the form

$$\frac{x_i\, d\ln f_i^\circ}{dp}\frac{dp}{dx_i}$$

permits (2-13) to be rewritten as

$$(V - x_1 V_1^\circ - x_2 V_2^\circ)\frac{1}{RT}\left(\frac{\partial p}{\partial x_1}\right)_T = x_1\frac{d\ln \gamma_1}{dx_1} + x_2\frac{d\ln \gamma_2}{dx_1} \tag{2-14}$$

The term $(V - x_1 V_1^\circ - x_2 V_2^\circ)$ is the change in total volume which occurs when x_1 and x_2 moles of components 1 and 2, respectively, are mixed to give 1 mole of mixture which has a total volume $V = x_1\bar{V}_1 + x_2\bar{V}_2$. Comparison of Eq. (2-14) with the binary form of (2-10) shows that the left side of (2-14) is the correction term which takes into account the variation of pressure with composition for isothermal data. The volume change of the liquids upon mixing is usually much smaller numerically than RT, and therefore, the correction for the nonconstancy of pressure is usually negligible.

Isobaric Binary Form. The constant-pressure binary form of Eq. (2-4) when written for 1 mole of liquid solution is

$$-S\,dt = x_1\,d\mu_1 + x_2\,d\mu_2$$

Since

$$\mu_i = RT \ln \bar{f}_i - RT \ln f_i^\circ + \mu_1^\circ \tag{2-15}$$

then

$$\left(\frac{\partial \mu_i}{\partial T}\right)_p = RT\left(\frac{\partial \ln \bar{f}_i}{\partial T}\right)_p - RT\left(\frac{\partial \ln f_i^\circ}{\partial T}\right)_p + R \ln \frac{\bar{f}_i}{f_i^\circ} + \left(\frac{\partial \mu_i^\circ}{\partial T}\right)_p$$

Also

$$-S = \frac{\bar{F} - \bar{H}}{T} = x_1 \frac{\bar{F}_1}{T} + x_2 \frac{\bar{F}_2}{T} - x_1 \frac{\bar{H}_1}{T} - x_2 \frac{\bar{H}_2}{T}$$

Substitution for $-S$, $d\mu_1$, and $d\mu_2$ gives

$$x_1 \frac{\bar{F}_1}{T} + x_2 \frac{\bar{F}_2}{T} - x_1 \frac{\bar{H}_1}{T} - x_2 \frac{\bar{H}_2}{T}$$

$$= x_1 RT \left(\frac{\partial \ln \bar{f}_1}{\partial T}\right)_p - x_1 RT \left(\frac{\partial \ln f_1^\circ}{\partial T}\right)_p + x_1 R \ln \frac{\bar{f}_1}{f_1^\circ} + x_1 \left(\frac{\partial \mu_1^\circ}{\partial T}\right)_p$$

$$+ x_2 RT \left(\frac{\partial \ln \bar{f}_2}{\partial T}\right)_p - x_2 RT \left(\frac{\partial \ln f_2^\circ}{\partial T}\right)_p + x_2 R \ln \frac{\bar{f}_2}{f_2^\circ} + x_2 \left(\frac{\partial \mu_2^\circ}{\partial T}\right)_p$$

From (2-15),

$$x_i \frac{\bar{F}_i}{T} - x_i R \ln \frac{\bar{f}_i}{f_i^\circ} = x_i \frac{F_1^\circ}{T}$$

Also

$$\left(\frac{\partial \mu_i^\circ}{\partial T}\right)_p = \left(\frac{\partial F_i^\circ}{\partial T}\right)_p = -S_i^\circ = \frac{F_i^\circ - H_i^\circ}{T}$$

The equation can now be rewritten as

$$x_1 \frac{F_1^\circ}{T} + x_2 \frac{F_2^\circ}{T} - x_1 \frac{\bar{H}_1}{T} - x_2 \frac{\bar{H}_2}{T}$$

$$= x_1 RT \left(\frac{\partial \ln \bar{f}_1}{\partial T}\right)_p - x_1 RT \left(\frac{\partial \ln f_1^\circ}{\partial T}\right)_p + x_1 \frac{F_1^\circ}{T} - x_1 \frac{H_1^\circ}{T}$$

$$+ x_2 RT \left(\frac{\partial \ln \bar{f}_2}{\partial T}\right)_p - x_2 RT \left(\frac{\partial \ln f_2^\circ}{\partial T}\right)_p + x_2 \frac{F_2^\circ}{T} - x_2 \frac{H_2^\circ}{T}$$

Making the substitution $\bar{f}_i = \gamma_i x_i f_i^\circ$ and using the fact that

$$\frac{\partial \ln x_1}{\partial \ln x_1} = \frac{\partial \ln x_2}{\partial \ln x_2} = 1.0$$

give, after cancellation of like terms,

$$[(x_1 H_1^\circ + x_2 H_2^\circ) - H]\frac{1}{RT^2}\left(\frac{\partial T}{\partial x_1}\right)_p = x_1 \frac{d \ln \gamma_1}{dx_1} + x_2 \frac{d \ln \gamma_2}{dx_2} \tag{2-16}$$

The term $(x_1H_1^\circ + x_2H_2^\circ)$ represents the enthalpy of x_1 and x_2 moles of 1 and 2, respectively, in the standard state (pure liquid component at temperature and pressure of the system). The term $H = x_1\bar{H}_1 + x_2\bar{H}_2$ represents the enthalpy of the 1 mole of mixture which is formed when x_1 moles of 1 and x_2 moles of 2 are mixed. The difference between the two terms as written in Eq. (2-16) is the negative of the integral heat of solution q_s per mole of mixture. Comparison of (2-16) with the binary form of (2-10) shows that the left side of (2-16) is the correction term which takes into account the variation of temperature with composition for isobaric data. The magnitude of the correction term is determined by the slope of the T vs x_1 curve (a function of the spread between the pure component boiling points) and by the size of the heat of mixing. Generally, the correction term will be important if the boiling-point spread is more than a few degrees unless the two components are from the same homologous series and therefore have a very low heat of mixing.

2-2. Thermodynamic Consistency of Equilibrium Data

The Gibbs-Duhem equation is used to check phase-equilibrium data for thermodynamic consistency. It is important that the consistency of the data be checked against the Gibbs-Duhem equation and not one of the integrated forms described in the next section. All the integrated forms will produce log γ vs x curves which are consistent with the Gibbs-Duhem data. However, a particular set of data may be thermodynamically consistent (that is, it agrees with the Gibbs-Duhem equation) and still not agree with any of the suggested integrated forms of the Gibbs-Duhem equation. Agreement with one of the integrated forms is sufficient to show that a set of data is consistent, but disagreement is not sufficient to show that the data are inconsistent. In the case of disagreement a consistency test against the Gibbs-Duhem equation will show whether the disagreement is due to inadequacy on the part of the integrated solution or to inherent inaccuracy in the data.

Since equilibrium data are always obtained at either constant temperature or constant pressure, a rigorous check on the data can be made through application of Eq. (2-14) for isothermal data or Eq. (2-16) for isobaric data. Both of these possibilities are discussed below.

Isothermal Binary Data. As pointed out in the previous section in the discussion of Eq. (2-14), the correction term

$$(V - x_1V_1^\circ - x_2V_2^\circ) \frac{1}{RT}\left(\frac{\partial p}{\partial x_1}\right)_T$$

which takes into account the effect of pressure change on the activity coefficients is entirely negligible in most cases. Therefore, the isothermal, isobaric form of the Gibbs-Duhem equation

$$x_1 \, d \log \gamma_1 + x_2 \, d \log \gamma_2 = 0 \qquad (2\text{-}10)$$

can be applied with little error to isothermal binary data. The most convenient form of (2-10) for this purpose has been presented by Redlich and Kister (20) and Herington (8).

$$\int_{x_1=0}^{x_1=1.0} \log \frac{\gamma_1}{\gamma_2} \, dx_1 = 0 \qquad (2\text{-}17)$$

Equation (2-17) can be derived by writing Eq. (2-10) for a binary and collecting terms to give

$$x_1 \, (d \log \gamma_1 - d \log \gamma_2) = - \, d \log \gamma_2$$

The left side of this equation is part of the derivative of $x_1(\log \gamma_1 - \log \gamma_2)$. The derivative of this product can be used to rewrite the equation as

$$\log \frac{\gamma_1}{\gamma_2} \, dx_1 = d \log \gamma_2 + d \, [x_1 \, (\log \gamma_1 - \log \gamma_2)]$$

Integrating between the limits of $x_1 = 0$ and $x_1 = 1.0$ and noting that $\log \gamma_2 = 0$ at $x_1 = 0$ and $\log \gamma_1 = 0$ at $x_1 = 1.0$ give Eq. (2-17).

Equation (2-17) is applied to binary systems by plotting $\log (\gamma_1/\gamma_2)$ vs x_1 and drawing a curve through the experimental data points. Such plots are shown in Figs. 2-6 through 2-9. If the data are correct (and if the effect of variable pressure is negligible), the curve which best fits the points will also give equal areas above and below the

$$\log \frac{\gamma_1}{\gamma_2} = 0$$

line. A curve which meets the equal-area test is consistent with the Gibbs-Duhem equation. Besides furnishing a quick test of the data, the $\log (\gamma_1/\gamma_2)$ curve can be used to provide constants for the Redlich-Kister correlating equation as explained in Sec. 2-3.

Dodge (5) has suggested the following integrations as a way of deriving a pair of mutually consistent $\log \gamma$ curves which represent the experimental points as well as possible. Since $\log \gamma_1 = 0$ at $x_1 = 1.0$, an integration of the binary form of Eq. (2-10) gives

$$(\log \gamma_1)_{x_1=x} = - \int_{x=1.0}^{x_1=x} \frac{1 - x_1}{x_1} \, d \log \gamma_2 \qquad (2\text{-}18)$$

Similarly, since $\log \gamma_2 = 0$ at $x_1 = 0$,

$$(\log \gamma_2)_{x_1=x} = - \int_{x_1=0}^{x_1=x} \frac{x_1}{1 - x_1} \, d \log \gamma_1 \qquad (2\text{-}19)$$

The experimental data are plotted as $\log \gamma$ vs x_1, and one or the other of the log γ curves drawn in a position which seems to fit the data best. This "assumed" curve is then used to calculate the other log γ curve by a graphical integration of the correct one of Eq. (2-18) or (2-19). If the calculated curve does not represent the data well, it is shifted to a position halfway between the points and its original position and is then used to calculate the originally assumed curve. This back-and-forth procedure is continued until a pair of mutually consistent curves are obtained which fit the experimental data points as well as possible. If the calculated curves cannot be made to fit the data, either the data are incorrect or the pressure variation was sufficiently large to make the isothermal, isobaric form of the Gibbs-Duhem equation inapplicable.

Redlich, Kister, and Turnquist (21) have presented a sensitive method for the selection of the best of two sets of data when Eqs. (2-17) to (2-19) fail to distinguish between the two. The method is based upon the prediction of the slope of the dew-point curve on the t-x diagram.

Equation (2-10) can be applied directly to binary data without integration. Since both γ_1 and γ_2 are functions of x, the equation can be divided by dx_1 and rearranged in the following "slope" form:

$$\frac{d \log \gamma_1}{dx_1} = - \frac{x_2}{x_1} \frac{d \log \gamma_2}{dx_1} \qquad (2\text{-}20)$$

The consistency of the data at any point can be checked by this relationship between the slopes of the log γ vs x_1 curves. This method depends upon graphical differentiation and is subject to the same assumptions and restrictions as the preceding equations.

The Gibbs-Duhem equation written in terms of the activity coefficients cannot be applied to systems where one component is above its critical temperature. Calculation of the activity coefficients requires a knowledge of the vapor pressures, and extrapolation of the vapor-pressure curve past the critical point is risky. In such cases the Gibbs-Duhem equation must be written in terms of the fugacities or, more conveniently, in terms of the distribution coefficients. This procedure has been described by Adler et al. (1).

Isothermal Ternary Data. A convenient way of presenting activity coefficient data for ternary systems is to plot constant-activity-coefficient curves on triangular composition diagrams. Equations similar to Eq. (2-17) which are useful for checking the thermodynamic consistency of the ternary experimental coefficients have been presented by Herington (9) and can be derived as follows: Writing equation (2-10) for three components and then eliminating one of the x's (x_2 will be eliminated for the purpose of illustration) give

$$x_1(d \log \gamma_1 - d \log \gamma_2) + d \log \gamma_2 + x_3(d \log \gamma_3 - d \log \gamma_2) = 0$$

Inspection of the first and third terms in this equation shows that they are parts of the derivatives of $[x_1(\log \gamma_1 - \log \gamma_2)]$ and $[x_3(\log \gamma_3 - \log \gamma_2)]$, respectively. Substitution for the first and third terms from these derivatives gives

$$\log \frac{\gamma_2}{\gamma_3} dx_3 - \log \frac{\gamma_1}{\gamma_2} dx_1 = -d \log \gamma_2 + d\, [x_3(\log \gamma_2 - \log \gamma_3)]$$
$$-d\, [x_1(\log \gamma_1 - \log \gamma_2)]$$

A choice must now be made as to the path of integration across the ternary diagram. Two choices are convenient. The integration can be at constant x_2, and in this case $dx_1 = -dx_3$. Or the integration can be at a constant ratio of x_1 to x_3, $r = x_1/x_3$, and in this case $dx_1 = r\, dx_3$.

For each of the two choices described, three equations can be derived depending upon whether x_1, x_2, or x_3 is eliminated in the original equation. These six equations can be expressed in terms of two cyclic general equations, one for each type of integration path. Along a path of constant x_i,

$$\int_{x_j=0}^{x_k=0} \log \frac{\gamma_j}{\gamma_k} dx_j = (x_i \log \gamma_i + x_j \log \gamma_j)_{x_k=0}$$
$$- (x_i \log \gamma_i + x_k \log \gamma_k)_{x_j=0} \quad (2\text{-}21)$$

Or along a path where the ratio $r = x_j/x_i$ is constant,

$$\int_{x_k=0}^{x_k=1.0} \log \frac{\gamma_k^{1+r}}{\gamma_i \gamma_j^{r}} dx_i = (x_i \log \gamma_i + x_j \log \gamma_j)_{x_k=0} \quad (2\text{-}22)$$

Often, the use of Eq. (2-22) over the entire range from $x_k = 0$ to $x_k = 1.0$ will involve such a large temperature change as to cast doubt on the validity of the results. In such cases, the temperature change can be reduced by integrating only to the mid-point of the ternary diagram along a line where $x_j/x_i = 1.0$. The form of Eq. (2-22) for this special purpose is

$$\int_{x_k=0}^{x_k=\frac{1}{3}} \log \frac{\gamma_k^{2}}{\gamma_i \gamma_j} dx_i = \frac{1}{2} [\log \gamma_i \gamma_j]_{x_k=0} - \frac{1}{3} [\log \gamma_i \gamma_j \gamma_k]_{x_k=\frac{1}{3}} \quad (2\text{-}23)$$

The application of Eqs. (2-21) to (2-23) involves plotting all the experimentally determined ternary activity coefficients within the ternary composition diagrams and all the limiting binary coefficients along the sides. The "best" constant-activity-coefficient curves are drawn, and their consistency with the Gibbs-Duhem equation checked by graphically evaluating Eq. (2-21), (2-22), or (2-23) along various paths across the composition diagrams. The equations can be applied rigorously only along constant-temperature and -pressure paths, but in the case of iso-thermal data the pressure correction is normally negligible. Even for isobaric data it is often possible in ternary systems to select paths that closely approximate isothermal conditions and the equations can be used to check portions of the diagrams in essentially a rigorous manner. If the data scatter badly, the equations are an invaluable aid in the construction of consistent activity-coefficient curves that fit the data as well as possible.

Schuhmann (23) has shown how the mutual consistency of the constant-activity- or activity-coefficient curves can be checked by an inspection of their slopes at some point in the composition triangle. The necessary graphical construction is shown in Fig. 2-4. The tangents to each

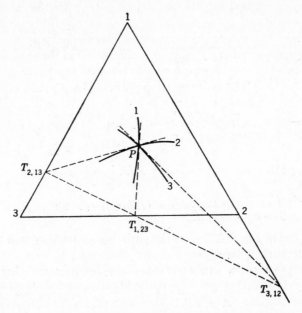

Fig. 2-4. Geometric relationship of the intercepts of tangents to the constant-activity-coefficient curves at point P. [R. Schuhmann, Acta Met., **3**, 219 (1955).]

of the constant-activity curves are drawn at some point P. The tangents are extended to the intercepts $T_{1,23}$, $T_{2,13}$, and $T_{3,12}$. The first subscript on the T's refers to the curve to which the tangent belongs, and the second pair to the side intercepted. It can be shown that the three intercepts should fall on a straight line. The construction can be used to check the mutual consistency of the three curves or to indicate the direction of an unknown curve at a point when the other two curves are known. The method provides a sensitive check on the slopes of the three curves but is subject to the same restrictions as the preceding equations.

Isobaric Binary Data. Equation (2-16) can be written as

$$x_1 \frac{d \ln \gamma_1}{dx_1} = x_2 \frac{d \ln \gamma_2}{dx_2} - [H - (x_1 H_1^\circ + x_2 H_2^\circ)] \frac{1}{RT^2} \left(\frac{\partial T}{\partial x_1}\right)_p$$
$$(2\text{-}16)$$

Ibl and Dodge (12) have shown that in the acetone–carbon disulfide system the correction term for the effect of temperature can be as much as 13 per cent of the $x_1 d \ln \gamma_1/dx_1$ term. Even in systems such as heptane and toluene the correction term is not always negligible when compared with the $x_i d \ln \gamma_i/dx_i$ terms as will be shown in Example 2-1. Whenever the correction term is not negligible, the binary data cannot be expected to satisfy Eq. (2-17). The rigorous equation for isobaric data which is analogous to (2-17) has been presented by Chao (3) and can be derived in the same manner as (2-17) except that the derivation is started with Eq. (2-16) rather than (2-10).

$$\int_{x_1=0}^{x_1=1.0} \left[\log \frac{\gamma_1}{\gamma_2} - \frac{q_s}{RT^2}\left(\frac{\partial T}{\partial x_1}\right)_p\right] dx_1 = 0 \qquad (2\text{-}24)$$

where $q_s = H - (x_1 H_1^\circ + x_2 H_2^\circ) =$ integral heat of solution per mole of solution. It can be seen from (2-24) that

$$\int_{x_1=0}^{x_1=1.0} \frac{q_s}{RT^2}\left(\frac{\partial T}{\partial x_1}\right)_p dx_1$$

represents the numerical value of the integral when Eq. (2-17) is applied to isobaric data and is therefore a measure of how far the system deviates from the "equal-area" requirement of (2-17) owing to the change in temperature with composition.

As in the case of isothermal data, the procedure of Adler et al. (1) should be used in systems where one component is above its critical temperature.

Example 2-1. Mathieson and Thynne [*J. Chem. Soc. (London)*, 1956, p. 3713] have presented the following correlation for the integral heat of solution in Btu per pound mole of solution for the heptane (1)–toluene (2) binary.

$$q_s = x_1 x_2 [953 - 19.3(x_1 - x_2) + 62.3(x_1 - x_2)^2]$$

Use this correlation to check the error involved in the application of the constant-temperature–constant-pressure form of the Gibbs-Duhem equation to the isobaric vapor-liquid equilibrium data for this system presented by Hipkin and Myers [*Ind. Eng. Chem.*, **46**, 2524 (1954)].

Solution. The rigorous form of the Gibbs-Duhem equation for isobaric data is

$$x_1 \frac{d \log \gamma_1}{dx_1} = x_2 \frac{d \log \gamma_2}{dx_2} - \frac{q_s}{2.3RT^2}\left(\frac{\partial T}{\partial x_1}\right)_p \qquad (2\text{-}16)$$

The error involved if the last term is neglected will be expressed as a ratio of the last term to the first term.

The $x_1 \, d \log \gamma_1/dx_1$ term can be estimated from Fig. 2-5, which shows the experimental $\log_{10} \gamma$ values calculated from the data of Hipkin and Myers. The $(\partial T/\partial x_1)_p$ term can be obtained from a plot of the equilibrium temperatures vs x_1. The following tabulation shows the values of the error obtained at various compositions:

x	$\left(\dfrac{\partial T}{\partial x_1}\right)_p$, °R	T, °R	q_s	$\dfrac{q_s}{2.3RT^2}\left(\dfrac{\partial T}{\partial x_1}\right)_p$	$\dfrac{x_1 \, d \log \gamma_1}{dx_1}$	% error
0.1	−44.1	685	88	−18.1 × 10⁻⁴	−0.021	8.6
0.3	−27.5	678	201	−26.3 × 10⁻⁴	−0.050	5.3
0.5	−19.5	674	238	−22.3 × 10⁻⁴	−0.058	3.8
0.7	−13.0	671	204	−13.0 × 10⁻⁴	−0.035	3.7

Despite the relatively large percentage errors incurred at certain compositions when the correction term is neglected, the effect on the $\log (\gamma_1/\gamma_2)$ vs x_1 plot is almost negligible. Figure 2-6 shows the uncorrected experimental $\log (\gamma_1/\gamma_2)$ points (round dots) and the corrected points (winged dots) calculated from Eq. (2-24). The curve represents values of $\log (\gamma_1/\gamma_2)$ predicted with a three-constant Redlich-Kister equation (see Sec. 2-3) obtained from a least-squares fit of the experimental data. Since the Redlich-Kister equation is a valid solution of the Gibbs-Duhem equation, the curve meets the equal-area requirement. By comparison with the curve, it can be seen that the experimental points, both corrected and uncorrected, deviate slightly from the equal-area requirement, particularly at high heptane concentrations.

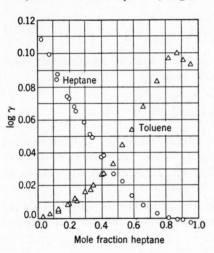

FIG. 2-5. Experimental activity coefficients for the heptane-toluene system at 1 atm. [*Hipkin and Myers, Ind. Eng. Chem.*, **46**, 2524 (1954).]

2-3. Integrated Binary Forms of the Gibbs-Duhem Equation

A general solution to the Gibbs-Duhem differential equation as written in (2-10) involves C relationships which relate the individual $\ln \gamma$'s to the liquid composition. Any set of equations which relate the activity coefficients to composition and which satisfy Eq. (2-10) comprises a solution of the Gibbs-Duhem equation. Many such solutions have been suggested for binary systems. Margules (14) was the first to express the activity coefficients as a power-series function of composition. The Margules equations so widely used at the present time are special cases of his original series. Van Laar (27) derived relationships between the binary activity coefficients and composition based on a crude model of binary molecular "clusters." Wohl (30, 29) has shown that the Margules and Van Laar equations as well as other equations of less or greater complexity can be derived from an empirical equation which relates the excess free energy of a nonideal mixture to the hypothesized interaction of molecules in aggregates of two, three, four, or more. Treybal (26) has presented a summary of Wohl's derivations. Wohl's approach to the problem is summarized below after the manner of Treybal.

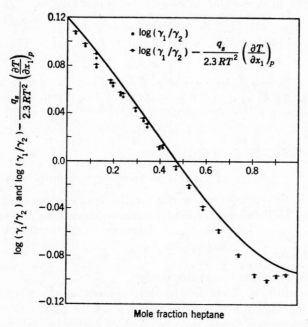

Mole fraction heptane

FIG. 2-6. Results of Example 2-1. The curve is calculated from the three-constant Redlich-Kister equation with $B_{H-T} = 0.1168$, $C_{H-T} = -0.01328$, and $D_{H-T} = -0.01144$.

The excess free energy of n moles of a nonideal mixture is defined as

$$\mathbf{F}^E = \mathbf{F}_{\text{nonideal}} - \mathbf{F}_{\text{ideal}}$$

and it can be related to the activity coefficients as follows: The free-energy change for component 1 which occurs when n_1 moles of pure 1 are put into a mixture containing components 1, 2, . . . , C is given by

$$n_1(\bar{F}_1 - F_1^\circ) = n_1 RT \ln \frac{\bar{f}_1}{f_1^\circ}$$

The standard state is taken as the pure liquid component at the temperature and pressure of the mixture. If the mixture is ideal, $\gamma_1 = 1.0$,

$\bar{a}_1 = \bar{f}_1/f_1^\circ = x_1$, and

$$n_1\bar{F}_1 = n_1RT \ln x_1 + n_1F_1^\circ$$

Similar expressions can be written for each of the C components. Addition then yields the total free energy of the ideal mixture.

$$
\begin{aligned}
\mathbf{F}_{\text{ideal}} &= n_1\bar{F}_1 + n_2\bar{F}_2 + \cdots + n_C\bar{F}_C \\
&= n_1RT \ln x_1 + n_2RT \ln x_2 + \cdots + n_CRT \ln x_C \\
&\quad + n_1F_1^\circ + n_2F_2^\circ + \cdots + n_CF_C^\circ \quad (2\text{-}25)
\end{aligned}
$$

For a nonideal solution, $\gamma \neq 1.0$ and $\bar{f}_1/f_1^\circ = \gamma_1 x_1$. The analog to Eq. (2-25) is

$$
\begin{aligned}
\mathbf{F}_{\text{nonideal}} &= n_1\bar{F}_1 + n_2\bar{F}_2 + \cdots + n_C\bar{F}_C \\
&= n_1RT \ln x_1 + n_2RT \ln x_2 + \cdots + n_CRT \ln x_C \\
&\quad + n_1RT \ln \gamma_1 + n_2RT \ln \gamma_2 + \cdots + n_CRT \ln \gamma_C \\
&\quad + n_1F_1^\circ + n_2F_2^\circ + \cdots + n_CF_C^\circ \quad (2\text{-}26)
\end{aligned}
$$

Subtraction of (2-25) from (2-26) gives

$$F^E = n_1RT \ln \gamma_1 + n_2RT \ln \gamma_2 + \cdots + n_CRT \ln \gamma_C$$

where $\mathbf{F}^E$ is the excess free energy for the entire n moles of solution. Dividing by nRT and converting to common logarithms provide

$$\frac{F^E}{2.303RT} = x_1 \log \gamma_1 + x_2 \log \gamma_2 + \cdots + x_C \log \gamma_C \quad (2\text{-}27)$$

where F^E is the molar excess free energy.

The procedure of Wohl in the development of the various Gibbs-Duhem integrations is to equate the left side of (2-27) to a series of empirical interaction terms, the size of which is determined by the number of molecules which are considered to form aggregates or complexes in the solution. Consider a binary solution of components 1 and 2 and assume that the following molecular "clusters" form in the solution: 1-2, 1-1-2, and 1-2-2. The effects of these three interactions in all their possible groups are then related empirically to the excess-free-energy function by

$$
\begin{aligned}
\frac{F^E}{2.303RT} &= x_1x_2k_{12} + x_2x_1k_{21} + x_1x_1x_2k_{112} \\
&\quad + x_1x_2x_1k_{121} + x_2x_1x_1k_{211} + x_1x_2x_2k_{122} \\
&\quad + x_2x_1x_2k_{212} + x_2x_2x_1k_{221}
\end{aligned}
$$

where the k's are constants whose subscripts reflect the order in which the concentration terms are written. Order of multiplication is unimportant, so $x_1x_2k_{12} = x_2x_1k_{21}$, etc., and

$$\frac{F^E}{2.303RT} = 2k_{12}x_1x_2 + 3k_{112}x_1^2x_2 + 3k_{122}x_1x_2^2$$

Since $x_1 + x_2 = 1.0$, the equation can be rewritten without modifying the equality as

$$\frac{F^E}{2.303RT} = 2k_{12}x_1x_2(x_1 + x_2) + 3k_{112}x_1^2x_2 + 3k_{122}x_1x_2^2$$

The constants are grouped by the following definitions:

$$A_{12} = 2k_{12} + 3k_{122}$$
$$A_{21} = 2k_{12} + 3k_{112}$$

Substitution of these relations gives

$$\frac{F^E}{2.303RT} = x_1x_2(x_1A_{21} + x_2A_{12})$$

Replacing x_1 and x_2 with the equalities

$$\frac{n_1}{n_1 + n_2} = x_1 \quad \text{and} \quad \frac{n_2}{n_1 + n_2} = x_2$$

provides

$$\frac{(n_1 + n_2)F^E}{2.303RT} = \frac{n_1n_2}{n_1 + n_2}\left(\frac{n_1A_{21}}{n_1 + n_2} + \frac{n_2A_{12}}{n_1 + n_2}\right) \tag{2-28}$$

Differentiation of

$$\mathbf{F}^E = n_1RT \ln \gamma_1 + n_2RT \ln \gamma_2$$

with respect to n_1 and n_2 gives

$$\frac{\partial \mathbf{F}^E}{\partial n_1} = \frac{\partial (n_1 + n_2)F^E}{\partial n_1} = RT \ln \gamma_1 \tag{2-29}$$

$$\frac{\partial \mathbf{F}^E}{\partial n_2} = \frac{\partial (n_1 + n_2)F^E}{\partial n_2} = RT \ln \gamma_2 \tag{2-30}$$

Differentiation of (2-28) with respect to n_1 and n_2 followed by the appropriate substitution of (2-29) or (2-30) provides the well-known three-suffix Margules equations.

$$\log \gamma_1 = (2A_{21} - A_{12})x_2^2 + 2(A_{12} - A_{21})x_2^3 \tag{2-31}$$
$$\log \gamma_2 = (2A_{12} - A_{21})x_1^2 + 2(A_{21} - A_{12})x_1^3 \tag{2-32}$$

Exercise 2-1. Show that Eqs. (2-31) and (2-32) are indeed solutions of the differential equation shown in Sec. 2-1 as Eq. (2-10).

The constants in Eqs. (2-31) and (2-32) are related to the terminal values of the activity coefficients as follows:

$$\lim_{x_1 \to 0} \log \gamma_1 = A_{12} \tag{2-33}$$

$$\lim_{x_2 \to 0} \log \gamma_2 = A_{21} \tag{2-34}$$

Determination of the constants by extrapolation to the infinite dilution values of the activity coefficients is uncertain, since the data points usually show considerable scatter in these regions. The best way to obtain the constants is by a least-squares fit of the equations to all the data points. Obviously incorrect points should be omitted (given a weight of zero). The least-squares procedure is illustrated in Appendix A.

Only one equilibrium datum point is sufficient for the determination of the constants in any two-constant binary equation. This datum point may be for either vapor-liquid or liquid-liquid equilibrium. Once the constants are known, the entire log γ curves or the binary equilibrium curve can be generated. The compilation of Horsley (11) provides azeotropic data for many systems which can be used to obtain the constants. If azeotropic data or at least one other vapor-liquid equilibrium point is not available, and if the binary system happens to be only partially miscible at some temperature not too far removed from the desired temperature, it is possible to calculate the constants from solubility data. Since activity coefficients are seldom reported for liquid systems, it is necessary to ratio the activity coefficients for a given component to give the distribution coefficient as shown by Eq. (2-55). If Eq. (2-55) is written for both components and Eqs. (2-31) and (2-32) substituted for the various γ's, two simultaneous equations are obtained which can be solved for the two constants. This procedure is illustrated in Example 2-2.

In the absence of any experimental data, the empirical equation of Pierotti et al. (18), which is described in detail in Sec. 11-1, can be used to predict infinite dilution activity coefficients which are related to the constants by Eqs. (2-33) and (2-34). The equilibrium data generated from these constants may or may not be accurate.

The largest "cluster" considered in this derivation of the Margules equation involved three molecules. Therefore, a three-suffix (or a third-degree) equation results. If the largest "cluster" contains only two molecules, two-suffix equations such as the following are obtained.

$$\log \gamma_1 = A(1 - x_1)^2 \qquad (2\text{-}35)$$
$$\log \gamma_2 = Ax_1^2 \qquad (2\text{-}36)$$

These equations are called the two-suffix Margules equations. The constant A is the same in both equations, and the log γ vs x curves must be symmetrical if the equations are to fit.

As mentioned previously the Margules equations presented above are special cases of Margules's original expression. The terms three-suffix and two-suffix arise from the manner in which Wohl carried out his derivations and have no relation to Margules's original derivation, which was based on a power series and not on a molecular model. On the other hand, Van Laar did hypothesize a molecular model (binary clusters). By appropriate arrangement of two-suffix interaction terms Wohl was also able to derive the Van Laar equations which are presented below.

$$\log \gamma_1 = \frac{A_{12}}{(1 + A_{12}x_1/A_{21}x_2)^2} \tag{2-37}$$

$$\log \gamma_2 = \frac{A_{21}}{(1 + A_{21}x_2/A_{12}x_1)^2} \tag{2-38}$$

The relationships between the constants in the Van Laar equations and the terminal values of the log γ terms are as shown in Eqs. (2-33) and (2-34) for the three-suffix Margules equations. Also the constants can be obtained from a single vapor-liquid or liquid-liquid datum point as described above.

Wohl also derived four-suffix equations. For the complete derivations of the various Wohl equations, their uses in the correlation and extrapolation of equilibrium data, and their limitations, the reader is referred to Wohl (30, 29), Treybal (26), and Perry (17).

Equations of Redlich and Kister. Redlich and Kister (20) have suggested solutions to the Gibbs-Duhem equation which are based upon the power series of Margules and therefore differ somewhat in approach from the Wohl equations. The Redlich-Kister correlations are particularly suited for extension to multicomponent systems, and time will now be taken to present the equations in some detail.

For the sake of convenience the excess-free-energy function $F^E/2.303RT$ is replaced by Q. Equation (2-27) can then be written for a binary system as follows:

$$Q = x_1 \log \gamma_1 + (1 - x_1)\log \gamma_2 \tag{2-39}$$

Whereas Wohl equated Q to a series of interaction terms, Redlich and Kister equate the excess-free-energy function for the binary 1-2 to a symmetrical, infinite power-series function of composition.

$$Q_{12} = x_1(1 - x_1)[B_{12} + C_{12}(2x_1 - 1) + D_{12}(2x_1 - 1)^2 + \cdots] \tag{2-40}$$

Other types of series could be used. Properties which recommend the $(2x_1 - 1)$ term are its symmetry with respect to the two components (substitution of x_2 for x_1 changes the sign but not the numerical value) and the fact that it meets the necessary requirement that $Q = 0$ at $x_1 = 0$ and $x_1 = 1.0$.

It was pointed out in Sec. 2-2 that a log (γ_1/γ_2) vs x_1 plot, besides checking the thermodynamic consistency of binary data, was useful in the determination of the Redlich-Kister constants. The form of the equation which is most convenient for the determination of the constants is

$$\frac{dQ_{12}}{dx_1} = \log \frac{\gamma_1}{\gamma_2} = B_{12}(1 - 2x_1) + C(6x_1x_2 - 1)$$
$$+ D(1 - 2x_1)(1 - 8x_1x_2) + \cdots \quad (2\text{-}41)$$

Equation (2-41) is obtained by differentiating the binary forms of Eqs. (2-39) and (2-40) and equating the results. The constants in Eq. (2-41) can be obtained from characteristic points on the log (γ_1/γ_2) curve as outlined in Table 2-1. It should be noted in Table 2-1 that the absolute values of log (γ_1/γ_2) must be equal at points 3 and 7 for all systems which can be represented by a three-constant equation. Therefore, a comparison of the log (γ_1/γ_2) values at $x_1 = 0.2113$ and $x_1 = 0.7887$ provides a quick check on the adequacy of a three-constant equation and, since three constants are usually more than sufficient, it normally provides a check on the data.

TABLE 2-1. CHARACTERISTIC POINTS FOR THE DETERMINATION OF THE BINARY REDLICH-KISTER CONSTANTS

No.	x	$\log \dfrac{\gamma_1}{\gamma_2}$
1	0	$B - C + D$
2	0.1464	$0.7071B - C/4$
3	0.2113	$0.5773(B - D/3)$
4	0.2959	$0.4082(B - 2D/3) + C/4$
5	0.5	$C/2$
6	0.7041	$-0.4082(B - 2D/3) + C/4$
7	0.7887	$-0.5773(B - D/3)$
8	0.8536	$-0.7071B - C/4$
9	1.0	$-B - C - D$

Redlich and Kister have suggested five types of solutions. The type to which a solution belongs depends upon whether the constants B, C, and D take on zero or nonzero values at the characteristic points listed in Table 2-1. The five solution types and convenient methods to obtain values of the constants for each are as follows:

Type 1. The simplest case is the ideal solution where $\gamma_1 = \gamma_2 = 1.0$ and log $(\gamma_1/\gamma_2) = 0$ for all values of x_1.

Type 2. The next simplest case can be represented by a straight, nonhorizontal line on the log (γ_1/γ_2) vs x_1 plot as shown in Fig. 2-7. If the equal-area requirement of Eq. (2-17) is to be met, the straight line must pass through $\log (\gamma_1/\gamma_2) = 0$ at $x_1 = 0.5$. For these systems, $B \neq 0$ and $C = D = \cdots = 0$.

Type 3. The majority of binary systems can be correlated with a two-constant equation such as the Van Laar or the three-suffix Margules equation. If $B \neq 0$, $C \neq 0$, and $D = \cdots = 0$, the Redlich-Kister equation reduces to the three-suffix Margules equation for binaries with $A_{12} = B_{12} - C_{12}$ and $A_{21} = B_{12} + C_{12}$. Since $D = 0$, the B constant

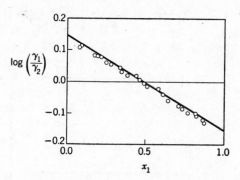

FIG. 2-7. n-Hexane-toluene system. $p = 1.0$ atm; $B = 0.153$. [*O. Redlich, A. T Kister, and C. E. Turnquist, Chem. Eng. Progr. Symposium Ser.,* **48**(2), 49 (1952).]

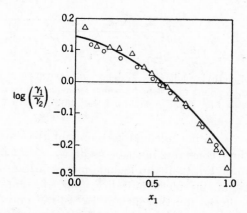

FIG. 2-8. Benzene-2,2,3-trimethylbutane system. $B = 0.118$; $C = 0.045$. $\triangle$ points from Harrison and Berg. $\circ$ points from Jost and Sieg. [*Redlich, Kister, and Turnquist, Chem. Eng. Progr. Symposium Ser.,* **48**(2), 49 (1952).]

can be determined directly from points 3 and 7. The C constant is obtained at point 5. The quantities $(B - C)$ and $(-B - C)$ must represent reasonable values at points 1 and 9. Figure 2-8 shows the $\log (\gamma_1/\gamma_2)$ vs x plot for a typical type 3 system.

Type 4.   Some alcohol-hydrocarbon systems can be approximated by an equation where $B \neq 0$, $D \neq 0$, and $C = 0$. The plot of log (γ_1/γ_2) in these cases is characterized by an S-shape curve. Since $C = 0$, the curve must pass through zero at $x = 0.5$ and the absolute values at points 1 and 9 must be equal. The values of B and D are obtained from points 2, 4, 6, and 8.

Type 5.   If three constants are required, C is obtained from point 5, B from points 2 and 8, and D from points 4 and 6 or points 3 and 7.

Inspection of the log (γ_1/γ_2) vs x plot will classify the system immediately if it belongs to type 1, 2, or 4. If it does not have the characteristics of one of these three, it belongs to either type 3 or 5. If two

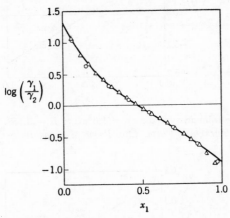

FIG. 2-9. Ethanol-methylcyclohexane system. $B = 0.920$, $C = 0.104$, $D = 0.162$, and $E = -0.075$. Δ points from Isii. ○ points from Kretschmer. [*Redlich, Kister, and Turnquist, Chem. Eng. Progr. Symposium Ser.*, **48**(2), 49 (1952).]

constants (type 3) do not adequately represent the data, the system is of type 5 and three constants must be used.

The extension of Eq. (2-40) to include more constants is obvious. If the binary exhibits extreme nonideality, four constants may be necessary. The use of the fourth constant is realistic only if the data are very precise. Figure 2-9 shows the diagram for such a system. Note the magnitude of the log (γ_1/γ_2) term and the S-shape curve typical of alcohol-hydrocarbon systems.

Any set of constants in Eq. (2-41) will generate a log (γ_1/γ_2) curve which satisfies the equal-area requirement of Eq. (2-17). The problem, therefore, is not in obtaining a consistent curve but in getting a consistent curve which fits the data. The characteristic points in Table 2-1 will provide quick approximations of the best set of constants, but unless the data are very inaccurate, it is usually more satisfactory to obtain the constants by a least-squares fit of Eq. (2-41) to the binary data points. If the data points scatter badly, the least-squares procedure may give unrealistic results unless some of the obviously incorrect points are omitted (given a weight of zero). The application of the least-squares technique is illustrated in Appendix A. As described previously in this section for Eqs. (2-31) and (2-32) the constants in a two-constant equation can be determined from a single equilibrium measurement, either vapor-liquid or liquid-liquid. It is sometimes necessary (despite the uncertainty involved) to obtain the constants for a binary from liquid-liquid solubility data (see Example 2-2) and then use the constants to generate vapor-liquid equilibrium data. Also, the equation of Pierotti et al. (18) can be used to obtain the constants as described previously for the Margules equation.

Example 2-2. Calculate the constants in the two-constant Redlich-Kister equation from solubility data at 70°C for the water (1)–trichloroethylene (2) system.

Solution. Let the symbols L and x refer to the trichloroethylene-rich phase and V and y to the water-rich phase. The solubility data at 70°C in terms of mole fractions are as follows:

$$y_1 = 0.9998 \qquad x_1 = 0.007463 \qquad K_1 = 134$$
$$y_2 = 0.0001848 \qquad x_2 = 0.9925 \qquad K_2 = 0.0001862$$

In the water-rich phase,

$$\log \gamma_1^V = B_{12}y_2^2 + C_{12}y_2^2(4y_1 - 1)$$
$$\log \gamma_2^V = B_{12}y_1^2 + C_{12}y_1^2(1 - 4y_2)$$

In the trichloroethylene-rich phase,

$$\log \gamma_1^L = B_{12}x_2^2 + C_{12}x_2^2(4x_1 - 1)$$
$$\log \gamma_2^L = B_{12}x_1^2 + C_{12}x_1^2(1 - 4x_2)$$

Since
$$K_1 = \frac{y_1}{x_1} = \frac{\gamma_1^L}{\gamma_1^V}$$

and
$$K_2 = \frac{y_2}{x_2} = \frac{\gamma_2^L}{\gamma_2^V}$$

we have

$$\log K_1 = B_{12}(x_2^2 - y_2^2) + C_{12}[x_2^2(4x_1 - 1) - y_2^2(4y_1 - 1)]$$
$$\log K_2 = B_{12}(x_1^2 - y_1^2) + C_{12}[x_1^2(1 - 4x_2) - y_1^2(1 - 4y_2)]$$

Substitution of numerical values into these equations followed by simultaneous solution for the constants provides $B_{12} = 2.93$ and $C_{12} = 0.798$.

Once the constants B_{12}, C_{12}, D_{12}, . . . have been obtained, Eq. (2-41) can be used to predict $\log(\gamma_1/\gamma_2)$ as a function of composition. The use of the Redlich-Kister equation in the form of Eq. (2-41) is often desirable for two reasons. First, the differentiation of Eqs. (2-39) and (2-40) to give (2-41) reduces the degree of the equation by one. Second, the relative volatility α in binary or multicomponent mixtures can be obtained easily from log (γ_r/γ_s) as follows:

$$\alpha_{r\text{-}s} = \frac{K_r}{K_s} = \frac{\gamma_r p_r^*}{\gamma_s p_s^*} \tag{2-42a}$$

or, from Eq. (1-15) where the gas law deviations and the effect of pressure on the liquid fugacities are taken into account,

$$\alpha_{r\text{-}s} = \frac{K_r}{K_s} = \frac{\gamma_r p_r^* \exp\{[(B_s - V_s{}^L)/RT](p - p_s^*)\}}{\gamma_s p_s^* \exp\{[(B_r - V_r{}^L)/RT](p - p_r^*)\}} \tag{2-42b}$$

When it is necessary to predict the individual activity coefficients rather than their ratio, the Redlich-Kister equations for binary systems can be used in the following forms:

$$\log \gamma_1 = x_2{}^2[B_{12} + C_{12}(4x_1 - 1) + D_{12}(x_1 - x_2)(6x_1 - 1) + \cdots]$$
$$\tag{2-43}$$
$$\log \gamma_2 = x_1{}^2[B_{12} + C_{12}(1 - 4x_2) + D_{12}(x_2 - x_1)(6x_2 - 1) + \cdots]$$
$$\tag{2-44}$$

It will be noted that (2-44) can be obtained from (2-43) by interchanging the subscripts 1 and 2 with the exception of a sign change in the coefficient of C_{12}. The sign change results from the fact that the composition term $(2x_1 - 1)$ in Eq. (2-40) changes sign when x_2 is substituted for x_1 or vice versa. Therefore, reversal of the subscripts on those constants whose coefficients in Eq. (2-40) are raised to an odd power requires an accompanying sign change.

$$B_{12} = B_{21}$$
$$C_{12} = -C_{21}$$
$$D_{12} = D_{21}$$
$$E_{12} = -E_{21}$$

Exercise 2-2. Show that when only two binary constants, B_{ij} and C_{ij}, are included in the Redlich-Kister binary equations, the equations are identical with the three-suffix Margules equations and that the relations between the constants are

$$A_{12} = B_{12} - C_{12}$$
$$A_{21} = B_{12} + C_{12}$$

Equation (2-43) can be written in a more symmetrical form, which makes obvious the form of additional terms if they are required.

$$\log \gamma_1 = B_{12}x_2(x_1 - x_2)^{-1}[x_1(1 - x_1 + x_2) - x_2]$$
$$+ C_{12}x_2(x_1 - x_2)^0[2x_1(1 - x_1 + x_2) - x_2]$$
$$+ D_{12}x_2(x_1 - x_2)^1[3x_1(1 - x_1 + x_2) - x_2] \tag{2-45}$$

The derivation of Eqs. (2-43) to (2-45) will be obvious from the following discussion of the multicomponent forms of the Redlich-Kister equations.

2-4. Multicomponent Equations

Both the Wohl and the Redlich-Kister equations can be extended to systems which contain as many components as desired. The multicomponent forms of the Wohl equations are presented in Wohl's original papers (30, 29), in Treybal (26), and in Perry (17). Numerous investigators (7, 13, 15, 16, 24) have used the Wohl equations to correlate and predict ternary and quaternary data with varying degrees of success. The use of the Redlich-Kister equations for this task will be briefly reviewed here.

The excess-free-energy function $Q_{123...c}$ of a multicomponent system can be represented by the sum of the binary functions plus terms which take into account the interactions between the various binaries. The following representation for a ternary system involves three ternary constants.

$$Q_{123} = Q_{12} + Q_{13} + Q_{23} + x_1x_2x_3[C + D_1(x_2 - x_3)$$
$$+ D_2(x_3 - x_1) + \cdots] \quad (2\text{-}46)$$

where Q_{12}, Q_{13}, and Q_{23} are the binary excess-free-energy functions which are correlated by Eq. (2-40), and C, D_1, and D_2 are ternary constants. Expansion of Eq. (2-46) to a quaternary system gives

$$Q_{1234} = Q_{12} + Q_{13} + Q_{14} + Q_{23} + Q_{24} + Q_{34}$$
$$+ x_1x_2x_3[C + D_1(x_2 - x_3) + D_2(x_3 - x_1) + \cdots]$$
$$+ x_1x_2x_4[E + F_1(x_2 - x_4) + F_2(x_4 - x_1) + \cdots]$$
$$+ x_1x_3x_4[G + H_1(x_3 - x_4) + H_2(x_4 - x_1) + \cdots]$$
$$+ x_2x_3x_4[I + J_1(x_3 - x_4) + J_2(x_4 - x_2) + \cdots]$$
$$+ x_1x_2x_3x_4[K + L_1(x_2 - x_3) + L_2(x_3 - x_4)$$
$$+ L_3(x_4 - x_1) + \cdots] \quad (2\text{-}47)$$

Equation (2-27) can be rearranged to express the activity coefficient γ_r of any given component r in terms of the excess-free-energy functions involved.

$$\log \gamma_r = Q + \frac{\partial Q}{\partial x_r} - \sum_{i=1}^{C} x_i \frac{\partial Q}{\partial x_i} \quad (2\text{-}48)$$

The partial derivatives in (2-48) are not true derivatives in the physical sense but are simply mathematical operations on Q where Q is expressed by Eq. (2-40), (2-46), or (2-47). For example, $\partial Q/\partial x_1$ indicates the derivative of Q when x_2, x_3, $\ldots$, x_c and γ_1, γ_2, $\ldots$, γ_c are considered as constants regardless of the fact that x_1 cannot be varied in a physical system unless at least one other x also varies. The reason for this special interpretation of the $\partial/\partial x_i$ operator is explained in the following derivation of Eq. (2-48).

For the purposes of this derivation let $\partial_p/\partial x_i$ denote the "physical" partial derivative, that is, the one obtained when two mole fractions are allowed to vary. To obtain such a derivative of Eq. (2-27), it is necessary to eliminate the x for one component from the equation. Let r refer to the component whose mole fraction is eliminated. Since

$$x_r = 1 - x_1 - x_2 - \cdots - x_C$$

Eq. (2-27) can be rewritten as

$$Q_{12\ldots c} = \log \gamma_r + x_1 (\log \gamma_1 - \log \gamma_r) + \cdots + x_C (\log \gamma_C - \log \gamma_r) \tag{2-49a}$$

or

$$Q_{12\ldots c} = \sum_{\substack{j=1 \\ j \neq r}}^{C} x_j \log \gamma_j + \log \gamma_r \left(1 - \sum_{\substack{j=1 \\ j \neq r}}^{C} x_j\right) \tag{2-49b}$$

where x_j can be any x except x_r. Differentiation of (2-49a) with respect to x_j gives

$$\left(\frac{\partial_p Q}{\partial x_j}\right)_{\substack{pTx_i \\ i \neq j \text{ or } r}} = \frac{\partial \log \gamma_r}{\partial x_j} + x_1 \left(\frac{\partial \log \gamma_1}{\partial x_j} - \frac{\partial \log \gamma_r}{\partial x_j}\right) + \cdots$$

$$+ x_j \left(\frac{\partial \log \gamma_j}{\partial x_j} - \frac{\partial \log \gamma_r}{\partial x_j}\right) + (\log \gamma_j - \log \gamma_r) + \cdots$$

$$+ x_C \left(\frac{\partial \log \gamma_C}{\partial x_j} - \frac{\partial \log \gamma_r}{\partial x_j}\right)$$

Collection of terms gives

$$\left(\frac{\partial_p Q}{\partial x_j}\right)_{\substack{pTx_i \\ i \neq j \text{ or } r}} = (\log \gamma_j - \log \gamma_r) + \sum_{\substack{i=1 \\ i=j \text{ and } r}}^{C} x_i \frac{\partial \log \gamma_i}{\partial x_j} \tag{2-50}$$

The last term in this equation is the constant-temperature–constant-pressure form of the Gibbs-Duhem equation and therefore zero if the restrictions of constant temperature and pressure are placed upon the equation. Making these restrictions and noting from Eq. (2-27) that

$$\frac{\partial Q}{\partial x_j} = \log \gamma_j$$

and

$$\frac{\partial Q}{\partial x_r} = \log \gamma_r$$

permit (2-50) to be written as

$$\log \gamma_j = \left(\frac{\partial_p Q}{\partial x_j}\right)_{\substack{pTx_i \\ i \neq j \text{ or } r}} + \log \gamma_r = \left(\frac{\partial Q}{\partial x_j} - \frac{\partial Q}{\partial x_r}\right) + \log \gamma_r$$

Substitution of this equation for log γ_j in Eq. (2-49b) gives

$$Q = \sum_{\substack{j=1 \\ j \neq r}}^{C} x_j \left(\frac{\partial Q}{\partial x_j} - \frac{\partial Q}{\partial x_r} \right) + \sum_{\substack{j=1 \\ j \neq r}}^{C} x_j \log \gamma_r + \log \gamma_r \left(1 - \sum_{\substack{j=1 \\ j \neq r}}^{C} x_j \right)$$

Since

$$\sum_{\substack{j=1 \\ j \neq r}} x_j \log \gamma_r = \log \gamma_r \sum_{\substack{j=1 \\ j \neq r}} x_j$$

the equation reduces to

$$\log \gamma_r = Q - \sum_{\substack{j=1 \\ j \neq r}}^{C} x_j \left(\frac{\partial Q}{\partial x_j} - \frac{\partial Q}{\partial x_r} \right)$$

Since

$$\frac{\partial Q}{\partial x_r} - \frac{\partial Q}{\partial x_r} = 0$$

the $j \neq r$ restriction can be dropped and j replaced by i. Splitting the summation term into two parts and noting that

$$\sum_{i=1}^{C} x_i \frac{\partial Q}{\partial x_r} = \frac{\partial Q}{\partial x_r}$$

permit the equation to be written as shown in Eq. (2-48).

The expression for log γ_r as a function of composition in a system of any number of components is obtained by the substitution for Q and the indicated derivatives of Q in Eq. (2-48). For a binary system, Q and the partials of Q with respect to x_1 and x_2 are obtained from Eq. (2-40) and the binary forms shown as Eqs. (2-43), (2-44), and (2-45) result. The use of Eqs. (2-46) and (2-47) provides the ternary and quaternary equations for log γ_r. Fortunately, many systems can be correlated adequately with only binary constants. This means that all ternary and higher constants can be considered to be zero and that the multicomponent excess-free-energy function can be represented as

$$Q_{12 \ldots c} = \sum_{i<j} Q_{ij} \tag{2-51}$$

where the form of the Q_{ij} expressions is given by the following equation, which is a more convenient form for this purpose of Eq. (2-40).

$$Q_{12} = x_1 x_2 [B_{12} + C_{12}(x_1 - x_2) + D_{12}(x_1 - x_2)^2 + \cdots)] \tag{2-52}$$

Performing on (2-52) the operations indicated by Eq. (2-48) provides the log γ expressions for the various components. These expressions can be represented by the followinge quation which, includes some extraneous terms to illustrate the symmetry.

$$\log \gamma_r = \sum_i B_{ri}x_i(x_r - x_i)^{-1}[x_r(1 - x_r + x_i) - x_i] - \sum_{\substack{i \neq j \\ i \neq r}} B_{ij}x_ix_j(x_i - x_j)^0$$

$$+ \sum_i C_{ri}x_i(x_r - x_i)^0[2x_r(1 - x_r + x_i) - x_i] - 2\sum_{\substack{i \neq j \\ i \neq r}} C_{ij}x_ix_j(x_i - x_j)^1$$

$$+ \sum_i D_{ri}x_i(x_r - x_i)^1[3x_r(1 - x_r + x_i) - x_i] - 3\sum_{\substack{i \neq j \\ i \neq r}} D_{ij}x_ix_j(x_i - x_j)^2$$

$$(2\text{-}53)$$

As an example, for $r = 2$, the B_{ri} terms become B_{21}, B_{23}, . . . , B_{2C}. The B_{ij} terms become B_{13}, B_{14}, . . . , B_{1C}; B_{34}, B_{35}, . . . , B_{3C}; B_{45}, B_{46}, . . . , B_{4C}; . . . $B_{(C-1)C}$. The B_{rr}, C_{rr}, and D_{rr} terms are zero and need not be included.

It can be seen that even when the ternary and higher constants are excluded the number of constants required in Eq. (2-53) is very large if the number of components C is large. The relation between the number of components C and the number of constants or terms in Eq. (2-53) is as follows:

	Terms in Eq. (2-53)
B and C constants	$C(C - 1)$
B, C, and D constants	$1.5C(C - 1)$

For cases where a suitable correlation of the data is impossible without including the ternary constants, the operations indicated by Eq. (2-48) on the ternary term in Eq. (2-46) provide the following terms for the 1-2-3 ternary which would be added to Eq. (2-53):

For $r = 1$: $+ Cx_2x_3(1 - 2x_1)$
$+ D_1x_2x_3(1 - 3x_1)(x_2 - x_3)$ $(2\text{-}54a)$
$+ D_2x_2x_3[x_1(-2 - 3x_3 + 3x_1) + x_3]$

For $r = 2$: $+ Cx_1x_3(1 - 2x_2)$
$+ D_1x_1x_3[x_2(2 - 3x_2 + 3x_3) - x_3]$ $(2\text{-}54b)$
$+ D_2x_1x_3(1 - 3x_2)(x_3 - x_1)$

For $r = 3$: $+ Cx_1x_2(1 - 2x_3)$
$+ D_1x_1x_2[x_3(-2 - 3x_2 + 3x_3) + x_2]$ $(2\text{-}54c)$
$+ D_2x_1x_2[x_3(2 - 3x_3 + 3x_1) - x_1]$

Terms similar to Eqs. (2-54a), (2-54b), and (2-54c) would be added to Eq. (2-53) for each ternary in the system for which ternary constants are necessary. The increased accuracy obtained by the addition of the ternary constants seldom justifies the additional complexity. This is usually true of the binary D constants also.

Equation (2-53) can be used to correlate liquid-liquid equilibrium data where only the distribution coefficients were measured and the activity coefficients are unknown. Since both liquid phases must be in equilibrium with the same vapor phase, the distribution coefficient between the two liquid phases is related to the liquid-phase activity coefficients by

$$K_i = \frac{y_i}{x_i} = \frac{\gamma_i{}^L}{\gamma_i{}^V} \tag{2-55}$$

where y and V refer to the raffinate phase while x and L refer to the extract phase. Combination of Eqs. (2-55) and (2-53) provides the following equation for the correlation of distribution coefficients in liquid-liquid systems:

$$\log K_r = [\text{right side of (2-53)}] - [\text{right side of (2-53) with}$$
$$y\text{'s substituted for } x\text{'s}] \tag{2-56}$$

Equation (2-56) is linear in the binary constants. It is neither linear nor explicit in the x's and y's. If the composition of one phase is known, the calculation of the other equilibrium-phase composition (other end of the tie line) involves the solution of C simultaneous, nonlinear algebraic equations which contain logarithmic terms.

As mentioned in the previous section it is sometimes more convenient to correlate the ratio of two activity coefficients because of the relation of the ratio to the relative volatility as shown in Eq. (2-42). It can be seen from Eq. (2-48) that the expression for $\log(\gamma_r/\gamma_s)$ can be obtained from

$$\log \frac{\gamma_r}{\gamma_s} = \frac{\partial Q}{\partial x_r} - \frac{\partial Q}{\partial x_s} \tag{2-57}$$

The multicomponent form with two binary constants is

$$\log \frac{\gamma_r}{\gamma_s} = \sum_i x_i(B_{ri} - B_{si}) + \sum_{i \neq r} C_{ri}x_i(2x_r - x_i) - \sum_{i \neq s} C_{si}x_i(2x_s - x_i) \tag{2-58}$$

Exercise 2-3. Show that Eq. (2-58) is cyclic; that is, the equation for $\log(\gamma_{r+1}/\gamma_{s+1})$ can be obtained from the equation for $\log(\gamma_r/\gamma_s)$ by replacing r with $r + 1$ and s with $s + 1$ according to a cyclic rotation which for a ternary can be represented by

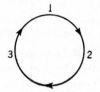

2-5. Prediction of Multicomponent Data

Analytical difficulties often restrict the number of components for which it is possible to obtain good phase-equilibrium data. In such cases it is often desirable to study only the binary systems involved and from a knowledge of the binary constants predict the multicomponent data. In the absence of ternary and quaternary data it is obviously necessary to assume that all ternary and higher constants are zero. The binary constants are found from a fit of Eq. (2-41) to each set of binary data. The binary constants are then inserted in (2-53) to predict the log γ's for any given liquid composition. If enough constants are used in (2-41) to correlate each binary adequately, and if all the binaries deviate in the same direction from Raoult's law, the predicted multi-component data will usually agree well with experimental results. See Steinhouser and White (25) for an example. If some of the binaries show positive deviations while others exhibit negative deviations, the correlation task is more difficult and the results obtained with any multi-component equation will be less accurate.

In building up a set of constants to represent a multicomponent mixture, it must be kept in mind that the various constants in the Redlich-Kister equations have no theoretical significance and their values depend upon the particular way in which the equations are fitted to the experi-mental data. For example, the binary constants obtained from a direct fit of the ternary equation to ternary equilibrium data will differ in general from the binary constants obtained when each binary is fitted individually. If a ternary constant is to be compatible with binary constants determined from binary systems, the binary constants must be inserted in the ternary equation before the ternary data are fitted. This not only produces a compatible set of constants but also reduces the number of unknown constants to be determined in the ternary fit. If three constants were to be used for each binary and three ternary constants were to be determined from the ternary data, the number of unknown constants (and the number of simultaneous equations to be solved) would be reduced from 12 to 3 if the binary constants were deter-mined previously from binary data. However, the availability of electronic computers to solve large sets of simultaneous equations makes the direct fit of available multicomponent data more practical than the laboratory determination of any binary data which may be lacking.

2-6. Effect of Temperature on Activity Coefficients

The liquid activity coefficient is defined in terms of fugacity as

$$\gamma_i = \frac{\bar{f}_i}{f_i^\circ x_i}$$

Taking the logarithm of this equation and differentiating with respect to temperature at constant pressure and composition provide

$$\left(\frac{\partial \ln \gamma_i}{\partial T}\right)_{p,x} = \left(\frac{\partial \ln \bar{f}_i}{\partial T}\right)_{p,x} - \left(\frac{\partial \ln {}_if^\circ}{\partial T}\right)_{p,x} \tag{2-59}$$

Writing the definition of fugacity in the integrated form

$$\bar{F}_i - F_i^\circ = RT \ln \bar{f}_i - RT \ln f_i^\circ$$

and differentiating with respect to temperature at constant pressure and composition give

$$\left(\frac{\partial \bar{F}_i}{\partial T}\right)_{p,x} - \left(\frac{\partial F_i^\circ}{\partial T}\right)_{p,x} = RT\left(\frac{\partial \ln \bar{f}_i}{\partial T}\right)_{p,x} + R \ln \bar{f}_i$$
$$- RT\left(\frac{\partial \ln f_i^\circ}{\partial T}\right)_{p,x} - R \ln f_i^\circ$$

Since $(\partial \bar{F}_i/\partial T)_p = -\bar{S}_i = (\bar{F}_i - \bar{H}_i)/T$, the equation can be rewritten as

$$(\bar{F}_i - F_i^\circ) - (\bar{H}_i - H_i^\circ) = RT^2\left(\frac{\partial \ln \bar{f}_i}{\partial T}\right)_{p,x} - RT^2\left(\frac{\partial \ln f_i^\circ}{\partial T}\right)_{p,x}$$
$$+ RT \ln \frac{\bar{f}_i}{f_i^\circ}$$

The first and last terms cancel. Combination of the remainder of the equation with Eq. (2-59) provides the following relationship between the activity coefficient and temperature:

$$\left(\frac{\partial \ln \gamma_i}{\partial T}\right)_{p,x} = -\frac{\bar{H}_i - H_i^\circ}{RT^2} \tag{2-60}$$

The standard state used in this text for the liquid phase is the pure component at the temperature and pressure of the system. The component in question may or may not be a liquid at these conditions, but in any case, the $\bar{H}_i - H_i^\circ$ term represents the difference in enthalpy per mole of the material between the pure (or standard) state and the solution state. Since the restriction of constant composition was used in the derivation, the $\bar{H}_i - H_i^\circ$ term must be evaluated at constant solution composition and is therefore termed the partial molar heat of solution and represented by the symbol L_i.

Equation (2-60) can be rewritten as

$$\left(\frac{\partial \ln \gamma_i}{\partial (1/T)}\right)_{p,x} = \frac{L_i}{R} \tag{2-61}$$

A plot of $\ln \gamma_i$ vs $1/T$ at constant pressure and composition provides straight lines with a slope of L_i/R. A typical plot of this sort is shown in Fig. 2-10. The top curve in Fig. 2-10 is at $x_1 = 0$, and the slope is related to the heat of solution of component 1 at infinite dilution in component 2. At $x_1 = 0$, $\ln \gamma_1 = A_{12}$ where A_{12} is a constant in the Van Laar and three-suffix Margules equations. Therefore, the top curve shows how the constant A_{12} varies with temperature.

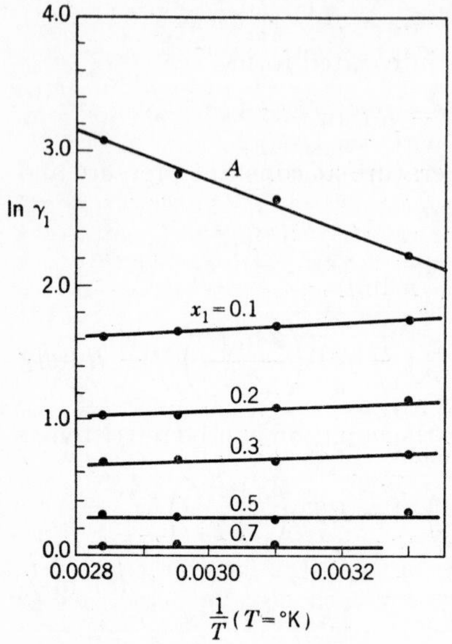

It can be seen from Eq. (2-61) that the error involved in the application of the Gibbs-Duhem equation to isobaric data depends upon the magnitude of the partial molar heat of solution. If the heat of solution is positive (heat absorbed), the activity coefficients will decrease with increasing temperature. Solutions with positive deviations from Raoult's law have positive heats of solution. If the heat of solution is negative (heat evolved), the activity coefficients will increase with increasing temperatures. Solutions which exhibit negative deviations from Raoult's law evolve heat when mixed.

FIG. 2-10. Effect of temperature on activity coefficients. Normal propanol (1)–water (2) system. [*Yu and Coull, Chem. Eng. Progr. Symposium Ser.*, **48**(2), 38 (1952).]

Modifications of some of the integrated forms of the Gibbs-Duhem equation have been suggested to take into account the variation of γ with temperature. White (28) and Yu and Coull (31) have suggested modified forms of the Van Laar equation. Chao and Hougen (4) have presented a modified form of the Redlich-Kister equation.

NOMENCLATURE

$a_i = f_i/f_i^\circ$ = activity of pure component i.

$\bar{a}_i = \bar{f}_i/f_i^\circ$ = activity of component i in solution.

A_{12}, A_{21} = constants in Van Laar or three-suffix Margules equations for binaries.

B_{12}, C_{12}, D_{12} = constants in the Redlich-Kister power series expression for the binary excess-free-energy function Q_{12}. Subscripts identify binary to which constants belong. See Eq. (2-40).

C = number of components in the system.

C, D_1, D_2, E, F_1, F_2, etc. = ternary constants in Redlich-Kister equations. See Eqs. (2-46) and (2-47).

d = differential operator.

E = internal energy. Also used as a superscript to denote excess-energy function.

f_i = fugacity of pure component i.

$\bar{f}_i$ = fugacity of component i in a solution.

f_i° = fugacity of component i in the standard state.

F_i = molar Gibbs free energy of pure component i.

$\bar{F}_i$ = partial molar free energy of component i.

F_i° = molar free energy of component i in the standard state.

$\mathbf{F}$ = free energy of entire system containing n moles.

H_i = molar enthalpy of pure component i.

$\bar{H}_i$ = partial molar enthalpy of component i.

H_i° = molar enthalpy of component i in the standard state.

$L_i = \bar{H}_i - H_i^\circ$ = partial molar heat of solution of component i. H in this chapter refers to both vapor and liquid.

K, L_1, L_2, L_3 = quaternary constants in Redlich-Kister equations. See Eq. (2-47).

k = constants in Wohl's empirical expression for the molar excess-free-energy function $F^E/2.303RT$.

n = number of moles in system.

p = total pressure.

p_i^* = vapor pressure of component i.

p_i = internal pressure of component i.

$Q = F^E/2.303RT$ = molar excess-free-energy function.

$r = x_j/x_i$. Used to define a straight-line integration path across a ternary composition diagram. Also used to denote any desired constituent in a multicomponent mixture.

R = gas law constant.

$\bar{S}_i$ = partial molar entropy of component i.

$\mathbf{S}$ = entropy of entire system containing n moles.

T = absolute temperature. Also used in Fig. 2-3 to denote an intercept.

V = molar volume.

$\mathbf{V}$ = volume of entire system containing n moles.

x = concentration in the heavy or extract phase.

y = concentration in the light or raffinate phase.

Greek Symbols

γ_i = liquid activity coefficient of component i. γ_i^L and γ_i^V refer to the activity coefficients in the extract and raffinate phases, respectively.

∂ = partial differential operator.

Δ = increment in a thermodynamic property. Calculated as state 2 minus state 1.

μ_i = chemical potential of component i.

μ_i° = chemical potential of component i in the standard state.

Subscripts and Superscripts

C = refers to the last component in a system which contains C components.

E = denotes excess-energy function.

i = refers to component i.

i, j = general subscripts in the multicomponent, Redlich-Kister equations. $i < j$.

i, j, k = cyclical subscripts used in the ternary equations (2-21), (2-22), and (2-23). The order of counting is $i\,j\,k\,i\,j\,\cdots\cdot$.

r, s = general subscripts to denote any two components in the system. $r \neq s$.

PROBLEMS

2-1. Fit a set of consistent log γ curves to the data of Richards and Hargraves [*Ind. Eng. Chem.*, **36**, 805 (1944)] for the system benzene-cyclohexane by each of the following methods: (Plot the curves vs mole fraction benzene.)

(a) Integrations suggested by Dodge.

(b) Van Laar equation.

(c) Three-suffix Margules equations.

(d) Two-constant Redlich-Kister equation.

(e) Three-constant Redlich-Kister equation.

(f) Use the values predicted by the correlation of part (b), (c), (d), or (e) to construct a plot of log (γ_1/γ_2) vs mole fraction benzene. Are equal areas obtained? Compare this log (γ_1/γ_2) vs x curve with the one obtained from the experimental values.

(g) Choose the best set of log γ curves, and check the slopes for consistency at two or three compositions.

2-2. (a) Use the equations and constants presented by Pierotti, Deal, and Derr [*Ind. Eng. Chem.*, **51**, 95 (1959)] along with the azeotropic conditions to determine the Redlich-Kister constants B_{12}, C_{12}, and D_{12} and predict the x-y curve for the ethanol (1)-water (2) system at a constant temperature of 60.65°C. The vapor-liquid equilibrium data for the ethanol-water system at this temperature have been reported by Jones et al. [*Ind. Eng. Chem.*, **35**, 666 (1943)]. The equations and constants of Pierotti et al. also appear in Sec. 11-1 of this text.

x_1	y_1	t, °C	p, mm Hg
0.051	0.316	60.60	219
0.086	0.393	60.65	249
0.197	0.517	60.70	298
0.375	0.596	60.65	325
0.509	0.648	60.70	342
0.527	0.660	60.65	344
0.545	0.671	60.50	343
0.808	0.826	60.65	363
0.851	0.862	60.65	364
0.860	0.867	60.80	366
0.972	0.972	60.65	362

(b) Use only the azeotropic conditions to determine the constants in the Margules three-suffix equations or the two-constant Redlich-Kister equations. Predict the equilibrium curve and compare the results to those obtained in (a) with the three-constant equation.

2-3. Isobaric vapor-liquid equilibrium data for the normal heptane (1)–cyclohexane

(2)–toluene (3) ternary and the various binary systems are presented in the following references:

n-C$_7$ (1)–CyC$_6$ (2): Myers, *Petrol. Refiner*, **36**(3), (1957).

n-C$_7$ (1)–toluene (3): Hipkin and Myers, *Ind. Eng. Chem.*, **46**, 2524 (1954).

CyC$_6$ (2)–toluene (3): Myers, *Ind. Eng. Chem.*, **48**, 1106 (1956).

Ternary: Myers, *A.I.Ch.E. J.*, **3**, 468 (1957).

Least-squares fit the three-constant Redlich-Kister equation to each of the binaries. Use the constants thus obtained to predict relative volatilities in the ternary system. Compare the predicted values with the experimental results.

2-4. Isothermal vapor-liquid equilibrium data at 50°C were obtained for the acetone (1)–chloroform (2)–methanol (3) system by Sesonske (Ph.D. Thesis, Delaware University). Fit the binary data with the three-constant Redlich-Kister equation and use the binary constants thus obtained to predict the ternary data. Assume that the gas phase forms an ideal solution ($\bar{V}_i = V_i$) but make corrections for the departure from ideal gas behavior and the effect of pressure on the liquid fugacities. The vapor pressures, molar liquid volumes, and second virial coefficients for the three components at 50°C are shown below:

p_1^*	612 mm Hg
p_2^*	521 mm Hg
p_3^*	416 mm Hg
$V_1{}^L$	76.4 ml/g mole
$V_2{}^L$	83.4 ml/g mole
$V_3{}^L$	42.1 ml/g mole
B_1	−1368 ml/g mole
B_2	−1216 ml/g mole
B_3	−2760 ml/g mole

2-5. Plot the activity coefficients reported by Myers [*A.I.Ch.E. J.*, **3**, 467 (1957)] for the normal heptane (1)–cyclohexane (2)–toluene (3) system on three triangular composition diagrams. Check the data for thermodynamic consistency along essentially constant-temperature paths.

REFERENCES

1. Adler, S. B., L. Friend, R. L. Pigford, and G. M. Rosselli: *A.I.Ch.E. J.*, **6**, 104 (1960).
2. Bondi, A., and D. J. Simkin: *A.I.Ch.E. J.*, **3**, 473 (1957).
3. Chao, K. C.: *Ind. Eng. Chem.*, **51**, 93 (1959).
4. Chao, K. C., and O. A. Hougen: *Chem. Eng. Sci.*, **7**, 246 (1958).
5. Dodge, B. F.: "Chemical Engineering Thermodynamics," 1st ed., pp. 555–562, McGraw-Hill Book Company, Inc., New York, 1944.
6. Ewell, R. H., J. M. Harrison, and L. Berg: *Ind. Eng. Chem.*, **36**, 871 (1944).
7. Griswold, J., and S. Y. Wong: *Chem. Eng. Progr. Symposium Ser.*, **48**(3), 18 (1952).
8. Herington, E. F. G.: *Nature*, **160**, 610 (1947).
9. Herington, E. F. G.: *J. Appl. Chem.*, **2**, 11 (1952).
10. Hildebrand, J. H., and R. L. Scott: "The Solubility of Nonelectrolytes," 3d ed., Reinhold Publishing Corporation, New York, 1950.
11. Horsley, L. H.: "Azeotropic Data," Advances in Chemistry Series, American Chemical Society, Washington, D.C.
12. Ibl, N. V., and B. F. Dodge: *Chem. Eng. Sci.*, **2**, 120 (1958).

13. Jordan, D., J. A. Gerster, K. Wohl, and A. P. Colburn: *Chem. Eng. Progr.*, **46**, 601 (1950).
14. Margules: *Sitzber. math.-naturw. Kl. Kaiserlichen Akad. Wiss.* (*Vienna*), **104**, 1243 (1895).
15. Murti, P. S., and M. Van Winkle: *A.I.Ch.E. J.*, **3**, 517 (1957).
16. Myers, H. S.: *A.I.Ch.E. J.*, **3**, 467 (1957).
17. Perry, J. H.: "Chemical Engineers' Handbook," 3d ed., p. 528, McGraw-Hill Book Company, Inc., New York, 1950.
18. Pierotti, G. J., C. H. Deal, and E. L. Derr: *Ind. Eng. Chem.*, **51**, 95 (1959).
19. Prigogine, I.: "The Molecular Theory of Solutions," 1st ed., chap. XV, Interscience Publishers, Inc., New York.
20. Redlich, O., and A. T. Kister: *Ind. Eng. Chem.*, **40**, 341, 345 (1958).
21. Redlich, O., A. T. Kister, and C. E. Turnquist: *Chem. Eng. Progr. Symposium Ser.*, **48**(2), 49 (1952).
22. Scatchard, G.: *Chem. Rev.*, **8**, 321 (1931).
23. Schuhmann, R.: *Acta Met.*, **3**, 219 (1955).
24. Severns, W. H., Jr., A. Sesonske, R. H. Perry, and R. L. Pigford: *A.I.Ch.E. J.*, **1**, 401 (1955).
25. Steinhouser, H. H., and R. R. White: *Ind. Eng. Chem.*, **41**, 2912 (1949).
26. Treybal, R. E.: "Liquid Extraction," chap. 3, McGraw-Hill Book Company, Inc., New York, 1951.
27. Van Laar: *Z. physik. Chem.*, **72**, 723 (1910); **83**, 599 (1913).
28. White, R. R.: *Trans. A.I.Ch.E.*, **41**, 539 (1945).
29. Wohl, K.: *Chem. Eng. Progr.*, **49**, 218 (1953).
30. Wohl, K.: *Trans. A.I.Ch.E.*, **42**, 215 (1946).
31. Yu, K. T., and J. Coull: *Chem. Eng. Progr. Symposium Ser.*, **48**(2), 38 (1952).

CHAPTER 3

DESIGN VARIABLES

The solution of a multistage, multicomponent separation problem is complete when the designer knows the composition, temperature, pressure, and flow rate of each stream associated with the separation unit. An operating unit gives such a solution continuously and therefore can be thought of as a computing machine. A feed of given composition, temperature, pressure, and rate is fed to the machine. The machine is restricted by certain physical characteristics such as diameter, number and efficiency of stages, feed-plate location, condensing and reboil capacities, efficiency of insulation, safe working temperatures and pressures, etc. In addition to these physical restrictions, the operator (or designer) imposes other arbitrary restrictions by specifying certain stream rates, ratios of rates, stream concentrations, recovery of certain components, or other desired operating restrictions. Subject to the restrictions, the machine produces a unique set of results. These results are in the form of stream compositions, temperatures, pressures, and flow rates.

The purpose of any design method is to approximate the answers which the operating unit will produce. The degree of accuracy required in the approximation will determine the design method used. None of the methods available at the present time will reproduce exactly the results provided by the operating unit, but several come sufficiently close to serve the purposes of the design engineer.

Just as the operating unit will require a certain fixed number of restrictions to force it to produce the desired separation, so will the design method require a certain fixed number of specifications if the solution is to converge to the desired set of answers. Too few restrictions will allow the unit or the calculation method to converge to some solution other than the one desired. Too many restrictions cannot be tolerated by the unit or the calculation method. If too many are specified, some must be ignored if convergence to any solution is to be obtained.

The ability to determine the exact number of restrictions which must be arbitrarily imposed by the designer has become more important with the advent of high-speed computers. Electronic computation makes possible duplication of the operating unit to almost any degree of exactness. It will eventually permit the calculation as one problem of whole processes which involve several units. However, since the computer lacks intuition, the problem must be defined exactly; that is, exactly the correct number of restrictions (or design specifications) must be known and fed to the computer. As the unit or process to be designed grows in complexity, the determination of the correct number of variables to specify becomes difficult. This problem has been studied by Dunstan (1), by Gilliland and Reed (2), and most recently by Kwauk (3). The nomenclature and method of analysis used in this chapter are essentially those of Kwauk. Several systems which range from the simplest to ones of moderate complexity will be analyzed in order to develop and illustrate a set of rules which can be applied to systems of any degree of complexity.

3-1. Types of Variables

The variables with which the designer of a separation unit must concern himself are as follows:

1. Stream concentrations
2. Temperatures
3. Pressures
4. Rates
5. Repetition variables N_r

The first three variables in the list are intensive variables; that is, they are independent of the quantity of material present. It would be permissible, of course, to substitute for any of the three listed other intensive properties such as molar enthalpy, molar entropy, etc., but this is seldom convenient.

The fourth variable on the list, rate, is an extensive property; that is, it depends on the amount of material present. Another extensive property, total stream enthalpy, is used in the calculations but only as a means of obtaining the stream rates. The term rate is used to describe the magnitude of both material streams and heat streams. Examples of the latter are the heat input to a reboiler and the heat removed through a condenser.

The fifth variable on the list is neither an intensive nor extensive variable. It is the single degree of freedom which the designer utilizes when he specifies how often a particular element will be repeated in a unit. For example, a distillation section is composed of a series of equilibrium stages, and when the designer specifies the number of stages which the section will contain, he utilizes the single degree of freedom represented by the repetition variable ($N_r = 1.0$). If the distillation column contains more than one section (such as above and below the feed stage), the number of stages in each section must be specified and as many repetition variables exist as there are sections; that is, $N_r = 2$.

The first step in the analysis of any system is to count the total number of variables N_v. The number N_v is analogous to the number of unknowns in a system of simultaneous algebraic equations. The second step is to count all the restricting conditions or relationships existing in the system. The number of such restrictions will be denoted as N_c. These restrictions are analogous to the independent equations which can be written in an algebraic system. If the number of equations equals the number of unknowns, a unique solution is possible. Likewise, if the number of restrictions N_c existing in a system equals the total number of variables N_v, then the system is completely defined. Such an equality does not often exist in the typical design problem. Then, just as in the case of an algebraic system, the designer must arbitrarily specify certain variables. The number which he can specify is referred to as the *degrees of freedom* in the system and can be calculated by the following simple equation:

$$N_i = N_v - N_c \tag{3-1}$$

The term *variance* is often used instead of degrees of freedom when referring to the N_i variables. In this book the N_i variables will often be called the *design variables*, since these are the variables which the designer must specify to define the design problem completely.

3-2. Restricting Relationships N_c

As will be seen later, the determination of N_v for any system is quite simple and straightforward. It is not difficult to count all stream concentrations, temperatures, pressures, and rates which exist in a system and add to their number the required repetition variables. Unfortunately, the restricting relationships N_c are not so easy to count. To avoid omission or duplication of restrictions it is necessary to follow some arbitrary but consistent procedure which reduces the chances for error.

It is convenient to divide all the possible restrictions into the following types:

1. Inherent restrictions
2. Material-balance restrictions
3. Energy-balance restrictions
4. Phase-distribution restrictions
5. Chemical-equilibrium restrictions

It is often arbitrary whether a given restriction is counted as an inherent restriction or as a material- or energy-balance restriction. Care must be taken to avoid redundancy in such cases.

Inherent Restrictions. Certain restricting relationships or conditions are often inherent in the particular system under consideration. These restrictions usually take the form of identities between two or more variables. For example, the concept of the equilibrium stage involves the inherent restrictions that the temperature and pressure of one equilibrium stream leaving the stage must be identical with the temperature and pressure of the other leaving stream. Figures 3-1 and 3-2 illustrate other common inherent restrictions. In Fig. 3-1, it is obvious that all the intensive properties of D and L_{N+1} must be equal. In Fig. 3-2, it is obvious that there are $C - 1$ independent concentration identities and one rate identity between A and A'. For each in-

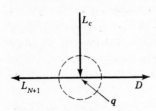

FIG. 3-1. Stream divider.

dependent identity, a restricting relationship can be counted and each such relationship (or its equivalent) must be subtracted from the total number of variables N_v in the calculation of the number of independent or arbitrary variables N_i.

Material-balance Restrictions. An over-all material balance can be written for each of the components† present. This furnishes C restricting relationships which must be subtracted from N_v when calculating N_i. Instead of C component balances, it is permissible, of course, to write $C - 1$ component balances plus one over-all quantity balance.

Care must be taken that the material-balance restrictions used are independent of the inherent restrictions listed previously. For example, in situations such as those shown in Fig. 3-2 where A and A' are actually one stream, the C-component-balance restrictions which might be applied are not independent of the $C - 1$ concentration

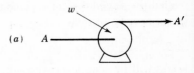
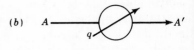

FIG. 3-2. (a) Pump, (b) heater, and (c) cooler.

identities and the one rate identity which may have been counted previously under inherent restrictions.

Energy-balance Restrictions. An over-all energy balance furnishes another restricting relationship. It will involve any heat streams as well

---

† The number of independent components C in a mixture where chemical reactions are not involved is identical with the number of chemical species which are present. This book deals exclusively with nonreactive mixtures, and therefore, no distinction will be made between phase-rule components and chemical species or constituents.

as all material streams. In some cases the energy balance may not be independent of the identities listed under inherent restrictions.

Phase-distribution Relationships. Each component in a system containing more than one phase will distribute itself between the various phases in a characteristic manner. The distribution of a component between two phases is described by its distribution coefficient K. A component which is distributed between three phases will possess three distribution coefficients, but only two will be independent. In general, if all components exist in all phases, the number of restricting relationships due to the distribution phenomenon will be $C(N_p - 1)$, where N_p is the number of phases present.

Chemical-equilibrium Relationships. In chemically reactive systems, the various chemical constituents will be related by chemical-equilibrium relationships. The number of such relationships is equal to the minimum number of stoichiometric equations which must be written to form all the chemical species assumed present from the independent components selected. A chemical-equilibrium relationship relates two or more constituents in the same phase or different phases, whereas a distribution relationship describes the distribution of one constituent between two phases.

Chemical-equilibrium relationships will not be a factor in any of the systems subsequently analyzed. They are listed here only for the sake of completeness.

Form of Restricting Relationships. The form of the various restricting relationships is immaterial in this type of analysis. Only the number of such relationships is important. For example, it makes no difference whether a distribution coefficient is expressed in terms of mole, weight, or volume fractions or whether it is even known. It is sufficient to know that the distribution relationship does or does not exist.

3-3. Analysis of Typical Elements

The direct application of Eq. (3-1) to a complex process would be confusing and time consuming. A more efficient approach is to break the process units down into parts or elements. The elements can then be analyzed individually to determine their respective degrees of freedom from

$$N_i^e = N_v^e - N_c^e \qquad (3\text{-}2)$$

where the superscript e denotes that the numbers refer to an element of the unit and not to the entire unit. Once a sufficient number of elements has been analyzed, any process unit can be constructed by connecting together the proper elements. The degrees of freedom for the unit

$N_i{}^u$ can then be calculated very simply from the summation of the $N_i{}^e$'s for the various elements involved.

The next section will deal with the combination of elements to form process units. Before we concern ourselves with units, however, it will be necessary to develop the rules of analysis by the consideration of several of the more common elements. As the rules are developed, a catalog of results for common elements will also be accumulated for subsequent use in the next section.

The so-called rules of analysis are rather arbitrary and actually are nothing more than a consistent method of counting variables and restrictions. Some educational benefit may be derived from the development of the rules, but the reader is reminded that the counting done in this chapter is analogous to counting in arithmetic. It is a tool which provides a result needed in subsequent analysis. The results in this chapter will take the form of expressions which represent the number of variables subject to control by the designer (degrees of freedom). Calculation of the degrees of freedom is little more than a mechanical exercise. It is not until the reader translates the degrees of freedom into convenient specifications that they begin to shed some light upon the workings of a separation process and the interrelationships among variables.

Single Stream. The simplest element which the process designer must design is a single homogeneous stream. When the stream is considered at one point, the variables involved are as follows:

	$N_v{}^e$
Composition	$C - 1$
Temperature	1
Pressure	1
Rate	1
	$\overline{C + 2}$

A single homogeneous stream will contribute $C + 2$ variables to any element or unit of which it is a part. There are no restricting relationships when a one-phase stream is considered only at a given point, so $N_c{}^e = 0$ and $N_i{}^e = N_v{}^e$. Once the designer has specified these $C + 2$ variables, the stream can be reproduced exactly by some other person; that is, the stream is completely defined.

Stream Divider. A divider simply splits a stream into two or more product streams. Figure 3-1 shows a divider with two product streams. Three material streams and one heat stream are involved, so

$$N_v{}^e = 3(C + 2) + 1 = 3C + 7$$

Each material stream contributes $C + 2$ variables, whereas a heat stream has only rate as a variable to contribute to $N_v{}^e$.

It will be arbitrarily assumed that the streams leaving any element are at the temperature and pressure of the element; that is, the $C + 2$ variables which each leaving stream possesses will be assigned before the stream leaves the element and before there is opportunity for pressure drop or temperature change. On the other hand, entering streams are considered to possess $C + 2$ variables which are in general independent of conditions within the element; that is, the variables are specified before the streams enter the elements.

The element in Fig. 3-1 can be considered to be inside the dotted circle. A set of independent restrictions is as follows:

Inherent	
t and p identities between L_{N+1} and D	2
Concentration identities between L_{N+1} and D	$C - 1$
Concentration identities between L_c and D	$C - 1$
Material balances	
One over-all quantity balance	1
Energy balance	1
Distribution restrictions	0
	$2C + 2$

Certain restrictions can be switched from one type to the other. For example, the $C - 1$ concentration identities between L_c and D (or L_{N+1}) and the over-all quantity balance could be replaced with C component balances. Also, the energy balance would be unnecessary (that is, not independent) if a temperature identity between L_c and D (or L_{N+1}) were listed as an inherent restriction for this element. These examples illustrate that it is the number of restrictions which is important and not their form.

The degrees of freedom (or the variables which the designer must specify) to define one unique operation of a stream divider are given by

$$N_i^e = N_v^e - N_c^e = (3C + 7) - (2C + 2) = C + 5$$

Specification of the feed stream L_c, the ratio L_{N+1}/D, the heat leak q, and the pressure of the divider utilizes these degrees of freedom.

Simple stream dividers are generally essentially adiabatic. Also, the pressure of the element could be taken as the pressure of the single feed stream. If these two restrictions were listed as inherent restrictions, N_i^e would be reduced to $C + 3$ for a simple, adiabatic stream divider. This would eliminate the necessity of listing q and p specifications for the divider. However, for the sake of conformity it is preferable (at least while learning) to treat the divider as all other elements and list the q and p as specifications rather than as inherent restrictions.

Mixer. A stream mixer is schematically represented in Fig. 3-3. Three material streams and one energy stream are involved, so

$$N_v^e = 3(C + 2) + 1 = 3C + 7$$

The restricting conditions include C component balances and one energy balance to give $N_c^e = C + 1$. There are no inherent or distribution restrictions. The number of variables over which the designer has control is

$$N_i^e = N_v^e - N_c^e = (3C + 7) - (C + 1) = 2C + 6$$

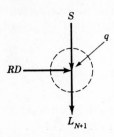

RD

Specification of the two feed streams takes care of $2C + 4$ variables. The magnitude of q and the pressure of the mixer can be specified to utilize the remaining two degrees of freedom.

FIG. 3-3. Mixer. In this particular case q would probably include energy input due to shaft work in addition to ordinary heat leak.

Pumps, Heaters, and Coolers. The flow diagrams for these elements are shown in Fig. 3-2. In each case,

$$N_v^e = 2(C + 2) + 1 = 2C + 5$$

Since only two material streams are involved, the application of C component balances takes care of the composition and rate identities between A and A'. The C component balances plus one over-all energy balance gives $N_c^e = C + 1$ and $N_i^e = C + 4$. Specification of stream A utilizes $C + 2$ variables, and usually the designer will utilize the other two degrees of freedom to specify the pressure and temperature of A'. The energy input q or w can then be calculated with the use of these specifications.

Total Condenser or Total Reboiler. A condenser is described as total when all the vapor feed is condensed to a liquid. Likewise, a reboiler could be termed a total reboiler if all the liquid feed were vaporized to vapor. These devices are shown schematically in Fig. 3-4. A total condenser is a practical device, whereas in most systems a total reboiler as defined above is not. Com-

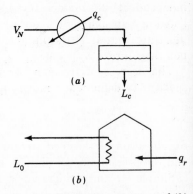

FIG. 3-4. (*a*) Total condenser and (*b*) a total reboiler.

plete vaporization usually results in fouling problems due to deposition of high-boiling impurities on the reboiler tubes. Also, in the case of a fired furnace, complete vaporization can cause overheating of the tubes due to the high heat-transfer resistance of vapor films. For these reasons, the type of reboiler shown in Fig. 3-4 is normally designed to vaporize only part of the stream passing through it. Despite this incomplete vaporization, the term *total* will be used to describe a reboiler where all the feed to the reboiler is returned to the distillation column without any part being withdrawn as the bottom product.

The analyses for total condensers and total reboilers are identical. As can be seen from Fig. 3-4, two material streams and one heat stream are involved and

$$N_v^e = 2(C + 2) + 1 = 2C + 5$$

The C component balances and one energy balance provide $C + 1$ restrictions. (As mentioned previously, the composition and rate identities are not independent of the component balances when only two material streams are involved.) The number of variables which must be specified is

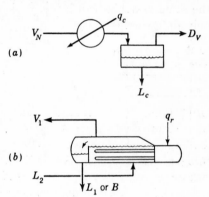

(a)

$$N_i^e = (2C + 5) - (C + 1) = C + 4$$

Convenient specifications would define the entering stream ($C + 2$ variables) and fix the temperature and pressure of the leaving stream. In some cases, it might be more convenient to specify q rather than the outlet temperature. For example, q might be assumed to be equal to the latent heat of vaporization or condensation. If this condition, q_c or q_r = latent heat, were made part of the definition of a total condenser or total reboiler, then it would be counted as a restricting condition to give

(b)

FIG. 3-5. (a) Partial condenser and (b) a partial reboiler.

$$N_c^e = C + 2 \quad \text{and} \quad N_i^e = C + 3$$

Partial Condenser or Reboiler. If the condenser or reboiler causes only part of the entering stream to change phase, it is termed a *partial* condenser or reboiler. In view of the discussion above for total reboilers it is necessary to extend this definition by adding the stipulation that part of the feed to a partial reboiler be removed as product instead of returning in total to the column. Partial condensers or partial reboilers are always assumed to be equilibrium stages in so far as the separation achieved is concerned. As in the case of total condensers and reboilers, the analysis of these elements is the same regardless of whether material is condensed or vaporized. From Fig. 3-5, it can be seen that three material streams and one heat stream are involved to make

$$N_v{}^e = 3(C + 2) + 1 = 3C + 7$$

The restricting relationships can be tabulated as follows:

Type of restriction	$N_c{}^e$
Inherent (two equilibrium streams at the same temperature and pressure)	2
Component balances	C
Over-all energy balance	1
Distribution relationships	C
	$\overline{2C + 3}$

So $\qquad N_i{}^e = N_v{}^e - N_c{}^e = (3C + 7) - (2C + 3) = C + 4$

This is the same result as obtained for total condensers or reboilers, and the same variables can be conveniently specified by the designer. The only difference is that the magnitude of q must be less than the latent heat of condensation or evaporation of the entering stream if the adjective "partial" is to apply.

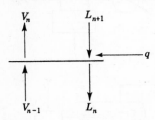

FIG. 3-6. Simple equilibrium stage.

Simple Equilibrium Stage. A schematic representation of a simple equilibrium stage (no fresh feed or side stream) appears in Fig. 3-6. Four material streams and one heat stream provide

$$N_v{}^e = 4(C + 2) + 1 = 4C + 9$$

Streams V_n and L_n are in equilibrium with each other by definition and, therefore, are at the same temperature and pressure. These two inherent identities when added to C component balances, one energy balance, and C distribution relationships give

$$N_c{}^e = 2C + 3$$

Then $\qquad N_i{}^e = N_v{}^e - N_c{}^e = (4C + 9) - (2C + 3) = 2C + 6$

The variables most commonly specified by the designer of a simple equilibrium stage are as follows:

Specifications	$N_i{}^e$
Specification of L_{n+1}	$C + 2$
Specification of V_{n-1}	$C + 2$
Stage pressure	1
Heat leak q	1
	$\overline{2C + 6}$

The stage pressure is not fixed by the pressure specifications for L_{n+1} and V_{n-1} and must be counted as a separate variable.

Feed Stage. A feed stage differs from a simple equilibrium stage in that a fifth material stream F is involved (see Fig. 3-7). In general, no identities exist between F and any of the other four material streams; that is, the composition, temperature, pressure, and rate of F are not identical with those of any of the other streams.

With five material streams and one heat stream,

$$N_v^e = 5(C + 2) + 1 = 5C + 11$$

The restricting relationships are the same as those listed for a simple equilibrium stage where

$$N_c^e = 2C + 3$$

The number of independent design variables is then

$$N_i^e = (5C + 11) - (2C + 3) = 3C + 8$$

This number exceeds the N_i^e for a simple equilibrium stage by $C + 2$,

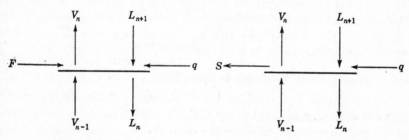

FIG. 3-7. Feed stage.   FIG. 3-8. Side-stream stage.

which is the number of variables necessary to define the additional stream F.

Kwauk (3) assumes that F and L_{n+1} are mixed in a separate mixer before entering the equilibrium stage to contact V_{n-1}. The pressure and heat loss for this extra mixer must be specified. This makes N_i^e for Kwauk's arrangement equal $3C + 10$, and two more variables, the mixer pressure and heat loss, must be specified by the designer. The author's analysis, in effect, assumes the inherent restrictions that the mixer pressure equals the pressure of the stage and that the heat leak in the mixer is zero.

Side-stream Stage. A side-stream stage is an intermediate stage in a series of simple equilibrium stages from which a product stream is withdrawn. The side stream may be returned to another stage after cooling or heating, but this is immaterial at this stage of the analysis. The term *side-stream stage* can be applied to any stage similar to the one pictured in Fig. 3-8 regardless of the subsequent disposal of the stream S.

Since five material streams and one heat stream are involved, the total number of variables $N_v{}^e$ is the same as for a feed stage, namely, $5C + 11$. The number of restricting relationships N_c is not the same as for a feed stage. The stream S must be identical in composition with either V_n or L_n and also have the same temperature and pressure as L_n and V_n. Therefore, $C + 1$ identities or restricting relationships must exist between S and L_n or V_n. Which one is unimportant, since the number and not the nature of the restrictions is important. The addition of these $C + 1$ identities to those present in a simple equilibrium stage gives

$$N_c{}^e = (C + 1) + (2C + 3) = 3C + 4$$

for a side-stream stage. The number of variables subject to the designer's control then becomes

$$N_i{}^e = N_v{}^e - N_c{}^e = (5C + 11) - (3C + 4) = 2C + 7$$

This number exceeds that for a simple equilibrium stage by one, and this additional degree of freedom would probably be used to specify the rate of S. The composition, temperature, and pressure of S are fixed by the specifications usually made for the equilibrium stage.

3-4. Combination of Elements to Form Units

A process unit is a combination of elements. In analyzing a unit, it is not necessary to count every variable and every restriction existing in the unit in order to calculate the degrees of freedom available to the unit designer. Instead, the analysis is simplified by analyzing the elements separately and then combining the results to furnish the answer for the unit just as the elements are combined to form the unit itself.

The total number of intensive and extensive variables which must be considered in a combination of elements is equal to the sum of the independent variables $N_i{}^e$ associated with each of the single elements. In addition to the intensive and extensive variables, one repetition variable must be counted for each decision which the designer must make concerning the number of repetitions of any particular type of element within the unit. These statements can be expressed mathematically as

$$N_v{}^u = N_r + \Sigma N_i{}^e \tag{3-3}$$

where the superscript u refers to the unit or combination of elements.

The number of independent variables $N_i{}^u$ associated with the combination can be calculated by

$$N_i{}^u = N_v{}^u - N_c{}^u \tag{3-4}$$

where $N_c{}^u$ refers to the *new* restricting relationships which arise when the elements are combined. $N_c{}^u$ does not include any of the restricting relationships considered in calculating the $N_i{}^e$'s for the various elements. The new restrictions denoted by $N_c{}^u$ are the stream identities which exist in each interstream between two elements. The interstream variables $(C + 2)$ were counted in each of the two elements when their respective $N_i{}^e$'s were calculated. Therefore, $(C + 2)$ new restricting relationships must be counted for each interstream in the combination of elements in order to prevent redundancy.

3-5. Analysis of Typical Units

Simple Absorption or Extraction Unit. Figure 3-9 pictures a simple absorption or extraction unit. The stream L_{N+1} is the lean absorber oil in an absorption unit or the fresh solvent in an extraction unit. The unit consists of a combination of N simple equilibrium stages. Stage 1 is not termed a feed stage nor stage N a side-stream stage, since each involves only four material streams. The heat-leak streams for each stage are not shown in Fig. 3-9, but they exist and their magnitude must be specified by the designer.

The decision as to the number of stages to be used rests with the designer, and therefore, the specification of N constitutes a single degree of freedom and $N_r = 1.0$. The total number of variables $N_v{}^u$ to be considered is given by

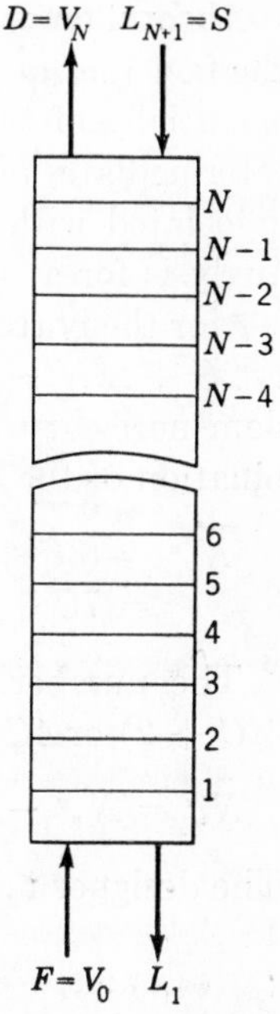

$$N_v{}^u = N_r + \Sigma N_i{}^e = 1 + N(2C + 6)$$

since $N_i{}^e = 2C + 6$ for a simple equilibrium stage.

There are $2(N - 1)$ interstreams, and therefore $2(N - 1)(C + 2)$ new identities (not previously counted) come into existence when the elements are combined. Subtraction of these restrictions from $N_v{}^u$ gives $N_i{}^u$, the degrees of freedom available to the designer.

$$N_i{}^u = N_v{}^u - N_c{}^u = [1 + N(2C + 6)]$$
$$- [2(N - 1)(C + 2)] = 2C + 2N + 5$$

The designer might utilize these $(2C + 2N + 5)$ degrees of freedom as follows:

FIG. 3-9. Simple absorption or extraction unit.

Specifications	$N_i{}^u$	
Pressure in each stage	N	
Heat leak in each stage	N	
V_0 or F	C	$+2$
L_{n+1} or S	C	$+2$
Number of stages N		1
	$2C + 2N + 5$	

Absorption or Extraction Unit with Two Feeds.   Figure 3-10 illustrates the fact that an absorption or extraction unit with two feeds is composed essentially of two simple absorption or extraction units connected to an intermediate feed stage.   The number of independent variables $N_i{}^e$ associated with a feed stage has been shown to be $3C + 8$.   From the analysis for a simple absorption or extraction unit, it is known that the $N_i{}^e$ for the bottom section is $2C + 2M + 5$ and the $N_i{}^e$ for the top section is $2C + 2(N - M - 1) + 5$.   These sums include the degrees of freedom necessary to specify the number of stages in each section.   Combination of the three elements gives

$$N_v{}^u = (3C + 8) + (2C + 2M + 5) + [2C + 2(N - M - 1) + 5]$$
$$= 7C + 2N + 16$$

The number of restricting relationships due to new interstreams is $4(C + 2)$ or $4C + 8$.   Therefore,

$$N_i{}^u = N_v{}^u - N_c{}^u = (7C + 2N + 16) - (4C + 8) = 3C + 2N + 8$$

The designer might specify the following variables:

Specifications	$N_i{}^u$	
Pressure in each stage	N	
Heat leak in each stage	N	
S or L_{N+1}	C	$+2$
F	C	$+2$
F' or V_0	C	$+2$
Total number of stages, N		1
Number of stages below feed, M		1
	$3C + 2N + 8$	

Extraction Unit with Extract Reflux.   An extraction column with one feed and operating with extract reflux is shown in Fig. 3-11.   If V_0 is substituted for F', it can be seen that the largest element in Fig. 3-11 is identical with the unit shown in Fig. 3-10 and therefore possesses $3C + 2N + 8$ independent variables when considered alone.   The element labeled divider in Fig. 3-11 has been analyzed previously and shown to have an $N_i{}^e = C + 5$.

The element labeled special divider is in reality not a divider at all. An extraction column which operates with extract reflux requires the use of some kind of separation device to recover the solvent from the extract stream leaving stage 1. A distillation column might be used, or a decanter which operates at some temperature above or below that of stage 1 might suffice to recover the solvent for recycle to stage N. Regardless of the type of separation device utilized, it is sufficient for the

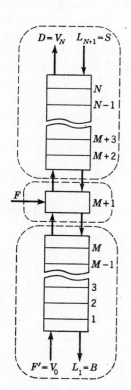

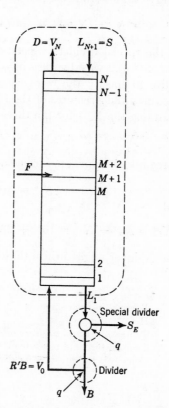

FIG. 3-10. Absorption or extraction unit with two feeds.

FIG. 3-11. Extraction column with one feed and extract reflux.

purpose of this analysis to consider the device as a special divider which splits each component in L_1 between S_E and $(L_1 - S_E)$ according to recovery factors specified by the designer. The streams S_E and $(L_1 - S_E)$ need not, in general, be in equilibrium with each other, and therefore, equilibrium distribution relationships cannot be assumed in the general case. Also, the streams S_E and $(L_1 - S_E)$ need not be at the same temperature and pressure.

The three material streams and one heat stream associated with the special divider give

$$N_v^e = 3(C + 2) + 1 = 3C + 7$$

Since there are no inherent identities or distribution relationships, the restricting relationships consist only of C component balances and one energy balance and $N_c^e = C + 1$. Then

$$N_i^e = N_v^e - N_c^e = (3C + 7) - (C + 1) = 2C + 6$$

for the special divider. Specification of the feed stream L_1 would take care of $C + 2$ variables leaving $C + 4$ variables which the designer could utilize to specify the recovery of each component in S_E and the temperatures and pressures of the two product streams.

Adding the N_i^e's for the three elements in Fig. 3-11 gives

$$N_v^u = \Sigma N_i^e = (3C + 2N + 8) + (C + 5) + (2C + 6) = 6C + 2N + 19$$

Three interstreams are formed when the three elements are combined so

$$N_c^u = 3(C + 2) = 3C + 6$$

and

$$N_i^u = N_v^u - N_c^u = (6C + 2N + 19) - (3C + 6) = 3C + 2N + 13$$

The designer might utilize these degrees of freedom as follows:

Specifications	N_i^u	
Pressure in each stage	N	
Heat leak in each stage	N	
Temperature and pressure of stream S_E		2
Temperature and pressure of stream $(L_1 - S_E)$		2
Heat leak and pressure for divider		2
S	C	$+ 2$
F	C	$+ 2$
Total number of stages, N		1
Number of stages below the feed stage, M		1
Extract reflux ratio $R' = V_0/B$		1
Recovery of each component in the special divider	C	
	$3C + 2N + 13$	

Exercise 3-1. Show that the extraction column given in Fig. 3-12 with two feeds and raffinate reflux has $3C + 2N + 13$ degrees of freedom. List a reasonable set of specifications.

Distillation Unit with One Feed, Total Condenser, and Partial Reboiler. Figure 3-13 pictures the various elements that comprise a distillation unit which has one feed, a total condenser, and a partial reboiler. The

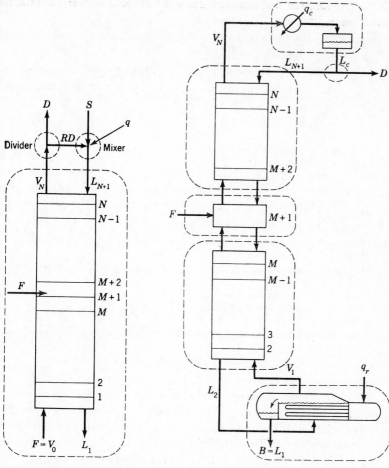

FIG. 3-12. Extraction column with two feeds and raffinate reflux.

FIG. 3-13. Distillation column with one feed, total condenser, and partial reboiler.

number of variables $N_v{}^u$ which must be considered in the analysis of the entire unit is the sum of the $N_i{}^e$'s for the six elements involved.

Element	$N_v{}^u = \Sigma N_i{}^e$	
Total condenser	C	$+ \ 4$
Divider (reflux)	C	$+ \ 5$
$N - (M + 1)$ simple equilibrium stages	$2C + 2(N - M - 1)$	$+ \ 5$
Feed stage	$3C$	$+ \ 8$
$M - 1$ simple equilibrium stages	$2C + 2(M - 1)$	$+ \ 5$
Partial reboiler	C	$+ \ 4$
	$10C$	$+ \ 2N + 27$

The $N_i{}^e$'s listed for the sections of simple equilibrium stages were obtained from the analysis of a simple absorption or extraction unit.

Nine interstreams are created by the combination of elements, and therefore

$$N_c{}^u = 9(C + 2) = 9C + 18$$

The degrees of freedom available to the designer are then

$$N_i{}^u = N_v{}^u - N_c{}^u = (10C + 2N + 27) - (9C + 18) = C + 2N + 9$$

One set of possible design specifications is as follows:

Specifications	$N_i{}^u$	
Pressure in each stage (including reboiler)	N	
Pressure in condenser		1
Pressure in reflux divider		1
Heat leak in each stage, excluding reboiler	$N - 1$	
Heat leak in reflux divider		1
F	C	$+ 2$
Reflux temperature		1
Total number of stages, N		1
Number of stages below feed stage, M		1
Distillate rate or D/F		1
Maximum allowable vapor rate or V/F		1
	$C + 2N + 9$	

The first six items listed above ($C + 2N + 4$ variables) are almost invariably specified by the designer. The remaining five variables can be utilized in a number of ways. The five specifications listed above are convenient when it is desired to calculate the performance of an existing column on a new feed. In such a situation, N and M are fixed and the maximum allowable vapor rate is known from experience or can be closely estimated. The composition of D and B can then be calculated as a function of overhead rate. Other specifications may be more convenient in other problems, and one or more of the following may be substituted for any of the last five listed above.

1. Reflux ratio R/D
2. Condenser load q_c
3. Reboiler load q_r
4. Recovery of one or two components in either D or B
5. Concentration of one or two components in either D or B

Some of the above variables cannot be specified indiscriminately. Some are restricted to narrow ranges. For example, the value of q_c specified must result in a condensate temperature somewhere between the bubble point and the freezing point. It is usually much easier to specify the condensate temperature instead of q_c. Some of the variables are so interrelated that it is difficult to specify reasonable values if they are used simultaneously. The q_r and q_c are so closely related that it is literally impossible to specify both before the complete enthalpy balance is known. Similarly, the maximum vapor rate is so closely related to q_r that an impossible situation will probably be created if a specification of both is attempted.

Distillation Unit with One Feed, Partial Condenser, and a Total Reboiler. The total number of variables N_v^u for the unit pictured in Fig. 3-14 is given by the following summation of the N_i^e's for the seven elements involved.

Element	$N_v^u = \Sigma N_i^e$	
Partial condenser	C	$+\ 4$
Divider (reflux)	C	$+\ 5$
$N - (M + 1)$ simple equilibrium stages	$2C + 2(N - M - 1)$	$+\ 5$
Feed stage	$3C$	$+\ 8$
M simple equilibrium stages	$2C + 2M$	$+\ 5$
Total reboiler	C	$+\ 4$
Divider (bottom)	C	$+\ 5$
	$11C + 2N$	$+\ 34$

The 10 interstreams make

$$N_c^u = 10(C + 2) = 10C + 20$$

Therefore, the number of design variables becomes

$$N_i^u = N_v^u - N_c^u = (11C + 2N + 34) - (10C + 20)$$
$$= C + 2N + 14$$

Convenient design specifications are as follows:

Specifications	N_i^u	
Pressure on each stage	N	
Pressure in condenser		1
Pressure in reboiler		1
Pressure in the two dividers		2
Heat leak in each stage	N	
Heat leak in the two dividers		2
F	C	$+\ 2$
Product ratio D_V/D_L		1
Circulation rate L_0 through reboiler		1
Total number of stages, N		1
Number of stages below feed stage, M		1
Maximum vapor rate, or V/F		1
Total distillate rate $D = D_V + D_L$, or D/F		1
	$C + 2N + 14$	

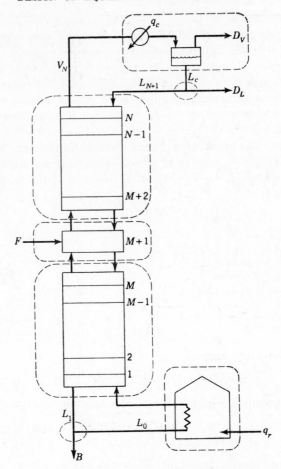

FIG. 3-14. Distillation column with one feed, partial condenser, and total reboiler.

Alternatives to the last six items might be the following:

1. Reflux ratio $R = L_{N+1}/D$
2. Reboiler load q_r
3. Condenser load q_c
4. Recovery of one or two components in either D or B
5. Concentration of one or two components in either D or B

As explained above, common sense must be used in the selection of the last six variables to avoid an impractical set of specifications.

Distillation Column with One Feed, One Side Stream, Total Condenser, and Partial Reboiler. The addition of a side stream to the unit shown in Fig. 3-13 increases the number of elements involved from six to eight and results in the unit pictured in Fig. 3-15. The $N_v{}^u$ for the

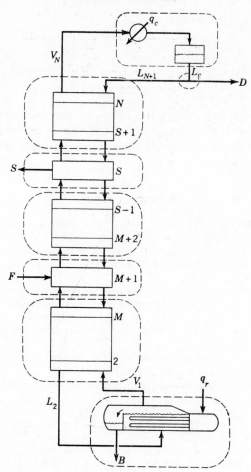

Fig. 3-15. Distillation column with one feed, one side stream, a total condenser, and a partial reboiler.

entire unit in Fig. 3-15 is found by summing the $N_i{}^e$'s for the various elements as follows:

Element	$N_v{}^u = \Sigma N_i{}^e$	
Total condenser	C	$+\ 4$
Divider (reflux)	C	$+\ 5$
$(N - S)$ simple equilibrium stages	$2C + 2(N - S)$	$+\ 5$
Side-stream stage	$2C$	$+\ 7$
$(S - 1) - (M + 1)$ simple equilibrium stages	$2C + 2(S - M - 2)$	$+\ 5$
Feed stage	$3C$	$+\ 8$
$(M - 1)$ simple equilibrium stages	$2C + 2(M - 1)$	$+\ 5$
Partial reboiler	C	$+\ 4$
	$14C + 2N$	$+\ 37$

Thirteen interstreams give an

$$N_c = 13(C + 2) = 13C + 26$$

Then,

$$N_i{}^u = N_v{}^u - N_c{}^u = (14C + 2N + 37) - (13C + 26)$$
$$= C + 2N + 11$$

The addition of a side stream has increased the degrees of freedom by two compared with the unit in Fig. 3-13. These two degrees could be used to specify the rate of S and the number of stages between the side stream and feed stages.

NOMENCLATURE

A = general stream designation. Stream A' is identical with A in composition and rate but different in temperature or pressure.

B = bottom or extract product rate; moles, weight, or volume per unit time. B' refers to extract product before extract reflux is removed but after solvent recovery.

C = number of components. For the chemically nonreactive systems discussed in this book, C is identical with the number of constituents in the system.

D = overhead or raffinate product; moles, weight, or volume per unit time. In extraction, D refers to raffinate after raffinate reflux is removed but before solvent recovery.

D_L = $L_c - L_{N+1}$ = liquid product rate; moles, weight, or volume per unit time. If a total condenser is used, $D_V = 0$ and $D_L = D$.

D_V = vapor product rate from a partial condenser; moles, weight, or volume per unit time.

F = upper feed rate; moles, weight, or volume per unit time.

F' = lower feed rate; moles, weight, or volume per unit time. If only one feed is used, $F' = 0$ and $M = 0$.

K_i = y_i/x_i = equilibrium distribution coefficient for component i.

L_c = liquid overflow from condenser; moles, weight, or volume per unit time.

L_{N+1} = RD = liquid or extract phase entering stage N; moles, weight, or volume per unit time.

M = number of theoretical stages below the feed stage. Feed stage is $M + 1$.

n = subscript referring to any stage. It may take on any value from 1.0 to N.

N = total number of theoretical stages including the reboiler (if it is a theoretical stage) and the feed stage but excluding the condenser.

N_c = independent restricting relationships.

N_i = degrees of freedom; variance; number of design variables which the designer must arbitrarily specify.

N_p = number of phases present in a multiphase system.

N_r = number of repetition variables.

N_v = total number of variables which the designer must consider.

q = general designation for a heat stream, Btu per unit time. q_c refers to the heat removed in a condenser and q_r to the heat input to a boiler.

R = external reflux ratio at top or raffinate end of column; amount of reflux to amount of product.

R' = external reflux ratio at bottom or extract end of column; amount of reflux to amount of product.

S = fresh solvent rate or side-stream rate; moles, weight, or volume per unit time. Also used in Fig. 3-15 to denote number of the stage from which the side stream S is withdrawn.

S_E = solvent-rich material recovered in the extract solvent-recovery equipment; mo'es, weight, or volume per unit time.

V = light or raffinate phase rate; moles, weight, or volume per unit time. Subscript denotes stage number.

V_0 = light or raffinate phase rate entering stage 1; moles, weight, or volume per unit time.

w = work.

x = concentration in heavy or extract phase. Units consistent with units on rates.

y = concentration in light or raffinate phase. Units consistent with units on rates.

Superscripts

e = denotes element where element is some part of a process unit.

u = denotes unit which is a combination of elements.

PROBLEMS

3-1. How many variables must be specified to define a partially flashed hydrocarbon stream flowing through a pipe? Consider each phase to be a separate stream in the analysis.

3-2. Derive the phase rule $N_i = C + 2 - N_p$ for a chemically nonreactive system.

3-3. Streams A and B flow countercurrently through a heat exchanger. Each stream exists in one phase only. How many variables can the designer specify?

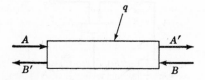

3-4. Count the design variables in the two-stage flash process shown below. Do not consider the valves as separate elements. They are shown on the diagram to denote the difference between the feed pressure and the system pressure in each element.

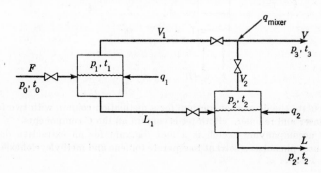

3-5. Analyze the stripping column shown in the adjacent sketch. Calculate $N_v{}^u$, $N_c{}^u$, and $N_i{}^u$. Tabulate a list of reasonable design specifications to utilize the available degrees of freedom.

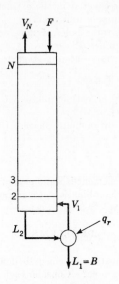

3-6. Calculate the degrees of freedom for the rectifying column shown in the adjacent sketch. Tabulate a list of reasonable design specifications.

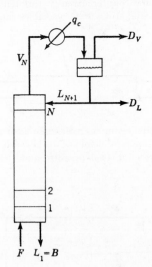

3-7. Calculate the degrees of freedom for a distillation column with two feeds and total condenser and reboiler. Both feeds contain all the C components.

3-8. The accompanying sketch is a flow diagram for an extractive distillation unit which uses phenol as a solvent to separate toluene and methylcyclohexane. The make-up phenol contains no toluene or methylcyclohexane, and the fresh feed contains no phenol. All three components exist to some extent in all other streams. Both columns have total condensers and reboilers. Calculate the number of design

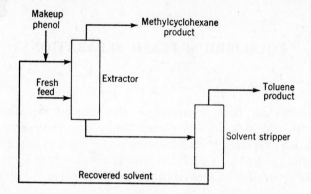

variables which must be specified and list a set of reasonable specifications. *Note:* In the analysis consider all streams to contain all three components and then utilize the information given above concerning stream purities to make the specifications.

REFERENCES

1. Dunstan, A. W., et al.: "The Science of Petroleum," p. 1563, Oxford University Press, London, 1938.
2. Gilliland, E. R., and C. E. Reed: *Ind. Eng. Chem.*, **34,** 551 (1942).
3. Kwauk, M.: *A.I.Ch.E. J.*, **2,** 240 (1956).

CHAPTER 4

EQUILIBRIUM FLASH SEPARATIONS

The equilibrium flash separator is the simplest equilibrium-stage process with which the designer must deal. The process involves (*a*) the regulation of a binary or multicomponent stream to the desired temperature and pressure, (*b*) a sudden reduction in pressure across a valve, and (*c*) the separation of the resulting liquid and vapor. Figure 4-1 is a schematic representation of the process. Normally, t_1 and p_1 are controlled to maintain the feed in the liquid state. The pressure reduction from p_1 to p_2 across the valve causes part of the feed to flash from the liquid to the vapor state. Since the flash occurs under essentially adiabatic conditions, the latent heat of vaporization must come from the sensible-heat content of the total material, and as a consequence the temperature falls when the flash occurs. The product liquid and vapor streams can be considered to be in equilibrium with each other owing to the intimate contact which prevails during the flashing operation.

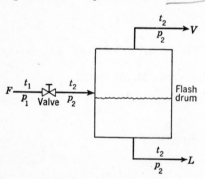

FIG. 4-1. Equilibrium flash separator.

The number of design variables or degrees of freedom in a flash process can be determined by the methods of Chap. 3. If the process is considered to be adiabatic, only the three material streams need be considered and

$$N_v = 3(C + 2) = 3C + 6$$

The vapor and liquid products are in equilibrium with each other, and as a consequence there are two inherent restrictions ($p_V = p_L$ and $t_V = t_L$) and C distribution relationships. In addition, there are C component balances and one over-all enthalpy balance to bring the number of total restrictions to

$$N_c = 2C + 3$$

The number of variables over which the designer has control is then

$$N_i = N_v - N_c = (3C + 6) - (2C + 3) = C + 3$$

One way to utilize the $C + 3$ degrees of freedom is to specify the feed rate, composition, temperature, and pressure ($C + 2$ variables) and the flash drum pressure p_2. The flash temperature t_2 must then be calculated along with the product compositions. Another method is to specify the feed rate, composition, and pressure ($C + 1$ variables) and then utilize the other two degrees of freedom to specify both t_2 and p_2. Once the separation at the specified t_2 and p_2 has been calculated, an over-all enthalpy balance will provide the t_1 required to produce the desired drum conditions.

Despite the fact that only one stage is involved, the calculation of the compositions and the relative amounts of the liquid and vapor phases

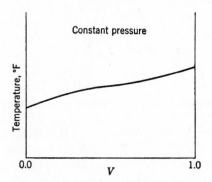

Fig. 4-2. Equilibrium flash curve.

at any given pressure and temperature usually involves a tedious trial-and-error solution. There are many different ways of combining the basic balance and equilibrium equations, and consequently, many different solution methods are possible. Whatever method is used, the basic purpose of the calculations is to establish a point on a flash curve such as the one shown in Fig. 4-2. A flash curve presents V as a function of the flash drum temperature. The basis of calculation is usually 1 mole of feed, and on this basis V represents that fraction of the feed which leaves the drum as vapor. The flash curve is always drawn at a constant pressure. Repetition of the calculations at a different pressure will result in another flash curve either above or below the original one depending upon whether the pressure was increased or decreased.

The separation between light and heavy components is relatively poor in a flash separation, since only one equilibrium stage is involved. The entire vapor is in equilibrium with the entire liquid, and therefore, the flash curve is relatively flat compared with the curves obtained by any multistage distillation process.   Figure 4-3 shows the position of the flash curve relative to the curves obtained in ASTM 10 per cent and true-boiling-point laboratory distillations.   More than one equilibrium stage is involved in the ASTM 10 per cent distillation.   A temperature difference exists between the boiling liquid-vapor interface and the point where the vapor is withdrawn.   As the vapors rise up the neck of the flash, some condensation occurs, improving the degree of separation between the light and heavy components.   Because of this, the initial boiling point is below the bubble point (left end of the flash curve), the final boiling point is above the dew point (right end of the flash curve), and therefore the general slope of the ASTM distillation curve is greater than that of the flash curve.

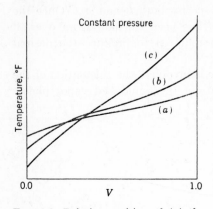

FIG. 4-3. Relative position of (a) the flash curve, (b) the ASTM 10 per cent distillation curve, and (c) the true-boiling-point distillation curve for a given feed.

In a true-boiling-point distillation, many equilibrium stages and a high reflux rate are employed and the degree of separation is high.   Actually, a true-boiling-point distillation on anything but very light, simple mixtures is almost an impossibility.

However, if a true-boiling-point distillation were available, the initial boiling point would be that of the lightest component present and the final boiling point that of the heaviest component present.   Therefore, the general slope of the true-boiling-point curve is greater than that of the ASTM distillation.

HYDROCARBON SYSTEMS

4-1. Binary Mixtures

Specification of the flash drum temperature and pressure for any given feed fixes the equilibrium-phase compositions.   In a binary system, the phase compositions are related to the distribution coefficients as follows:

$$x_1 = \frac{1 - K_2}{K_1 - K_2} \tag{4-1}$$

$$y_1 = \frac{K_1 K_2 - K_1}{K_2 - K_1} \tag{4-2}$$

These two equations can be combined with the over-all material-balance equation for component 1

$$Fz_1 = Vy_1 + Lx_1$$

to give the following expression for V when $F = 1.0$ mole:

$$V = \frac{z_1(K_1 - K_2)/(1.0 - K_2) - 1.0}{K_1 - 1.0} \tag{4-3}$$

It can be seen from Eqs. (4-1) to (4-3) that the calculation of the product rates and compositions from a binary flash involves no trial and error if the flash drum temperature and pressure are specified. Note that for a basis of $F = 1.0$, V is equivalent to the fraction flashed.

Example 4-1. Calculate the flash curve for a 50-50 mole per cent mixture of ethane and propane at 250 psia.

Solution. The pressure is constant at 250 psia, but the drum temperature will vary from one end of the flash curve to the other. Equation (4-3) must be evaluated at several temperatures. The results, when plotted, will constitute the flash curve. The calculations are best done in tabular form as follows:

Temp, °F	K_1	K_2	$K_1 - 1$	$K_1 - K_2$	$1 - K_2$	$z_1 \dfrac{K_1 - K_2}{1 - K_2} - 1.0$	V
50	1.5	0.462	0.50	1.038	0.538	−0.035	−0.070
55	1.57	0.496	0.57	1.074	0.504	0.083	0.146
60	1.63	0.523	0.63	1.107	0.477	0.160	0.254
65	1.70	0.550	0.70	1.150	0.450	0.278	0.397
70	1.77	0.585	0.77	1.185	0.415	0.429	0.558
75	1.85	0.617	0.85	1.233	0.383	0.610	0.718
80	1.94	0.655	0.94	1.285	0.345	0.861	0.917
85	2.04	0.700	1.04	1.340	0.300	1.235	1.187
90	2.11	0.735	1.11	1.375	0.265	1.595	1.438

A plot of the V's vs temperature gives the flash curve. The temperatures read from the curve for $V = 0$ and $V = 1.0$ are the bubble and dew points, respectively. The values of V below zero and above 1.0 are fictitious and serve only to locate the ends of the flash curve.

4-2. Multicomponent Mixtures

When more than two components exist in the system, a direct solution for V is not feasible and it becomes necessary to resort to a trial-and-error solution. This section describes convenient methods of calculation for miscible multicomponent systems.

Distribution Coefficients. When a component analysis of the feed stream is available, the K values for the individual components can be obtained by the methods described in Chap. 1. The convergence-pressure charts (5) or the DePriester (1) pressure-temperature-composition charts should be used for pressures much above 150 psia. Below 150 psia, ideal-solution K values can be obtained from Perry (8), from Maxwell (7), and from the DePriester (1) nomographs (Fig. B-3a and b) or calculated through the use of the equation

$$K_i = \frac{f_{i,p_i{}^*}}{f_{i,p}}$$

and the fugacity coefficient charts (Fig. B-2a and b).

If the final boiling point of the feed is above 150 to 160°F, a component analysis becomes expensive and will probably not be available. Above 200°F, a component analysis becomes impossible because of the large number of components present. The only information available on such complex feeds will usually be a laboratory distillation, and the designer

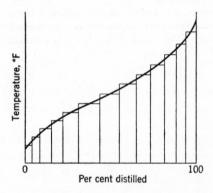

FIG. 4-4. Breakdown of a complex mixture into pseudocomponents.

of the flash separator must use the laboratory-distillation curve in lieu of a component analysis. Empirical methods have been developed to estimate the atmospheric flash curve from an ASTM 10 per cent or a crude assay distillation curve. The atmospheric flash curve can then be corrected to the desired operating pressure. For a discussion of these empirical methods and the charts necessary for their use, the reader is referred to Maxwell (7), Edmister and Bowman (3), Edmister (2), and Edmister and Okamoto (4).

An alternative to the empirical methods for complex mixtures is the pseudocomponent method. The laboratory-distillation curve can be used to separate the feed into pseudocomponents as shown in Fig. 4-4. Each volume fraction corresponding to the base of one of the rectangles is treated as a component with a boiling point corresponding to the intersection of the distillation curve with the top of the rectangle. The number of pseudocomponents marked off is arbitrary, as are their respective volume per cents and boiling points. The only restriction is that the sum of the areas enclosed in the rectangles must equal the area under the distillation curve. If the laboratory distillation involved enough stages and reflux, the curve may exhibit flattened sections, or plateaus. Advantage should be taken of any such plateaus in marking off the pseudocomponents, since a flat portion corresponds to a single component or to a constant-boiling mixture. Increasing the number of pseudocomponents improves the accuracy of the calculations but at the same time increases the calculation time.

Calculation of the K values for the various pseudocomponents involves a series of approximations. The molecular weight and specific gravity of each of the pseudocomponents must be known or estimated in order to convert the volume per cents to mole per cents. Maxwell (7) and Edmister (2) provide charts useful for the estimation of molecular weights and specific gravities as a function of boiling point. A Cox chart can be used to estimate the vapor pressures of the pseudocomponents at the flash drum temperature from their respective boiling points at the pressure of the laboratory distillation. Ideal gases can then be assumed and the K values calculated from $K_i = p_i^*/p$, or the pseudocritical properties can be estimated for each component and the K values calculated from $K_i = f_{i,p_i^*}/f_{i,p}$. Maxwell (7) presents charts for the estimation of pseudocritical properties.

Once the mole fractions and K values have been estimated for each of the pseudocomponents, the calculations proceed in exactly the same way as if a component analysis had been available. It can be seen that considerable calculation effort is required by the pseudocomponent method and that many estimates and approximations are needed. For these reasons, it is generally advisable to resort to the empirical methods mentioned above for complex mixtures.

Bubble and Dew Points. A pure liquid when heated will begin to boil when its vapor pressure becomes equal to the confining pressure. Likewise, a liquid mixture of several components will begin to boil when the sum of the vapor pressures exerted by the individual components becomes equal to the confining pressure. Boiling is characterized by the formation of vapor bubbles within the liquid phase. Since the bubbles in boiling are formed within the confines of the liquid phase, their vapor can contain only those components present in the liquid phase; that is, the Σy_i for the liquid components must equal 1.0 in the vapor bubble. This fact differentiates the vapor formed at the boiling temperature from vapor formed at the liquid-gas interface by evaporation at temperatures below the boiling point. Below the boiling point, the Σy_i for the liquid-phase components in the vapor formed by evaporation is less than 1.0.

The boiling point of a multicomponent liquid mixture at a given pressure is not constant as is the boiling point of a pure liquid. The lighter, more volatile components boil away more rapidly than the heavier components, and consequently, the boiling point of the remaining mixture increases as the vaporization proceeds. Because of its indefinite nature, the term *boiling point* is not used for the temperature at which the first bubble of vapor forms. Instead, the term *bubble point* is used to describe the temperature of incipient boiling in a multicomponent mixture.

The term *dew point* is used for the condensation of a multicomponent vapor mixture in a manner analogous to the use of bubble point for vaporizing liquid mixtures. The dew-point temperature is that temperature at which the first drop of liquid appears as the vapor mixture is cooled. The liquid drop is in equilibrium with the entire vapor mixture, and the Σx_i in the liquid drop must be 1.0. As condensation proceeds, the vapor phase is more rapidly depleted in the heavier components, since they tend to condense before the lighter ones. Consequently, the temperature must drop as condensation proceeds. The dew point for the original mixture refers only to the temperature of incipient condensation.

It can be seen from these definitions of the bubble and dew points that they represent the points at $V = 0$ and $V = 1.0$, respectively, on the flash curve. The mathematical relationships by which the dew- and bubble-point temperatures can be calculated are as follows: For a liquid mixture at its bubble point,

$$x_1 + x_2 + x_3 + \cdots + x_C = 1.0 \tag{4-4}$$
$$y_1 + y_2 + y_3 + \cdots + y_C = 1.0 \tag{4-5}$$

where the subscripts refer to the respective components and the y's refer to the first small bubble of vapor formed within the liquid phase. At the bubble point and before an appreciable amount of vapor has escaped, $x_1 = z_1$, $x_2 = z_2$, etc., where the z's refer to the total feed concentrations. Since $y_1 = K_1 x_1 = K_1 z_1$, $y_2 = K_2 x_2 = K_2 z_2$, etc., it can be seen that $\Sigma K_i z_i = 1.0$ at the bubble point.

If Eqs. (4-4) and (4-5) are written for a vapor mixture at its dew point, the x's refer to the first drop of liquid formed. At the dew point it is still true that $y_1 = z_1$, $y_2 = z_2$, etc. If the relations $x_1 = y_1/K_1$, $x_2 = y_2/K_2$, etc., are substituted in Eq. (4-4), it can be seen that $\Sigma z_i/K_i = 1.0$ at the dew point.

When a vapor exists at a temperature above its dew point, $\Sigma z_i/K_i < 1.0$ and the vapor is said to be superheated. When a liquid exists at a temperature below its bubble point, $\Sigma K_i z_i < 1.0$ and the liquid is said to be subcooled. The calculation of a dew or bubble point at any given pressure involves a trial-and-error search for the temperature at which $\Sigma z_i/K_i$

or $\Sigma K_i z_i = 1.0$.   In some situations, the temperature will be specified and the dew- or bubble-point pressure must be found by trial and error.

Example 4-2. Calculate the bubble and dew points at 200 psia for a feed containing 8 per cent ethane, 22 per cent propane, 53 per cent n-butane, and 17 per cent n-pentane.

Solution. The bubble point will be calculated first. Ideal-solution K values will be assumed for the sake of simplicity.

Component	z	$t = 100°F$		$t = 151°F$		$t = 150°F$	
		K	Kz	K	Kz	K	Kz
C_2	0.08	2.70	0.216	3.92	0.314	3.910	0.313
C_3	0.22	0.945	0.208	1.52	0.334	1.500	0.330
$n\text{-}C_4$	0.53	0.339	0.180	0.605	0.321	0.597	0.317
$n\text{-}C_5$	0.17	0.111	0.019	0.240	0.041	0.238	0.040
	1.00		0.623		1.010		1.000

The calculation of the dew point proceeds as follows:

Component	z	$t = 200°F$		$t = 205°F$	
		K	$\dfrac{z}{K}$	K	$\dfrac{z}{K}$
C_2	0.08	5.50	0.015	5.60	0.014
C_3	0.22	2.16	0.102	2.24	0.098
$n\text{-}C_4$	0.53	0.96	0.552	1.01	0.525
$n\text{-}C_5$	0.17	0.45	0.378	0.47	0.362
	1.00		1.047		0.999

Intermediate Points. The calculation of an intermediate point on a flash curve should not be attempted until the ends of the curve have been fixed by the calculation of the dew and bubble points. The feed may not form two phases at the specified drum temperature and pressure. Calculation of the dew and bubble points will show this discrepancy, and more suitable conditions can be chosen.

The balance and equilibrium relationships can be combined in many different ways for the calculation of intermediate points on the flash curve. Two different ways of using the equations, each best suited for a particular application, will now be described.

Classical Method. This method was first suggested by Katz and Brown (6) in 1933. It is best suited for the calculation of one specific point on the flash curve, that is, the V to be obtained at one specific temperature. The working equations can be derived as follows. If

1 mole of feed is taken as the basis of calculation, and if y_i/K_i is substituted for x_i, the over-all material balance for component i can be written as

$$z_i = Vy_i + \frac{Ly_i}{K_i}$$

Solving for y_i and multiplying both sides of the equation by V provide

$$y_iV = \frac{z_i}{1.0 + L/VK_i} \tag{4-6}$$

Since $\Sigma y_i = 1.0$, then

$$\sum_{i=1}^{i=C} y_iV = V \sum_{i=1}^{i=C} y_i = V$$

and

$$V = \sum_{i=1}^{i=C} \frac{z_i}{1.0 + L/VK_i} \tag{4-7}$$

The component concentrations in the vapor phase can be calculated from the following combination of Eqs. (4-6) and (4-7):

$$y_i = \frac{z_i/(1.0 + L/VK_i)}{\displaystyle\sum_{i=1}^{i=C} z_i/(1.0 + L/VK_i)} = \frac{v_i}{V} \tag{4-8}$$

The amount of liquid per mole of feed is given by

$$L = 1.0 - V \tag{4-9}$$

and the liquid mole fractions by

$$x_i = \frac{z_i - Vy_i}{L} = \frac{z_i - v_i}{L} \tag{4-10}$$

Solution by the classical method is accomplished by assuming a V, evaluating Eq. (4-6) for each component at the specified flash drum conditions, and summing the y_iV terms to get a calculated V which can be compared with the assumed V. If $\Sigma y_iV >$ assumed V, assume a higher V and repeat. If $\Sigma y_iV <$ assumed V, assume a lower V and repeat. Convergence is slow, especially near the ends of the curve. The agreement between the assumed and calculated V's should be within five in the fourth decimal place to ensure that convergence has been obtained. Because of the slow convergence, the classical method is inconvenient when more than one point on the flash curve is to be calculated.

Example 4-3. Calculate by the classical method the separation which will be obtained at 200 psia and 180°F for the feed composition given in Example 4-2.

Solution:

Component	z	$K^{180°F}_{200\,psia}$	$V = 0.5;\ \dfrac{L}{V} = 1.0$		$V = 0.4;\ \dfrac{L}{V} = 1.5$	
			$1.0 + \dfrac{L}{VK}$	$\dfrac{z}{1.0 + L/VK}$	$1.0 + \dfrac{L}{VK}$	$\dfrac{z}{1.0 + L/VK}$
C_2	0.08	4.78	1.209	0.0665	1.314	0.0609
C_3	0.22	1.87	1.535	0.1432	1.801	0.1220
$n\text{-}C_4$	0.53	0.805	2.242	0.2365	2.863	0.1851
$n\text{-}C_5$	0.17	0.35	3.860	0.0440	5.290	0.0322
	1.00			0.4902		0.4002

Note the relatively close check obtained between the calculated and assumed V's in the first trial and how much the assumed V had to be changed to obtain convergence.

Smith-Wilson Method. A rearrangement of the equilibrium and material-balance relationships has been suggested by D. A. Smith and G. W. Wilson (9) to obtain a faster convergence than that obtained by the classical method. The Smith-Wilson method converges rapidly and should be used when an entire flash curve is to be calculated. However, since the method usually converges to a temperature other than the one originally assumed, it is not convenient when the separation at one specific temperature is desired.

The Smith-Wilson method selects one component as the reference component and specifies the split on this component in place of specifying the operating temperature in the flash drum. Equations relating the reference component to any other component in the feed can be derived as follows:

$$\frac{K_i}{K_r} = \frac{y_i x_r}{y_r x_i} \qquad (4\text{-}11)$$

where the subscript r refers to the selected reference component. If l_i, l_r, v_i, and v_r are defined as the moles of components i and r in the liquid and vapor phases, the mole fractions in Eq. (4-11) can be replaced to give

$$\frac{K_i}{K_r} = \frac{(v_i/V)(l_r/L)}{(v_r/V)(l_i/L)} = \frac{v_i l_r}{v_r l_i} \qquad (4\text{-}12)$$

This equation can be rearranged to

$$\frac{v_i}{l_i} = K_i \frac{v_r}{l_r K_r} \qquad (4\text{-}13)$$

which is used to calculate the split for component i between V and L for any specified split on component r. The over-all balance, $z_i = v_i + l_i$ when $F = 1.0$, can be rearranged to

$$l_i = \frac{z_i}{v_i/l_i + 1.0} \tag{4-14}$$

and used to calculate l_i and v_i from the v_i/l_i ratio.

The Smith-Wilson method specifies the flash drum pressure and the split on the reference component, v_r/l_r. The reference component should have an appreciable concentration in the feed, and it should be in the middle of the feed mixture so that its K value will be close to unity. The flash drum temperature is the major iteration variable. It can be seen from Eqs. (4-13) and (4-14) that for any specified v_r/l_r and pressure the assumption of a drum temperature will permit the calculation of l_i and v_i for every component. Smith and Wilson have suggested the following guides for the initial assumption of the temperature corresponding to the specified v_r/l_r:

At 50 per cent vaporized,

$$t_{50\%} = 0.5(t_{\text{D.P.}} + t_{\text{B.P.}})$$

At 25 per cent vaporized,

$$t_{25\%} = t_{50\%} - 0.35(t_{50\%} - t_{\text{B.P.}})$$

At 75 per cent vaporized,

$$t_{75\%} = t_{50\%} + 0.35(t_{\text{D.P.}} - t_{50\%})$$

The subscripts D.P. and B.P. refer to the dew and bubble points, respectively. Once the respective l_i's and v_i's have been calculated, they are summed to give V and L. The following equation is then used to check the correctness of the assumed temperature.

$$K_r = \frac{v_r/\Sigma v_i}{l_r/\Sigma l_i} = \frac{v_r/V}{l_r/L} \tag{4-15}$$

The temperature which corresponds to the K_r calculated from Eq. (4-15) with the calculated v_i's and l_i's is compared with the assumed temperature. Agreement within 5°F is usually satisfactory unless the flash curve is very flat. If the agreement is not satisfactory, the following equation can be used to approximate the new temperature for the next trial

$$t_{\text{new}} = t_{\text{old}} + 0.9(t_{\text{calc.}} - t_{\text{old}})$$

where $(t_{\text{calc.}} - t_{\text{old}})$ is the difference between the calculated and assumed temperature from the previous trial.

Example 4-4. Use the Smith-Wilson method to calculate a point on the flash curve at 200 psia for the feed composition given in Example 4-2.

Solution. The n-butane is the logical choice for the reference component, and the point on the flash curve corresponding to a 50-50 split on the reference component will be calculated. Since the butane is in the middle of the feed mixture, the split on the total feed will be approximately 50 50 also. The dew and bubble points were calculated in Example 4-2 as 150 and 205°F, respectively. An arithmetic average indicates an initial temperature assumption in the region of 180°F. At 180°F, $K_r = 0.81$ and $v_r/l_rK_r = 0.5/(0.5)(0.81) = 1.237$.

Component	z	$K_{200 \text{ psia}}^{180°F}$	$\dfrac{v}{l} = 1.237K$	$l = \dfrac{z}{v/l + 1.0}$	v
C_2	0.08	4.78	5.90	0.0116	0.0684
C_3	0.22	1.87	2.31	0.0665	0.1535
n-C_4	0.53	0.81	1.00	0.2650	0.2650
n-C_5	0.17	0.35	0.432	0.1187	0.0513
	1.00			0.4618	0.5382

From Eq. (4-15)   Calculated $K_r = \dfrac{0.265/0.5382}{0.265/0.4618} = 0.857$

Corresponding $t = 188°F$

$$\text{Next } t = 180 + 0.9 \, (188 - 180) = 187°F$$
$$\text{New } K_r = 0.85 \quad \text{and} \quad v_r/l_rK_r = 1.179$$

Component	z	$K_{200 \text{ psia}}^{187°F}$	$\dfrac{v}{l} = 1.179K$	$l = \dfrac{z}{v/l + 1.0}$	v
C_2	0.08	5.00	5.89	0.0116	0.0684
C_3	0.22	1.95	2.30	0.0666	0.1535
n-C_4	0.53	0.85	1.00	0.2650	0.2650
n-C_5	0.17	0.38	0.447	0.1176	0.0524
	1.00			0.4608	0.5393

Calculated $K_r = \dfrac{0.265/0.5393}{0.265/0.4608} = 0.856$ Corresponding $t = 188°F$

4-3. Enthalpy Balances

After the flash curve has been calculated, the fraction flashed can be read directly for any desired temperature and the specified pressure. When both temperature and V are known, the calculation of the vapor and liquid compositions is straightforward, involving no trial and error. However, the temperature of the flash often is not specified directly. Instead, the feed temperature and pressure are specified (along with feed composition), and this in turn fixes the flash temperature. Combination of an enthalpy balance with the material balance and the equilibrium calculations described previously yields the unique solution pertaining to a given feed composition and enthalpy.

The enthalpy balance is written as follows:

$$Fh_F = VH_V + Lh_L$$

where h_F = enthalpy of 1 mole of feed
H_V = enthalpy of 1 mole of the vapor product
h_L = enthalpy of 1 mole of the liquid product

Substituting $L = F - V$ and setting $F = 1.0$ give

$$V = \frac{h_F - h_L}{H_V - h_L} \qquad (4\text{-}16)$$

To calculate V from this expression the temperature and the compositions of the liquid and vapor phases must be known in order to calculate the molar enthalpies of the vapor and liquid products. The liquid and vapor compositions at various temperatures will be available from the

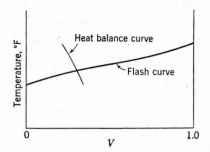

Fig. 4-5. Diagram showing the location of the actual operating conditions in a flash separator by means of the intersection of a fictitious heat-balance curve with the flash curve.

previous calculation of the flash curve. These sets of temperature and compositions can be used to calculate corresponding V's by means of the enthalpy-balance equation. These V's can be plotted to give an enthalpy-balance curve which intersects the flash curve as shown in Fig. 4-5. The V's on the enthalpy-balance curve are fictitious except at the point of intersection. Note that each calculated fictitious V plotted on the enthalpy-balance curve is always on the other side of the intersection from the true V corresponding to the temperature and compositions used.

The enthalpy-balance curve is not a straight line. If it is necessary to pinpoint the intersection exactly, three points are required. Two points are calculated using two points from the calculation of the flash curve. A straight line is drawn between these points. Then a third point is calculated using the V and temperature where the straight line intersects the flash curve.

HYDROCARBON-WATER SYSTEMS

4-4. Binary Mixtures

Hydrocarbons and water usually can be assumed to be completely immiscible for the purposes of flash calculations.† Each of the two immiscible phases exerts its own vapor pressure, and the total vapor pressure is the sum of the two individual vapor pressures.

$$p_t^* = p_{HC}^* + p_{H_2O}^*$$

where the subscripts t and HC refer to the total and hydrocarbon vapor pressures, respectively. If the temperature is such that $p_t^* < p$, then no vapor phase can exist unless a fixed gas such as air is present to make up the difference between the vapor pressure of the mixture and the total system pressure.

Bubble Point. If the temperature is increased until $p_t^* = p$, vaporization will start and continue as long as $p_t^* \geqq p$ and as long as heat is supplied to furnish the heat of vaporization. The point of incipient vaporization (where p_t^* just becomes equal to p) is the bubble point of the mixture.

Example 4-5. Calculate the bubble point of a mixture of water and n-heptane at 100 psia.

Solution. A bubble-point calculation for a pair of immiscible liquids is simply a trial-and-error search for the temperature at which

$$p_t^* = p = p_{HC}^* + p_{H_2O}^*$$

Component	p_i^*, psia		
	250°F	290°F	288.1°F
H_2O	29.8	57.6	55.9
$n\text{-}C_7$	32.0	45.0	44.1
	61.8	102.6	100.0

The sum of the individual vapor pressures equals 100 psia at 288.1°F. Note that the relative amounts of H_2O and $n\text{-}C_7$ are unimportant in the calculation of the

† In high-temperature cases where the amount of water is so small that it is completely dissolved in the hydrocarbon, the assumption of immiscibility cannot be made. If the distribution of this small amount of water between the vapor and liquid phases is important, a K value can be assigned to the water and the methods of the previous sections used. The water is treated as just another component.

bubble point. If a phase is present, it will provide its vapor pressure whether it is present to the extent of only one small drop or many barrels.

Dew Point. The partial pressure of either component in the vapor phase cannot be greater than its vapor pressure. When a binary hydrocarbon-water vapor mixture is cooled, a temperature will eventually be reached where the vapor pressure of one or the other of the components becomes equal to its partial pressure. That component will begin to condense, and the point of incipient condensation is called the dew point of the mixture. It is important to realize that, in general, only one of the components will condense initially and that each must be checked separately when determining the dew point. Which one condenses first will depend upon the individual partial pressures and the individual vapor pressures of the two. (It is, of course, possible for the composition to be such that both phases begin to condense at the same time.) Since the dew point of an immiscible binary pair depends upon the partial pressures, the dew point is not independent of the feed composition as was the bubble point.

Example 4-6. Calculate the dew point of a 60 mole per cent water and a 40 mole per cent n-heptane vapor mixture at 100 psia.

Solution. The temperatures at which the individual vapor pressures equal the partial pressures are given below:

Component	$z = y$	Partial pressure, psia	p^*, psia 293°F	p^*, psia 280°F
H_2O	0.6	60	60	<60
n-C_7	0.4	40	>40	40
	1.0			

At 293°F, the vapor pressure of the water becomes equal to its partial pressure. Therefore, water will begin to condense, and this is the dew point of the mixture. The mixture must be cooled further to 280°F before the original partial pressure of the n-C_7 becomes equal to the vapor pressure, but by then some of the water has already condensed and the partial pressure of the n-C_7 is no longer 40 psia. So the numbers at 280°F are meaningless. If the over-all composition is changed as follows a dew point is obtained where the n-C_7 condenses first.

Component	$z = y$	Partial pressure, psia	p^*, psia 314°F	p^*, psia 267°F
H_2O	0.4	40	>40	40
n-C_7	0.6	60	60	<60

Here the n-C_7 begins to condense at 314°F.

Intermediate Points. It must always be remembered that in the flash operation all the vapor formed is in equilibrium with all the remaining liquid. The vapor is not removed as fast as it is formed as it is in a differential distillation such as an ASTM 10 per cent or other distillations. The fact that all the vapor remains in contact with all the liquid (until separated in the drum) is helpful in the calculation of points on the flash curve for a two-liquid-phase mixture. As the flash curve is constructed by calculating a series of points from $V = 0$ toward $V = 1.0$, a point is reached at which one of the two liquid phases disappears. Which one disappears depends entirely on how much of each was originally present and on their respective vapor pressures. Since all the vapor remains in contact with the liquid, the amount in the vapor phase of the component whose liquid phase has disappeared is equal to the amount

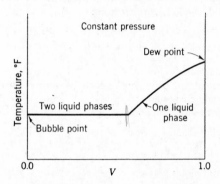

FIG. 4-6. Typical flash curve for an immiscible binary.

originally present in the feed. Knowing the amount of one component present in the vapor, the other component can be calculated by utilizing the vapor pressures.

The phase rule states that for a two-component system containing three phases the variance is one. Therefore, if the separator pressure is specified, no control can be exercised over the temperature or phase compositions. This means that from $V = 0$ (bubble point where the third phase appears) out to the point where one liquid phase disappears and the number of phases returns to two, the binary flash curve will be a horizontal line at the bubble-point temperature. The vapor composition remains unchanged as V increases in this region. When one liquid phase disappears, the variance increases to two and another variable must be specified to define the system. In this region the flash curve is a curved line extending upward from the end of the horizontal line to the temperature of the dew point as shown in Fig. 4-6.

Example 4-7. Calculate the entire flash curve for a 60 per cent H_2O-40 per cent n-C_7 mixture at 100 psia.

Solution. From Examples (4-5) and (4-6) the bubble point is 288.1°F and the dew point is 293°F.

The initial vapor composition (at the bubble point) is given by the ratio of the respective vapor pressures at 288.1°F.

$$\frac{\text{Moles } H_2O}{\text{Moles } n\text{-}C_7} = \frac{55.9}{44.1}$$

As long as two liquid phases are present, this will be the vapor composition. Since this ratio is smaller than the ratio of moles of H_2O to $n\text{-}C_7$ in the original feed, the liquid $n\text{-}C_7$ will be depleted first as V is increased and its liquid phase will be the one that disappears. At the point of disappearance the amount of water which has been vaporized can be calculated as follows:

$$\text{Moles } H_2O \text{ in vapor} = 0.4 \frac{55.9}{44.1} = 0.506$$

where the 0.4 is the amount of $n\text{-}C_7$ present in the feed. Also at the point of disappearance of $n\text{-}C_7$ liquid, the following are true:

$$\text{Total moles of vapor} = 0.4 + 0.506 = 0.906$$
$$\text{Total moles of liquid water} = 0.6 - 0.506 = 0.094$$

Once the end of the flat portion has been found, a point on the curved portion of the flash curve can be calculated by choosing a temperature between the bubble and dew points and proceeding as follows:

Let $t = 290°F$. Then $p^*_{H_2O} = 57.6$ psia.

$$\text{Partial pressure of } n\text{-}C_7 = 100 - 57.6 = 42.4 \text{ psia}$$
$$\text{Moles } H_2O \text{ in vapor} = 0.4 \frac{57.6}{42.4} = 0.545$$
$$\text{Total moles vapor} = V = 0.4 + 0.545 = 0.945$$

It may be necessary to calculate more than one point on the curved portion in order to define the curve adequately.

4-5. Multicomponent Hydrocarbon–Water Mixtures

Calculations for multicomponent hydrocarbon-water mixtures are based on either of two assumptions. (1) The hydrocarbons and water are assumed to be completely immiscible as was done above for binary mixtures, or (2) the slight solubility of water in the hydrocarbon can be recognized and the water considered to have a K value just as all the hydrocarbons do. Unfortunately, it is more convenient to use one assumption in certain situations and the other assumption in other situations. Description of both calculation methods serves little academic purpose, but the practicing engineer may find an orderly presentation useful. Therefore, both methods will be described and illustrated below. For the sake of convenience the two methods will hereafter be described as Method 1 and Method 2.

Method 1. Method 1 is based upon the assumption of complete immiscibility of the two liquid phases. This method is essentially that used in Sec. 4-4 for binary mixtures, and the relationships between the total pressure and the vapor pressures described there are equally applicable to multicomponent systems. Each liquid phase is assumed to act independently of the other. The partial pressure of the hydrocarbon portion of the system is calculated by subtracting the vapor pressure of the water from the total pressure. Calculations are then made with the hydrocarbon portion as if no water were present except that the K values are read from the charts at the hydrocarbon partial pressure instead of the total system pressure.

Bubble and Dew Points. The procedure for calculating bubble and dew points by Method 1 is as follows: Convert the feed composition to a water-free basis and assume the bubble-point or dew-point temperature. Calculate the partial pressure of the hydrocarbon portion of the system from

$$p_{HC} = p - p^*_{H_2O}$$

Obtain K values at the assumed temperature and the hydrocarbon partial pressure p_{HC}. Calculate $\Sigma y_i = \Sigma K_i z'_i$ or $\Sigma x_i = \Sigma z'_i / K_i$, where z'_i refers to a concentration on the water-free basis. If Σy_i or Σx_i is not 1.0, a new temperature must be assumed and the procedure repeated.

The temperature at which $\Sigma y_i = 1.0$ for the hydrocarbon is always the bubble point of the hydrocarbon-water mixture. However, the temperature at which $\Sigma x_i = 1.0$ for the hydrocarbons is not always the dew point of the total mixture. The water will begin to condense when $p_{H_2O} = p^*_{H_2O}$, and this equality may be reached at a temperature above that at which $\Sigma x_i = 1.0$ for the hydrocarbon portion alone.

Example 4-8. Calculate the bubble and dew points at 25 psia for the hydrocarbon-water mixture whose feed composition is shown below in the solution.

Solution. The bubble point will be calculated first.

Component	z	z'	$t = 160°F$ $p^*_{H_2O} = 4.7$ $p_{HC} = 20.3$		$t = 163.4°F$ $p^*_{H_2O} = 5.2$ $p_{HC} = 19.8$	
			$K^{160°F}_{20.3}$	Kz'	$K^{163.5°F}_{19.8}$	Kz'
n-C$_5$	0.225	0.250	1.94	0.485	2.05	0.511
n-C$_6$	0.450	0.500	0.77	0.385	0.815	0.408
n-C$_7$	0.225	0.250	0.30	0.075	0.324	0.081
H$_2$O	0.100					
	1.000	1.000		0.945		1.000

Note that the water enters into the calculation of the bubble point only when the hydrocarbon partial pressure is being determined from $p_{HC} = p - p^*_{H_2O}$. When the dew point is being calculated, the partial pressure of the hydrocarbon portion is independent of the temperature and is given by $p_{HC} = (0.9)(25) = 22.5$ psia. The dew-point calculation then proceeds as follows:

| Com- | | | $t = 190°F$ | | $t = 195°F$ | |
ponent	z	z'	$K_{22.5}^{190°F}$	$\dfrac{z'}{K}$	$K_{22.5}^{195°F}$	$\dfrac{z'}{K}$
$n\text{-}C_5$	0.225	0.250	2.5	0.100	2.71	0.092
$n\text{-}C_6$	0.450	0.500	1.08	0.463	1.17	0.423
$n\text{-}C_7$	0.225	0.250	0.46	0.543	0.515	0.485
H_2O	0.100					
	1.000	1.000		1.106		1.000

If oil condenses first, the dew point is 195°F. A check must be made to see if this temperature is higher than the temperature at which water will begin to condense if it condenses first.

$$p_{H_2O} = 0.1(25) = 2.5 \text{ psia}$$

The vapor pressure of water will exceed 2.5 psia until 134°F is reached. Therefore, the oil condenses first and 195°F is the dew point of the total mixture. ✓

Method 2. Method 2 recognizes that water is present in the hydrocarbon liquid to a slight extent and has its own K value. However, the dissolved water has such a high K value that it is impractical to use it to calculate the concentration of water in the vapor phase or in the liquid phase. It is easier to assume that the concentration in the liquid phase is so low that it will not appear in calculations carried to three or four digits. Also, it is easier to get the concentration of the water in the vapor phase by using the vapor pressure if a liquid water phase is present or by using the amount of water originally in the feed if the water phase has already disappeared. This method is presented in addition to Method 1 because it is useful in calculating the one-liquid-phase portion of the flash curve when the liquid phase is that of the hydrocarbons.

Bubble and Dew Points. The procedure for a bubble-point calculation by Method 2 is as follows: Calculate the composition of the liquid hydrocarbon phase. Assume only a trace of water present in the hydrocarbon; that is, the hydrocarbon phase is essentially water free. Assume a bubble-point temperature, and obtain K values at this assumed temperature and the total system pressure. Obtain the vapor pressure of water at the assumed temperature. Calculate $y_i = K_i x_i$ for each hydrocarbon using the composition of the hydrocarbon liquid phase, but calculate the vapor concentration for water from $y_{H_2O} = p_{H_2O}^*/p$. Calculate the Σy_i from

$$\Sigma y_i = y_{H_2O} + \Sigma(K_i x_i)_{HC}$$

If Σy_i is not 1.0, assume a new temperature and repeat the calculations.

The procedure for the calculation of the hydrocarbon dew point by Method 2 is the same as for a bubble-point calculation except that the

z_i's used in $\Sigma x_i = \Sigma z_i/K_i$ are for the total feed. The K for water is so large that z/K for water is essentially zero. As before, the dew-point calculation must always be followed by a check on the temperature at which the water vapor would start condensing if it condenses before the hydrocarbons.

Example 4-9. Rework Example 4-8 using Method 2.

Solution. The bubble point will be calculated first. The water concentration in the vapor is obtained from $y_{H_2O} = p^*_{H_2O}/p$.

Component	z	x_{HC}	$t = 163°F$ $p^*_{H_2O} = 5.1$ $y_{H_2O} = 0.204$		$t = 162°F$ $p^*_{H_2O} = 5.0$ $y_{H_2O} = 0.200$	
			$K_{25}^{163°F}$	y	$K_{25}^{162°F}$	y
$n\text{-}C_5$	0.225	0.250	1.65	0.412	1.62	0.405
$n\text{-}C_6$	0.450	0.500	0.67	0.335	0.66	0.330
$n\text{-}C_7$	0.225	0.250	0.26	0.065	0.26	0.065
H_2O	0.100	Trace		0.204		0.200
	1.000	1.000		1.016		1.000

Note that when water is included in Σy_i, the K values of the hydrocarbons must be evaluated at the total system pressure.

The dew-point calculation proceeds as shown below. No K value for water need be listed because z/K for water is essentially zero.

Component	z	$t = 196°F$		$t = 195°F$	
		$K_{25}^{196°F}$	$\dfrac{z}{K}$	$K_{25}^{195°F}$	$\dfrac{z}{K}$
$n\text{-}C_5$	0.225	2.50	0.090	2.45	0.092
$n\text{-}C_6$	0.450	1.09	0.413	1.06	0.424
$n\text{-}C_7$	0.225	0.475	0.474	0.465	0.484
H_2O	0.100		0		0
	1.000		0.977		1.000

Now a check must be made to see if the water will condense above 195°F. This is done in the same way as in Method 1.

$$p_{H_2O} = 0.1(25) = 2.5 \text{ psia}$$
$$p^*_{H_2O} = 2.5 \text{ psia at } 134°F$$

The hydrocarbon condenses first.

General Considerations. The following general comments are helpful in the calculation of the entire flash curve for a multicomponent hydrocarbon-water feed.

1. Two liquid phases will be present for the initial part of the curve,

and usually only one liquid phase will be present over the latter part of the curve. Neither part of the curve will be a straight line.

2. The dew-point calculation establishes which liquid phase disappears first.

3. Calculations for the two-liquid-phase region are the same no matter which phase disappears first.

4. The calculation method used in the one-phase region will depend on which phase remains. If the liquid phase is water, use Method 1. If the liquid phase is hydrocarbon, use Method 2.

5. The intersection point of the two portions of the flash curve can be calculated, but the method of calculation depends on which phase disappears.

6. It is usually most convenient to calculate the various parts of the flash curve in the following order:

  (a) Calculate dew and bubble points.

  (b) Locate the point of intersection of the two parts of the curve.

  (c) Calculate the one-liquid-phase curve.

  (d) Calculate the two-liquid-phase curve.

One-liquid-phase Region. The one liquid phase which exists at the higher values of V may be either water or hydrocarbon depending on the relative amounts of each in the original feed. The most convenient calculation method to be used depends on the nature of the remaining liquid.

Liquid Phase Is Water. If the single liquid phase is water, then all the hydrocarbon portion of the feed is in the vapor phase. The calculation of the dew point of these hydrocarbon vapors will locate the point of intersection of the two parts of the flash curve. The dew point can be calculated by either Method 1 or 2, but the former is more convenient when the one liquid phase is water.

Example 4-10. A dew-point calculation at 25 psia for the feed composition shown below has shown that the water condenses first and therefore is the liquid phase in the one-liquid-phase region. Locate the point of intersection of the two parts of the flash curve at 25 psia.

Component	z	z'	$t = 180°F$ $p^*_{H_2O} = 7.51$ $p_{HC} = 25 - 7.5$		$t = 178°F$ $p^*_{H_2O} = 7.18$ $p_{HC} = 25 - 7.18$	
			$K^{180°F}_{17.5}$	$\dfrac{z'}{K}$	$K^{178°F}_{17.8}$	$\dfrac{z'}{K}$
$n\text{-}C_5$	0.025	0.25	2.93	0.035	2.80	0.089
$n\text{-}C_6$	0.050	0.50	1.20	0.416	1.15	0.432
$n\text{-}C_7$	0.025	0.25	0.52	0.480	0.48	0.520
H_2O	0.900					
	1.000	1.00		0.981		1.041

By interpolation, the point of intersection is approximately 179.3°F. Also, at the point of intersection the vapor is composed as follows:

$$\text{Moles hydrocarbon} = 0.10$$
$$\text{Moles H}_2\text{O} = 0.1 \frac{7.4}{17.6} = 0.042$$
$$V = 0.10 + 0.042 = 0.142 \text{ mole flashed}$$

Once the dew point and point of intersection have been determined, the calculation of a point in the one-liquid-phase region is quite simple and involves no trial and error when the one liquid phase is water. The procedure is as follows: A value of V is selected to the right of the point of intersection. Since all the hydrocarbons are in the vapor phase, the moles of water in the vapor are easily found from

$$v_{\text{H}_2\text{O}} = V - v_{\text{HC}}$$

where $v_{\text{H}_2\text{O}}$ and v_{HC} refer to the moles of water and hydrocarbons in the vapor. The mole fraction and partial pressure of water in the vapor phase are given by

$$y_{\text{H}_2\text{O}} = \frac{v_{\text{H}_2\text{O}}}{V}$$
$$p_{\text{H}_2\text{O}} = p_{\text{H}_2\text{O}}^* = y_{\text{H}_2\text{O}}p$$

The temperature corresponding to this vapor pressure of water also corresponds on the flash curve to the V originally selected.

Liquid Phase Is Hydrocarbon. If the dew-point calculation shows that the hydrocarbon portion of the feed will condense first, then the liquid in the one-liquid-phase region will be hydrocarbon. The calculation of the point of intersection of the one-liquid-phase portion with the two-liquid-phase part of the flash curve is aided by the following two facts: At the intersection, all the water is in the vapor phase but it is also true that $p_{\text{H}_2\text{O}} = p_{\text{H}_2\text{O}}^*$. These two facts permit the calculation of a line which is the locus of all possible intersection points. The calculation of a point on this locus is accomplished as follows: A temperature somewhere between the bubble and dew points is selected, and the vapor pressure of water obtained at this temperature. A fictitious V corresponding to the selected temperature can be related to the vapor pressure of water by the following equations:

$$p_{\text{H}_2\text{O}} = p_{\text{H}_2\text{O}}^*$$
$$p_{\text{HC}} = p - p_{\text{H}_2\text{O}}^*$$
$$\frac{V_{\text{HC}}}{V_{\text{H}_2\text{O}}} = \frac{p_{\text{HC}}}{p_{\text{H}_2\text{O}}} = \frac{p - p_{\text{H}_2\text{O}}^*}{p_{\text{H}_2\text{O}}^*}$$
$$V = z_{\text{H}_2\text{O}} \left(1 + \frac{p - p_{\text{H}_2\text{O}}^*}{p_{\text{H}_2\text{O}}^*} \right) = z_{\text{H}_2\text{O}} \frac{p}{p_{\text{H}_2\text{O}}^*} \qquad (4\text{-}17)$$

A V calculated in this manner is, of course, fictitious unless the temperature assumed is the actual intersection point. Nevertheless, a series of such V's when plotted on a flash-curve diagram provides a locus of all possible intersections and reduces the number of points which must be calculated on each portion of the flash curve. Figure 4-7 shows a typical locus on a multicomponent hydrocarbon-water flash curve.

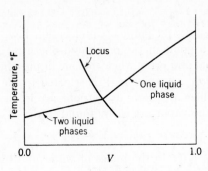

Fig. 4-7. Typical flash curve for a multicomponent hydrocarbon-water system where the hydrocarbon phase condenses first. The calculated locus of possible intersection points is shown.

Example 4-11. Calculate the locus of possible intersections of the two portions of the flash curve for the feed of Example 4-8. The pressure is 25 psia, and the mole fraction of water in the feed is 0.1.

Solution. The bubble and dew points were calculated in Example 4-8 to be 163.4 and 195°F. The solution of Eq. (4-17) for three temperatures in this range is shown below.

Assumed t	$p_{H_2O}^*$	$V = \dfrac{z_{H_2O}p}{p_{H_2O}^*}$
180	7.51	0.333
170	5.99	0.418
160	4.74	0.528

A plot of these three points on the flash-curve diagram gives a locus of all possible intersections.

After the locus of possible intersections has been plotted, one can proceed with the calculation of the one-liquid-phase curve. Either the classical or the Smith-Wilson method can be used. If the classical method is used, a temperature must be selected which falls between the dew point and the point of intersection. Sometimes, when the curve is almost flat, the temperature first selected will fall outside the one-liquid-phase region and the calculation will be wasted.

If the Smith-Wilson method is used, a percentage flashed must be selected for the reference component which will give a V and a temperature in the one-liquid-phase region. If the reference component is in the middle of the feed mixture (as it usually must be if its K value is close to 1.0), then the fraction of the reference component flashed will be roughly equal to V. It is quite easy to pick a V which will lie to the right of the locus of intersections. As in the case of the classical method, the calculations are wasted if the point does not fall within the one-liquid-phase region.

The water in either the classical or the Smith-Wilson method is best treated according to the assumptions of Method 2; that is, the K value for water is assumed to be very large and all the water goes to the vapor phase. The K values for the hydrocarbons are evaluated at the total system pressure, and the calculations proceed as shown in Example (4-3) or (4-4).

A point calculated by either the classical or Smith-Wilson method must be checked to ensure that it actually lies within the one-liquid-phase region. This can be done, of course, by plotting the point on the flash-curve diagram and noting whether it lies to the right of the locus of possible intersections. If the locus has not been calculated, the apparent partial pressure of water can be calculated from Eq. (4-17) as

$$p_{H_2O} = p \frac{y_{H_2O}}{V} = p \frac{z_{H_2O}}{V}$$

and checked against the vapor pressure. If $p_{H_2O}^* > p_{H_2O}$, all the water is in the vapor phase.

Two-liquid-phase Region. Points in the two-liquid-phase region can be calculated by either the classical or Smith-Wilson methods. If the Smith-Wilson method is used, the agreement between the assumed and calculated temperatures should be within 1 or 2°F, since the two-liquid-phase portion of the flash curve is often quite flat in hydrocarbon-water systems.

The assumptions of Method 1 are usually most convenient for the two-liquid-phase portion of the flash curve. The calculations by either the classical or Smith-Wilson method are performed on the basis of 1 mole of water-free material, and the K values are evaluated at the partial pressure of the hydrocarbons. Since 1 mole of hydrocarbons is used as a calculation basis and since 1 mole of total feed contains less than 1 mole of hydrocarbons, the V and L calculated for the hydrocarbons must be corrected to correspond to the amount of hydrocarbons actually present in the feed. This correction is shown in the following example, which illustrates both the classical and Smith-Wilson methods.

Example 4-12. Calculate a point on the flash curve at 25 psia in the two-liquid-phase region for the feed mixture shown below in the solution. Calculate a point by both the classical and Smith-Wilson methods.

Solution. The classical method will be illustrated first, and a point at 166°F will be calculated using the assumptions of Method 1. At 166°F, $p_{H_2O}^* = 5.5$ and

$$p_{HC} = 25 - 5.5 = 19.5 \text{ psia}$$

Com-ponent	z	z'	$K_{19.5}^{166°F}$	$\dfrac{L}{V} = 4.0$		$\dfrac{L}{V} = 5.7$	
				$1.0 + \dfrac{L}{VK}$	$\dfrac{z'}{1.0 + L/VK}$	$1.0 + \dfrac{L}{VK}$	$\dfrac{z'}{1.0 + L/VK}$
n-C$_5$	0.225	0.250	2.17	2.84	0.0880	3.63	0.0688
n-C$_6$	0.450	0.500	0.87	5.60	0.0894	7.55	0.0663
n-C$_7$	0.225	0.250	0.345	12.60	0.0197	17.50	0.0143
H$_2$O	0.100						
	1.000				0.1971		0.1494

$$V = 0.1971 \qquad V = 0.1494$$
$$L = 0.8029 \qquad L = 0.8506$$
$$\frac{L}{V} = 4.09 \qquad \frac{L}{V} = 5.7$$

Since 1 mole of total feed contains only 0.9 mole of hydrocarbons, the moles of hydrocarbons flashed on a basis of 1 mole of total feed is

$$v_{HC} = 0.9(0.1494) = 0.1346$$

Then
$$v_{H_2O} = v_{HC}\frac{p_{H_2O}}{p_{HC}} = (0.1346)\frac{5.5}{19.5} = 0.0379$$
$$V = v_{H_2O} + v_{HC} = 0.1346 + 0.0379 = 0.1725$$

A check must be made to be sure the point calculated falls in the two-liquid-phase region. This can be done by calculating the moles of water flashed and comparing with the moles originally in the feed.

$$v_{H_2O} = V - v_{HC} = 0.1725 - 0.1346 = 0.0379$$
$$l_{H_2O} = 0.100 - 0.0379 = 0.0621$$

Some of the water remains in the liquid phase.

The example will now be worked by the Smith-Wilson method. The n-C$_6$ is the logical choice for the reference component. The temperature of 166°F will again be assumed, and in order to reduce the trial and error, the split on the n-C$_6$ calculated by the classical method will be assumed. From the calculations above, $p_{HC} = 19.5$ psia and $v_r/l_r = 0.153$. The K_r at 166°F and 19.5 psia is 0.87, and $v_r/l_rK_r = 0.1759$.

Com-ponent	z	z'	$K_{19.54}^{166°F}$	$\dfrac{v}{l} = 0.1759K_i$	$l = \dfrac{z'}{v/l + 1.0}$	v
n-C$_5$	0.225	0.250	2.17	0.3820	0.1808	0.0692
n-C$_6$	0.450	0.500	0.87	0.1529	0.4330	0.0670
n-C$_7$	0.225	0.250	0.352	0.0619	0.2350	0.0150
H$_2$O	0.100					
	1.000	1.000			0.8488	0.1512

$$\text{Calculated } K_r = \frac{0.067/0.1512}{0.433/0.8488} = 0.868 \qquad \text{Corresponding } t = 166°F$$

Fraction of hydrocarbon flashed $= 0.1512$

Moles of hydrocarbon in vapor $= v_{HC} = 0.9(0.1512) = 0.1361$

$$\text{Total moles of vapor} = V = \frac{v_{FC}}{y_{HC}} = \frac{v_{HC}}{p_{HC}/p} = \frac{0.1361}{19.5/25} = 0.1745$$

This 0.1745 corresponds to the 0.1725 calculated by the classical method.

NOMENCLATURE

C = number of components in system.

$f_{i,p}$ = fugacity of component i at total pressure.

$f_{i,p_i}{}^{*}$ = fugacity of component i at vapor pressure.

F = feed to separator. Usually taken as 1.0 mole.

h = molar enthalpy of liquid. Subscript denotes stream.

H = molar enthalpy of vapor. Subscript denotes stream.

$K_i = y_i/x_i$ = equilibrium-distribution coefficient of component i. Subscript r refers to component r.

l_i = moles of component i in L.

L = fraction of feed which leaves separator as a liquid. Range of L is from 0.0 to 1.0.

N_c = independent restricting relationships.

N_i = degrees of freedom; variance; number of design variables which the designer must arbitrarily specify.

N_v = total number of variables which the designer must consider.

p = total pressure.

p_i^{*} = vapor pressure of pure component i.

p_t^{*} = total vapor pressure exerted by a mixture.

p_{HC}^{*} = vapor pressure exerted by the hydrocarbon portion of a mixture.

$p_{H_2O}^{*}$ = vapor pressure of water.

p_{HC} = partial pressure of hydrocarbons in vapor phase.

p_{H_2O} = partial pressure of water in vapor phase.

r = subscript denoting reference component.

t = temperature, °F. Subscripts V and L refer to vapor and liquid phases, respectively.

v_i = moles of component i in V.

V = fraction of feed which is flashed. Range of V is from 0.0 to 1.0.

x = mole fraction in liquid phase.

y = mole fraction in vapor phase.

z = mole fraction in feed to flash separator.

z' = mole fraction in feed on water-free basis.

PROBLEMS

4-1. A 50-plate distillation column is charged with the feed given below:

	Mole %
C_3	6.0
$i\text{-}C_4$	65.0
$n\text{-}C_4$	26.0
$n\text{-}C_5$	3.0
	100.0

An overhead fraction containing no more than 5 per cent n-butane and a bottoms fraction containing no more than 5 per cent isobutane are desired. Assume complete separation of propane and pentane. Assuming a bubble-point condensate, an accumulator temperature of 100°F, and a feed at its bubble point, calculate:

(a) Accumulator pressure.

(b) Top vapor temperature.

(c) Feed temperature.

(d) Reboiler temperature.

4-2. A feed containing 50 mole per cent ethane and 50 mole per cent propane enters at 70°F and 250 psia. Calculate the quantity and composition of overhead and bottoms for each of the following situations:

(a) Distillation column with infinite number of plates but no reflux.

(b) Flash drum, no reflux.

(c) Flash drum with 90% of the liquid stream returned to the flash drum as reflux.

(d) Distillation column with infinite number of plates and an almost infinite reflux ratio.

4-3. There is insufficient high-pressure tankage to store excess propylene. An alternate being considered is storage at 100°F and 30 psia in saturated solution in a solvent having the properties of n-octane. It is recovered as required in the system shown below. Calculate:

(a) Quantity and composition of feed.

(b) Quantity of liquid product.

(c) Temperature in flash drum.

(d) Quantity and composition of vapor product.

(e) Quantity and composition of recycle stream.

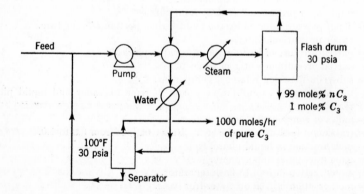

4-4. The following analysis was obtained on the total vapor overhead from a distillation column. If the material is cooled to 130°F and the accumulator pressure is 150 psig, what is the per cent condensed?

Component	Mole %
C_2	8.7
C_3	20.2
i-C_4	69.6
n-C_4	1.5
	100.0

If the pressure is reduced to 105 psig, to what temperature will the material have to be cooled to give the same percentage condensed?

4-5. Calculate the temperature at the top of the tower shown in the following diagram. Assuming that all the data above are reliable, can you conclude that something is wrong with the operation of the tower? If so, what is wrong?

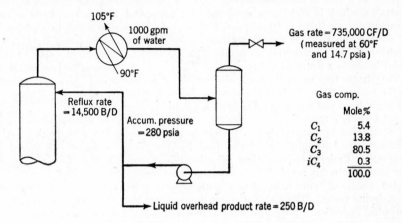

Gas comp.	
	Mole%
C_1	5.4
C_2	13.8
C_3	80.5
iC_4	0.3
	100.0

4-6. A gas stream having the composition given below is compressed and then cooled before being sent to a gas separator drum operating at 100°F and 100 psia.

Component	Moles/hr
C_1	50
C_2	200
C_3	300
$n\text{-}C_4$	450
Total	1000

(a) Calculate the compositions and quantities of the gas and liquid streams leaving the separator at 100°F and 100 psia.

(b) In order to decrease the volume of gas leaving the drum, a proposal was made to recycle 500 moles/hr of the liquid to the cooler. If this liquid is mixed with the compressor effluent just upstream of the cooler and if the compressor effluent temperature is 200°F, how would you go about calculating the compositions and quantities of the gas and liquid produced at steady state? Assume that the heat transferred in the cooler is the same before and after recycle begins.

4-7. The following mixture flows through a control valve into a vessel where a portion is flashed.

Component	Mole %
C_2	5.0
C_3	8.0
$i\text{-}C_4$	32.0
$n\text{-}C_4$	30.0
$i\text{-}C_5$	15.0
$n\text{-}C_5$	10.0
	100.0

(a) Calculate the entire flash curve at 155 psia.

(b) Calculate the amounts of liquid and vapor leaving the vessel, their compositions, and the vessel temperature if the temperature and pressure ahead of the control valve are 200 F and 285 ps'a and the vessel pressure is 155 psia.

4-8. Light-hydrocarbon-rich absorber oil is charged to an open steam stripping column at a rate of 10,000 moles/hr. The overhead hydrocarbon product is 1000 moles/hr, which is withdrawn as 60 per cent vapor and 40 per cent liquid at the accumulator condition of 50 psia and 140°F. In addition, 800 moles/hr of reflux is in the top vapor, where the temperature is 180°F. Superheated steam is admitted underneath the bottom tray at a rate of 13,500 lb/hr. The mixed vapor at this point indicates a water partial pressure of 20 psi, and the temperature here is 300°F. A dehydrator tray up in the tower will remove any liquid water which accumulates in the tower. There is also a liquid water draw-off in the accumulator. Neglect all pressure drops and any water dissolved in a liquid hydrocarbon phase.

(a) Calculate the amount of water removed from each of the possible places.

(b) If the dehydrator plate becomes plugged, additional heat is added either to the feed or to the bottom of the column (none by open steam) to remove all the water overhead. Under these conditions, all other things remaining constant, what must the minimum reflux rate be? (Assume top vapor temperature constant.)

4-9. Relatively pure isobutylene is withdrawn from storage at 50°F and charged to a drying tower. The material is wet due to carry-over from water washing in a preceding operation. Entrained water is removed in an "Excelso" drier (coagulating device), and the remaining material preheated to 135°F before charging to the drying tower. Water removed by distillation is accumulated in a "boot" on the reflux drum and periodically drained to the sewer. A pressure controller throttles the cooling water from the condenser and is set for a tower pressure of 110 psig. The control valve to the steam chest on the reboiler is connected to a 25-psig supply of saturated exhaust steam.

(a) What is the temperature of reflux returned to the tower?

(b) What is the top vapor temperature?

(c) What is the average water rate drained from the reflux drum?

(d) What is the reflux rate (bbl/day)?

Note: Neglect tower and condenser pressure drop and heat loss. Assume 15-psia drop across steam control valve.

Additional information	
Feed rate to tower	1200 bbl/day
Steam to reboiler	1600 lb/hr
Cooling water in	85°F
Cooling water out	120°F

4-10. Calculate the entire flash curve at 25 psig for the following mixture:

Component	Mole %
$n\text{-}C_5$	20
$n\text{-}C_6$	20
$n\text{-}C_7$	20
$n\text{-}C_8$	20
H_2O	20
	100

REFERENCES

1. DePriester, C. L.: *Chem. Eng. Progr. Symposium Ser.*, **49**(7), 1 (1953).
2. Edmister, W. C.: *Ind. Eng. Chem.*, **47**, 1685 (1955).
3. Edmister, W. C., and J. R. Bowman: *Chem. Eng. Progr. Symposium Ser.*, **48**(2), 112 (1952).
4. Edmister, W. C., and K. K. Okamoto: *Petrol. Refiner*, **38**(8), 117 (1959); **38**(9), 271 (1959).
5. "Engineering Data Book," 7th ed., p. 161, Natural Gasoline Supply Men's Association, Tulsa, Okla., 1957.
6. Katz, D. L., and G. G. Brown: *Ind. Eng. Chem.*, **25**, 1373 (1933).
7. Maxwell, J. B.: "Data Book on Hydrocarbons," 1st ed., D. Van Nostrand Company, Inc., Princeton, N.J., 1958.
8. Perry, J. H.: "Chemical Engineers' Handbook," 3d ed., pp. 568–572, McGraw-Hill Book Company, Inc., New York, 1950.
9. Smith, D. A., and G. W. Wilson: *Trans. A.I.Ch.E.*, **42**, 927 (1946).

CHAPTER 5

BINARY DISTILLATION

x-y Diagrams

Binary-distillation problems arise rather infrequently in plant practice, and the engineer rarely finds occasion to use the graphical calculation methods by which such problems are most conveniently solved. The availability of electronic computers further reduces the possibility of graphical solution of binary problems. Despite their limited practical use, the diagrams used to solve binary problems are valuable instructional tools since they permit the visualization of the distillation process as calculated through the concept of the equilibrium stage. The diagrams also make possible simple definitions of the concepts of total reflux, minimum reflux, and plate efficiency and therefore would serve a useful academic purpose even if they were never used in plant practice.

The separation of binary mixtures by continuous distillation can be calculated rigorously by means of the enthalpy-concentration diagram originally proposed by Ponchon (5) and Savarit (6), or the solution can be approximated through the use of an x-y diagram as suggested by McCabe and Thiele (3). The use of the enthalpy-concentration diagram is described in the next chapter. This chapter is devoted to the x-y diagram, which, although less rigorous, gives a better pictorial view of the distillation process than does the enthalpy-concentration diagram.

5-1. Equilibrium Data

A complete experimental determination of the phase-equilibrium data for a binary system provides the equilibrium vapor and liquid compositions over the entire range of composition. The vapor compositions can then be plotted vs the corresponding liquid compositions as shown in Fig. 5-1. If the y and x for the more volatile component are plotted, the equilibrium curve will fall above the 45° diagonal, since $y > x$. Curve A in Fig. 5-1 shows a typical curve for a system in which one component remains more volatile over the entire concentration range. Curve B is an illustration of those systems (ethanol-water, for example) where the component which is more volatile at low values of x becomes less volatile than the other component at high values of x. At the point where curve B crosses the 45° diagonal, the vapor and liquid compositions are identical. The term *azeotrope* is used to describe this condition. If only one liquid phase exists at the azeotropic composition, the azeotrope is a *homogeneous* azeotrope and the equilibrium curve will be similar to curve B. If a second liquid phase appears at the azeotrope (two liquid phases in equilibrium with the same vapor phase), the azeotrope is *heterogeneous* and the equilibrium curve has the general shape of curve C in Fig. 5-1. The normal butanol-water system is an example of a heterogeneous azeotrope. Regardless of whether an azeotrope is

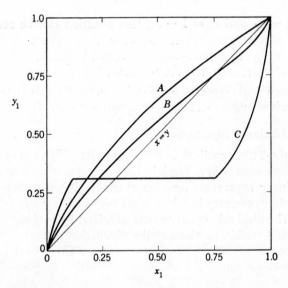

Fig. 5-1. Typical binary equilibrium curves. Curve A, system with normal volatility. Curve B, system with homogeneous azeotrope (one liquid phase). Curve C, system with heterogeneous azeotrope (two liquid phases in equilibrium with one vapor phase).

homogeneous or heterogeneous, its presence in a system limits the separation which can be obtained between the components by simple distillation.

The phase rule states that only two variables can be arbitrarily specified in the experimental determination of binary equilibrium data for a two-phase system. If both the temperature and pressure are specified, only one set of vapor and liquid compositions is possible. If it is desired to vary the liquid and vapor compositions, then either the temperature or the pressure must be varied. Binary vapor-liquid equilibrium data over the entire composition range can be obtained either at constant temperature (isothermal) or at constant pressure (isobaric) but not both.

Equilibrium data can be reported and used as mole, weight, or volume fractions. The units for the phase rates must, of course, agree with the basis for the equilibrium data.

It is sometimes permissible to assume that the volatility of component 1 with respect to component 2 is constant over the range of column conditions. The relative volatility α of component 1 with respect to component 2 is defined as

$$\alpha_{1-2} = \frac{K_1}{K_2} = \frac{y_1 x_2}{x_1 y_2}$$

Since $y_2 = 1 - y_1$ and $x_2 = 1 - x_1$, this equation can be rewritten as

$$y_1 = \frac{x_1 \alpha}{1 + (\alpha - 1)x_1} \tag{5-1}$$

If α is assumed to be constant and this constant value is known, the entire equilibrium curve can be calculated from Eq. (5-1).

5-2. Material-balance Equations

The McCabe-Thiele method is based upon the representation of the material-balance equations as straight lines upon the x-y diagram. Solution of a binary separation problem therefore becomes an exercise in simple analytical geometry in which algebraic equations are transformed into lines. The algebraic equations are obtained by making component balances around certain portions of the distillation column. For example, a balance for any given component around envelope 1 in Fig. 5-2 gives the following equation:

$$V_n y_n = L_{n+1} x_{n+1} + D x_D$$

Division by V_n puts the component balance into the analytical form of the equation for a straight line

$$y_n = \frac{L_{n+1}}{V_n} x_{n+1} + \frac{D}{V_n} x_D \tag{5-2}$$

where L_{n+1}/V_n is the slope and $D x_D/V_n$ is the intercept. If the component balance is written around envelope 2, the following equation results:

$$y_{n-2} = \frac{L_{n-1}}{V_{n-2}} x_{n-1} + \frac{D}{V_{n-2}} x_D \tag{5-3}$$

The $D x_D$ term is the same in both Eqs. (5-2) and (5-3). Also, $L_{n+1} = L_{n-1}$ and $V_n = V_{n-2}$ if *constant molar overflow* is assumed. The assumption of constant molar overflow depends upon two prior assumptions. (a) The two components must be assumed to have identical molar heats of vaporization, and (b) all heat effects (heats of solution and heat leak to or from the column) must be assumed to be zero. It follows from these two assumptions that a mole of vapor must be formed for each mole of vapor that condenses within a stage. Therefore, within any section of the column where no material or energy is added or withdrawn, the moles of liquid overflow from stage to stage must be constant as are the

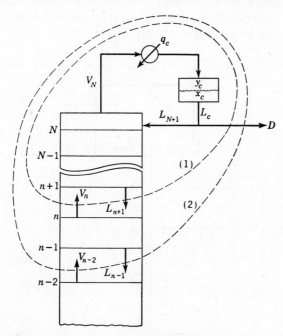

Fig. 5-2. Two material-balance envelopes in the top section of a distillation column.

moles of vapor ascending the column. Under these conditions of constant molar overflow, the subscripts on the phase rates can be dropped and Eqs. (5-2) and (5-3) written in the following general form for that section of the column:

$$y_n = \frac{L}{V} x_{n+1} + \frac{Dx_D}{V} \tag{5-4}$$

Before this equation can be plotted on the x-y diagram, the L/V and Dx_D/V terms must be known. If they are not specified in the problem statement, they must be assumed before the solution can begin.

The component balance may contain other external streams besides D.

For example, the component balance around the envelope in Fig. 5-3 is

$$y_n = \frac{L}{V}\, x_{n+1} + \frac{Dx_D + Sx_S}{V} \tag{5-5}$$

Or if the side stream had been a feed stream,

$$y_n = \frac{L}{V}\, x_{n+1} + \frac{Dx_D - Fx_F}{V} \tag{5-6}$$

Regardless of how many external streams are included in the intercept term of the equation, they must all be specified or assumed before the straight line can be located on the x-y diagram.

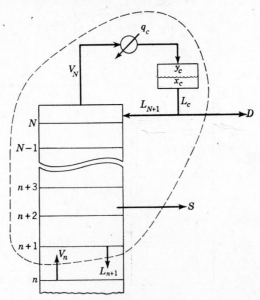

Fig. 5-3. Material-balance envelope which contains two external streams D and S, where S represents a side-stream product withdrawn above the feed plate.

Similar equations can be written around the bottom end of the column. Below the lowest feed or side stream, the component balance as outlined by the envelope in Fig. 5-4 is

$$y_m = \frac{L}{V}\, x_{m+1} - \frac{Bx_B}{V} \tag{5-7}$$

Or if the material-balance envelope around the bottom end of the column cuts across a bottom side stream,

$$y_m = \frac{L}{V}\, x_{m+1} - \frac{Bx_B + Sx_S}{V} \tag{5-8}$$

Equations (5-4) to (5-8) when plotted on the x-y diagram furnish lines which are called *operating lines*. It can be seen from an inspection of the equations that a point on an operating line represents two operating streams which have been cut by the material-balance envelope. In a countercurrent process the point relates two *passing* streams. In a concurrent process the point on the operating line relates two concurrent streams. One operating line will represent all pairs of passing (or concurrent) streams between points in the column where material or enthalpy is added or withdrawn. It can be seen from the equations that the

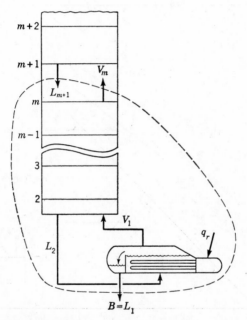

FIG. 5-4. Material-balance envelope around bottom end of column. The partial reboiler is equilibrium stage 1.

addition or withdrawal of material or energy at any point in the column will change both the slope and the intercept terms and necessitate the use of a new operating line as soon as the addition or withdrawal point is passed.

A material-balance or operating line can be located on the x-y diagram if (*a*) two points on the line are known or (*b*) one point and the slope are known. The known points on an operating line are usually its intersection with the 45° diagonal and/or its intersection with another operating line. The location of the lines in the various parts of the column is described in more detail in the following sections.

5-3. Equilibrium Stages

Before proceeding with the specific rules for the location of the various operating lines, it is advantageous to discuss the manner in which the operating lines are used in conjunction with the equilibrium curve to locate the equilibrium stages. Let the solid lines above the $y = x$ line in Fig. 5-5 represent those portions of the equilibrium curve and the operating line which cover the vapor- and liquid-composition ranges found

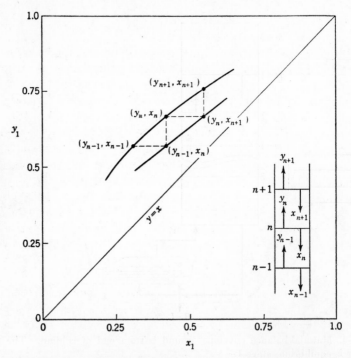

Fig. 5-5. Illustration of how equilibrium stages can be located on the x-y diagram through alternating use of the equilibrium curve and the operating line.

in a given section of a distillation column. If y_n and x_n represent the compositions (in terms of the more volatile component) of the equilibrium vapor and liquid leaving stage n, then point (y_n, x_n) on the equilibrium curve must represent the equilibrium stage n. The operating line is the locus for the compositions of all possible pairs of passing streams within the section, and therefore, a horizontal line (dotted) at y_n must pass through the point (y_n, x_{n+1}) on the operating line, since y_n and x_{n+1} represent passing streams. Likewise, a vertical line (dotted) at x_n must intersect the operating line at point (y_{n-1}, x_n). The equilibrium stages above and below stage n can be located by a vertical line through (y_n, x_{n+1}) to find (y_{n+1}, x_{n+1}) and a horizontal line through (y_{n-1}, x_n) to find (y_{n-1}, x_{n-1}). It can be seen that one can work up or down the column through alternating use of the equilibrium and operating lines.

5-4. Top Section of Column

It was shown in Sec. 5-2 that the operating-line equation for the top section of the column (above the uppermost feed or side stream) is

$$y_n = \frac{L}{V} x_{n+1} + \frac{Dx_D}{V} \qquad (5\text{-}4)$$

The Dx_D product in this equation represents the total overhead product withdrawn from the column. The overhead product may be withdrawn as (a) a liquid, (b) a vapor, or (c) a combination of liquid and vapor. The construction for each of these cases will be discussed separately.

Total Condenser. A condenser is a *total* condenser when all the overhead vapor is condensed to the liquid state. If the heat load on the condenser is exactly equal to the latent heat of the saturated (dew point) overhead vapor, the condensate will be a saturated (bubble point) liquid. If the condenser duty is greater than the latent heat of V_N, the condensate will be subcooled. In either case, the pressure in the condenser and the accumulator will be the total vapor pressure of the condensate unless some light gas is injected to increase the column pressure. Figures 5-2 and 5-3 show columns with total condensers and liquid overhead products. If the liquid overhead product is denoted by D_L, then $D = D_L$ and $x_c = x_D$ in these cases.

Equation (5-4) is the operating-line equation for the top section of the column. The point at which this operating line intersects the $y = x$ diagonal is found by setting $y = x$ in Eq. (5-4) to give

$$Vx = Lx + Dx_D \qquad (5\text{-}9)$$

Since $V - L = D$, then $x = x_D$ and the intersection occurs at $y = x = x_D$ on the diagonal line. The intersection will always be at x_D whether the condensate is at its bubble point or subcooled.

Either another point on the line or the slope must be known to locate the operating line. Setting $x = 0$, the intersection of the line with the left-hand side of the diagram is found to occur at $y = Dx_D/V$. If the reflux is a saturated liquid, then $V = V_N = L_{N+1} + D$. If the condensate is subcooled, an enthalpy balance must be made around stage N to calculate V_{N-1} and L_N, which can then be used as the V and L for the top section. The values of D and L_{N+1} are specified or assumed.

The slope of the line is found from the following equation when saturated reflux is used:

$$\frac{L}{V} = \frac{L_{N+1}}{V_N} = \frac{RD}{(1 + R)D} = \frac{R}{1 + R} \tag{5-10}$$

If the reflux is subcooled, $L \neq L_{N+1}$ and $V \neq V_N$ and the enthalpy balance around stage N must be used to calculate L/V. The L/V ratio is the *internal* reflux ratio and changes each time energy or material

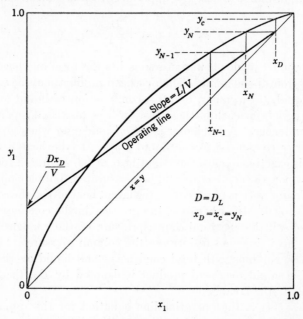

FIG. 5-6. Construction at top of column when a total condenser is used. $D_L = D$ and $x_c = x_D$.

is added to or withdrawn from the column. The *external* reflux ratio, $R = L_{N+1}/D$, is usually fixed for a particular problem by specification.

When the reflux is subcooled, L/V does not equal L_{N+1}/V_N and therefore the point which represents the two external passing streams V_N and L_{N+1} must lie on a different operating line from the one which represents pairs of passing streams below stage N. The line upon which V_N and L_{N+1} fall must have the lesser slope, since $L_{N+1}/V_N < L/V$. Both lines must intersect the $x = y$ diagonal at x_D. Since

$$y_N = x_{N+1} = x_D$$

the only point of interest on the operating line for streams V_N and L_{N+1}

is at x_D, and consequently, the line need not be drawn. Figure 5-6 illustrates the construction for either a saturated or subcooled reflux and overhead product. Note that the composition of the equilibrium vapor (not withdrawn) in the accumulator can be located on the equilibrium curve by a vertical line through x_D.

Partial Condenser. A condenser is a *partial* condenser when it condenses only part of the overhead vapor. The vapor and liquid produced can be considered to be in equilibrium, and therefore the point (y_c, x_c) must fall on the equilibrium curve. If all the condensed liquid is returned to the column while all the vapor is withdrawn for overhead product as shown in Fig. 5-7a, then $D_V = D$ and $y_c = x_D$, where D and x_D refer

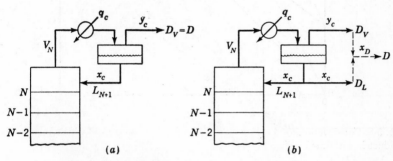

FIG. 5-7. (a) Column with partial condenser from which only a vapor product is withdrawn. (b) Column with partial condenser from which both a vapor and liquid product are withdrawn.

to the *total* overhead product. If both vapor and liquid are withdrawn as product as shown in Fig. 5-7b, then $D = D_V + D_L$ and

$$x_D = \frac{D_V y_c + D_L x_c}{D_V + D_L}$$

It can be shown in the same manner as for a total condenser that the top operating line must always intersect the $y = x$ diagonal at x_D regardless of the way the product is withdrawn. The first step, therefore, in the solution of a problem which involves both vapor and liquid overhead products is the calculation of x_D from the known values of x_c and y_c. The latter two quantities are related by the equilibrium curve, and only one need be specified. It should be recognized that D and x_D are only calculated quantities and represent no actual stream.

In the McCabe-Thiele method the reflux from a partial condenser is always assumed to be at its bubble point, and therefore L_{N+1}/V_N always equals L/V. The intercept of the operating line with the ordinate at $x = 0$ is still Dx_D/V, where D and x_D refer to the *total* overhead product

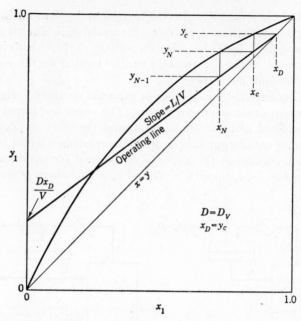

FIG. 5-8. Construction for a column with a partial condenser from which only a vapor product is withdrawn. See Fig. 5-7a.

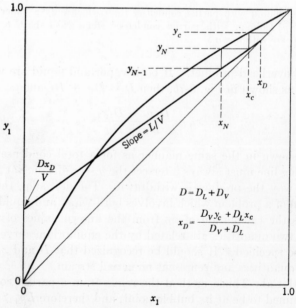

FIG. 5-9. Construction for a column with a partial condenser from which both a vapor and liquid product are withdrawn.

(see Fig. 5-7b). Figure 5-8 illustrates the construction when only a vapor product is withdrawn, while Fig. 5-9 illustrates a partial condenser with both a vapor and liquid product.

5-5. Bottom Section of Column

Figure 5-4 shows the bottom end of a distillation column with a partial reboiler. Equation (5-7) is the equation of the operating line which relates all pairs of passing streams (including L_2 and V_1) in the section of the column below the lowest feed or side stream. The intersection of the bottom operating line with the $y = x$ diagonal is found by setting $y = x$ in Eq. (5-7) to give

$$Lx = Vx + Bx_B$$

Since $L - V = B$, it can be seen that the intersection must occur at x_B.

The other point most often used to locate the bottom operating line is the point of intersection with the next operating line above. The possible intersection points are discussed later in the sections on feed plates and side streams.

In some problems it may be convenient to specify the reflux ratio at the bottom of the column instead of at the top. This is done by specifying the ratio $R' = V_1/B$. Normally this ratio is termed the *reboil* ratio rather than a reflux ratio. If the reboil ratio is known, it can be used to calculate the slope term in Eq. (5-7) and this slope used to locate the bottom operating line.

Open Steam. When the less volatile component in the binary mixture is water, it may be advantageous to eliminate the reboiler and add heat to the column by injecting open steam beneath the bottom stage. The flow diagram for this operation is shown in Fig. 5-10 along with the McCabe-Thiele construction. If the x-y diagram is to be used in the solution of such a problem, it is, of course, necessary to assume that the water has the same molar heat of vaporization as the other component and that the condition of constant molar overflow exists.

If the steam feed to the bottom stage is designated as V_0, the more volatile component balance around the envelope in Fig. 5-10 is

$$L_{m+1}x_{m+1} + V_0y_0 = V_my_m + Bx_B$$

Since $y_0 = 0$ for the more volatile component, the material balance can be written in the usual straight-line form as

$$y_m = \frac{L}{V}x_{m+1} - \frac{Bx_B}{V}$$

If y is set at zero in this equation, it can be seen that the operating line will intersect the base of the diagram at $x = x_B$, since $L = B$. The intersection with the $y = x$ diagonal is found by setting $y = x$ and solving to give

$$x = \frac{B}{B - V_0} x_B$$

The McCabe-Thiele construction is illustrated in Fig. 5-10.

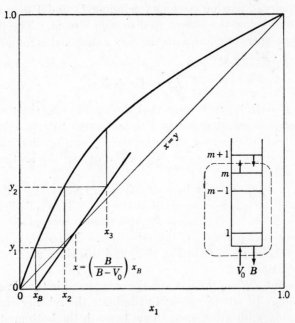

FIG. 5-10. Construction for a column with open steam. Water is the less volatile component. V_0 represents the open steam introduced below stage 1.

5-6. Feed Stage

The effect of a feed on the internal phase rates in the column depends upon the thermal condition of the material introduced. The feed may be anything from a subcooled liquid to a superheated vapor. As an aid in the analytical description of the feed stream, it is convenient to introduce the quantity q which is defined by the following equations:

$$L' = L + qF \qquad (5\text{-}11)$$
$$V = V' + (1 - q)F \qquad (5\text{-}12)$$

The L' and V' refer to the rates below and L and V to the rates above the feed stage. It can be seen that q represents the moles of saturated liquid formed in the feed stage per mole of feed. The thermal condition of the feed must be determined at the temperature and pressure of the feed stage. If this is done, and since equal molar heats of solution are assumed, the following expression for q is synonymous to Eqs. (5-11) and (5-12).

$$q = \frac{\text{heat to convert 1 mole of feed to a saturated vapor}}{\text{molar heat of vaporization}}$$

For a partially flashed feed at the feed-stage conditions, q is simply the fraction of the feed which is liquid. The following values of q result from these definitions.

Subcooled liquid feed: $q > 1.0$
Saturated liquid feed: $q = 1.0$
Partially flashed feed: $1.0 > q > 0.0$
Saturated vapor feed: $q = 0$
Superheated vapor feed: $q < 0$

The quantity q can be used to derive an equation whose line is the locus of all possible intersections of the two operating lines which pertain to the column sections just above and below the feed stage. Let the following equation represent the operating line just above the feed stage:

$$V y_n = L x_{n+1} + D x_D \tag{5-13}$$

Represent the operating line just below the feed stage with

$$V' y_m = L' x_{m+1} - B x_B \tag{5-14}$$

For the sake of simplicity, Eqs. (5-13) and (5-14) assume no other feed streams or side streams above or below the feed in question, and therefore the intercept terms in the equations involve only the end-product terms. The intercept terms cancel in this derivation, and therefore the simplification places no restrictions on the generality of the equation being derived.

At the point of intersection of the operating line just above the feed with the operating line just below the feed, the x's and y's in Eqs. (5-13) and (5-14) become identical. Subtracting (5-14) from (5-13) and substituting $(F x_F - D x_D)$ for $B x_B$ give

$$(V - V')y = (L - L')x + F x_F$$

at the point of intersection. Since $(V - V') = (1 - q)F$ and

$$(L - L') = -qF$$

this equation becomes

$$y = \frac{q}{q - 1} x - \frac{x_F}{q - 1} \tag{5-15}$$

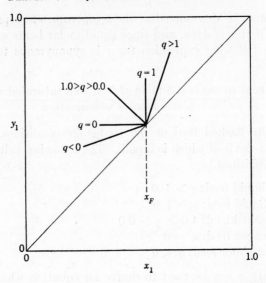

FIG. 5-11. Five possible locations for the q line at a feed stage.

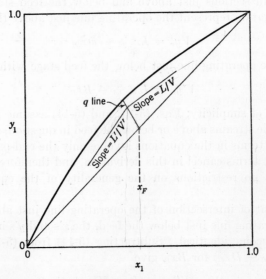

FIG. 5-12. Typical intersection of the two operating lines at the q line for a feed stage. The q line shown is for a partially flashed feed.

Equation (5-15) is known as the q-line equation. A plot of the equation on the x-y diagram generates a straight line with a slope of $q/(q-1)$. If y is set equal to x, it can be seen that the q line must intersect the $y = x$ diagonal at the feed composition. Figure 5-11 illustrates five possible slopes for the q line.

The two operating lines which come together must intersect each other on the q line at some point between the equilibrium curve and the $y = x$ diagonal. Figure 5-12 shows a typical construction with a partially flashed feed. The intersection can, of course, occur on the equilibrium curve or on the diagonal. These two possibilities are discussed further in the sections on minimum and total reflux.

5-7. Side-stream Stage

The withdrawal of a side stream will change the internal reflux ratio L/V in much the same manner as the introduction of a feed. The situation at a side-stream stage is simpler, however, because the side stream must be either a saturated vapor or a saturated liquid. If the quantity q is used to describe the thermal condition of a side stream, it can take on only the values of 1.0 (saturated liquid) or 0.0 (saturated vapor). This means that the q line at the side-stream stage must fall in one of two positions. It will be vertical for a liquid side stream and horizontal for a vapor side stream, and it must intersect the $x = y$ diagonal at the composition of the side stream. The operating line just above the side stream must intersect the operating line just below the side stream on the vertical line at x_S if the side stream is liquid or on the horizontal line at y_S if the side stream is vapor. The correctness of these statements can be demonstrated by repeating the derivation of the q line given in the preceding section with $-S$ substituted for F. The material balances analogous to Eqs. (5-13) and (5-14) should both be written around the top or the bottom of the column.

The change in the phase rates caused by withdrawal of the side stream is easily calculated from $L' = L - qS$ and $V = V' - (1 - q)S$, where L' and V' represent the rates just below and L and V represent rates just above the side-stream stage. So if one of the two operating lines has been previously located, the slope of the other can be calculated and used with the intersection point on the q line at the side stream to locate the new operating line.

If there are no other side streams or feeds above the side stream in question, the operating line just below the side stream will intersect the $y = x$ diagonal at

$$x = \frac{Sx_S + Dx_D}{S + D} \qquad (5\text{-}16)$$

This can be shown by setting $y = x$ in Eq. (5-5) and solving for x. If the same thing is done with Eq. (5-8), it can be shown that the operating line just above a bottom side stream (no other feeds or side streams below it) must intersect the $y = x$ diagonal at

$$x = \frac{Sx_S + Bx_B}{S + B} \qquad (5\text{-}17)$$

The graphical construction for both an upper and lower side stream is illustrated in Fig. 5-13. The lower side stream in this figure is assumed to be a vapor side stream, while the upper one is assumed to be liquid.

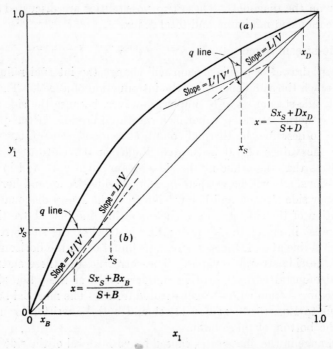

FIG. 5-13. Typical construction for a side stream showing the intersection of the two operating lines with the q line and with the $x = y$ diagonal. (a) Liquid side stream near the top of the column. (b) Vapor side stream near the bottom of the column.

5-8. Total Column

The rules of graphical construction for each of the various addition or withdrawal points in a distillation column have been illustrated in the previous four sections. The graphical solution of any binary separation problem merely involves the application of these rules to the process in question. The application will now be illustrated by example problems. The following processes will be considered.

Example 5-1: one feed and a total condenser
Example 5-2: one feed, one side stream, and a total condenser

The necessary problem specifications are discussed in each example. The problem of feed-stage location is discussed after the first example.

Example 5-1. A column with eight equilibrium stages (including the reboiler) is to be used to produce a 95 per cent concentrate of component 1 from 1000 moles/hr of a binary feed which is 40 per cent component 1. The construction of the column is such that the feed can be introduced only in the fourth stage from the bottom. The column has a total condenser and a partial reboiler. The maximum vapor capacity at the top of the column is approximately 2000 moles/hr at the normal operating condenser pressure of 40 psia. Make the necessary number of additional specifications and calculate the yield of overhead product.

Solution. In Chap. 3 it was shown that a column such as the one in this example has $C + 2N + 9$ design variables. Some of these variables have been fixed in the problem statement. Those items are marked with a dagger in the following list of convenient specifications.

Specifications		$N_i{}^u$
Pressure in each stage (including reboiler)		N
† Pressure in condenser		1
Heat leak in each stage (except reboiler)		$N - 1$
Pressure and heat leak in reflux divider		2
† Feed composition and rate	C	
Feed temperature and pressure		2
† Total number of stages, N		1
† Feed-stage location		1
† One overhead purity		1
Reflux temperature		1
External reflux ratio		1
		$\overline{C + 2N + 9}$

The pressures can be specified at any level below the safe working pressure of the column. The condenser pressure will be set at 40 psia, and the pressure drops within the column will be neglected. The equilibrium curve in Fig. 5-14 will be assumed to represent the equilibrium data at this pressure.

The heat leak in each stage (excluding the reboiler) and in the reflux divider must be assumed to be zero if the McCabe-Thiele method is to be used.

The feed temperature and pressure will be arbitrarily specified to give a $q = 1.0$ (saturated liquid) at the feed stage temperature and pressure.

The reflux temperature will be set at the bubble point. An external reflux ratio of 4.5 will be used to give an L/V of 0.818 above the feed. See Eq. (5-10) for the calculation of L/V.

Note that the maximum top vapor rate has not been used as a specification. The top vapor rate along with D is fixed by the purity and reflux specifications. When the solution has been completed and D is known, the vapor rate must be calculated and checked against the known column capacity.

The graphical solution is shown in Fig. 5-14. The feed q line is drawn vertically at x_F. The top operating line always passes through x_D on the diagonal and has been drawn from this point with a slope of $L/V = 0.818$. The stages above the feed stage are stepped off between this operating line and the equilibrium curve. Since a total condenser is used, $y_8 = x_D$ and a horizontal line from x_D on the diagonal locates y_8 on the equilibrium curve. A vertical line from y_8 on the equilibrium curve locates x_8 on the operating line. This point on the operating line also represents y_7, so a horizontal line across to the equilibrium curve locates y_7 and x_7. This alternating use of the equilibrium curve and top operating line is continued until the feed stage is reached. Below stage 4 the top operating line no longer relates the passing streams. Therefore the vertical line from the point (y_4, x_4) on the equilibrium curve must be drawn to the lower operating line to locate the point (y_3, x_4).

The location of the bottom operating line is a trial-and-error procedure. The bottom operating line must intersect the top operating line on the q line, and this fixes one point on the bottom line. The slope L'/V' cannot be calculated, since L and V are not known individually at this time. It is known, however, that the column has eight equilibrium stages, and therefore, the lower operating line must be drawn in such a manner as to give exactly eight steps between x_D and x_B. The bottom line is located by assuming an x_B, drawing the operating line between the

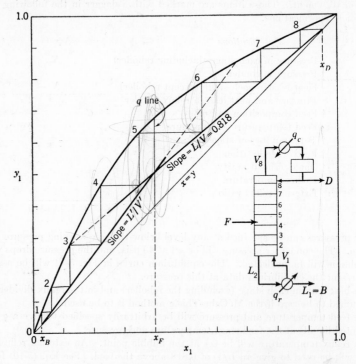

FIG. 5-14. Graphical solution of Example 5-1.

assumed x_B and the known point on the q line, and stepping off three stages below the feed stage. If the vertical line on the last step does not coincide with the assumed x_B, a new value of x_B is assumed and the procedure repeated until a match is obtained. This trial-and-error procedure is always necessary when the number of stages is specified.

The bottoms composition can now be read from Fig. 5-14 as $x_1 = 0.026$. Using 100 moles of feed as a basis, the yield of overhead product can be calculated with a component 1 balance.

$$F x_F = B x_B + D x_D = (F - D) x_B + D x_D$$

$$D = F \frac{x_F - x_B}{x_D - x_B} = 100 \frac{0.40 - 0.026}{0.95 - 0.026} = 40.5$$

The maximum top vapor rate which the column will handle is 200 moles per 100 moles of feed. The required V_8 is as follows:

$$V_8 = L_{N+1} + D = D(R + 1) = 40.5(4.5 + 1.0) = 223$$

The vapor-handling capacity of the column has been exceeded by 11.5 per cent. This means that the 40.5 per cent yield of overhead product cannot be obtained at a purity of 95 per cent. Either yield or purity must be sacrificed to fit within the limitation of the column. If the 95 per cent specification is retained, the reflux ratio must be reduced to unload the column. With the lower reflux ratio, the operating lines will lie closer to the equilibrium curve and the eight stages will not reach to so low an x_B as previously. The use of a larger value of x_B in the material balance reduces D, which together with the lower value of R results in a lower required V_8. It can be seen that the calculation of the yield when the column is fully loaded is a trial-and-error calculation.

Feed-stage Location. The optimum feed-stage location is that location which, with a given set of other operating specifications, will result in the widest separation between x_D and x_B with the given number of stages. Or if the number of stages is not specified, the optimum feed location is the one which requires the fewest number of stages to accomplish a specified separation between x_D and x_B. Either of these criteria will always be satisfied if the operating line farthest from the equilibrium curve is used in stepping off the stages. On the x-y diagram, this means that the feed should be inserted and the operating line changed when the q line for that feed is crossed. This should always be done, of course, in the design of a new column where the feed-stage location is under the control of the designer.

The calculation of the separation which can be obtained in an existing column requires that the diagram correspond to the existing physical realities. The switch from one operating line to the other must occur on that step which corresponds to the feed stage in the existing column. This may or may not correspond to the optimum location. In Fig. 5-14, it can be seen that stage 5 would be the optimum feed location. If the feed were introduced in stage 5, the vertical line from that stage could be drawn down to the lower operating line and stage 4 along with the other three lower stages would be moved down the diagram. A wider separation between x_D and x_B would result. However, since the feed was actually introduced on the fourth stage, the upper operating line must be used to locate the point (y_4, x_5) and the lower line cannot be used until the point (y_3, x_4).

Only certain stages are operable feed stages for any given location of the operating lines. If the two operating lines in Fig. 14 are extended to the equilibrium line, it can be seen that only those stages between the two intersections thus formed are operable feed stages. For example, if the feed were introduced on stage 7, the vertical from point (y_7, x_7) to the operating line to locate point (y_6, x_7) would have to go up and would intersect the lower operating line above the equilibrium curve. The horizontal line from (y_6, x_7) on the operating line would intersect the equilibrium curve above stage 7, and the steps would move back to the right rather than down the diagram toward the x_B which was used to locate the lower operating line in the first place.

Example 5-2. Calculate the number of stages required to produce a 95 per cent overhead product and a 5 per cent bottom product from 1000 moles/hr of 40 per cent feed when the internal reflux ratio L/V at the top of the column is 0.818 and a liquid side stream is withdrawn from the second stage from the top at a rate equal to that of the overhead product. The column will operate at a condenser pressure of 40 psia, and the interstage pressure drops can be neglected. The equilibrium curve is the same as that shown in Fig. 5-14. A total condenser with saturated reflux will be used, and the feed stage will be located at the optimum position. The feed will be assumed to be 50 per cent flashed at the feed-stage conditions.

Solution. It was shown in Chap. 3 that the addition of a side stream to the process pictured in Fig. 5-14 increased the number of design variables by 2 to a total of $C + 2N + 11$. The two additional variables are most conveniently used to specify the rate and the location of the side stream. A complete set of specifications for this example problem is shown below. Those items specified in the problem statement are marked with a dagger.

Specifications	$N_i{}^u$	
† Pressure in each stage (including reboiler)	N	
† Pressure in condenser		1
Heat leak in each stage (excluding reboiler)	$N - 1$	
Pressure and heat leak in the reflux divider		2
† Feed composition and rate	C	
† Feed temperature and pressure		2
† Side-stream-stage location		1
† Feed-stage location (optimum)		1
† Two product purities		2
† Reflux temperature (saturated liquid)		1
† One internal reflux ratio		1
† Side-stream rate		1
	$C + 2N + 11$	

The assumption of zero heat leak in the reflux divider and in all the stages except the reboiler is inherent in the McCabe-Thiele method.

The graphical solution for the example problem is shown in Fig. 5-15. Since the side-stream location is specified as the second stage from the top and the L/V is given at the top, it is more convenient to step off the stages from the top. The vertical line from the $N - 1$ stage locates x_S at 0.746. Now that all the product compositions are known, it would be possible to calculate D, S, and B and from the known L/V ratio at the top calculate the L's and V's above and below the side stream. Once the L/V ratio below the side stream is known, the second operating line can be drawn through the point where the top operating line intersects the vertical line from stage $N - 1$. This vertical line is the q line for the liquid side stream, and the two operating lines which come together at a side stream (or feed) must always intersect on the q line for that stream.

Rather than calculate the slope of the second operating line as described above, it is easier in this problem to calculate the point of intersection of the second operating line with the $x = y$ diagonal. Since $D = S$, both can be set equal to 1.0 in Eq. (5-16) to give

$$x = \frac{(1.0)(0.746) + (1.0)(0.95)}{1.0 + 1.0} = 0.848$$

for the point of intersection on the diagonal. The operating line below the side stream can now be drawn through the two known points. The intersection of this

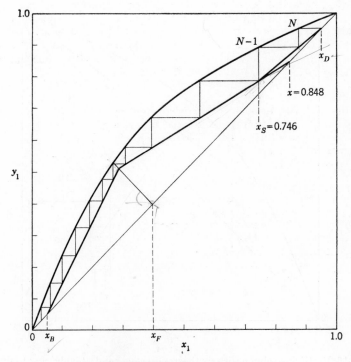

Fig. 5-15. Graphical solution of Example 5-2.

operating line with the feed q line together with the specified x_B furnish two points on the bottom operating line. Once the operating lines are located, the stages can be stepped off. The feed is introduced on the optimum stage (sixth from the top), and slightly more than 11 stages are required to reach the specified x_B of 0.05.

5-9. Rectifying and Stripping Columns

That part of a distillation column above the feed stage is called the *rectifying section*, while the part below is termed the *stripping section*. Each of these sections (plus the feed stage) can be used alone to form either a rectifying column as shown in Fig. 5-16 or a stripping column as pictured in Fig. 5-17. The feed to a rectifying column must contain some vapor, while the feed to a stripping section must be at least partly liquid if the columns are to be operable. Usually a superheated vapor is fed to a rectifier and a subcooled liquid to a stripper. In any case, the feed condition is represented by the slope of the q line as described in Sec. 5-6.

The McCabe-Thiele diagram for a single-section column involves only one operating line. This line can be located as described previously

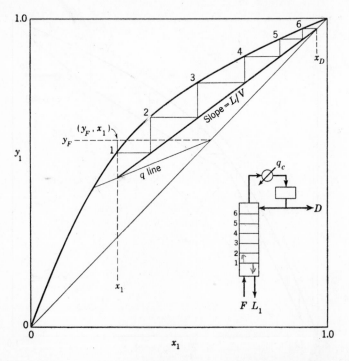

FIG. 5-16. Rectifying column with six theoretical stages and a highly superheated feed.

from a knowledge of its slope and the x_D point for a rectifier or the x_B point for a stripping column. One other point on the operating line is particularly useful in its location, and that is its intersection with the feed q line. The intersection must always occur at x_1 for a rectifier and at y_N for a stripper. This can be proved by equating the operating-line and q-line equations and solving for the value of x or y at the intersection. For a rectifier, y is eliminated and the equations written as

$$\frac{q}{q-1}x - \frac{y_F}{q-1} = \frac{L}{V}x + \frac{Dx_D}{V}$$

olving for x and substituting for $(q - 1)$ and q from $V = (1 - q)F$ give

$$x(F + L - V) = Fy_F - Dx_D$$

since $F + L - V = L_1$, then x must equal x_1 and the point of intersection of the q line and operating line must occur at x_1 regardless of the value of q. The proof that the intersection occurs at y_N in a stripper is similar.

Figures 5-16 and 5-17 illustrate the flow diagrams and graphical construction for rectifying and stripping columns. The feed to the rectifying column in Fig. 5-16 is assumed to be highly superheated in order to show

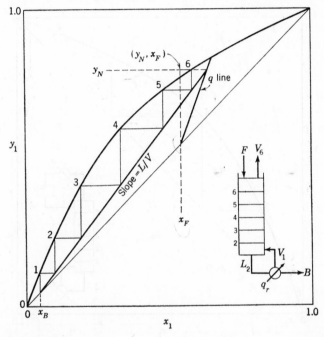

FIG. 5-17. Stripping column with a highly subcooled feed and six theoretical stages ncluding the reboiler.

that the vapor rising from stage 1 (the feed stage) may contain a lower concentration of the more volatile component than the feed. Likewise, Fig. 5-17 shows that a highly subcooled feed to a stripping column may cause the liquid overflow from the top stage to contain a higher concentration of the more volatile component than does the feed. These irregularities in the concentration profiles in the columns are a result of the change in each of the phase rates across the feed stage when the feed is not saturated. A superheated vapor feed will vaporize a considerable amount of the liquid entering the feed stage and make $V_1 > F$ and $L_1 < L_2$ in the rectifier. Since L_1/F does not equal L/V, the pair of external passing streams cannot lie on the same operating line as do the internal pairs. Therefore, the vertical leg on step 1 in Fig. 5-16 must turn upward to the point (y_F, x_1) which represents the pair of external streams and which has been located by the intersection of the operating line with the q line. An analogous construction is shown in Fig. 5-17 for a stripping column.

5-10. Total Reflux

The condition of total reflux exists when the entire overhead vapor stream from the top stage is condensed and returned to that stage as

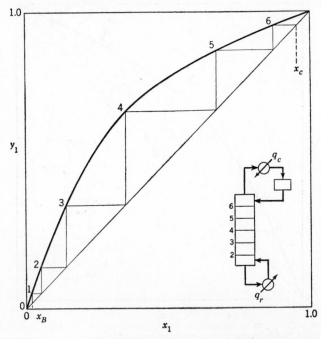

FIG. 5-18. Diagram for a column operating with no feed and under total reflux.

liquid reflux. No overhead product is withdrawn, and $L_{N+1} = V_N$. A distillation column can be operated under conditions of total overhead reflux with or without a feed stream. Each of these situations will be discussed separately.

Total Reflux—No Feed. Assume that the feed and product streams on a continuous column which has been operating at steady state are simultaneously blocked. The reflux is adjusted to maintain the level in the accumulator, and the reboiler and condenser loads are adjusted to maintain the enthalpy balance. When the column regains the steady state, it is operating at total reflux throughout its length. All the vapor entering the condenser is condensed and returned to the column as reflux.

All the liquid entering the reboiler is vaporized and returned to the column. A material balance around either end of the column and cutting between any two stages will show that the two streams in any pair of passing streams in the column are equal in magnitude and identical in composition. Their temperatures and pressures are not identical, but $L_{n+1} = V_n$, $L/V = 1.0$, and $x_{n+1} = y_n$, where n is any stage in the column. One operating line suffices for the entire column, and since this line must pass through x_B and x_D, it must coincide with the $y = x$ diagonal. This position of the operating line provides the maximum possible

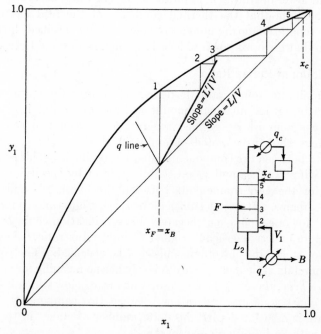

FIG. 5-19. Flow diagram and McCabe-Thiele construction for a column with total reflux and continuous feed and bottom streams.

slope for the operating line in the top part of the column (normally the rectifying section) and the minimum possible slope for the operating line in the lower part (normally the stripping section). These limiting slopes require the minimum number of stages to effect a specified separation between x_D and x_B, or they provide the maximum possible separation with a specified number of stages. Figure 5-18 represents the McCabe-Thiele construction for a column which is operating under total reflux over its entire length.

Total Reflux with Feed. The flow diagram for a column which is operating under total reflux conditions in only the rectifying section is shown in Fig. 5-19. The bottoms stream is the only product stream and therefore must be identical with the feed stream in rate and composition. The withdrawal of a bottoms stream requires that $L'_{m+1} \neq V'_m$, $L'/V' \neq 1.0$, and $y_m \neq x_{m+1}$, where m refers to any stage in the stripping section and the primes refer to rates below the feed stage. Above the feed stage $L_{n+1} = V_n$, $L/V = 1.0$, $y_n = x_{n+1}$, and the operating line coincides with the diagonal.

The McCabe-Thiele construction for this type of operation is shown in Fig. 5-19. Since $x_B = x_F$, the intersection of the lower operating line with the $y = x$ diagonal must occur at x_F. The intersection of the upper and lower operating lines on the q line must also occur at x_F on the diagonal. Note that the thermal condition of the feed has no effect upon the separation or the number of stages at any specified slope L'/V'. Note also that the optimum feed location would be in stage 1, the reboiler.

5-11. Minimum Reflux Ratio

The minimum reflux ratio is defined as that reflux ratio which if decreased by an infinitesimal amount will require an infinite number of stages to accomplish a specified separation between two components. Two things should be noted carefully in this definition. First, the separation between *two* components must be specified. This can be done by specifying both x_D and x_B on the diagram or by specifying the split on each of the two components between the overhead- and bottom-product streams. It is not sufficient to specify only one concentration or a split for one component because this specification could be met by taking more or less overhead product at any reflux ratio. Second, the number of stages in the column must *not* be specified. The concept of minimum reflux as defined above obviously has no meaning for an existing column. True, there will be a certain reflux rate below which it is impossible to effect a desired separation between two components, but this reflux rate is associated with the finite number of stages in the column and not with an infinite number of stages.

It can be seen from the preceding paragraph that the discussion of the concept of minimum reflux for binary systems must be restricted to those problems where both x_D and x_B are specified and the number of stages is not specified. Also, it is necessary to specify that the feed be introduced in the optimum location (stage that straddles the q line), since otherwise it would be possible to require an infinite number of stages at any reflux by simply not introducing the feed until an infinite number of stages had been stepped off on the first operating line.

Consider a hypothetical column with a variable number of stages which is producing fixed product compositions of x_D and x_B with a finite number of stages and one feed. Hold the product composition constant while slowly reducing the external reflux ratio L_{N+1}/D. The reduction in the reflux ratio will cause the internal reflux ratio L/V at the top of the column to decrease, and the top operating line will pivot upward around the fixed point x_D. The lower operating line will also pivot upward around the fixed point x_B in order to maintain its intersection with the upper operating line on the q line of the feed. As the operating lines move closer to the equilibrium curve, more steps will be required to pass from x_D to x_B. The feed is always introduced at the optimum point, and therefore stages must be added to both the rectifying and

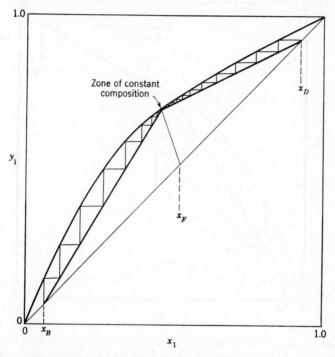

Fig. 5-20. McCabe-Thiele diagram for minimum reflux with a system of normal volatility.

stripping sections. Continue the reduction in the reflux ratio until any further reduction will cause the two operating lines to intersect on the equilibrium curve. The L_{N+1} at this point is the *true* minimum reflux if the binary mixture exhibits "normal" volatility characteristics. The McCabe-Thiele construction for this situation is shown in Fig. 5-20. Note that the zone of constant composition (no change in composition from stage to stage) is centered on the feed stage. The reflux ratio pictured in Fig. 5-20 is the true minimum reflux because further reduction in the reflux ratio would create two zones of constant composition which do not touch at the feed stage and this is an inoperable situation. It should be noted that each set of x_D, x_B, and q-line locations has its own unique minimum reflux ratio.

The equilibrium curve may exhibit the "abnormal" shape shown in Fig. 5-21. With such a system, a zone of constant composition will develop in the rectifying section at a higher reflux ratio than the one associated with the development of such a zone around the feed stage. That reflux ratio which, with the optimum feed location, first produces a

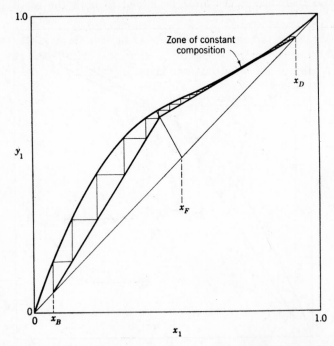

FIG. 5-21. McCabe-Thiele diagram at minimum reflux with a system of abnormal volatility where the zone of constant composition does not include the feed stage.

zone of constant composition anywhere in the column is always the *true* minimum reflux ratio.

5-12. Optimum Reflux

Somewhere between the limiting conditions of total and minimum reflux lies one reflux ratio which is the optimum for the specified separation and feed. When a new column is being designed, it can be seen that an increase in reflux will decrease the required stages and vice versa. The number of stages affects only the height of the column, whereas a change in reflux causes a change in the amount of vapor and liquid traffic in the column. The diameter of the column, the heat-transfer surfaces in the condenser and reboiler, the pumps, and all the other accessories to the column must be sized to handle the vapor and liquid circulated within the column to provide reflux. The designer can work through several cases at various refluxes and then make cost estimates for each of the resulting columns. A plot of construction cost vs reflux ratio would then pinpoint that case with the lowest investment cost. The investment cost would decrease from infinity at minimum reflux (infinite number of stages), pass through a minimum, and then increase again due to increased column diameter and accessory costs at the higher reflux ratios.

Since the steam (or fuel) and water rates to the reboiler and condenser and the power consumption of the circulating pumps vary with reflux ratio, the point of lowest initial investment in general is not the optimum

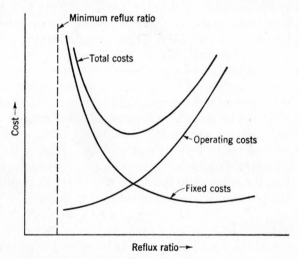

Fig. 5-22. Location of the optimum reflux ratio for a given feed and specified separation.

design. The designer must figure the costs on a per unit basis, a unit of time or a unit of material (feed or product). Some yearly rate of return on the initial investment is assumed, manpower and maintenance requirements are estimated, allowances are made for depreciation and taxes, operating costs are calculated, etc., to provide total costs per unit of time or per unit of material. This cost when plotted vs the assumed reflux ratios locates the optimum design for the specified separation and feed. The total cost curve goes from infinity at minimum reflux (owing to the fixed costs on an infinite investment at this point), passes through a minimum as the investment decreases, and then begins to rise again owing to the increase in operating (utility) costs at the higher refluxes. Studies of this sort have been presented by Bolles (1). Typical curves are shown in Fig. 5-22.

It is important to realize the restricted nature of the optimum reflux ratio described above.   It applies only as long as the feed and the specified separation remain unchanged.   It often happens that within the year or more in which the unit is engineered and constructed, one or the other will change and the column constructed will no longer be the optimum one.   The product specifications and values often change, requiring a change in the specified separation.   Or the unit or outside supplier from which the feed stock is obtained changes operation in such a manner as to affect the rate or composition of the assumed feed.   Market demands may change so drastically that the operation of the column at start-up time bears little resemblance to that visualized in the original design.   Because of these uncertainties, the wise designer will attempt to build as much flexibility as possible into the column (consistent with reasonable investment costs) so that the column can operate efficiently under any foreseeable eventuality.   Time spent in saving a few dollars by optimization for one particular case will generally be wasted unless the designer is absolutely sure the column will operate as specified.

Optimum reflux for an operating column depends upon the product values and the degree of separation desired.   Only utility (steam, power, and water) costs can be saved by a decrease in reflux ratio on an existing column.   Obviously no more reflux should be used than necessary to obtain the desired product yields and purities.   More often than not, however, the credit for additional separation far overshadows the utilities savings.   If so, the reflux ratio should be set as high as possible without causing flooding or excessive entrainment in the column.

5-13. Stage Efficiency

The concept of the equilibrium stage is based on the assumption that all the vapor leaving the stage is in equilibrium with all the liquid leaving the same stage.   A real stage that actually produces vapor and liquid streams which are in equilibrium with each other is said to have an efficiency of 100 per cent.   Real stages may have efficiencies which range from values close to zero to values of more than 100 per cent.   Efficiencies greater than 100 per cent are possible because in large-diameter columns all the vapor entering a stage is not of the same composition and all the vapor does not contact all of the liquid.   The estimation of plate efficiency and the factors affecting efficiency will be discussed in Chap. 16.   Two simple types of efficiency will be discussed here in order to utilize the x-y diagram in their illustration.

The simplest type of stage efficiency to apply is the *over-all* efficiency.   The over-all efficiency is defined as the number of equilibrium stages required to accomplish a given separation divided by the number of actual stages required to accomplish the same separation.   For example, if the over-all stage efficiency were 80 per cent in the system represented in Fig. 5-15, the number of actual stages required would be approximately 11.3/0.8 = 14.1 or 15 stages. It is reasonable to use fractions of an equilibrium stage but not fractions of an actual stage. The over-all efficiency factor to be applied usually comes from the designer's experience with other columns in similar services.

Instead of applying an over-all efficiency, the designer may choose to apply the same or a different efficiency factor to each individual stage

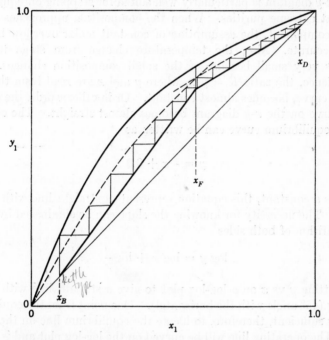

FIG. 5-23. Application of a 50 per cent Murphree vapor-phase efficiency to each stage (excluding the reboiler) in the column. Each step in the diagram corresponds to an actual stage.

in the column. The Murphree (4) efficiency is convenient for this purpose. The Murphree efficiency is defined for the vapor phase as

$$\frac{y_n - y_{n-1}}{y_n^* - y_{n-1}}$$

where y_n^* is the composition of the vapor which would be in equilibrium with the liquid leaving stage n, and y_n is the actual composition of the vapor. A similar efficiency can be defined for the liquid phase.

Figure 5-23 illustrates the application of a Murphree vapor-phase efficiency of 50 per cent to each stage on an x-y diagram. A "pseudo"-

equilibrium curve is drawn halfway (on a vertical line) between the operating lines and the true equilibrium curve. The true equilibrium curve is used for the first stage, since the reboiler is usually assumed to be an equilibrium stage. For every other stage above the reboiler (excluding a partial condenser), the 50 per cent efficiency factor is applied to each stage through use of the pseudoequilibrium curve.

5-14. Extreme Purities

The x-y diagram is particularly well suited for stepping off equilibrium stages at extreme purities. When the composition approaches that of a pure component, the assumption of constant molar overflow becomes quite accurate. Also, the temperature change from stage to stage becomes very small because of the small composition changes. As a consequence, the ratio $K = y/x$, where y and x are read from the equilibrium curve, becomes almost constant. Or in other words, the equilibrium curve on the x-y diagram becomes almost straight. The equation for the equilibrium curve can be written as

$$y = \frac{y}{x} x + 0.0$$

If y/x is a constant, this equation represents a straight line with a slope of y/x. The necessity for knowing the slope can be eliminated by taking the logarithm of both sides

$$\log y = \log x + \log \frac{y}{x}$$

and plotting y vs x on a log-log plot to give a straight line with a slope of unity (45° angle with the horizontal). One point from the equilibrium curve is sufficient, therefore, to locate the equilibrium line on the log-log plot. The operating line will be curved on the log-log plot and is located by plotting the appropriate material-balance equation. Both the equilibrium and operating lines can be extended to any purity desired.

5-15. Curved Operating Lines

It is, of course, possible to modify the graphical results which are based on the assumption of constant molar overflow by the subsequent trial-and-error application of enthalpy balances. The approximate stage compositions as determined with straight operating lines can be used to make enthalpy balances which provide better values of the interstage phase rates than those based on the assumption of constant molar overflow. The new phase rates can be used to plot new curved operating lines which can be used to step off the stages again and provide still better values of the stage compositions. The calculations are repeated until the true operating lines are located. Rather than use this cumbersome procedure, it is better to use the graphical method of Ponchon (5) and Savarit (6), which includes the enthalpy balances in the graphical construction. The Ponchon-Savarit method is described in the next chapter, and it should be used in binary calculations where the assumption of constant molar overflow is not sufficiently accurate.

5-16. Smoker Equation

The McCabe-Thiele or any other graphical method becomes inconvenient if a large number of stages are required to effect the desired separation. A large number of stages will be required if the relative volatility α of one component with respect to the other is close to one. The relative volatility in such cases can often be assumed to be constant over the concentration range.

Smoker (7) has utilized the assumption of a constant α along with the usual McCabe-Thiele assumption of constant molar overflow to derive an equation which presents the graphical construction in analytical form. The Smoker equation is convenient to use when a large number of stages are required and the assumptions of constant α and constant molar overflow are valid.

Since a constant α is assumed, the equilibrium curve can be represented analytically by Eq. (5-1).

$$y = \frac{\alpha x}{1 + (\alpha - 1)x} \tag{5-1}$$

The assumption of constant molar overflow permits the representation of any operating line (in either section of the column) by

$$y = mx + b \tag{5-18}$$

where m is the slope and b the intercept at $x = 0$. Alternate use of these two equations in a stage-to-stage calculation through the column is analogous to the construction of steps on the McCabe-Thiele diagram. The Smoker equation, however, eliminates the need for the stage-to-stage calculation by relating the two purities at the ends of a column section to the number of stages within the section. The derivation proceeds as follows:

The intersection of the operating line (within any given section of the column) with the equilibrium curve (see Fig. 5-24) is given by the combination of Eqs. (5-1) and (5-18).

$$m(\alpha - 1)x^2 + [m + b(\alpha - 1) - \alpha]x + b = 0 \tag{5-19}$$

The assumption of a constant α eliminates the possibility of an "abnormally" shaped equilibrium curve such as the one shown in Fig. 5-21. If, in addition, the analysis is limited to single feed columns, it can be seen that only one intersection [or root of Eq. (5-19)] of the operating line and equilibrium curve will occur between $x = 0$ and $x = 1.0$. This root will be denoted by $x = k$, and the other root will be disregarded.

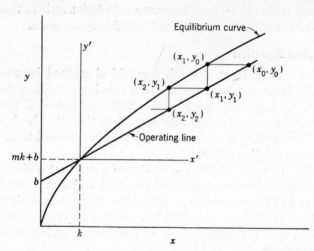

FIG. 5-24. Diagram used in the derivation of the Smoker equation. Note that the coordinate points on the steps are not numbered according to stage number.

At this point it is convenient to shift the origin of the x and y axes from $x = y = 0$ to the point of intersection of the operating and equilibrium lines where $x = k$ and $y = mk + b$. The transformation equations between the two sets of coordinate axes are then

$$x' = x - k \qquad (5\text{-}20)$$
$$y' = y - (mk + b) \qquad (5\text{-}21)$$

The operating-line and equilibrium-curve equations when transformed to the new coordinates are

$$y' = mx' \qquad (5\text{-}22)$$

and

$$y' + mk + b = \frac{\alpha(x' + k)}{1 + (\alpha - 1)(x' + k)} \qquad (5\text{-}23)$$

At $x' = y' = 0$, this last equation becomes

$$mk + b = \frac{\alpha k}{1 + (\alpha - 1)k}$$

This form can be combined with (5-23) to eliminate b and give

$$y' = \frac{\alpha x'}{c^2 + c(\alpha - 1)x'} \qquad (5\text{-}24a)$$

or

$$x' = \frac{c^2 y}{\alpha - c(\alpha - 1)y'} \qquad (5\text{-}24b)$$

where

$$c = 1 + (\alpha - 1)k \qquad (5\text{-}25)$$

Equation (5-24) is the equation for the equilibrium curve with respect to the coordinate axes whose origin is at the intersection of the equilibrium curve with the operating line.

Now start at some point x_0, y_0 (say the overhead product purity) on the operating line in Fig. 5-24, and by alternate use of Eqs. (5-22) and (5-24) calculate down the column to x_n, where x_n represents some point short of the new origin.

$$y_0' = mx_0'$$

and

$$x_1' = \frac{c^2 y_0'}{\alpha - c(\alpha - 1)y_0'}$$

Elimination of y_0' between these two equations relates x_1' to x_0'.

$$x_1' = \frac{mc^2 x_0'}{\alpha - mc(\alpha - 1)x_0'}$$

Since $y_1' = mx_1'$,

$$x_2' = \frac{c^2 y_1'}{\alpha - c(\alpha - 1)y_1'} = \frac{mc^2 x_1'}{\alpha - mc(\alpha - 1)x_1'}$$

Substitution for x_1' from the previous equation relates x_2' to x_0'.

$$x_2' = \frac{m^2 c^4 x_0'}{\alpha^2 - mc(\alpha - 1)(\alpha + mc^2)x_0'}$$

Alternating use of the material-balance and equilibrium equations until stage n is reached gives the following relation between x_n' and x_0'.

$$x_n' = \frac{m^n c^{2n} x_0'}{\alpha^n - mc(\alpha - 1)[(\alpha^n - m^n c^{2n})/(\alpha - mc^2)]x_0'}$$

which can be rearranged to

$$n = \frac{\log \dfrac{x_0'\{1 - [mc(\alpha - 1)x_n'/(\alpha - mc^2)]\}}{x_n'\{1 - [mc(\alpha - 1)x_0'/(\alpha - mc^2)]\}}}{\log (\alpha/mc^2)} \qquad (5\text{-}26)$$

Equation (5-26) must be applied individually to the rectifying and stripping sections. A summary of the equations for the rectifying section is as follows:

$$m = \frac{R}{R + 1} \qquad \text{where } R = \frac{L_{N+1}}{D} \qquad (5\text{-}27)$$

$$b = \frac{x_D}{R + 1} \qquad \text{since } y = x \text{ at } x = x_D \qquad (5\text{-}28)$$

$$x_0 = x_D$$
$$x_0' = x_D - k$$
$$x_n = \text{intersection of operating line with } q \text{ line}$$
$$x_n' = x_n - k$$
$$c = 1 + (\alpha - 1)k \tag{5-25}$$

The k is that root of Eq. (5-19) between 0.0 and 1.0.

For the stripping section the expression for m can be obtained from the known reflux ratio, feed thermal condition, and feed and product compositions as follows:

$$m = \frac{L'}{V'} = \frac{L + qF}{V - (1 - q)F} = \frac{R + q + q(B/D)}{R + q - (1 - q)(B/D)}$$

From an over-all component balance

$$\frac{B}{D} = \frac{x_D - x_F}{x_F - x_B}$$

Substitution for B/D gives

$$m = \frac{Rx_F + qx_D - (R + q)x_B}{(R + 1)x_F + (q - 1)x_D - (R + q)x_B} \tag{5-29}$$

The x_F in Eq. (5-29) is the feed composition regardless of the thermal condition of the feed. Substitution for m in $y = mx + b$ from Eq. (5-29) and use of the fact that $y = x$ at $x = x_B$ give

$$b = \frac{(x_F - x_D)x_B}{(R + 1)x_F + (q - 1)x_D - (R + q)x_B} \tag{5-30}$$

When a stripping column is used, the slope of the operating line is simply

$$m = \frac{L'}{V'} = \frac{F}{F - B} = \frac{F/B}{F/B - 1}$$

and

$$b = \frac{x_B}{1 - F/B}$$

The other relationships needed for the stripping section are

$$x_0 = \text{intersection of operating line with } q \text{ line}$$
$$x_0' = x_0 - k$$
$$x_n = x_B$$
$$x_n' = x_B - k$$
$$c = 1 + (\alpha - 1)k$$

The k is obtained from Eq. (5-19) and the m and b for the stripping section.

At total reflux, $R = \infty$, $m = 1$, $b = 0$, $k = 0$, and $c = 1$. Equation (5-26) then reduces to

$$n = \frac{\log\,[x_0(1 - x_n)]/[x_n(1 - x_0)]}{\log \alpha} \qquad (5\text{-}31)$$

No primes are needed, since there is no shift in the coordinate axes.

Example 5-3. Smoker (7) illustrated the correctness of Eq. (5-26) by comparison of results with a graphical solution. He took as an example a continuous still, a bubble-point feed of 40 per cent benzene and 60 per cent toluene, an external reflux ratio $R = 3$, and product purities of 99 per cent. An average α of 2.5 was used. For the rectifying section,

$$m = \frac{3}{3 + 1} = 0.75$$

$$b = \frac{0.99}{3 + 1} = 0.2475$$

$$m + b(\alpha - 1) - \alpha = 0.75 + 0.2475(1.5) - 2.5 = -1.38$$
$$m(\alpha - 1) = 0.75(1.5) = 1.125$$

$$k = \frac{1.38 \pm \sqrt{(1.38)^2 - 4(1.125)(0.2475)}}{2(1.125)} = 1.01 \text{ or } 0.2184$$

$$c = 1 + (1.5)(0.2184) = 1.328$$
$$mc^2 = (0.75)(1.328)^2 = 1.321$$

$$\frac{mc(\alpha - 1)}{\alpha - mc^2} = \frac{(0.75)(1.328)(1.5)}{2.5 - 1.321} = 1.265$$

$$x_0' = 0.99 - 0.2184 = 0.7716$$
$$x_n' = x_F' = 0.40 - 0.2184 = 0.1816$$

$$n = \frac{\log \dfrac{0.7716[1 - (1.265)(0.1816)]}{0.1816[1 - (1.265)(0.7716)]}}{\log\,(2.5/1.321)} = 7.9$$

For the stripping section,

$$m = \frac{3(0.4) + 1.0(0.99) - (3 + 1)(0.01)}{(3 + 1)(0.4) + 0(0.99) - (3 + 1)(0.01)} = 1.378$$

$$b = \frac{(0.4 - 0.99)(0.01)}{(3 + 1)(0.4 - 0.01)} = -0.00378$$

$$k = 0.5486 \text{ [from Eq. (5-19)]}$$

$$c = 1 + (1.5)(0.5486) = 1.823$$
$$x_0' = x_F' = x_F - k = 0.40 - 0.5486 = -0.1486$$
$$x_n' = x_B' = x_B - k = 0.01 - 0.5486 = -0.5386$$
$$mc^2 = (1.378)(1.823)^2 = 4.58$$

$$\frac{mc(\alpha - 1)}{\alpha - mc^2} = \frac{(1.378)(1.823)(1.5)}{2.5 - 4.58} = -1.812$$

$$n = \frac{\log \dfrac{(-0.1486)[1 - (-1.812)(-0.5386)]}{(-0.5386)[1 - (-1.812)(-0.1486)]}}{\log\,(2.5/4.58)} = 7.8$$

The total number of equilibrium stages required is $7.9 + 7.8 = 15.7$, which checks a graphical solution.

Example 5-4. Separation of oxygen isotopes by the distillation of ordinary water requires such a large number of stages to accomplish any significant separation that a graphical solution is impractical. Dodge and Huffman (2) discuss an example in which ordinary water with an H_2O^{18} concentration of 0.002 mole fraction was fed to a stripping column which produced a bottom product with 0.0254 mole fraction H_2O^{18}. At the temperature of operation the relative volatility of H_2O^{16} to H_2O^{18} was 1.006. Calculate the number of stages required to accomplish this separation at total reflux (no bottom product withdrawn but with feed introduced continuously and the top vapor product withdrawn continuously). What is the effect of reducing the F/B ratio from infinity to 3920?

Solution. Equation (5-31) is applicable at total reflux. Since the H_2O^{16} is the more volatile component, the concentrations should be expressed in terms of H_2O^{16}.

$$x_0 = x_0' = x_F = 1.0 - 0.002 = 0.998$$
$$x_n = x_n' = x_B = 1.0 - 0.0254 = 0.9746$$
$$n = \frac{\log\,[0.998(1 - 0.9746)]/[0.9746(1 - 0.998)]}{\log 1.006} = 430 \text{ at total reflux}$$

More stages will be required when product is withdrawn continuously. When $F/B = 3920$,

$$m = \frac{L'}{V'} = \frac{F}{F - B} = \frac{F/B}{F/B - 1} = \frac{3920}{3919} = 1.000255$$

$$b = \frac{x_B}{1 - F/B} = \frac{0.9746}{-3919} = -0.0002487$$

Solution of Eq. (5-19) provides $k = 0.9991$. From these values of m, b, and k it can be seen that the operating line deviates only slightly from the $x = y$ diagonal.

$$c = 1 + (\alpha - 1)k = 1 + (0.006)(0.9991) = 1.00599$$
$$mc^2 = (1.000255)(1.00599)^2 = 1.01227$$
$$\frac{mc(\alpha - 1)}{\alpha - mc^2} = \frac{(1.000255)(1.00599)(0.006)}{1.006 - 1.0123} = -0.9582$$
$$x_0' = x_0 - k = x_F - k = 0.998 - 0.9991 = -0.0011$$
$$x_n' = x_n - k = x_B - k = 0.9746 - 0.9991 = -0.0245$$
$$n = \frac{\log\,\dfrac{(-0.0011)[1.0 - (-0.9582)(-0.0245)]}{(-0.0245)[1.0 - (-0.9582)(-0.0011)]}}{\log\,(1.006/1.01227)} = 503$$

Slide-rule accuracy is not sufficient when n is very large. Note that the value of n obtained is quite sensitive with respect to the α used.

NOMENCLATURE

b = intercept of operating line at $x = 0$.

B = bottoms product rate; moles, weight, or volume per unit time.

c = subscript to denote concentrations leaving the condenser. Also used in Sec. 5-15 as defined by Eq. (5-25).

D = total overhead product rate; moles, weight, or volume per unit time. D_L and D_V refer to the liquid and vapor overhead product rates, and $D = D_L + D_V$. For total condenser, $D_V = 0$ and $D = D_L$.

F = feed rate; moles, weight, or volume per unit time.

k = root between 0.0 and 1.0 in Eq. (5-19).

$K_i = y_i/x_i$ = equilibrium-distribution coefficient.

L = heavy phase rate; moles, weight, or volume per unit time. Primes are sometimes used below a feed or side-stream stage to distinguish between the liquid rates above and below the stage.

m = subscript denoting any stage in the stripping section. Also used in Sec. 5-15 to denote slope of the operating line.

n = subscript denoting any stage in the rectifying section. If the column contains only one section, n denotes any stage in the column. Also used to denote the number of stages calculated with the Smoker equation.

N = total number of stages in the column including partial reboiler, if any, but excluding partial condenser, if any.

$N_i{}^u$ = independent design variables. See Chap. 3.

q = analytical function defined by Eqs. (5-11) and (5-12) to describe thermal condition of a feed stream. Also used to describe thermal condition of a side stream (see Sec. 5-7).

q_c = condenser heat load, Btu/unit time.

q_r = reboiler heat load, Btu/unit time.

$R = L_{N+1}/D$ = external reflux ratio at top of column.

$R' = V_1/B$ = reboil ratio at bottom end of column with partial reboiler.

S = side-stream product rate; moles, weight, or volume per unit time.

V = light phase rate; moles, weight, or volume per unit time. Primes are sometimes used below a feed or side-stream stage to distinguish between the vapor rates above and below the stage.

x = concentration in heavy phase rate. Units consistent with units on rates. Subscript refers to stage or stream in which concentration appears. Subscripts 1 and 2 also used to refer to components 1 and 2.

y = concentration in light phase rate. Units consistent with units on rates. Subscript refers to stage or stream in which concentration appears. Subscripts 1 and 2 also used to refer to components 1 and 2.

Greek Symbol

$\alpha_{1-2} = K_1/K_2 = y_1 x_2/y_2 x_1$ = relative volatility of component 1 to component 2.

PROBLEMS

5-1. A column is to be designed to separate a mixture of benzene and toluene. The feed, at its bubble point and containing 50 mole per cent of benzene, is to be fractionated to produce an overhead product containing 90 mole per cent benzene and a bottoms product containing 10 mole per cent benzene. If the reflux ratio L/V is 0.6, would plate 11 counting from the top be an operable feed plate? What would be the highest operable feed stage? Vapor-liquid equilibrium data for this system are given in Perry, "Chemical Engineers Handbook," 3d ed., page 578. Make the usual simplifying assumptions.

5-2. It has been decided to use an eight-plate column with partial reboiler to separate materials A and B at atmospheric pressure. The feed stock consists of an equimolar mixture of A and B at its bubble point, and it is desired to obtain an overhead product containing 95 mole per cent A. It is possible to introduce the feed on either the sixth or eighth plate from the top. Calculate the yield of distillate product per mole of feed when the feed is introduced on (a) the sixth plate and (b) the eighth plate from the top. Which of the two would you use and why? Is either one the optimum feed tray? Assume the relative volatility constant at 2.8, and use a reflux ratio L/V above the feed of 0.6.

5-3. A mixture containing 50 mole per cent A, which is at its bubble point, is introduced on the second tray of a column containing three trays. The unit is equipped with a reboiler equivalent to a theoretical tray and a total condenser. The reboil ratio V_1/B is 2.0. Assume that each tray has an efficiency of 100 per cent and that the usual simplifying assumptions hold. The relative volatility α is 4.0.

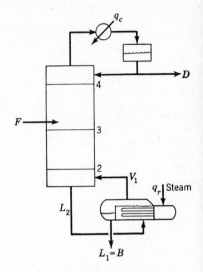

(a) If the bottoms contain 10 mole per cent A, what would be the composition of the overhead product?

(b) What would be the composition of the vapor leaving the top tray if the withdrawal of overhead product is stopped and all the condensate from the condenser is returned to the top tray? The reboil ratio remains at $V_1/B = 2.0$.

5-4. In the following, assume that the feeds are saturated liquids and the usual simplifying assumptions hold. In all cases show the McCabe-Thiele diagram. Use a separate x-y diagram for each part. The relative volatility is 2.5. A is the more volatile.

(a) At the present time, a column separating A and B has two feeds, $x_F = 0.4$ and $x_F = 0.2$. The feeds are supplied to the column in the proportion of 25 moles of the feed richer in A and 75 moles of the feed leaner in A. When the reflux ratio $L_{N+1}/D = 3.0$, the column produces an overhead product containing 0.7 mole fraction of A and a bottoms product containing 0.1 mole fraction of A. On what stages should the feeds be introduced and how many stages are required? Number from the bottom.

(b) If the two feeds were switched so that the lean feed came in on the stage where the rich feed was introduced in part (a) and the rich feed came in on the stage where the lean feed was previously brought in, how many stages would be required?

(c) A feed which has the same average composition as the total feed in part (a) is fed to the same feed stages as determined in part (a). The feed is introduced to the column in two streams so that 25 moles/unit time goes to the upper feed stage and 75 moles/unit time goes to the lower feed stage. How many stages would be required to effect the specified separation?

5-5. Consider a continuous binary distillation column with five equilibrium stages and a bubble-point feed on the middle (optimum) stage operating at steady state. Describe by a series (three are enough) of x-y diagrams the changes that occur as the reflux L_{N+1} is slowly reduced while the overhead product rate remains constant. The diagrams should each represent the steady state at the assumed reflux. The first diagram should show the original steady state before the reflux reduction is started, and the last diagram the steady state at the minimum reflux.

5-6. An ethanol-water mixture containing 40 mole per cent ethanol and which is 50 mole per cent vapor is fed into a packed column. Saturated steam enters at the bottom of the column, and pure liquid ethanol at the boiling point enters at the top. The feed enters at the optimum point. The liquid leaving the bottom of the column contains 5 mole per cent ethanol. At one point in the column the liquid contains 20 mole per cent ethanol while the vapor contains 30 mole per cent ethanol. At another point in the column the liquid contains 60 mole per cent ethanol while the vapor contains 65 mole per cent ethanol.

Make the assumptions which are necessary for the use of the Mc-Cabe-Thiele method.

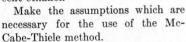

(a) What is the composition of the vapor leaving the column?

(b) If the feed rate is 100 lb moles/hr, what are the flow rates of the other entering and leaving streams?

(c) Determine the number of theoretical plates to which the column is equivalent.

5-7. A column supplied with a saturated liquid feed containing 33 mole per cent ethane and 67 mole per cent propane is to produce an overhead product containing 97 per cent ethane and a bottoms product containing 4 per cent ethane. The operating pressure is to be 170 psia and the reflux will be returned to the column as a saturated liquid.

(a) Using the fugacity function charts in Maxwell's "Data Book on Hydrocarbons," construct the x-y diagram.

(b) Using the McCabe-Thiele diagram, determine the minimum refluxes L/V and L/D, the minimum plates (total reflux), and the theoretical plates required for a reflux rate of $1.3(L/D)_{min}$.

REFERENCES

1. Bolles, W. L.: *Petrol. Refiner*, **25**(12), 613 (1946).
2. Dodge, B. F., and J. R. Huffman: *Ind. Eng. Chem.*, **29**, 1434 (1937).
3. McCabe, W. L., and E. W. Thiele: *Ind. Eng. Chem.*, **17**, 605 (1925).
4. Murphree, E. V.: *Ind. Eng. Chem.*, **17**, 747 (1925).
5. Ponchon, M.: *Tech. moderne*, **13**, 20, 55 (1921).
6. Savarit, R.: *Arts et métiers*, 1922, pp. 65, 142, 178, 241, 266, 307.
7. Smoker, E. H.: *Trans. A.I.Ch.E.*, **34**, 165 (1938).

CHAPTER 6

BINARY DISTILLATION
Enthalpy-Concentration Diagrams

Binary-distillation problems can be solved graphically in a rigorous manner by means of the enthalpy-concentration diagram. The McCabe-Thiele method described in the previous chapter utilized only the equilibrium and material-balance relationships to approximate the separation

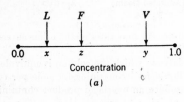

between two components in a distillation column. The enthalpy-concentration diagram adds the third basic relationship, the energy balance. If enough basic calorimetric data are available to permit the calculation of the equilibrium-phase enthalpies, the binary-distillation problem can be solved on the enthalpy-concentration diagram without any simplifying assumptions.

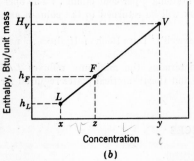

FIG. 6-1. (a) Component material balance on a concentration scale. (b) Simultaneous enthalpy and component material balance on an enthalpy-concentration diagram.

6-1. Lever-arm Rule

The use of the enthalpy-concentration diagram to solve binary-distillation problems was originally proposed by Ponchon (2) and Savarit (3). The graphical construction is based upon the lever-arm rule, which is a technique for making material- and energy-balance calculations with straight lines. Figure 6-1a illustrates how component material balances can be made on a concentration scale in which concentrations are expressed in ratio form as mass/unit mass, moles/mole, or volume/unit volume. Consider as an example the component-i balance in an equilibrium flash separation,

$$Vy_i + Lx_i = Fz_i$$

164

The streams V and L are located on the concentration scale according to their respective concentrations of component i. The point which represents their sum Fz_i must fall on the concentration scale somewhere between V and L. Its location depends upon the relative amounts of V and L as well as their concentrations. This fact is illustrated by the following rearrangements of the component-i balance (the i's are omitted for simplicity):

$$\frac{V}{L} = \frac{z - x}{y - z}$$

$$\frac{L}{F} = \frac{y - z}{y - x}$$

$$\frac{V}{F} = \frac{z - x}{y - x}$$

It can be seen that the relative amount of stream V is represented by the line segment $\overline{xz}$ while the relative amount of stream L is represented by the segment $\overline{zy}$. The *addition point* F is represented by the entire line segment between L and V.

Figure 6-1a can also be used to illustrate the subtraction of one stream from another. If L is subtracted from F, the *difference point* must lie on the other side of F at V. The exact location of V is determined by the relative amounts of L and F along with their concentrations as shown by the ratios in the previous paragraph. If L is larger than F (not possible in the case of the flash separator), the difference point may fall so far to the right that $y > 1.0$. Or if V is subtracted from F when V is larger than F, then the difference point L may fall so far to the left that $x < 0$. In such cases, the stream which the difference point represents can have only algebraic meaning; that is, it cannot exist physically. Nevertheless, difference points with negative concentrations (and enthalpies) or with concentrations greater than 1.0 arise often in the graphical solution of equilibrium processes. These algebraic quantities are handled exactly as any physical stream in the application of the lever-arm rule.

Enthalpy balances can be made in a manner analogous to the component balances if the concentration scale in Fig. 6-1a is replaced with an enthalpy scale whose divisions represent Btu per unit mass of mixture instead of mass per unit mass of mixture where mass refers to weight, moles, or volume.

The enthalpy and component balances can be made simultaneously on the two-dimensional diagram shown in Fig. 6-1b. The enthalpy scale is used for the ordinate and the concentration scale as the abscissa. The streams V and L are plotted on this two-dimensional diagram and connected by a straight line. It can be shown that the addition point F must lie on this straight line. This will be done by showing that the slope of the line segment $\overline{FV}$ is identical with the slope of the entire line $\overline{LV}$.

$$\text{Slope of } \overline{FV} = \frac{H_V - h_F}{y - z}$$

$$\text{Slope of } \overline{LV} = \frac{H_V - h_L}{y - x}$$

The subscript i is omitted for simplicity. From the component i and the enthalpy balances,

$$z = \frac{Vy + Lx}{V + L}$$

$$h_F = \frac{VH_V + Lh_L}{V + L}$$

Substituting for z and h_F in the expression for the slope of $\overline{FV}$ gives

$$\text{Slope of } \overline{FL} = \frac{H_V - h_L}{y - x}$$

which is the expression for the slope of $\overline{LV}$.

The position of the addition point F on the straight line between L and V is determined (as in the case of a single dimension) by the relative amounts of V and L. The line segments can be measured off on the line $\overline{LV}$ or on its projection on either the abscissa or the ordinate. Subtraction of one stream from another is accomplished as shown previously on a single scale.

The solution of binary distillation problems on the enthalpy-concentration diagram involves nothing other than the application of the lever-arm rule to the material and energy balances in the column.

6-2. Equilibrium Data

The most convenient way to handle the equilibrium-composition data is by means of the x-y plot described in Sec. 5-1. The x-y equilibrium curve is superimposed on the enthalpy-concentration diagram as shown in Fig. 6-2. Once the saturated vapor and liquid curves have been located as described below, points from the x-y equilibrium curve can be transferred to the enthalpy-concentration diagram as needed. Each point on the x-y curve represents a pair of equilibrium-phase compositions. Therefore, each point on the x-y curve becomes a pair of points on the enthalpy-concentration diagram, one of which is located on the saturated vapor curve and the other on the saturated liquid curve. A straight line which connects the equilibrium-phase compositions is a *tie line*. The tie lines which are determined experimentally can, of course, be plotted directly on the enthalpy-concentration diagram without recourse to the x-y diagram. However, the x-y diagram is the best method for smoothing the data and for interpolation.

If liquid heat of solution data are available as a function of
composition at some convenient temperature, and if the heat capacity
of the binary liquid mixtures are known as a function of temperature
and composition, it is possible to locate the saturated vapor and liquid
enthalpy curves exactly.   Reference states are chosen for each of the
two components.   In most cases, the pure liquid components at some
convenient temperature are satisfactory reference states.   Arbitrary
values of enthalpy (usually zero) are assigned to the reference states.

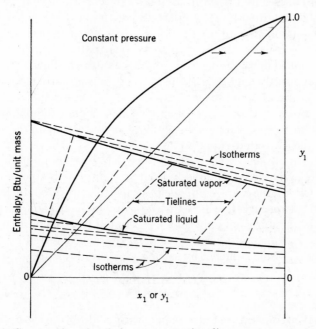

Fig. 6-2. Construction of enthalpy-concentration diagram at constant pressure.

A base isotherm is constructed at the temperature for which the heat-
of-solution data are available (not necessarily the reference temperature)
by means of the equation

$$H = x_1 H_1 + x_2 H_2 + q_s \qquad (6\text{-}1)$$

where H = enthalpy of a unit of solution
H_1, H_2 = enthalpies of units of pure components 1 and 2, respectively
   q_s = heat of solution per unit of solution at the composition x_1
       and x_2
Once the base isotherm is established, the liquid-heat-capacity data for
both the pure components and various solution compositions can be used
to locate other isotherms above and below the base isotherm.   The
saturated-liquid curve can then be located by spotting the bubble points

(equilibrium temperatures) at various liquid compositions on the isotherms. The ends of the saturated-liquid curve are fixed by the boiling points of the pure components at the system pressure.

If the binary liquid solutions are ideal, the q_s term in Eq. (6-1) is zero. The enthalpy along an isotherm then becomes a straight-line function of x_1. It is usually necessary to assume ideal-liquid-solution behavior and draw the isotherms as straight lines because no heat-of-solution data are available for the system of interest. The assumption of ideal solutions does not require the saturated-liquid (or vapor) curve to be straight.

Once the saturated-liquid curve is fixed, the heats of vaporization of the pure components are used to locate the ends of the saturated-vapor curve. It is usually satisfactory to assume that the vapors form ideal solutions and to draw the vapor isotherms as straight lines. If the pressure is sufficiently high (>200 psia), generalized correlations (1) can be used to estimate the heat effects of mixing. Once a sufficient number of vapor isotherms are established, the dew points (equilibrium temperatures) at various vapor compositions can be spotted on the isotherms to locate the saturated-vapor curve.

Other assumptions even less rigorous than the assumption of ideal-solution behavior are often used to locate the saturated-vapor and -liquid curves. Often, the pure component properties are used to locate the ends of the two curves, which then are drawn as straight lines. This construction assumes not only that q_s is zero but that the pure component enthalpies are constant over the temperature range covered by the equilibrium data. This is shown by setting $q_s = 0$ and rewriting Eq. (6-1) as

$$H = (H_1 - H_2)x_1 + H_2 \qquad (6\text{-}2)$$

Most diagrams of practical interest are drawn at constant pressure, and therefore the temperature must change with x_1 along the saturated-vapor and -liquid curves. Therefore an equation such as (6-2) cannot represent the saturation curves unless the change in H_1 and H_2 with temperature is ignored.

If the saturation curves are drawn to be parallel as well as straight, the assumption of equal molar heats of vaporization is involved and the results become identical with those obtained with the McCabe-Thiele diagram described in the previous chapter.

6-3. Material- and Enthalpy-balance Equations

The over-all balances for a distillation column with one feed (see Fig. 6-3) are

$$F = D + B$$
$$Fx_{i,F} = Dx_{i,D} + Bx_{i,B}$$
$$Fh_F + q_r = Dh_D + Bh_B + q_c$$

where q_r is the total energy input (Btu per hour) to the reboiler and q_c is the total energy removed (Btu per hour) by the condenser. The enthalpy balance assumes adiabatic operation except in the condenser and reboiler. Representation of the energy streams q_r and q_c on the enthalpy-concentration diagram requires that they be expressed as Btu per unit mass. (The term "unit mass" as used here can be a unit weight, a mole, or a unit volume.) The change from Btu per hour to Btu per unit mass is accomplished most conveniently as follows:

$$Q_D = \frac{q_c}{D} = \text{Btu/unit mass of } D$$

$$Q_B = \frac{q_r}{B} = \text{Btu/unit mass of } B$$

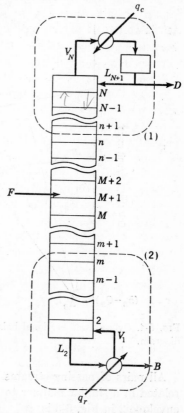

FIG. 6-3. Distillation column with one feed, total condenser, and partial reboiler.

The condenser and reboiler loads can, of course, be related to any stream in the column if desired.

The over-all enthalpy balance can now be written as

$$Fh_F = D(h_D + Q_D)$$
$$+ B(h_B - Q_B) \quad (6\text{-}3a)$$
or $\quad Fh_F = \Delta_D + \Delta_B \quad (6\text{-}3b)$

where the stream Δ_D has the coordinates $(x_D, h_D + Q_D)$ and the stream Δ_B has the coordinates $(x_B, h_B - Q_B)$. Equation (6-3b) is represented in Fig. 6-4 as a straight line after the method described in Sec. 6-1.

All pairs of passing streams within the column can be related to the points which represent Δ_D and Δ_B. The enthalpy balance for envelope 1 in Fig. 6-3 is

$$V_n H_n = L_{n+1} h_{n+1} + D h_D + q_c$$
or $\quad V_n H_n - L_{n+1} h_{n+1} = D(h_D + Q_D) = \Delta_D \quad (6\text{-}4)$

Equation (6-4) shows that the stream Δ_D is the difference point for all pairs of passing streams above the uppermost side stream or feed stream. The point which represents the liquid overflow from stage $n + 1$ has the coordinates (x_{n+1}, h_{n+1}) and falls on the saturated liquid curve. Likewise, the equilibrium vapor rising from stage n has the coordinates (y_n, H_n) which define a point on the saturated vapor curve. The subtraction of L_{n+1} from V_n is represented on the diagram (Fig. 6-4) by a straight line from L_{n+1} through V_n to the difference point Δ_D.

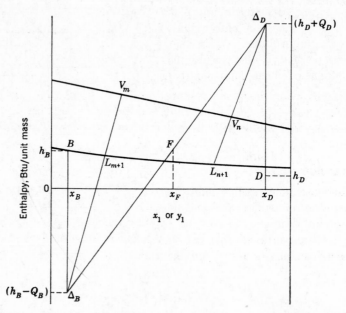

FIG. 6-4. Over-all and sectional balances represented on an enthalpy-concentration diagram.

All pairs of passing streams below the lowest side stream or feed are related in a similar manner to the point Δ_B. The enthalpy balance for envelope 2 in Fig. 6-3 is

$$L_{m+1}h_{m+1} = V_m H_m + B h_B + q_r$$

or $\qquad L_{m+1}h_{m+1} - V_m H_m = B(h_B - Q_B) = \Delta_B \qquad (6\text{-}5)$

The points representing L_{m+1} and V_m fall on the saturated-liquid and -vapor curves and have the coordinates (x_{m+1}, h_{m+1}) and (y_m, H_m), respectively. The balance represented by Eq. (6-5) is made on the diagram (Fig. 6-4) by drawing a straight line from V_m through L_{m+1} to the difference point Δ_B.

A balance which passes between two equilibrium stages and then around the end of the column may cut other external streams besides q_c, D, q_r, and B. For example, assume that a side stream S were withdrawn above stage n in Fig. 6-3. The enthalpy-balance equation would then be

$$V_n H_n = L_{n+1} h_{n+1} + S h_S + D h_D + q_c$$

or

$$V_n H_n - L_{n+1} h_{n+1} = S h_S + \Delta_D = \Delta' \qquad (6\text{-}6)$$

The difference point Δ' now obtained when L_{n+1} is subtracted from V_n is the addition point obtained when S, D, and q_c are added. Equation (6-6) illustrates the fact that whenever energy is added to or withdrawn from the column, pairs of passing streams on opposite sides of the addition or withdrawal points are not related to the same difference point.

The straight lines drawn through pairs of passing streams to the respective difference points are called *operating lines*. Enough information must be included in the specification of the distillation problem to permit the location of the various difference points. The operating lines can then be used along with the equilibrium tie lines to step off the equilibrium stages as shown in the next section.

6-4. Equilibrium Stages

Figure 6-5 illustrates the equilibrium and balance relationships involved in an equilibrium stage. Assume that by some means the composition of the liquid overflow L_{n+1} to the equilibrium stage n is known. As shown in the previous section, L_{n+1} is related by an enthalpy balance to its passing stream V_n and a difference point Δ. The location of Δ is

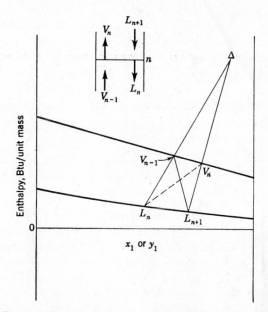

FIG. 6-5. Representation of a single equilibrium stage.

fixed by certain specifications in the distillation problem (condenser load, overhead product rate, etc.). A straight operating line from the known L_{n+1} to the fixed Δ locates V_n on the saturated-vapor curve. By definition, V_n is in equilibrium with L_n and L_n can be located by means of the equilibrium tie line (dotted) which passes through V_n. If stage n is assumed to be adiabatic, and since no material is added or withdrawn through the column wall, no change in the difference point will occur across stage n Therefore, V_{n-1} can be located by means of a straight operating line from L_n to Δ.

The over-all balance around stage n is

$$L_{n+1} + V_{n-1} = L_n + V_n$$

The sum of L_n and V_n must lie on the tie line which connects them, and the addition point is found at the intersection of the tie line with the straight line between V_{n-1} and L_{n+1}.

6-5. Top Section in Column

It has been shown that the difference point for that section of the column above the uppermost feed or side stream is given by

$$\Delta_D = D(h_D + Q_D)$$

The D in this expression refers to the total overhead product withdrawn from the column. Depending upon the type of condenser employed, the overhead product may be withdrawn as (a) liquid, (b) a vapor, or (c) a combination of liquid and vapor. The construction for each of these cases will be discussed separately.

Total Condenser. A *total* condenser condenses all the overhead vapor V_N to liquid. The condensate will be a saturated liquid if the condenser load q_c is exactly equal to the latent heat of V_N, and in this case h_D will fall on the saturated-liquid curve at x_D. If the condenser load is somewhat higher than the latent heat of V_N, the condensate will be subcooled and h_D will fall below the saturated-liquid curve.

A balance around the condenser gives

$$V_N H_N = L_{N+1} h_{N+1} + D h_D + q_c$$

or $$V_N H_N - L_{N+1} h_{N+1} = D(h_D + Q_D) = \Delta_D \tag{6-7}$$

Since $y_N = x_D = x_{N+1}$ for a total condenser, the operating line which represents this balance on the enthalpy-concentration diagram is a vertical line at x_D as shown in Fig. 6-6a. The various segments of this vertical balance line can be used to represent the reflux ratios. The V_N can be eliminated from Eq. (6-7), and the equation rearranged to give the external reflux ratio.

$$\frac{L_{N+1}}{D} = \frac{(h_D + Q_D) - H_N}{H_N - h_{N+1}} \tag{6-8}$$

If D is eliminated,

$$\frac{L_{N+1}}{V_N} = \frac{(h_D + Q_D) - H_N}{(h_D + Q_D) - h_{N+1}} \qquad (6\text{-}9)$$

Equation (6-9) can be written for any stage in the top section as

$$\frac{L_{n+1}}{V_N} = \frac{(h_D + Q_D) - H_n}{(h_D + Q_D) - h_{n+1}} \qquad (6\text{-}10)$$

Equation (6-10) describes the *internal* reflux in terms of the various segments of the operating lines.

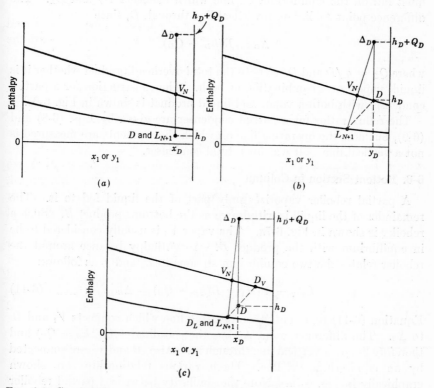

Fig. 6-6. Construction on enthalpy-concentration diagram for (a) total condenser with subcooled reflux, (b) partial condenser with vapor product only, and (c) partial condenser with both vapor and liquid products withdrawn.

Partial Condenser. A *partial* condenser condenses only part of the overhead vapor V_N. The vapor and liquid from the condenser can be considered to be at equilibrium. Normally the vapor is withdrawn as the overhead product D while the equilibrium liquid is returned to the column as the reflux L_{N+1} (see Fig. 5-7a). The enthalpy balance around the condenser is again given by Eq. (6-7), but now $y_N \neq y_D \neq x_{N+1}$ and the three streams do not all fall on the same straight line. Figure 6-6b shows the construction for a partial condenser with only a vapor product withdrawn. The streams D and L_{N+1} are connected by the equilibrium tie line which passes through the saturated-vapor curve at y_D. The streams L_{N+1} and V_N are related by the operating line which corresponds to Eq. (6-7).

If both a vapor product D_V and a liquid product D_L are withdrawn, the total overhead product D is given by $D = D_V + D_L$ (see Fig. 5-7b). In this case D is the addition point obtained by adding D_V and D_L and must fall on the equilibrium tie line which connects D_V and D_L. The difference point Δ_D lies on a vertical line through D, since

$$\Delta_D = D(h_D + Q_D)$$

where $Q_D = q_c/D$ and D refers to the total overhead product whether it is liquid, vapor, or a combination of both. The construction for a partial condenser with both a vapor and liquid product is shown in Fig. 6-6c.

The reflux ratios for a partial condenser are given by Eqs. (6-8) and (6-9), but now the operating line on which the segments are measured is not a vertical line as it was for a total condenser.

6-6. Bottom Section in Column

A partial reboiler vaporizes only part of the liquid fed to it. The remainder of the liquid is withdrawn as the bottoms product B. Such a reboiler is shown in Fig. 6-7a. The vapor V_1 is usually considered to be in equilibrium with the product B. An enthalpy balance around the reboiler relates the two equilibrium streams to L_2 and q_r as follows:

$$L_2 h_2 - V_1 H_1 = B(h_B - Q_B) = \Delta_B \qquad (6\text{-}11)$$

Equation (6-11) represents the operating line which connects V_1 and L_2 to Δ_B. The difference point Δ_B has the coordinates $(x_1, h_B - Q_B)$ and therefore lies on a vertical line through x_1. Also, B and V_1 are connected by an equilibrium tie line. These various relationships are shown graphically in Fig. 6-7a. Note the similarity between a partial reboiler as represented by Fig. 6-7a and the partial condenser represented in Fig. 6-6b.

Open Steam. When the less volatile component in a binary mixture is water, it may be advantageous to eliminate the reboiler and add heat to the column by injecting open steam beneath the bottom stage. The construction when open steam is used is shown in Fig. 6-7b. The enthalpy of the entering steam (denoted as V_0) is located on the ordinate at $x = 0$, where x refers to the more volatile component. From an

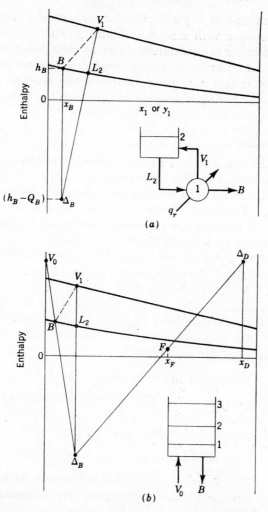

Fig. 6-7. Construction on an enthalpy-concentration diagram for (a) a partial reboiler and (b) a column with open steam where the steam V_0 is slightly superheated.

over-all enthalpy balance around the column,

$$Fh_F = D(h_D + Q_D) + Bh_B - V_0H_0$$

Defining $\Delta_B = Bh_B - V_0H_0$ for this case, the balance can be rewritten as

$$F = \Delta_D + \Delta_B$$

It can be seen that Δ_B lies at the intersection of a line through Δ_D and F with another line through V_0 and B.

The replacement of a partial reboiler with open steam reduces the separation obtained with a given column. Elimination of the reboiler reduces the number of equilibrium stages by one. Also, the steam acts as a diluent and a fraction of a stage is required to overcome this effect. The net effect is that one and a fraction stages must be added to the column when the switch to open steam is made if the same separation as that obtained with the partial reboiler is desired.

6-7. Feed Stages

The construction on the enthalpy-concentration diagram for a feed stage is much simpler than on an x-y diagram. The enthalpy or thermal condition of the feed stream is taken into account in the location of F on the diagram. The point F will always be located on a vertical line through x_F. If the feed is a subcooled feed, it will fall below the saturated-liquid curve as shown in Fig. 6-7b. If the feed is partially flashed, F will fall between the two saturation curves (see Fig. 6-4) on a tie line whose ends represent the equilibrium vapor and liquid portions of the feed stream. A saturated-vapor feed will fall on the saturated-vapor curve while a superheated-vapor feed will fall above the curve.

It has been shown that for a single feed column, all pairs of passing streams above the feed stage are related by an enthalpy balance around the top end of the column as follows:

$$V_n H_n - L_{n+1} h_{n+1} = \Delta_D \qquad (6\text{-}4)$$

Pairs of passing streams below the feed stage can also be related to each other by an enthalpy balance around the top end of the column. The enthalpy balance now includes the feed stream.

$$V_m H_m - L_{m+1} h_{m+1} = \Delta_D - F h_F = \Delta_B \qquad (6\text{-}5)$$

It can be seen that as soon as the feed is introduced, the difference point which relates passing streams immediately shifts from Δ_D to Δ_B in a single feed column.

Optimum Feed Location. The optimum feed-stage location is that one which, with a given set of other operating conditions, will result in the widest separation between x_D and x_B with the given number of stages. Or if the number of stages is not specified, the optimum location is the one which requires the fewest number of stages to accomplish a specified separation between x_D and x_B. Either of these criteria will always be satisfied if the feed is introduced on that stage whose tie line crosses the over-all material-balance line which passes through F, Δ_D, and Δ_B. The switch from one difference point to the other at that stage will always result in the maximum separation. This fact is illustrated by Fig. 6-8. Stage 4 (when numbering from the bottom) is the optimum feed stage, and solid operating lines are drawn on the assumption that the feed is actually introduced on the fourth stage.   Below stage 4, V_m and L_{m+1} are related by Δ_B, while above stage 4, V_n and L_{n+1} are related to Δ_D.   Now assume that the feed was introduced on stage 3.   Then V_3 and L_4 would lie above the feed stage and their difference point would be $V_3 - L_4' = \Delta_D$.   An operating line (dotted) through V_3 from Δ_D would locate L_4'.   Comparison of the locations of L_4' and L_4 show the deleterious effect of introducing the

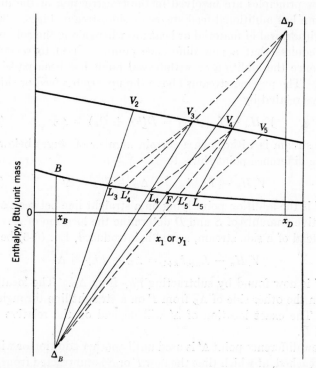

FIG. 6-8. Illustration of the concept of optimum feed-stage location. The solid operating lines correspond to the introduction of the feed on stage 4 (optimum). The dotted operating lines assume the introduction of the feed one stage below or one stage above the optimum location.

feed on stage 3 rather than stage 4.   The separation would suffer also if the feed were introduced on stage 5.   If stage 5 were the feed stage, V_4 and L_5 would lie below the feed stage and their difference point would be $L_5' - V_4 = \Delta_B$.   An operating line (dotted) from V_4 to Δ_B would locate L_5', which lies to the left of the L_5 obtained when the feed is introduced on stage 4.

In the design of a new column, the feed-stage location is under the control of the designer and should, of course, always be located in the optimum position.   In a study of an existing column, the feed-stage location is fixed and the switch from one difference point to the next one must occur at the actual feed stage which may or may not be the optimum feed stage for the separation under study.   Usually whenever a column is switched to a new service, the feed stage will not be the optimum one.

6-8. Other Feed or Side-stream Stages

No new principles are involved in the construction of the diagram if the column has additional feed stages or side-stream stages.   The addition or withdrawal of material or heat anywhere along the column length merely requires that a new difference point be used to relate passing streams once the addition or withdrawal point has been passed on the diagram.   The passing streams above the uppermost feed or side stream are always related by

$$V_n H_n - L_{n+1} h_{n+1} = D(h_D + Q_B) = \Delta_D \qquad (6\text{-}4)$$

If a side stream is withdrawn a certain number of stages below the top stage, the difference point shifts to

$$V_n H_n - L_{n+1} h_{n+1} = \Delta_D + S h_S = \Delta' \qquad (6\text{-}6)$$

where Δ' is the addition point lying on a straight line between Δ_D and S. The relative amounts of S and D determine the location of Δ'.

If, instead of a side stream, a feed is introduced, Eq. (6-6) becomes

$$V_n H_n - L_{n+1} h_{n+1} = \Delta_D - F h_F = \Delta'$$

where Δ' is now found by subtracting $F h_F$ from Δ_D.   The location of Δ' will lie on the other side of Δ_D from F on a straight line through the two points.   The exact location of Δ' will depend on the relative amounts of F and D.

The new difference point Δ' is used until another addition or withdrawal point is reached, at which time the new F or S is subtracted from or added to Δ' to obtain the next difference point Δ''.   If this latest addition or withdrawal point is the lowest feed or side stream in the column, then Δ'' must correspond to Δ_B.

6-9. Total Column

The construction for an entire column involves simply the application in turn of the rules developed in the previous sections for the construction at the various addition or withdrawal points in the column.   The application of these rules will now be illustrated by example problems. The following examples will be considered.

Example 6-1: design of a new column with one feed and a partial condenser and reboiler

Example 6-2: study of an existing column with two feeds, a total condenser, and a partial reboiler

The necessary problem specifications are discussed in each example.

Example 6-1. How many equilibrium stages will be required to produce 5 and 95 mole per cent bottom and overhead products, respectively, from a feed which contains 55 mole per cent of the more volatile component? The column will operate

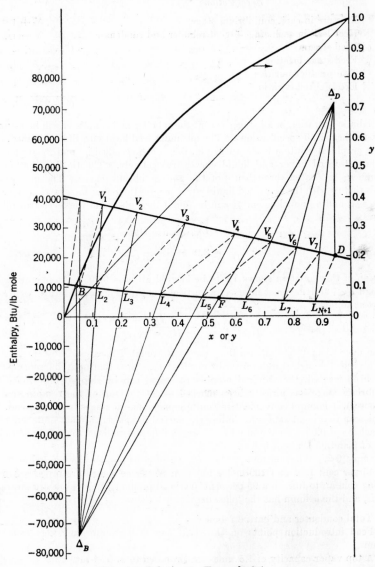

Fig. 6-9. Solution to Example 6-1.

at essentially atmospheric pressure and with an external reflux ratio of $L_{N+1}/D = 3.0$. The feed is a saturated liquid at column conditions. The equilibrium-composition and enthalpy data are as shown in Fig. 6-9.

Solution. From the principles presented in Chap. 3 it can be shown that a column such as the one in this example has $C + 2N + 6$ design variables. Some of these variables have been fixed in the problem statement. Those items are marked with a dagger in the following list of specifications.

<table>
<tr><td align="center">*Specifications*</td><td></td><td align="right">$N_i{}^u$</td></tr>
<tr><td>† Pressure in each equilibrium stage</td><td></td><td align="right">$N + 1$</td></tr>
<tr><td>Heat leak in each stage except reboiler and condenser</td><td></td><td align="right">$N - 1$</td></tr>
<tr><td>† Feed stream</td><td align="center">C</td><td align="right">$+ 2$</td></tr>
<tr><td>Feed-stage location</td><td></td><td align="right">1</td></tr>
<tr><td>† Two product purities</td><td></td><td align="right">2</td></tr>
<tr><td>† External reflux ratio</td><td></td><td align="right">1</td></tr>
<tr><td></td><td></td><td align="right">$\overline{C + 2N + 6}$</td></tr>
</table>

Adiabatic operation in all stages except the reboiler and condenser will be assumed to fix the heat-leak specifications. The optimum feed location will be specified.

The graphical solution is shown in Fig. 6-9. The specified y_D and x_B are located on the saturated vapor and liquid curves, respectively. The tie line that passes through y_D is located through use of the x-y equilibrium curve. The intersection of this tie line with the saturated-liquid curve locates the reflux L_{N+1} at $x_{N+1} = 0.877$. With x_{N+1} and y_D now known, y_N can be found through use of the lever-arm rule as follows:

$$V_N y_N = L_{N+1} x_{N+1} + D y_D$$

or
$$\frac{L_{N+1}}{D} = 3.0 = \frac{0.95 - y_N}{y_N - 0.877}$$

Solution of this equation shows V_N to be located at $y_N = 0.895$. An operating line through L_{N+1} and V_N must intersect a vertical line through y_D and locate Δ_D in such a way as to make

$$\frac{L_{N+1}}{D} = 3.0 = \frac{(h_D + Q_D) - H_N}{H_N - h_{N+1}}$$

The over-all material-balance line from Δ_D through F locates Δ_B by intersection with a vertical line drawn downward from B at x_B. Now that both difference points have been located, the required number of stages can be stepped off. Since one equilibrium stage has already been marked off at the top of the column (partial condenser), it is more convenient to continue stepping off stages from that end. It can be seen that eight stages (including the partial condenser) are not quite sufficient to reach $x_B = 0.05$. So the number of stages required to meet the purity specifications of 5 and 95 per cent is 8+.

Example 6-2. What bottom-product composition will be obtained if 700 moles/hr of a 45 per cent feed and 100 moles/hr of an 85 per cent feed are processed in an existing column to furnish a 95 per cent overhead product? The feeds are saturated liquids, and the column has the following characteristics:

1. Total condenser and partial reboiler
2. Feed introduction points on the third and fifth equilibrium stage from the bottom
3. A top vapor capacity of 1.5 times the proposed total feed rate
4. Seven equilibrium stages including the reboiler
5. Essentially atmospheric operating pressure

Solution. It can be shown by the methods of Chap. 3 that this column has $2C + N + 12$ design variables subject to control by the operator. A convenient list of specifications is given below. Those which have been specified in the problem statement are marked with a dagger.

Specifications	$N_i{}^u$
† Pressure in each stage and condenser	$N + 1$
† Pressure in reflux divider	1
Heat leak in each stage excluding reboiler	$N - 1$
Heat leak in reflux divider	1
† Feed streams	$2C + 4$
† N, total number of stages	1
† Feed-stage locations	2
† One product purity	1
Reflux temperature	1
External reflux ratio L_{N+1}/D	1
	$2C + 2N + 12$

The heat leaks in the reflux divider and in all stages except the reboiler will be assumed to be zero. The condenser effluent will be specified to be a saturated liquid, and a reflux ratio of $L_{N+1}/D = 2.5$ will be used. The top vapor capacity of the column will not be used as a specification. Instead, the problem will be solved at the specified L_{N+1}/D, and once D is known, the required V_N will be calculated and compared with the known capacity to see if the column can actually handle the calculated rates.

The graphical solution of the example is shown in Fig. 6-10. The solution begins with the drawing of a vertical line at $x_D = 0.95$. Since a total condenser is used, y_N and x_{N+1} also fall on the vertical at 0.95. The difference point Δ_D for the top part of the column is found by measuring distances on the vertical line so that

$$\frac{L_{N+1}}{D} = 2.5 = \frac{(h_D + Q_D) - H_N}{H_N - h_{N+1}} \tag{6-8}$$

The difference point Δ_B is found by adding F_1 and F_2 to give the addition point F' and then drawing an over-all enthalpy-balance line from Δ_D through F' to the point of intersection with a vertical line through x_B. However, x_B is not known at this point in the solution, and it is necessary to assume a value for x_B in order to locate the lower and intermediate difference points. These assumed points are then used to step off the required number of stages to go from the assumed x_B to the specified x_D. If the required number equals the specified number of stages, the assumed x_B is correct. This trial-and-error procedure is necessary whenever the number of stages is specified as it always will be when an existing column is under study.

Once a Δ_B has been located by the assumption of an x_B, the intermediate difference point Δ' for the column section between the two feed stages is located by drawing lines as indicated by the following rearrangement of the over-all balance:

$$F_1 h_{F_1} - \Delta_B = \Delta_D - F_2 h_{F_2} = \Delta'$$

The subtraction of Δ_B from F_1 and F_2 from Δ_D is accomplished by drawing straight lines which intersect at Δ'.

The stages are now stepped off, starting at the assumed x_B of 0.10. The bottom difference point Δ_B is used to relate V_1 to L_2 and V_2 to L_3. The lower feed is introduced on the third stage, and therefore Δ' must be used to locate L_4 from V_3. The point Δ' is also used to relate V_4 to L_5. The 85 per cent feed is introduced on stage

5, and this requires a change in the difference point from Δ' to Δ_D which is used for the remaining two stages. Note that the tie line for stage 7 overshoots the specified x_D slightly. This indicates that the assumed value of 0.1 for x_B was slightly high. A subsequent trial showed that $x_B = 0.096$ is the correct value. An over-all material

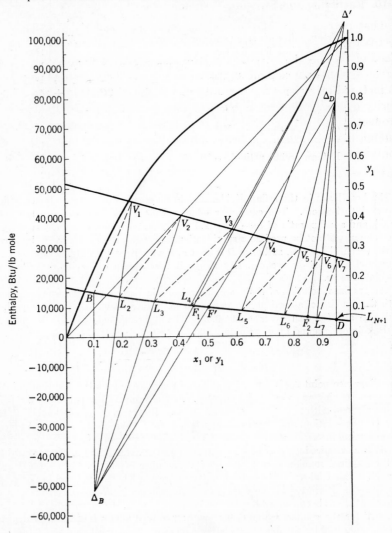

FIG. 6-10. Solution to Example 6-2.

balance with $x_D = 0.95$ and $x_B = 0.096$ shows that $D = 378.5$ and $B = 421.5$ on the basis of 800 moles/hr of total feed. The required top vapor rate is

$$V_N = D(R + 1) = 378.5(3.5) = 1325 \text{ moles/hr}$$

The allowable vapor rate is only $1.5(800) = 1200$ moles/hr. The solution obtained in Fig. 6-10 is therefore not possible in this particular column. It would be necessary to make other trials at lower reflux ratios until that reflux is found which loads the column to capacity but does not exceed the flooding point.

6-10. Rectifying and Stripping Columns

That part of a distillation column above the feed stage is called the rectifying section, while the part below is termed the stripping section. Each of these sections (plus the feed stage) can be used alone to form either a rectifying column as shown in Fig. 6-11 or a stripping column as pictured in Fig. 6-12. A rectifying column is practical when a high-purity overhead product is required and contamination of the bottoms product with light material can be tolerated. Conversely, a stripping column is useful when the heavy product must have a high purity and contamination of the overhead product with the heavier material can be

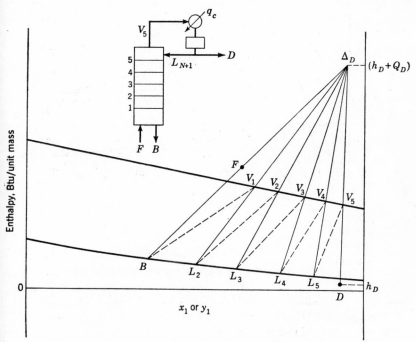

Fig. 6-11. Construction for a five-stage rectifying column.

tolerated. The feed to a rectifying column must contain some vapor, while the feed to a stripping section must be at least partially liquid if the columns are to be operable. Usually a superheated vapor is fed to a rectifier and a subcooled liquid to a stripper.

The construction for rectifying and stripping columns is relatively simple on an enthalpy-concentration diagram. The thermal condition of the feed determines the position of the point F on a vertical at x_F or y_F. There is only one column section and therefore only one difference point. The over-all enthalpy balances are

$$FH_F - Bh_B = Dh_D + q_c = D(h_D + Q_D) = \Delta_D \qquad (6\text{-}12)$$

for a rectifying column and

$$Fh_F - V_N H_N = Bh_B - q_r = B(h_B - Q_B) = \Delta_B \qquad (6\text{-}13)$$

for a stripping column. Equations (6-12) and (6-13) together with the individual stage balances and equilibrium relationships are illustrated in

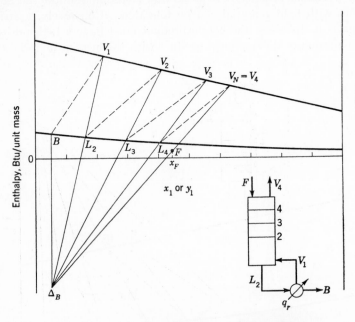

Fig. 6-12. Construction for a four-stage stripping column.

Figs. 6-11 and 6-12, respectively. Note in Fig. 6-11 that if the feed is superheated enough, it is possible for V_1 to have a lower concentration of the more volatile component than does F. Similarly, in Fig. 6-12, if the feed is subcooled enough, it is possible for L_4 to lie to the right of F. These irregularities in the concentration profiles would be caused by the large changes in the vapor and liquid rates across the feed stage when the thermal condition of the feed is extreme.

6-11. Total Reflux

The condition of total reflux exists when the entire overhead vapor stream from the top stage is condensed and returned to that stage as liquid reflux. No overhead product is withdrawn and $L_{N+1} = V_N$. A distillation column can be operated under conditions of total reflux with or without a feed stream. Each of these situations is discussed separately below.

Total Reflux—No Feed. Assume that the feed and product streams on a continuous column which has been operating at steady state are simultaneously blocked. The reflux is adjusted to maintain the level in the accumulator, and the reboiler and condenser loads are adjusted to maintain the enthalpy balance. When the column regains the steady

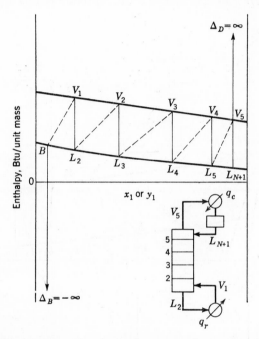

Fig. 6-13. Graphical construction for a five-stage column operating under total reflux throughout its length.

state, it is operating at total reflux throughout its length. All the vapor entering the condenser is condensed and returned to the column as reflux. All the liquid entering the reboiler is vaporized and returned to the column. A material balance around either end of the column and cutting between any two stages will show that the two streams in any pair of passing streams in the column are equal in magnitude and identical in composition. Their temperatures and pressures are not identical, but $L_{n+1} = V_n$ and $x_{n+1} = y_n$, where n is any stage in the column. The identity between x_{n+1} and y_n requires that L_{n+1} and V_n fall on the same vertical line, or in other words, the operating lines are vertical. This fact is also illustrated by an enthalpy balance which, since $D = 0$ and $Q_D = q_c/D = \infty$, reduces to

$$V_n H_n - L_{n+1} h_{n+1} = \infty$$

Since $F h_F$ is zero, the over-all balance reduces to

$$\Delta_D = -\Delta_B$$

and since $\Delta_D = \infty$, $\Delta_B = -\infty$. The graphical construction for a column under total reflux throughout its length is shown in Fig. 6-13.

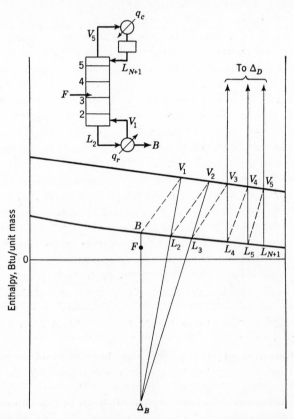

Fig. 6-14. Graphical construction for a column with total reflux and continuous feed and bottom streams.

Total Reflux with Feed. All the overhead vapor may be condensed and returned to the column as reflux ($D = 0$) while feed is continuously introduced at the feed stage. Since $D = 0$, the rate and composition identities must prevail between all passing streams above the feed stage, or in other words, the rectifying section is operating under total reflux conditions. Passing streams in the stripping section, however, are not identical. The withdrawal of a bottoms stream requires that $L_{m+1} \neq V_m$ and $y_m \neq x_{m+1}$, where m refers to any stage below the feed stage.

The graphical construction for this operation is shown in Fig. 6-14. Except for the fact that $x_B = x_F$, the construction below the feed is normal. Above the feed stage, the difference point Δ_D lies at $+\infty$ and the operating lines become vertical.

6-12. Minimum Reflux Ratio

The minimum reflux ratio is defined as that reflux ratio which, if decreased an infinitesimal amount, will require an infinite number of stages to accomplish a specified separation between two components. Two facts should be noted carefully in this definition. First, the separation between *two* components must be specified. This can be done by specifying the split on each of the two components between the overhead- and bottoms-product streams. It is not sufficient to specify only one concentration or the split on only one component because this specification could be met by taking more or less overhead product at any reflux ratio. Second, the number of stages in the column must *not* be specified. The concept of minimum reflux as defined above obviously has no meaning for an existing column. True, there will be a certain reflux rate below which it is impossible to effect a desired separation between two components, but this reflux rate is associated with the finite number of stages in the column and not with an infinite number of stages.

It can be seen from the preceding paragraph that the discussion of the concept of minimum reflux for binary systems must be restricted to those problems where both x_D and x_B are specified and the number of stages is not specified. Also, it is necessary to specify that the feed be introduced in the optimum location, since otherwise it would be possible to require an infinite number of stages at any reflux by simply not introducing the feed until an infinite number of stages had been stepped off with one difference point. If one difference point Δ_D or Δ_B is used for a sufficient number of stages, the operating lines will finally coincide with a tie line and no further progress can be made regardless of the number of stages.

The true minimum reflux for any specified x_D and x_B can be found by extrapolating the tie line which passes through the feed point F until it intersects the vertical line at x_D. In systems with normal volatility (curve A in Fig. 5-1) this intersection gives the lowest possible position for Δ_D, since at this point the operating lines will become coincident with the extrapolated tie line and a zone of constant composition (no change in composition from stage to stage) will develop around the feed stage. In systems with abnormal volatility (such as curve B in Fig. 5-1) some tie line other than the one through the feed will give the highest intersection on the vertical at x_D or the lowest intersection on the vertical at x_B and therefore control the position of Δ_D at which the operating lines first become coincident with a tie line. To find the controlling tie line, those tie lines to the right of the feed are extrapolated to x_D while the tie lines

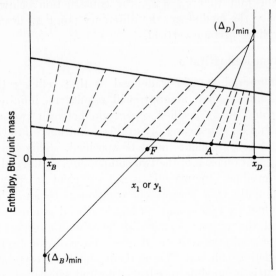

FIG. 6-15. Determination of the true minimum reflux by extrapolation of tie line. Tie line through A gives highest intersection at x_D and therefore determines the minimum reflux ratio.

to the left are extrapolated to x_B. Figure 6-15 illustrates the construction when a tie line to the right of the feed gives a higher intersection than the tie line through the feed.

6-13. Stage Efficiency

The manner in which the calculated number of equilibrium stages (100 per cent efficient) is converted to the required number of real stages through the concept of stage efficiency has been discussed briefly in Sec. 5-13 and is covered more thoroughly in the last chapter in the book. The concept of efficiency is mentioned now only to permit the use of the enthalpy-concentration diagram to illustrate the Murphree-stage efficiency.

The Murphree-stage efficiency can be defined in terms of either the vapor or the liquid phase. The vapor-phase efficiency is most convenient when the stage-to-stage calculations begin at the bottom. The Murphree-stage efficiency in terms of vapor-phase concentrations is

$$\frac{y_n - y_{n-1}}{y_n^* - y_{n-1}}$$

where y_n and y_{n-1} are the actual concentrations in the vapors rising from stages y_n and y_{n-1} while y_n^* represents the concentration which would exist in the vapor from stage n if the vapor were in equilibrium with the liquid overflow from that stage.   Figure 6-16 illustrates the application of a Murphree efficiency of 80 per cent to the first few stages above the reboiler.   Since the reboiler is usually considered to have an efficiency of 100 per cent, V_1 lies at the other end of the tie line which passes

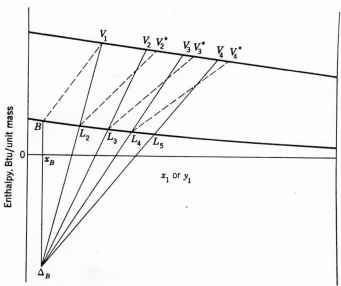

Fig. 6-16. Illustration of Murphree vapor-phase efficiency on enthalpy-concentration diagram.   Illustrated efficiency = 80 per cent.

through B.   An operating line from V_1 to Δ_B locates L_2.   If stage 2 were 100 per cent efficient, V_2 would coincide with y_2^*, which lies at the other end of the tie line through L_2.   The efficiency of 80 per cent is applied by locating y_2 (or V_2) at a point 80 per cent of the distance from y_1 to y_2^*.   The operating line is then drawn from the actual y_2 (not the equilibrium value) to Δ_B to locate L_3.   The 80 per cent efficiency factor is applied in this manner to each of the succeeding stages.

NOMENCLATURE

B = bottoms product rate; moles, weight, or volume per unit time.
C = number of components in system.
D = total overhead product rate; moles, weight, or volume per unit time. D_L
    refers to the liquid portion; D_V to the vapor portion.
F = feed rate; moles, weight, or volume per unit time.
H = vapor enthalpy, Btu/unit mass.
h = liquid enthalpy, Btu/unit mass.
i = subscript referring to component i.

L = heavy phase rate; moles, weight, or volume per unit time. L_{N+1} refers to reflux rate.

m = subscript referring to any stage below feed stage.

n = subscript referring to any stage above the feed stage. In some instances n may refer to any stage in the column.

N = total number of stages in the column including the partial reboiler if any but excluding the partial condenser if any.

$N_i{}^u$ = independent design variables in the process. See Chap. 3.

q_c = condenser heat load, Btu/unit time.

q_r = reboiler heat load, Btu/unit time.

q_s = heat of solution per unit mass of solution at the composition x_1 and x_2.

$Q_D = q_c/D$ = heat removed in the condenser per unit mass of D.

$Q_B = q_r/B$ = heat added in the reboiler per unit mass of B.

S = side-stream product rate; moles, weight, or volume per unit time.

V = light phase rate; moles, weight, or volume per unit time.

x = concentration in heavy phase. Units consistent with units on rates. Subscript refers to stage or stream in which concentration appears. Subscripts 1 and 2 also used to refer to components 1 and 2.

y = concentration in heavy phase. Units consistent with units on rates. Subscript refers to stage or stream in which concentration appears. Subscripts 1 and 2 also used to refer to components 1 and 2.

y^* = equilibrium vapor concentration. Used when y refers to actual concentration.

Greek Symbols

$\Delta_B = B(h_B - Q_B)$ = difference point for bottom section of column.

$\Delta_D = D(h_D + Q_D)$ = difference point for top section in column.

Δ' = difference point for intermediate section in column.

PROBLEMS

6-1. For the enriching column shown here determine the amount of heat which is removed from the partial condenser per mole of distillate product when the vapor product contains 0.9 mole fraction of A, the vapor into the condenser 0.75 mole fraction of A, and the saturated vapor feed to the bottom plate 0.47 mole fraction of A. How many theoretical trays are required in the column to effect the separation? What is the composition of the bottom product? Assume a relative volatility of 4.0. The enthalpies of the vapor and liquid phases can be represented by straight lines between 15,000 and 0 Btu/lb mole for the saturated vapor and liquid of pure B to 10,000 and 0 Btu/lb mole for the saturated vapor and liquid of pure A.

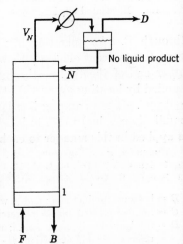

6-2. A column supplied with a saturated liquid feed containing 33 mole per cent ethane and 67 mole per cent propane is to produce an overhead product containing

97 per cent ethane and a bottoms product containing 4 per cent ethane. The operating pressure is to be 170 psia, and the reflux will be returned to the column as a saturated liquid.

(a) Using the enthalpy data in Maxwell's "Data Book on Hydrocarbons" and assuming ideal gas and liquid solutions, construct the enthalpy-concentration diagram.

(b) Determine the minimum refluxes L_{N+1}/V_N and L_{N+1}/D, the minimum stages (total reflux), and the number of equilibrium stages required for a reflux rate of 1.3 times the minimum.

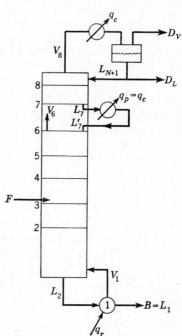

6-3. For a saturated feed of 52.5 per cent A and a D_V of 90 per cent A, calculate the following items when $D_V = D_L = 1.0$ and $L_{N+1} = D_V + D_L$. Use a relative volatility of 2.4, and assume that the saturated-vapor and -liquid curves can be represented by straight lines between 900 and 200 Btu/lb at 0 per cent A and 570 and 75 Btu/lb at 100 per cent A (see the accompanying figure).

(a) What is the condenser load per pound of total overhead product $D_V + D_L$?

(b) Calculate the L_7, L_7', and L_6 rates (basis is $D_V = 1.0$).

(c) What is x_B?

(d) What are the feed and bottom rates relative to $D_V = 1.0$?

(e) What is the reboiler load per pound of bottom product?

6-4. Rework Prob. 5-4 using enthalpy-concentration diagrams. Assume that the saturated-vapor and -liquid curves are straight and parallel.

REFERENCES

1. Maxwell, J. B.: "Data Book on Hydrocarbons," 1st ed., p. 92, D. Van Nostrand Company, Inc., Princeton, N.J., 1950.
2. Ponchon, M.: *Tech. moderne*, **13**, 20, 55 (1921).
3. Savarit, R.: *Arts et métiers*, 1922, pp. 65, 142, 178, 241, 266, 307.

CHAPTER 7

TERNARY EXTRACTION DIAGRAMS

Separation between components with dissimilar molecular structures can be accomplished without vaporization by means of liquid-liquid extraction. A new component (solvent) in which the original components exhibit different degrees of solubility is added to the mixture. The solvent must be partially immiscible with at least one of the original components. Two liquid phases are formed, and all the components (including the solvent) distribute themselves between the two phases in accordance with their respective distribution coefficients. One of the phases will contain a high concentration of the solvent, and this solvent-rich phase is termed the *extract*. The other phase is called the *raffinate*. Either the extract or the raffinate phase may contain that particular component which is most valuable.

Liquid-liquid extractions can be performed with any number of components and more than one solvent. If more than one solvent is used, they can be used simultaneously or in sequence. This chapter will be restricted to the separation of binary mixtures with a single solvent. Calculation methods for systems with more than three components are described in Chap. 12.

Liquid-liquid extraction is analogous to vapor-liquid processes such as distillation and absorption in that separation is achieved through the tendency of the individual components to distribute themselves in a unique manner between the two phases. Since the components remain in the liquid state in extraction, heats of vaporization are not involved. The only heat effects are heats of solution, and although these are much smaller than heats of vaporization, they may be sufficiently large to cause a temperature change as the liquid compositions vary. Solubilities are closely related to temperature, and therefore heat-of-solution effects can affect the distribution coefficients. However, it is usually satisfactory to neglect heat effects and assume that the extraction will be carried out isothermally. Or a temperature gradient can be arbitrarily imposed upon the extraction system in order to take full advantage of the solubility relationships. All the examples shown in this chapter assume an isothermal system. The extension of the illustrated principles to non-isothermal systems will be obvious.

Extraction processes involve one or more contacts between a solvent (or solvent-rich stream) and the raffinate material. Each contact will approach an equilibrium stage to a greater or lesser degree depending upon the degree of mixing, contact time, and physical properties such as viscosity, molecular weight, molar volumes, etc. It is therefore convenient to treat the extraction process in terms of equilibrium stages in a manner analogous to that followed in distillation. The principles developed for the vapor-liquid processes can be applied to the calculation of extraction problems. The transfer of principles from one process to the other is facilitated if a common nomenclature is used. Since the distillation nomenclature is well established, it is convenient to retain the same symbols for extraction. One arbitrary decision must be made in this transfer of nomenclature, and that is whether the extract or the raffinate phase in extraction is analogous to the liquid phase in distillation. As long as only binary extraction systems (or ternary systems if the solvent is included) are considered, it is convenient to consider the extract and vapor streams as analogous because of the rather complete analogy (4, 5) which can be built between the solvent in extraction and heat in distillation. This analogy is particularly strong when the enthalpy-concentration diagram for distillation (Chap. 6) is compared with the mass-ratio diagram for extraction which will be described later in this chapter. In these two graphical constructions, enthalpy and solvent are treated in an entirely analogous manner. The solvent-rich phase (extract) is analogous to the enthalpy-rich phase (vapor), and a similar analogy exists between the raffinate (extraction) and liquid (distillation) phases. For multicomponent systems (more than two components plus a solvent) for which graphical techniques are less useful, the analogy between solvent and heat is of little value. The analogy is not useful in any of the design calculations and adds very little in the understanding of the solubility relationships. When all the common equilibrium-stage processes such as distillation, absorption, extraction, stripping, washing, extractive distillation, etc., are considered, it is more reasonable to say that the extract phase in the liquid-liquid processes is analogous to the liquid phase in the vapor-liquid processes. The solvent in extraction is more analogous to the lean oil than to the fresh vapor feed in absorption. Likewise, the extract phase in extraction is more analogous to the solvent-rich phase in extractive distillation than to the vapor phase. Greater uniformity among the various processes is obtained if the symbols L and x are applied to the extract phase while V and y are used for the raffinate phase.

Since all the vapor-liquid processes are numbered from the bottom in this text, it will be necessary to number extraction processes from the extract end. The extract end may or may not be the bottom of the column (if a column is used), depending upon the relative densities of the extract and raffinate phases.

7-1. Equilibrium Data and Diagrams

Extraction systems can be characterized by the number of partially miscible pairs which they contain. Since this chapter will deal only with ternary liquid systems, the only possible systems are those which contain one, two, or three partially miscible pairs. These systems are known as type I, type II, and type III systems, respectively. Many extraction systems will change from one type to another as the temperature is raised or lowered. This phenomenon is discussed more fully in the following section. For the present it is sufficient to say that at any given temperature, a system can be conveniently characterized by saying it is of a certain type.

Type I systems are most often used, since it is usually desirable that the solvent be completely miscible with one component (or class of components) and as immiscible as possible with the raffinate material. Almost complete separation can be obtained in these systems with only a few stages, since the solvent has a high selectivity for one component and a low selectivity for the other. If the solvent is only partially miscible with both the raffinate and extract materials (type II), the difference in selectivity is much smaller and more stages will be required. Type II systems can often be converted to type I systems by a change in temperature. Likewise, type III systems can usually be converted to type II or even type I systems if desired.

Three graphical methods for the representation of ternary liquid-liquid equilibrium data and solution of ternary design problems are in widespread use. These are the triangular diagram (1), the mass-ratio diagram (2), and the distribution diagram. The first two involve the same construction principles used in the enthalpy-concentration diagram (Chap. 6), while the distribution diagram is more akin to the x-y diagram (Chap. 5). All three diagrams serve the same basic purposes, and the only excuse for the development of more than one method is the fact that the graphical construction in particular problems may be jammed into a small portion of one type of diagram whereas on another diagram the construction lines are spread and easier to handle.

Equilibrium diagrams for liquid-liquid systems are always drawn at constant temperature. The phase rule stipulates that for a three-component system with three phases (two liquid and one vapor) there are two degrees of freedom. Specification of the system temperature and one-phase concentration defines one unique state of the system. The system pressure will be the vapor pressure of the two liquid phases. Most equilibrium data in liquid systems, however, are not obtained at the system vapor pressure. If the vapor pressure of the system is less than 1 atm, the tie-line determinations are normally made in the presence of air at essentially atmospheric pressure. The air can be considered as a fourth component which gives the third degree of freedom and permits the experimenter to "fix" the pressure. Or, if the system pressure is maintained greater than the vapor pressure and an inert gas such as air is excluded, no vapor phase will exist and three degrees of freedom are still available to the experimenter. In cases where the vapor pressure is greater than atmospheric pressure, an air pocket can be maintained in the equilibrium cell to allow for thermal expansion of the liquid. If so, four components and three phases are present to give the experimenter

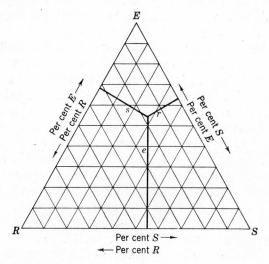

FIG. 7-1. Representation of ternary composition data on an equilateral triangle.

the same three degrees of freedom. The presence of air and moderate changes in system pressure usually have an entirely negligible effect on the compositions of the equilibrium liquid phases and are unimportant unless it is desired to determine the activity coefficients from measurements of the total pressure and equilibrium vapor compositions.

Equilateral-triangle Diagrams. Ternary liquid-liquid equilibrium data are conveniently plotted on an equilateral triangle such as the one shown in Fig. 7-1. The equilateral triangle has the property that the sum of the lengths of the three perpendiculars drawn from an interior point to the three sides must equal the altitude. Therefore, if each of the three altitudes is divided into 100 divisions and numbered from zero at the base to 100 at the apex, it is possible to represent the composition of any ternary mixture by a point within the diagram. An apex is assigned to each component. Then each of the perpendiculars to a given point represents the percentage concentration of a given component and the concentrations of the three components sum to 100. This fact is illustrated in Fig. 7-1. The raffinate component is designated by R and assigned the lower left-hand apex. The extract component is E and is represented by the upper apex. The letter S is used to denote the solvent, which will always be represented by the lower right-hand apex. Lines parallel to each of the sides are drawn at regular intervals

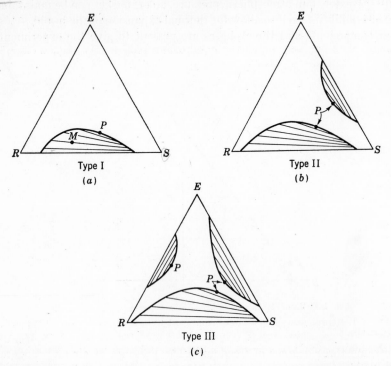

Fig. 7-2. Isothermal representation of types I, II, and III systems on the equilateral triangle.

to provide a grid on which distances can be easily measured. The grid can be used to determine the length of the three perpendiculars E, R, and S and show that the sum is 100.

Figure 7-2a, b, and c illustrates types I, II, and III systems on the equilateral-triangle diagram. The curved lines within the triangles are solubility envelopes and separate the one-liquid-phase regions from the regions of immiscibility. For systems of design interest a region of immiscibility will always open to at least one side of the triangle which represents a partially immiscible binary pair. Straight lines can be drawn across an immiscible region to connect points which represent equilibrium-phase compositions. These lines are called tie lines. A mixture represented by point M in the type I diagram will separate into two layers whose compositions at equilibrium are represented by the two intersections which the tie line through M makes with the solubility envelope. There are an infinite number of tie lines in any immiscible area, and they may have a negative or positive slope depending upon the particular system under consideration. One tie line will be coincident with the side of the triangle and represents the binary solubilities. On the other side of the immiscible area, the tie lines become shorter and shorter as the compositions of the two equilibrium phases approach each other. The point at which the equilibrium phases become coincident is

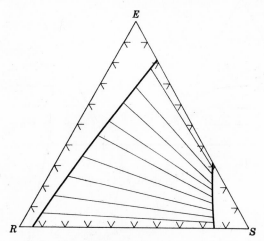

Fig. 7-3. Type II system with only one region of immiscibility.

called the *plait point* and is represented by P in Fig. 7-2. The position of the plait point depends upon the slope of the tie lines and may fall at the highest point on the solubility envelope or on either side of it.

Types II and III systems as shown in Fig. 7-2 are of little interest as far as the principles of graphical calculation are concerned. The graphical construction for any one separation device will be confined to one region of immiscibility, and the only effect of multiple regions on design lies in the fact that the designer may have a choice as to which area the process will utilize.

Type II systems which are important both industrially and in the development of graphical design procedures can be formed from the type II system in Fig. 7-2 by a change of temperature which causes the two immiscible regions to merge and form a band of immiscibility across the diagram as shown in Fig. 7-3. If the plait points do not come together, it is possible for a three-liquid-phase region to form in type II systems. Three-liquid-phase regions can also be formed in type III systems, but such systems are of little practical importance. Type I systems as depicted in Fig. 7-2 and type II systems as illustrated in Fig. 7-3 represent the vast majority of systems of practical importance, and the remainder of the chapter will be devoted to their consideration.

The properties of the triangle provide a convenient means of interpolating and extrapolating the equilibrium data when only a few tie lines have been determined experimentally. The construction is shown in Fig. 7-4 for a type I system. Lines parallel to one side of the triangle

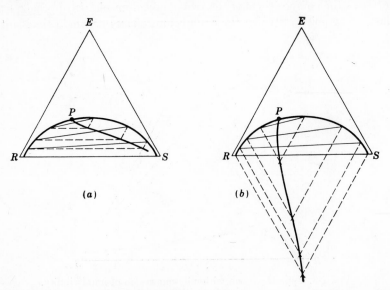

FIG. 7-4. Construction of conjugate line for interpolation and extrapolation of available tie-line data.

are drawn from the R-rich ends of the known tie lines to intersect similar lines which pass through the S-rich ends and are parallel to a different side. A curve drawn through the intersections thus formed is the locus of all possible intersections and is called a conjugate line, since the intersecting lines connect conjugate phases. The conjugate line, of course, passes through the plait point. By proper choice of the sides to which the lines are to be drawn parallel, the conjugate line can be made to fall outside or inside the triangle. That construction which produces the straightest conjugate line is preferred.

The simplest method for interpolating and extrapolating the tie-line data for both type I and II systems is probably the distribution diagram. The distribution diagram is similar to the equilibrium curve in the x-y

(McCabe-Thiele) diagram. An "equilibrium" curve for each of the three components can be produced by plotting the raffinate concentration vs the extract concentration or vice versa. Such a plot is given in Fig. 7-5, which shows the three distribution curves which correspond to the tie lines and solubility curve drawn in Fig. 7-4. Any one of these three curves can be used to find new tie lines in a manner similar to the way the equilibrium curve in the x-y diagram for distillation is often used to locate new tie lines on the enthalpy-concentration diagram.

Many other schemes for the extension of tie lines have been suggested. For a review of these the reader is referred to Treybal (5).

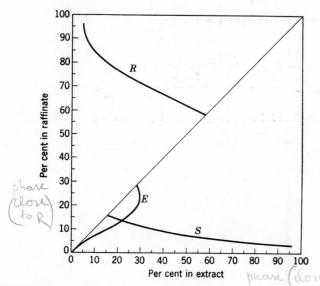

phase (close to R)

phase (close to S)

FIG. 7-5. Distribution curves for components R, E, and S in the typical type I system shown in Fig. 7-4.

Right-triangle Diagrams. The concentration scales on the equilateral-triangle diagram are not perpendicular to one another, and therefore it is somewhat tedious to locate compositions on such a diagram. The concentration scales can be made much easier to read if a right triangle is used. Figure 7-6 shows how the solubility envelope and tie lines of the type I system in Fig. 7-2 appear when plotted on rectangular coordinates. Figure 7-7 shows a replot of the type II system from Fig. 7-3. In each case, the right-triangle figure can be considered to be formed by warping the equilateral triangle to make the RE side vertical. Each component is still represented by an apex of the triangle, but now all compositions can be represented by two mutually perpendicular scales. Any two of the three components can be utilized for the two concentra-

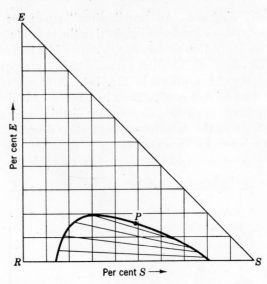

FIG. 7-6. Replot of the type I system in Fig. 7-2 on rectangular coordinates.

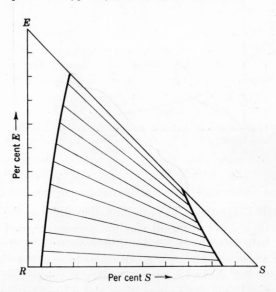

FIG. 7-7. Replot of the type II system in Fig. 7-3 on rectangular coordinates.

tion scales.   In this text the ordinate scale will always represent concentrations of the extract component E while solvent concentration is plotted on the abscissa.   The rules for the graphical solution of extraction problems apply equally well to both the right- and equilateral-triangle diagrams.   However, because of its greater convenience, only the right triangle will be used in the subsequent discussions.

Mass-ratio Diagrams. Janecke (2) suggested a diagram for the representation of liquid-liquid equilibrium data which is analogous to the enthalpy-concentration diagram described in Chap. 6. The amount of solvent per unit mass of nonsolvent material

$$Y_S \text{ or } X_S = \frac{\text{amount of } S}{\text{combined amounts of } R \text{ and } E}$$

is plotted as the ordinate on rectangular coordinates vs the composition on a solvent-free basis on the abscissa,

$$Y_E \text{ or } X_E = \frac{\text{amount of } E}{\text{combined amounts of } R \text{ and } E}$$

The composition of the solvent-free material could be expressed in terms of Y_R and X_R, but the extract component E will be used in this text. It can be seen that X_S will take on values from zero for a solvent-free stream to infinity for a pure solvent stream. Y_R, Y_E, X_R, and X_E can all take on values from 0.0 to 1.0, and it is always necessary that

$$Y_R + Y_E = 1.0$$
and
$$X_R + X_E = 1.0$$

The amounts of R, E, and S used to calculate the ratios Y and X can be measured in terms of weight, moles, or volume. The only restriction is that the stream rates be expressed in the same units. If volume units are used, the phase rates must be expressed as summations of pure component volumes. In other words, the volumetric flow rates are calculated from weight or molar rates through the use of the pure component densities at some selected temperature. The volumetric flow rates thus obtained will not correspond exactly to the actual flow rates, which are affected by the changes in volume due to mixing and temperature changes. The only time that the designer needs to convert his "ideal" volumes (which represent the sum of the individual mass rates multiplied by constant conversion factors) to actual volumes is in the sizing of mechanical equipment. For example, the actual volumetric flow rate must be known in sizing flow meters and the liquid-handling devices in a contact stage.

The appearance of a type I and II system on the mass-ratio diagram is shown in Fig. 7-8. The solubility envelope which encloses the immiscible region takes on the shape of a horizontal band which reaches all the way across the diagram in a type II system. Outside the band only one liquid phase exists. The top of the region where two liquid phases exist is bounded by a line which represents all possible solvent-rich phase (extract) compositions, while the bottom is bounded by a line which represents all possible raffinate compositions. Tie lines can be drawn across the immiscible region to connect extract and raffinate phases in equilibrium. The diagrams are always drawn at some specific temperature. At a given temperature, the system vapor pressure must change as the composition varies along the solubility envelope. However, as explained above under Equilateral-triangle Diagrams the equilibrium data are usually taken with a fourth component (air) present or without any vapor phase present, which permits the experimenter to hold the pressure constant.

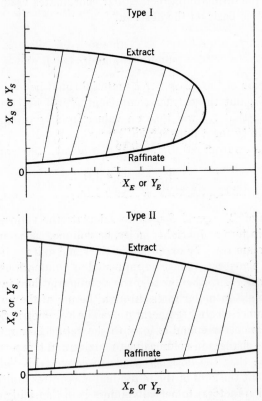

FIG. 7-8. Typical appearance of types I and II systems on the mass-ratio diagram.

The location of new tie lines from interpolation or extrapolation of a few experimentally determined tie lines is best accomplished through use of the conjugate line construction or by use of the distribution curves. Both of these methods have been described above under Equilateral-triangle Diagrams. It may be convenient to plot a distribution curve (on a solvent-free basis) directly on the mass-ratio diagram for interpolation purposes just as the x-y equilibrium curve is plotted in Fig. 6-2 for use with the enthalpy-concentration diagram in distillation. On a solvent-free basis, the distribution diagram for type II systems has the same form as the binary equilibrium curve in vapor-liquid systems.

Sources of Data. Extensive bibliographies of solubility and tie-line data sources have been compiled by the Petroleum Refinery Laboratory at Pennsylvania State University, University Park, Pa. (*Rept.* PRL-6-57, December, 1957) and by Himmelblau et al. (Bureau of Engineering Research, University of Texas, Austin 12, Tex., *Spec. Publ.* 30, 1959). References to data published since these reports can be found in the extraction section of the Unit Operations Review in *Industrial and Engineering Chemistry* each year.

The designer is often faced with the task of selecting a solvent for some particular separation for which no equilibrium data have been obtained. Treybal (5) discusses the use of the concept of internal pressure to estimate the relative attraction which a proposed solvent will have for each of the components to be separated. Pierotti et al. have presented an empirical equation for the prediction of infinite dilution activity coefficients in binary systems. A comparison of these activity coefficients for each of the components in the proposed solvent indicates which is more like the solvent and therefore more likely to concentrate in the solvent phase. This method is discussed more thoroughly in Sec. 11-1.

The integrated forms of the Gibbs-Duhem equation can be used to correlate experimental solubility in extraction systems containing three or more components. Example 2-2 illustrates how the constants in a two-constant equation can be obtained from binary solubility data. The constants for miscible binaries must be obtained from a fit of the ternary solubility data. The distribution coefficients are related to the binary constants and the equilibrium-phase compositions by Eq. (2-56). The application of this equation to the correlation and calculation of tie-line data is discussed briefly at the beginning of Chap. 12.

7-2. Effect of Temperature

A change in temperature will cause the two-liquid-phase region in the equilibrium diagrams to increase or decrease in size. A typical example is shown in Fig. 7-9. If the change in temperature is large enough, the system may change from one type to another as shown in Fig. 7-10. The slope of the tie lines often changes appreciably as the solubility envelope expands or contracts in size. These effects make temperature an important process variable in extraction. Ideally, an extraction process should be operated at that temperature where the solvent is almost completely immiscible with the raffinate material but still miscible with the extract material. Often, even though the system can be made to approach this condition at some temperature, that temperature

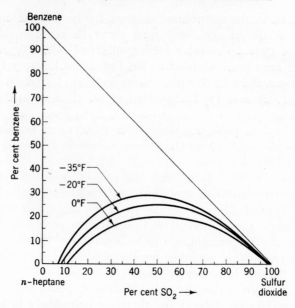

FIG. 7-9. Effect of temperature on the solubility envelope in the *n*-heptane–benzene–sulfur dioxide system. [*Data from Satterfield et al., Ind. Eng. Chem.*, **47**, 1458 (1955).]

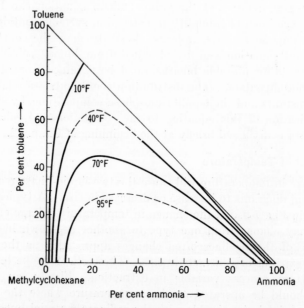

FIG. 7-10. Effect of temperature on the solubility envelope in the methylcyclohexane–toluene–ammonia system. System changes from type I to type II between 40 and 10°F. [*Data from Fenske et al., A.I.Ch.E. J.*, **1**, 335 (1955).]

may lie outside the range of economical operation. It may be so low as to require more expensive materials of construction and expensive cooling equipment which cannot be justified by the increased efficiency of the extraction process. Or the temperature may be so high as to require an excessive pressure to maintain the system in the liquid phase. However, the effect of temperature within the range of practical operation should be investigated thoroughly, since temperature can have a large effect upon ease of separation and recovery.

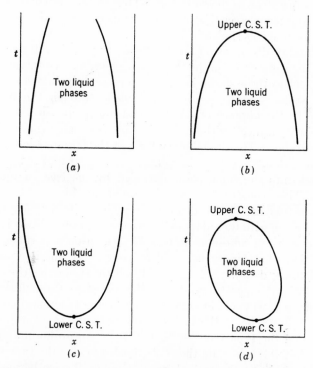

FIG. 7-11. Four general types of binary solubility curves in the stable liquid region, C.S.T. is critical solution temperature.

The effect of temperature on a ternary system is largely determined by the binary t-x diagrams for the partially miscible binary pairs. The solubility curves on a t-x diagram for a partially miscible binary can have any of the four general configurations shown in Fig. 7-11. Each of the four diagrams is concerned only with the temperature range in which both components exist as liquids. Figure 7-11a represents systems in which the partial immiscibility extends over the entire range. The curves as drawn show increasing solubility with increasing temperature and indicate that if the system pressure were high enough to maintain the liquid state, the components would become completely miscible at some high temperature. If the sign of the slope of each curve were changed, complete miscibility at some low temperature would be indicated if the transformation to the solid state could be prevented in some manner. The temperature at which the two solubility curves meet is known as the *critical solution temperature*. In many systems the solubility curves meet at a temperature whree the liquid states are stable to give an upper critical solution temperature as shown in Fig. 7-11b or a lower critical solution temperature as shown in Fig. 7-11c. A few systems exhibit both an upper and a lower critical solution temperature in the stable liquid temperature range as shown in Fig. 7-11d.

The solubility diagrams for the three binary pairs in a ternary system represent the three faces on the ternary t-x diagram. Isothermal ternary diagrams such as those shown in Fig. 7-2 represent the intersections of the solubility surfaces with an isothermal plane through the ternary t-x diagram. In the absence of ternary data much can be deduced about the solubility surfaces within the diagram from an inspection of the binary diagrams. The three binary faces of the ternary t-x diagram can be studied without resort to three-dimensional drawings if the faces are folded back into the same plane occupied by the base of the ternary diagram. This construction is illustrated in Fig. 7-12. Common isotherms across the binary diagrams represent the intersections of an isothermal plane with the sides of the ternary diagram. If the isothermal plane does not cut across an immiscible region on a particular binary diagram, then that binary pair will be completely miscible on the isothermal ternary diagram. In Fig. 7-12 all three binaries exhibit upper critical solution temperatures. An isothermal plane through the ternary t-x diagram at t_1 will cut through a binary region of immiscibility on the R-S side only. The isothermal ternary diagram at t_1 will have the appearance of the type I system in Fig. 7-2. An isothermal plane at t_2 cuts an immiscible region on the E-S side also to give the type II system in Fig. 7-2. The isothermal plane at t_3 gives the isothermal ternary diagram shown on the triangular base which is the same as the type III system shown in Fig. 7-2.

The isotherms through the diagram show only on what faces the immiscible regions will appear. It is quite possible that at t_2 or t_3 the two larger solubility envelopes will have merged to give a band of immiscibility which extends across the diagram as shown in Fig. 7-3. Another possibility on which the binary data shed no light is that an upper ternary critical solution temperature may exist at some point within the diagram. In such a situation, successively lower isothermal planes will first intersect the solubility surfaces entirely within the diagram. The corresponding isothermal ternary diagram will show a roughly circular region of immiscibility which does not open on any of the three sides.   The exact location of the solubility surfaces within the diagram is impossible without resort to experimental data.   The surfaces can be located approximately at some given temperature with one of the integrated forms of the Gibbs-Duhem equation (Chap. 2) if sufficient binary

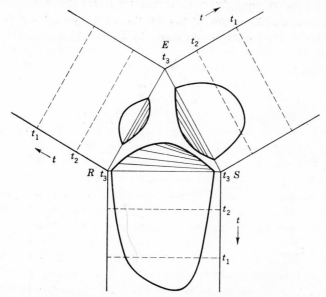

FIG. 7-12. Two-dimensional representation of the three binary faces of a ternary t-x diagram.   The triangular composition base represents the isotherm at t_3.   All three binaries exhibit upper critical solution temperatures.

equilibrium data are available to permit the evaluation of the equation constants at that temperature.

7-3. Lever-arm Rule

The solution of ternary extraction problems on the triangular or mass-ratio diagrams is based upon the *lever-arm rule*. This rule permits material and energy balances to be made with straight lines when a proper coordinate system is used. The application of the rule to enthalpy-concentration diagrams for distillation has been described in Sec. 6-1. The equations derived there apply directly to the triangular and mass-ratio diagrams after appropriate changes in nomenclature. The corresponding development in the mass-ratio nomenclature is shown below.

Mass-ratio Diagram. Consider any mixture F whose composition falls within the two-liquid-phase region. This mixture will divide into an extract phase L and a raffinate phase V. The composition of L and

V must be expressed on a solvent-free basis for representation on the abscissa of the mass-ratio diagram (see Fig. 7-8). Let $\hat{F}$, $\hat{L}$, and $\hat{V}$ represent the solvent-free portions of F, L, and V. A component E balance on a solvent-free basis can then be written as

$$\hat{L}X_E + \hat{V}Y_E = \hat{F}Z_E$$

where X_E, Y_E, and Z_E represent the fraction of E on a solvent-free basis and can be located on the concentration scale shown in Fig. 7-13a.

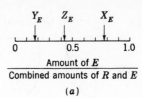

$$\frac{\text{Amount of } E}{\text{Combined amounts of } R \text{ and } E}$$

(a)

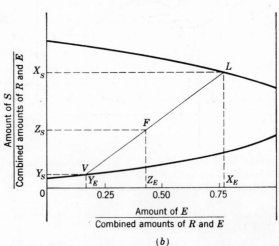

(b)

Fig. 7-13. (a) Component E material balance on a solvent-free concentration scale. (b) Simultaneous S and E component balances on a mass-ratio diagram.

Since $\hat{F}$ is the sum of $\hat{L}$ and $\hat{V}$, the addition point Z_E which represents its composition must fall between X_E and Y_E. The exact location of Z_E is related to the relative amounts of $\hat{L}$ and $\hat{V}$ by the following rearrangement of the component E balance:

$$\frac{\hat{V}}{\hat{L}} = \frac{X_E - Z_E}{Z_E - Y_E}$$

$$\frac{\hat{L}}{\hat{F}} = \frac{Z_E - Y_E}{X_E - Y_E}$$

$$\frac{\hat{V}}{\hat{F}} = \frac{X_E - Z_E}{X_E - Y_E}$$

It can be seen that the amount of $\hat{V}$ is proportional to the line segment $\overline{X_E Z_E}$, and the amount of $\hat{L}$ to the line segment $\overline{Z_E Y_E}$. $\hat{F}$ is represented by the entire line segment between X_E and Y_E.

The concentration scale in Fig. 7-13a can also be used to subtract one stream from another. Rearrangement of the component balance as

$$\hat{F} Z_E - \hat{V} Y_E = \hat{L} X_E$$

shows that when $\hat{V}$ is subtracted from $\hat{F}$, the difference point must lie on the other side of $\hat{F}$ at the concentration X_E. The line-segment ratios in the previous paragraph again give the relative amounts of $\hat{V}$, $\hat{F}$, and $\hat{L}$.

A solvent component balance can be accomplished simultaneously with the component E solvent-free balance if the two-dimensional diagram shown in Fig. 7-13b is used. A vertical concentration scale whose units represent the amount of solvent per unit of nonsolvent material is constructed at right angles to the concentration scale of Fig. 7-13a. Points which represent F, V, and L can now be located in the two-dimensional space. L and V fall on the saturated extract and raffinate curves, respectively. Before the lever-arm rule can be applied to the two-dimensional diagram, it must be shown that the addition point F falls on a straight line between L and V. This can be done by showing that the slope of the line segment $\overline{LF}$ (or $\overline{VF}$) is identical with the slope of the entire line from V to L. Expressions for the slopes of the line segments are as follows:

$$\text{Slope of } \overline{FL} = \frac{X_S - Z_S}{X_E - Z_E}$$

$$\text{Slope of } \overline{VL} = \frac{X_S - Y_S}{X_E - Y_E}$$

The component E and S balances can be written as

$$Z_E = \frac{\hat{L} X_E + \hat{V} Y_E}{\hat{L} + \hat{V}}$$

$$Z_S = \frac{\hat{L} X_S + \hat{V} Y_S}{\hat{L} + \hat{V}}$$

Substitution for Z_E and Z_S in the expression for the slope of $\overline{FL}$ gives

$$\text{Slope of } \overline{FL} = \frac{X_S - Y_S}{X_E - Y_E}$$

which is identical with the expression for the slope of $\overline{VL}$.

The position of the addition point F depends (as in the previous discussion for a single dimension) upon the relative amounts of V and L. The line segments can be measured along the line $\overline{VL}$ or from their projections on either the ordinate or abscissa.

Subtraction of one stream from another is accomplished as shown previously on a single scale. It is obvious that in some cases, subtraction will result in difference points which fall outside the range 0.0 to 1.0 on the abscissa and below 0.0 on the ordinate. Although ratios less than zero on either scale or greater than one on the abscissa are physically impossible, they have algebraic meaning and the use of the lever-arm rule proceeds as described above.

Triangular Diagrams.   The lever-arm rule can be demonstrated for the triangular diagram shown in Fig. 7-14 in exactly the same manner as shown above for the mass-ratio diagram. The only difference between

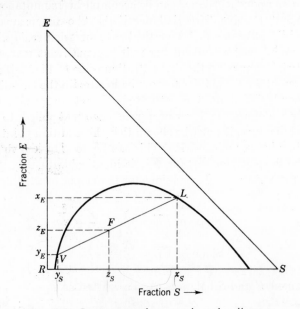

Fig. 7-14. Lever-arm rule on a triangular diagram.

the two diagrams is the basis for the concentration ratios. In the so-called mass-ratio diagram, the concentration ratios are based upon nonsolvent material, while in the triangular diagrams, the concentration ratios are based upon the entire material. The lever-arm rule works with either. The demonstration of its applicability to triangular diagrams (either right or equilateral) is left to the reader.

TRIANGULAR DESIGN DIAGRAMS

The triangular diagram is inherently more convenient than the mass-ratio diagram because the equilibrium data can be plotted directly as weight, mole, or volume per cent without the calculation of ratios. Therefore, the triangular diagram is generally used except in those cases where the graphical construction becomes overly crowded in some part of the diagram. In such cases it may be advisable to switch to the mass-ratio diagram or transfer the construction points to a distribution diagram where the construction lines may be less congested and easier to read.

Solution of an extraction problem on a triangular (or mass-ratio) diagram involves little more than repeated use of the lever-arm rule to perform the additions and subtractions indicated in the various material-balance equations. Some judgment must be exercised in the choice of the correct balances and the arrangement of the terms into the most convenient groupings, but beyond this the solution requires nothing more than the mechanical application of the simple rules described in the previous section.

The material-balance equations always are written as an equality between sums or as an equality between differences. Consider the extraction column (or train) shown in Fig. 7-15. The over-all balance around the extraction column can be written as

$$V_0 + L_{N+1} = L_1 + V_N \qquad (7\text{-}1)$$

or as

$$V_0 - L_1 = V_N - L_{N+1} \qquad (7\text{-}2)$$

The latter form equates the difference between the passing streams V_0 and L_1 to the difference between another pair of passing streams V_N and L_{N+1}. It is generally necessary to use both forms in the solution of a problem.

The graphical construction for Eq. (7-1) is shown in Fig. 7-16. All four streams are located on the diagram according to their

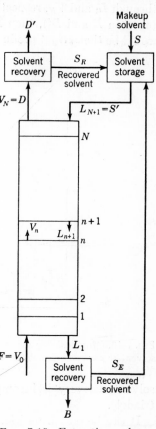

FIG. 7-15. Extraction column (or train) with one feed and no refluxes.

respective concentrations of components E and S. The feed is assumed to contain no solvent, but the entering solvent stream L_{N+1} is assumed to contain small amounts of R and E owing to incomplete separation in the solvent-recovery devices. The addition point M which represents the sum $V_0 + L_{N+1}$ must fall on a straight line between V_0 and L_{N+1} and, according to Eq. (7-1), also must represent the sum $L_1 + V_N$. It can be seen that if the composition of L_{N+1}, V_0, and either V_N or L_1 is specified along with the solvent treat L_{N+1}/V_0, the composition of the fourth stream and the relative amounts of V_N and L_1 can be found immediately.

The solvent in the extract L_1 and raffinate V_N normally must be recovered for recycle to the extraction column. The compositions of D' and B can be found by subtracting S_R from $V_N(=D)$ and S_E from L_1. This is done graphically by drawing lines from S_E and S_R to the RE side through L_1 and V_N, respectively. If the solvent recovery were complete (no S in B and D'), then $\hat{L}_1$ would be the raffinate product B and $\hat{V}_N$ would be the extract product D'. On the diagram it is assumed that the

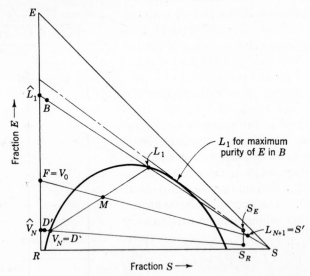

Fig. 7-16. Graphical representation of Eq. (7-1) and the material balances around the solvent-recovery equipment.

solvent recovery is incomplete and B and D' therefore lie inside the triangle.

It can be seen from Fig. 7-16 that, as far as the solubility envelope is concerned, the L_1 which gives the maximum concentration of E in B for the assumed S_E can be located by drawing a straight (dashed) line from S_E tangent to the solubility curve. It will be shown later in the discussions on minimum solvent treats that in some systems it may not be possible to reach the maximum concentration of E owing to the slope of the tie lines.

Equation (7-1) serves the usual purpose of an over-all balance in that it relates all the external streams and aids in locating them in the diagram. It does not shed any light, however, on internal (interstage) streams, and until all the internal streams can be located on the diagram, it is not possible to perform the required stage-to-stage calculations. For this purpose the designer must turn to the difference form of the over-all balance as written in Eq. (7-2). This equation provides a relationship between passing streams which makes easy the location of any possible pair of passing streams on the diagram. Comparison of Eq. (7-2) with the difference form of the balances which cut through the column in Fig. 7-15,

$$V_0 - L_1 = V_n - L_{n+1} \tag{7-3}$$

and

$$V_N - L_{N+1} = V_n - L_{n+1} \tag{7-4}$$

shows that the difference point for passing streams is identical throughout the column. The three material balances can be combined to emphasize this fact and to define the symbol Δ.

$$\Delta_L = V_0 - L_1 = V_n - L_{n+1} = V_N - L_{N+1} \tag{7-5a}$$

or

$$\Delta_R = L_1 - V_0 = L_{n+1} - V_n = L_{N+1} - V_N \tag{7-5b}$$

The symbol Δ is used to denote the hypothetical stream which represents the difference between two passing streams. Equations (7-5a) and (7-5b) define two hypothetical streams, one of which is the negative of the other. Instead of using Δ and $-\Delta$ to denote the two streams, the symbols Δ_L and Δ_R will be used for greater clarity. The Δ_L denotes the left-hand difference $(V_n - L_{n+1})$; that is, when the extract stream is subtracted from the raffinate stream, the difference must lie to the left of the diagram. Similarly, Δ_R denotes the right-hand difference obtained when the raffinate stream V_n is subtracted from the extract stream L_{n+1} to give a difference point to the right of the diagram. The locations of V_0, L_1, V_N, and L_{N+1} in the particular problem under consideration determine automatically which definition of Δ must be used. The Δ must be at the intersection of a straight line through V_0 and L_1 with another straight line through V_N and L_{N+1}. If the intersection occurs to the left, the Δ is defined by Eq. (7-5a); to the right, the Δ from Eq. (7-5b) is used.

It should be recognized that Δ has its own unique rate and composition and that it is treated as any other stream in the application of the lever-arm rule. The concentrations of the various components in Δ will not necessarily fall between 0.0 and 1.0 but may be negative or greater than 1.0. The sign on a concentration denotes the direction of net flow in the column for that particular component. For example, if a concentration of a component in Δ_L is negative, the net flow for that component is down (toward the extract end). The concentration of S in Δ_L will almost always be negative because L_{n+1} generally carries so much more solvent than V_n. Conversely, the concentration of S in Δ_R will generally be greater than 1.0.

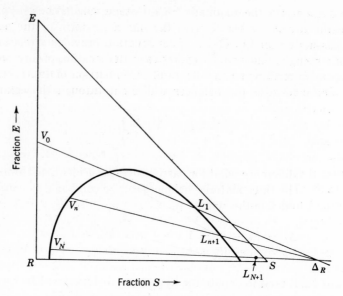

FIG. 7-17. Graphical representation of Eq. (7-2).

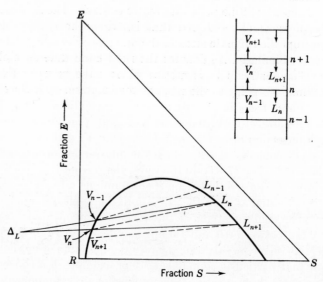

FIG. 7-18. Alternate use of the operating lines and the equilibrium tie lines in the stage-to-stage construction.

Once Δ has been located (usually by the intersection of the two straight lines drawn through L_1 and V_0 and through L_{N+1} and V_N), any possible pair of passing streams V_n and L_{n+1} within the column can be located by drawing a straight line from Δ across the solubility envelope. The construction is illustrated in Fig. 7-17. All the interstage streams must fall on the solubility curve (a consequence of the assumption of equilibrium stages), and therefore V_n and L_{n+1} must fall at the intersections of the straight line with the envelope, L_{n+1} on the solvent-rich side and V_n on the raffinate side.

In the graphical solution of extraction problems the designer is interested in the location of those specific pairs of passing streams which would exist if the equilibrium-stage column were in actual operation. The straight lines from Δ are drawn to connect these operating streams and are therefore termed *operating* lines. These lines perform the function of the material balance and when used alternately with the equilibrium tie lines permit the designer to make the stage-to-stage calculation by graphical means. The combined use of material balances (operating lines) and equilibrium relationships (tie lines) is illustrated in Fig. 7-18. The tie lines (dotted) connect streams which are in equilibrium with each other. The operating lines connect interstage streams which pass each other. Assume that by some means the composition of the stream V_n is known. L_n can be located by the tie line which passes through V_n. V_{n-1} is found by means of an operating line drawn from L_n to Δ_L. The tie line through V_{n-1} then locates L_{n-1}, etc.

The construction methods described above are the same for both type I and type II systems. The principles which have been developed are applied to some of the more common extraction processes in the following sections.

7-4. Single-stage Contact

The simplest extraction process is a one-stage contact between the feed and solvent. The feed and solvent are thoroughly mixed and then allowed to separate into two equilibrium phases which are withdrawn as the extract and raffinate products. The process may be either batch or continuous. A schematic diagram of a continuous single-stage process is shown in Fig. 7-19 along with the corresponding graphical construction. F and S when added give the addition point M. The material balance

$$F + S = L + V$$

requires that M also be the addition point for L and V. Since L and V represent equilibrium phases, they are connected by a tie line, and therefore L and V must lie on the tie line which passes through M.

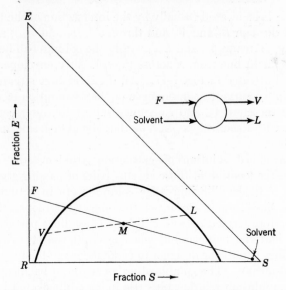

Fig. 7-19. Single-stage contact with impure solvent.

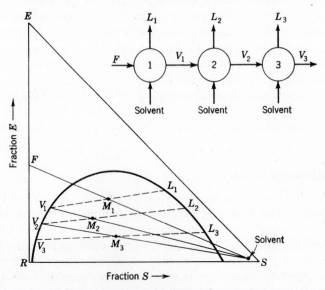

Fig. 7-20. Series of single-stage contacts.

A series of single-stage contacts can be used. In such a process, th raffinate from each stage becomes the feed to the next stage, where it i mixed with more solvent and further extracted. Figure 7-20 shows schematic diagram for this process and the corresponding graphica construction for three contacts.

5. Multistage Countercurrent Contact

A multiple stage extraction process in which the feed is contacted .ntercurrently with solvent is shown in Fig. 7-15. It has been shown Chap. 3 that the extraction column alone (excluding the solvent-:overy and storage devices) has $2C + 2N + 5$ independent design riables. The most convenient specifications for a graphical solution ₂ as follows:

Specifications	$N_i{}^u$	
Pressure in each stage	N	
Temperature in each stage	N	
V_0	C	$+ 2$
L_{N+1}	C	$+ 2$
One concentration in V_N or L_1		1
		$2C + 2N + 5$

The pressure specifications are relatively unimportant as far as the aphical construction is concerned. The only restriction on the pressure that it not deviate so widely from the pressure at which the equilibrium .ta were obtained as to affect the distribution coefficients.

Extraction processes are often essentially isothermal. In such cases e same solubility curve can be used for all the stages. If there is to be temperature gradient through the column, the tie line for each stage ould correspond to the temperature of that stage; that is, a series of >thermal equilibrium diagrams must be determined experimentally to •ver the expected temperature range.

Specification of V_0 and L_{N+1} not only locates these streams on the agram but also fixes their addition point M, since rate is one of the $+ 2$ variables specified for each stream. One specified concentration . V_N (usually a maximum concentration for the extract component E) sufficient to locate that stream on the solubility envelope. A line from $_N$ through M locates L_1. This construction has been illustrated in ig. 7-16.

Now that the problem has been completely defined by the specification : the required number of variables, the graphical construction can be erformed to determine the required number of stages. The stage-to-age calculations are illustrated in Fig. 7-21. Lines through F and L_1 nd through V_N and L_{N+1} locate Δ_R. The stages can be stepped off from .ther end. The construction in Fig. 7-21 begins with the location of the e line through L_1. V_1 must fall on the other end of this tie line. An perating line from V_1 to Δ_R locates L_2. V_2 lies on the other end of the .e line through L_2. Another operating line gives L_3, and another tie line ives V_3, etc. V_4 falls below the specified V_N, indicating that four equilibrium stages would more than accomplish the desired separation. Fractional equilibrium stages have significance, since an efficiency factor must be applied to convert them to actual stages. Consequently, a reasonable answer for this example would be 3.7 stages. After the efficiency factor has been applied, the number of actual stages must, of course, be rounded off.

The stage-to-stage calculation described above involves no trial and error. This would not be the case if the one degree of freedom utilized to fix a concentration in V_N or L_1 were used instead to specify the number of stages. If neither V_N nor L_1 is fixed by a concentration specification

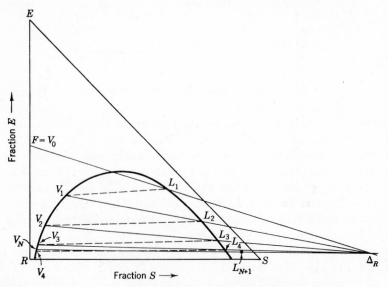

Fig. 7-21. Stage-to-stage construction for a multistage countercurrent process with one feed (see Fig. 7-15).

their location must be assumed before the stage-to-stage calculation is started. If the number of stages stepped off between the assumed V_N and L_1 does not equal the specified number, new values for V_N and L_1 must be assumed until a check is obtained. It can be seen that there are basically two types of problems. If the number of stages is specified, a trial-and-error solution is necessary. If not, the solution is straightforward without trial and error.

The over-all balances for the process pictured in Fig. 7-15 are shown in Fig. 7-22. The addition point M comes from the balance around the extraction column,

$$F + L_{N+1} = L_1 + V_N = M$$

he addition point M' comes from the balance around the entire process
cluding the solvent-recovery equipment,

$$F + S = B + D' = M'$$

here S represents the pure make-up solvent. The S_E and S_R points
present the composition of the solvent-rich streams from the solvent-
covery devices. Addition of these two streams with S gives L_{N+1}.

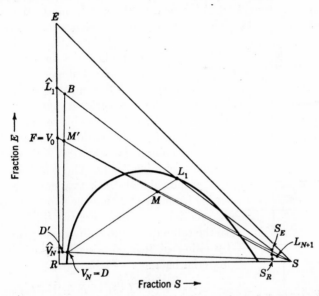

IG. 7-22. Over-all material-balance lines for the multistage, countercurrent process
ictured in Fig. 7-15.

ubtraction of S_E from L_1 gives B ($\hat{L}_1$ would result if the solvent removal
rom L_1 were complete). Subtraction of S_R from V_N gives D' ($\hat{V}_N$ would
esult if the solvent removal from V_N were complete).

-6. Two Feeds

Often the designer will have two feeds to process. One will be rela-
ively rich in the extract component E; the other relatively poor. Rather
han combine the two into a single feed and lose the separation already
chieved, it is usually preferable to design a process for two separate
eeds. Such a process is illustrated in Fig. 7-23. The feed stream F'
vhich has the higher concentration of component E is introduced at the
xtract end as V_0. The other feed stream F is introduced at some
ntermediate stage $M + 1$ in the column. This process has $3C + 2N + 8$

degrees of freedom or $C + 3$ more than the single-feed column discusse
in the previous section.   The specification of the additional feed strea
utilizes $C + 2$ variables.   The remaining variable is used to specify tl
intermediate feed location.   In computer solutions of design problem

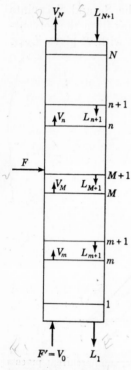

it is most convenient to specify the numeric
value of M, the number of stages below the fee
stage.   In graphical solutions it is convenient
utilize the degree of freedom by simply specifyii
that the feed-plate location be the optimum on
The *optimum location* is defined as that one whic
provides the greatest separation between R a
E with a given number of stages or as the o
which requires the fewest stages to obtain a sp
cified separation.   The manner in which the
criteria can be met in the graphical constructic
will be described later in this section.

The principles of graphical construction for
two-feed column are identical with those deve
oped for a single-feed.   The only differences i
the graphical solutions for the two processe
spring from the fact that the two-feed column h
two sections and therefore two difference point
In the raffinate end the difference point is define
by either of the following forms of the over-a
material balance:

$$\Delta_L = V_N - L_{N+1} = F + V_0 - L_1$$
$$= M_F - L_1 \qquad (7\text{-}6a$$
or $\quad \Delta_R = L_{N+1} - V_N = L_1 - F - V_0$
$$= L_1 - M_F \qquad (7\text{-}6b$$

Fig. 7-23. Two-feed proc-
ess.   The solvent-recov-
ery   equipment   (not
shown) would be similar
to that shown in Fig.
7-15.

where $V_0 = F'$ and $M_F = F + F'$.   It will b
noted that $\Delta_R = -\Delta_L$ and therefore Δ_R and Δ
represent the same difference point Δ.   How
ever, rather than use the sign to denote the two possible locations of th
point, the subscripts R and L will be used to show which form of Eq. (7-6
is used and whether the point falls to the right (Δ_R) or to the left (Δ_L) o
the solubility envelope.   In the extract end,

$$\Delta'_L = V_0 - L_1 = V_N - L_{N+1} - F \qquad (7\text{-}7a$$
or $\qquad \Delta'_R = L_1 - V_0 = F + L_{N+1} - V_N \qquad (7\text{-}7b$$

where Δ'_L and Δ'_R represent the left- and right-hand locations of Δ'.

Four possibilities exist for the locations of the Δ's for the two sections
The two difference points can both lie to the left, both can lie to the

ght, Δ can fall to the left while Δ' falls to the right, or Δ can fall to the ght while Δ' falls to the left. The relationships between the two differ-ace points and the intermediate feed F for all these possibilities follow om Eq. (7-7) and the fact that $\Delta_L = -\Delta_R$ and $\Delta'_L = -\Delta'_R$.

$$\Delta'_L = \Delta_L - F \tag{7-8a}$$
$$\Delta'_R = F + \Delta_R \tag{7-8b}$$
$$\Delta'_R + \Delta_L = F \tag{7-8c}$$
$$\Delta'_L + \Delta_R = -F \tag{7-8d}$$

hese expressions are four different forms of the over-all material balance, id all stipulate that the two difference points and the intermediate ed F must all lie on a straight line.

The difference points defined by Eqs. (7-6) and (7-7) in terms of ternal streams can be related to any pair of passing internal streams by alances which cut the column sections. In the raffinate section,

$$V_N - L_{N+1} = V_n - L_{n+1} = \Delta_L \tag{7-9a}$$
$$L_{N+1} - V_N = L_{n+1} - V_n = \Delta_R \tag{7-9b}$$

1 the extract section

$$V_0 - L_1 = V_m - L_{m+1} = \Delta'_L \tag{7-10a}$$
$$L_1 - V_0 = L_{m+1} - V_m = \Delta'_R \tag{7-10b}$$

'he stage-to-stage construction is started from either end of the column sing the appropriate difference point. The feed is introduced on the pecified stage or, if the optimum location is specified, on that stage whose e line crosses the line represented by Eq. (7-8). As soon as the feed is atroduced (optimum location or not), the material-balance lines must e switched to the other difference point. These principles are illustrated elow for both the optimum and nonoptimum feed locations.

The use of the over-all material balance to locate the two difference oints is shown in Fig. 7-24. Specification of the feed streams F' and ' fixes their rates and permits their addition to give M_F. The difference oint Δ for the raffinate end section is located by the application of Eq. 7-6a). Subtraction of L_{N+1} from V_N gives a straight line which passes hrough Δ. Subtraction of L_1 from M_F also gives a straight line which aust pass through Δ. Therefore, Δ can be located from the intersection f these two lines without any prior knowledge of the relative rates of $_{N+1}$, V_N, L_1, and M_F. Equations (7-7b) and (7-8c) show that Δ' must e at the intersection of lines through V_0 and L_1 and through Δ and F.

The stages will first be stepped off with the optimum feed location. Another diagram (Fig. 7-25) without the over-all balance lines will be used o avoid congestion on Fig. 7-24. The tie line through L_1 is located to tart the calculation. The difference point Δ' is used for the material

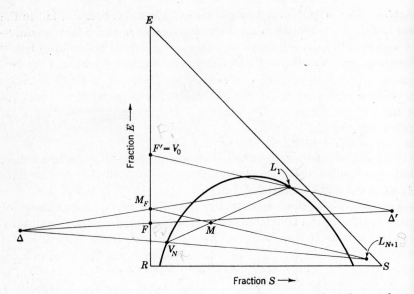

FIG. 7-24. Location of difference points for a two-feed process such as the one shown in Fig. 7-23.

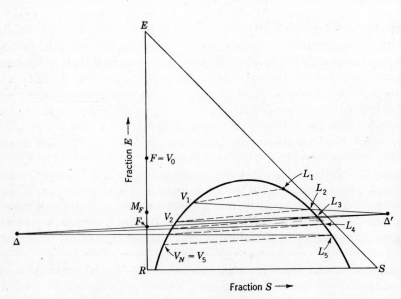

FIG. 7-25. Stage-to-stage construction for the two-feed process in Fig. 7-23 and the material balances in Fig. 7-24.

balance lines all the way through the extract end section. The tie line for stage 3 crosses the over-all material-balance line which connects Δ and Δ' and passes through F. It can be seen that the use of Δ to locate L_4 gives a greater difference between L_3 and L_4 than if Δ' were used again. The reader can determine also that the use of Δ one stage earlier would have decreased the difference between L_2 and L_3. Therefore, stage 3 is the optimum feed location. The optimum feed-introduction point will always be on that stage whose tie line crosses the over-all material-balance line which connects Δ and Δ'. *in Greas notation*

Five stages are required in Fig. 7-25 when the feed is introduced in an optimum manner. As mentioned earlier, the feed location may be fixed at some stage other than the optimum location. This will often be the case when existing equipment is used for separations or operating conditions other than those for which it was originally designed. The mislocation of the feed will always lower the amount of separation obtained in a fixed number of stages. In Fig. 7-25 where the number of stages is not fixed, mislocation of the feed increases the number of stages required to make the specified separation. For example, if the feed F were introduced on stage 5 instead of stage 3, Δ' would be used for two more stages and a total of approximately 6.7 stages would be required, an increase of 1.7 over the number required with optimum feed location.

7-7. Minimum Solvent Treat

The solvent treat is defined as the ratio L_{N+1}/F for a single-feed column and as $L_{N+1}/(F + F')$ for a column with two feeds. In either case, the solvent treat plays a major role in the location of the difference point for any column section since the treat largely determines the internal reflux ratios L_{n+1}/V_n and L_{m+1}/V_m. Increasing the solvent treat will, in general, make the tie line and operating line for any given stage more divergent, that is, increase the separation obtained in the stage. Conversely, decreasing the solvent treat causes the tie line and operating line in any given stage to approach each other, and as a result the effectiveness of the stage is decreased. If the solvent treat is decreased enough, a point will be reached where the tie line and operating line become coincident in some stage. The stage-to-stage construction cannot proceed past this point. This situation is described by saying that an infinite number of stages are required to make the separation between L_1 and V_N. The *minimum* solvent treat is defined as that treat which if decreased an infinitesimal amount will require an infinite number of stages to pass from L_1 to V_N. The minimum solvent treat is like the minimum reflux in distillation in that the concept is associated with an infinite number of stages. It has no meaning for a column with a fixed number of stages. It differs from minimum reflux in complete distillation columns in that it does not depend upon the prior specification of a separation between the two components. The degrees of freedom available for a single-feed column were listed in Sec. 7-5, and inspection of the tabulation there shows that only one product concentration can be specified at any given solvent treat. Therefore, it is not possible to locate both V_N and L_1 arbitrarily by specification and then find the solvent treat which requires an infinite number of stages to accomplish the specified separation. The solvent treat is fixed by the location of V_N and L_1, and there is only one set of V_N and L_1 locations (plus the corresponding solvent treat) which will require an infinite number of stages.

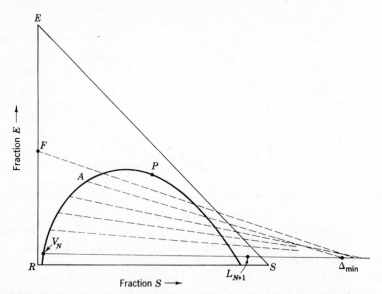

FIG. 7-26. Determination of minimum solvent treat in system where the tie lines have a negative slope. Tie line A gives the intersection closest to L_{N+1}. The operating Δ must lie between L_{N+1} and Δ_{min}.

The location of the minimum solvent treat is illustrated in Fig. 7-26 for a single-feed column. If the tie lines have a negative slope as shown in this figure, the minimum treat will occur while L_{N+1} is still larger than V_N and the difference point still lies to the right of the diagram. At infinite solvent treat, Δ is coincident with L_{N+1}. As the treat is decreased, Δ moves to the right on a straight line through V_N and L_{N+1}, since

$$\Delta_R = L_{N+1} - V_N$$

The minimum treat as defined above will occur when Δ reaches the first intersection which any extrapolated tie line within the diagram makes with the line through V_N and L_{N+1}. All tie lines, or rather tie lines from all parts of the diagram, must be extrapolated to find that one whose intersection lies closest to L_{N+1}. In Fig. 7-26, tie line A is the one with which the operating lines first become coincident and its intersection determines the minimum solvent treat. The operating Δ must lie between L_{N+1} and $\Delta_{\min}$.

If the tie lines have a positive slope, their intersections with the line through V_N and L_{N+1} must occur to the left of the diagram as shown in Fig. 7-27. The location of Δ at infinite solvent treat is still at L_{N+1}, and Δ still moves to the right as the treat is decreased. No intersections with

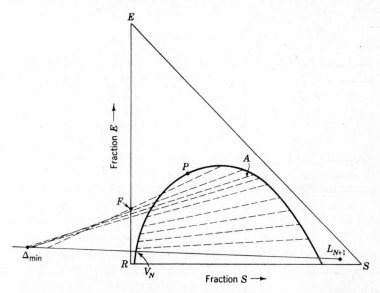

FIG. 7-27. Determination of minimum solvent treat in system where the tie lines have a positive slope. Tie line A gives intersection farthest to the left. The operating Δ must lie to the left of $\Delta_{\min}$.

the extrapolated tie lines occur to the right, and therefore the decrease in the treat can be continued until V_N equals L_{N+1} in rate. When $V_N = L_{N+1}$, Δ falls at an infinite distance to the right of the diagram. Further decrease in the treat causes Δ to reappear to the left of the diagram (Δ switches from Δ_R to Δ_L). Now Δ moves toward V_N as the treat is decreased and must eventually reach the first intersection with a tie line. Usually the tie line through F gives the intersection farthest from V_N, but the tie lines in Fig. 7-27 have been located in such a manner as to show that this need not always be true.

The feed point F in Fig. 7-27 may lie high enough so that even the tie line at the plait point will fall below it. If this is the case, and if the tie lines are regular (tie line through P gives the intersection farthest from V_N), then L_1 will coincide with P at the minimum solvent treat. If P falls below the material-balance line between F and L_{N+1}, then L_1 can go no farther in the stage-to-stage construction than the point of intersection of this line with the solubility envelope. L_1 then becomes coincident with $M = F + L_{N+1}$, which falls on the solubility envelope. The "minimum" treat must then be obtained from this location of M and will be less than that which corresponds to the intersection of the tie line through P with the line through V_N and L_{N+1}.

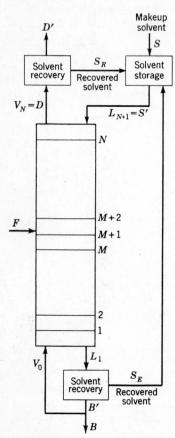

FIG. 7-28. Single-feed column with extract reflux.

7-8. Extract Reflux

In the processes described in the previous two sections and pictured in Figs. 7-15 and 7-23, the final equilibrium extract stream L_1 is removed from the same stage to which a feed stream is admitted. Even though L_1 is in equilibrium with V_1 rather than the entering feed, the introduction of the feed restricts the purity of L_1 which can be obtained. This is particularly so if the feed has a low concentration of the extract component E. The obvious remedy for the situation is to replace the feed with a stream which has a high concentration of E, thereby permitting a higher purity extract L_1 to be withdrawn. This can be accomplished by extract reflux as pictured in Fig. 7-28. The feed introduction point is moved from stage 1 to some intermediate stage $M + 1$. Part of the solvent-lean effluent B' from the solvent-recovery device is then returned to stage 1 as V_0. Enough solvent is removed from L_1 to move B' to the other side of the solubility diagram as shown in Fig. 7-29. If all the solvent were removed, B' would correspond to $\hat{L}_1$. Usually B' will fall somewhere between $\hat{L}_1$ and the solubility envelope.

A type II system is used to illustrate extract reflux in Fig. 7-29 for two reasons: first, to show that the principles of graphical construction are the same as for type I systems and, second, because extract reflux is generally more useful in type II systems. Reflux is more apt to be needed when the solvent is partially immiscible with both components and shows no great selectivity for one over the other.

The over-all material-balance lines and location of difference points proceed exactly as in the previous section for a two-feed column if V_0 is treated as the second feed. In the raffinate end section the difference point Δ is defined by

$$\Delta_L = V_N - L_{N+1} = F + V_0 - L_1 = M_F - L_1 \qquad (7\text{-}11a)$$

or
$$\Delta_R = L_{N+1} - V_N = L_1 - F - V_0 = L_1 - M_F \qquad (7\text{-}11b)$$

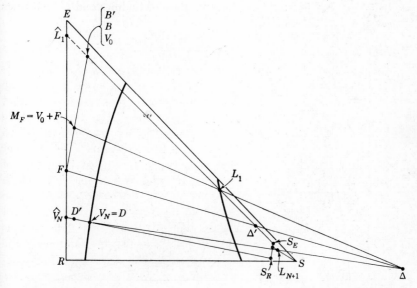

FIG. 7-29. Material-balance lines and location of Δ and Δ' for the process with extract reflux shown in Fig. 7-28.

which are identical with Eqs. (7-6a) and (7-6b). In the extract end, the Δ' must fall to the right, since L_1 must be greater than V_0.

$$\begin{aligned}
\Delta_R' = L_1 - V_0 &= F + L_{N+1} - V_N \\
&= F + \Delta_R \\
&= F - \Delta_L \qquad (7\text{-}12)
\end{aligned}$$

A balance around the extract-recovery device relates Δ' to B and S_E.

$$\Delta_R' = L_1 - V_0 = B + S_E \qquad (7\text{-}13)$$

Equations (7-11) to (7-13) plus the usual balances around the solvent-recovery and storage equipment are demonstrated graphically in Fig. 7-29. Subtraction of S_R and S_E from V_N and L_1 give D' and B', respectively. The point which represents B' also represents B and V_0. Addi-

tion of V_0 and F give M_F, and according to (7-11), Δ must lie at th
intersection of a line through M_F and L_1 with a line through V_N an
L_{N+1}. From Eqs. (7-12) and (7-13), Δ' can be located by drawing th
over-all material-balance line which connects F to the two differenc
points and noting its intersection with the line through L_1 and V_0.

After the two difference points have been located, the stage-to-stag
construction proceeds exactly as for any two-feed column. The stage
can be stepped off from either end. In Fig. 7-30 the construction i
started from L_1. The tie line through L_1 locates V_1, and an operatin
line from V_1 to Δ' locates L_2. Alternate use of tie lines and Δ' continue

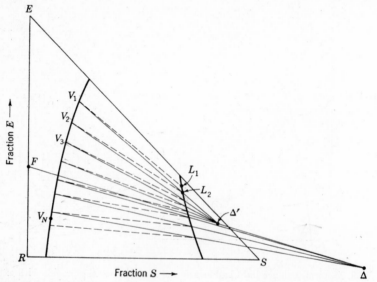

Fig. 7-30. Stage-to-stage construction for the process with extract reflux pictured in
Fig. 7-28. Location of Δ and Δ' is shown in Fig. 7-29.

until that stage whose tie line crosses the line through F, Δ, and Δ' is
reached. As explained in the previous section this stage is the optimum
feed stage. Introduction of the feed in this stage requires that the
difference point for the remaining stages be switched from Δ' to Δ.

Extract Reflux Ratio. In the foregoing discussion, it was presumed
that the rate of V_0 had been specified (as well as its composition) and that
the ratio V_0/F was therefore known. The construction proceeded then as
for a two-feed column. Specification of the ratio V_0/F is a legitimate and
often convenient specification. In some cases, however, the designer
may wish to specify the *external* reflux $R' = V_0/B$ instead. This ratio
can be used to locate Δ' on the line through V_0 and L_1. The over-all
material-balance line through F, Δ, and Δ' then locates Δ.

Before the external reflux ratio can be used to locate the difference points, it is necessary to express the reflux ratio in terms of line segments on the operating lines. The relationship is simple for the *internal* reflux ratio $\frac{L_{n+1}}{V_n}$ or $\frac{L_{m+1}}{V_m}$. The definition of Δ,

$$\Delta_R = L_{n+1} - V_n$$

or
$$\Delta_L = V_n - L_{n+1}$$

establishes that the line segment $\overline{V_n\Delta}$ represents the point L_{n+1} while the line segments $\overline{V_nL_{n+1}}$ and $\overline{L_{n+1}\Delta}$ represent the streams Δ and V_n, respectively. Therefore,

$$\frac{L_{n+1}}{V_n} = \frac{\overline{V_n\Delta_R}}{\overline{L_{n+1}\Delta_R}} \tag{7-14a}$$

or
$$\frac{L_{n+1}}{V_n} = \frac{\overline{V_n\Delta_L}}{\overline{L_{n+1}\Delta_L}} \tag{7-14b}$$

Similarly, in the extract end section where with extract reflux only the right-hand difference is used,

$$\frac{L_{m+1}}{V_m} = \frac{\overline{V_m\Delta_R'}}{\overline{L_{m+1}\Delta_R'}} \tag{7-14c}$$

Equation (7-14c) can be applied directly to the external reflux ratio V_0/L_1.

$$\frac{V_0}{L_1} = \frac{\overline{L_1\Delta_R'}}{\overline{V_0\Delta_R'}} \tag{7-14d}$$

If the streams V_0 and L_1 are specified sufficiently to permit their location on the diagram, then Δ_R' can be located immediately through use of (7-14d) if the reflux ratio V_0/L_1 is known. Rather than measure off the line segments on the line through V_0 and L_1, it is more convenient to project the line to one of the axes and use the concentration scales to measure distances. If the projection is made to the ordinate, the line segments can be expressed in terms of concentrations of E as follows:

$$\frac{V_0}{L_1} = \frac{\overline{L_1\Delta_R'}}{\overline{V_0\Delta_R'}} = \frac{(x_1)_E - (x_{\Delta'})_E}{(y_0)_E - (x_{\Delta'})_E} \tag{7-14e}$$

Or if the projection is made to the abscissa, the subscript E's are replaced with S's, since the line segments would then be expressed in terms of solvent concentrations. The line on which distances are being measured should always be projected to that axis which is most nearly parallel with the line.

The location of Δ' is not so straightforward when the external reflux ratio is expressed as V_0/B rather than as V_0/L_1. One method is as follows: The ratio L_1/S_E is determined from the diagram (see Fig. 7-29) by

$$\frac{L_1}{S_E} = \frac{\overline{B'S_E}}{\overline{B'L_1}} = \frac{\overline{V_0S_E}}{\overline{V_0L_1}} \tag{7-15}$$

The use of (7-15) depends upon the prior specification (or assumption if the number of stages is specified) of the location of the streams V_0, S_E, and L_1. The ratio S_E/B can now be expressed in terms of L_1/S_E and the external reflux ratio V_0/B in the following manner: A balance around the solvent-recovery device gives

$$L_1 = S_E + V_0 + B = S_E + \left(\frac{V_0}{B} + 1\right) B$$

Replacing L_1 with $(L_1/S_E)S_E$ and then rearranging give

$$\frac{S_E}{B} = \frac{V_0/B + 1}{L_1/S_E - 1} \tag{7-16}$$

The definition of Δ' when combined with the balance around the solvent-recovery device gives

$$\Delta'_R = L_1 - V_0 = S_E + B \tag{7-13}$$

which can be rearranged to

$$\frac{\Delta'_R}{B} = \frac{S_E}{B} + 1$$

From (7-13) it can be seen that the ratio Δ'_R/B can be expressed in terms of line segments as follows:

$$\frac{\Delta'_R}{B} = \frac{S_E}{B} + 1 = \frac{\overline{V_0S_E}}{\overline{\Delta'_RS_E}} \tag{7-17a}$$

Or in terms of the projected lengths on the ordinate expressed as concentrations of E,

$$\frac{\Delta'_R}{B} = \frac{S_E}{B} + 1 = \frac{(y_0)_E - (x_{S_E})_E}{(x_{\Delta'})_E - (x_{S_E})_E} \tag{7-17b}$$

In summary, the location of Δ' through the external reflux ratio V_0/B involves the use of Eqs. (7-15) and (7-16) to calculate S_E/B, which when substituted in (7-17b) permits the determination of $(x_{\Delta'})_E$. The y_0 and x_{S_E} must already be known from prior specifications or assumptions. The intersection of the horizontal line which corresponds to $(x_{\Delta'})_E$ with the straight line through L_1 and V_0 locates Δ', the difference point for the extract end section. The difference point for the raffinate end section Δ can then be found by drawing a line through F and Δ' to its intersection with the line through V_N and L_{N+1} as shown in Fig. 7-29.

Total Extract Reflux. The effect of returning all B' to the column as V_0 is best described with Eq. (7-13).

$$\Delta_R' = L_1 - V_0 = B + S_E \qquad (7\text{-}13)$$

At total extract reflux $B = 0$ and Δ_R' becomes coincident with S_E. As can be seen from Figs. 7-29 and 7-30, movement of Δ' downward to S_E increases the separation obtained in each stage of the extract end of the column. Also Δ must be shifted downward, since it must lie on a straight line through F and Δ'. Therefore increased reflux tends to reduce the

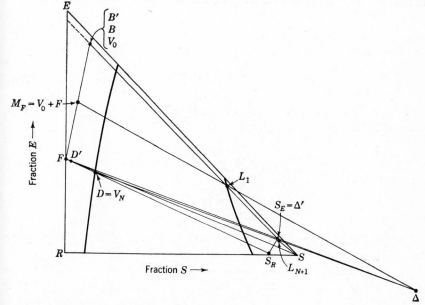

FIG. 7-31. Material balances for column operating at total extract reflux ($B = 0$) but with continuous introduction of feed and continuous withdrawal of raffinate product.

number of stages required. However, the effect of the over-all balance must be considered. As B is decreased, the raffinate product D becomes more and more like F. At $B = 0$, the balance around the entire process becomes

$$F + S = D'$$

where S is the makeup solvent. This material balance is shown in Fig. 7-31. The final raffinate product D' must also lie on a straight line through S_R and V_N. Addition of S_E, S_R, and S gives L_{N+1}. At total reflux, $\Delta' = S_E$. A line through Δ' and F must intersect a line through V_N and L_{N+1} at Δ. These two lines are almost coincident in Fig. 7-31, and therefore only a small fraction of an equilibrium stage would be required in the raffinate end section. The rest of the stages would be in the extract end section and with total reflux would have the maximum possible effectiveness. It can be seen that for a given number of stages N, total extract reflux would provide the maximum concentration of E in L_1. However, the extract product rate would be zero and the raffinate product on a solvent-free basis identical with the feed F.

Minimum Extract Reflux. The minimum extract reflux rate is defined as that rate of V_0 which if decreased an infinitesimal amount would require an infinite number of stages to accomplish a specified separation between components R and E. An infinite number of stages requires

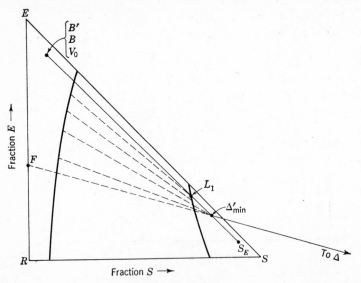

FIG. 7-32. Determination of minimum extract reflux rate by extrapolation of tie lines in the extract end section to their intersection with the line $\overline{L_1 S_E}$. The lowest intersection locates Δ'_{min}.

the establishment of a zone of constant composition somewhere within the column. A zone of constant composition is created when the material-balance (or operating) lines become coincident with a tie line. When this happens, no further progress can be made in the stage-to-stage construction.

The similarity between minimum extract reflux and the concept of minimum solvent treat discussed earlier in the chapter is obvious. It will be necessary to assume in the discussion of minimum reflux that the solvent treat L_{N+1}/F is great enough to make the column operable at some reflux less than total reflux; that is, the specified separation between R and E can be accomplished with a finite number of stages.

The minimum value of V_0 (or the maximum value of B) which can be used without causing the operating lines to become coincident with a tie line can be found by the extrapolation of all the tie lines in the extract end section to their point of intersection with the line from V_0 to S_E. These extrapolations are shown in Fig. 7-32. That tie line in the extract end section which intersects closest to S_E locates Δ' for the minimum reflux. Usually the tie line which extrapolates through the feed F is the one which has the lowest intersection, and Fig. 7-32 is drawn in that manner. Any location of Δ' between Δ'_{min} and S_E would not cause the operating lines to become coincident with any tie line in the extract end section.

The procedure described above assumes that the tie lines in the raffinate section do not reverse or change their slope in such a manner as to cause the development of a zone of constant composition in the raffinate end section at a higher "minimum" reflux than indicated by the construction in Fig. 7-32. Such a possibility is remote but nevertheless must always be checked in any actual system.

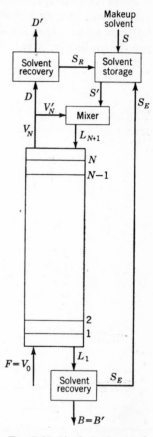

FIG. 7-33. Single-feed extraction column with raffinate reflux.

7-9. Raffinate Reflux

Part of the stream V_N is sometimes diverted from the raffinate solvent-recovery device and mixed directly with the incoming solvent stream to form L_{N+1} as shown in Fig. 7-33. A reflux ratio can be defined as the amount refluxed divided by D, or $R = V'_N/D$. However, this so-called reflux is nothing other than the recycle of an equilibrium phase to stage N. Since the intensive variables of a system at equilibrium are independent of the amounts of the equilibrium phases present, recycle of part of V_N cannot affect the separation in an equilibrium-stage system. Therefore, raffinate reflux has no effect upon the calculation of the number of equilibrium stages. In practical operations where the contact stages are not equilibrium stages (less than 100 per cent efficient) raffinate reflux may have some practical advantages. Presaturation of the solvent with raffinate material outside stage N in effect increases the size of the column, since the mixing no longer must occur in the contact stage.

Also, the degree to which equilibrium is approached is usually a function of time as well as the physical properties of the liquids. If part of the mass transfer is accomplished in the mixer, a closer approach to equilibrium may be possible in the end stage of the column or train, resulting in a higher stage efficiency.

Even though the raffinate reflux does not affect the stage-to-stage calculation, it does result in some new points in the over-all material-balance lines. The material-balance lines for the process in Fig. 7-33 are shown in Fig. 7-34. Two addition points can be located.

$$M = F + L_{N+1} = L_1 + V_N$$
$$M' = F + S' = L_1 + D$$

The addition of V_N' to S' locates L_{N+1} close to the saturation curve. The stream S' is located by the addition of S_E, S_R, and S, where S represents

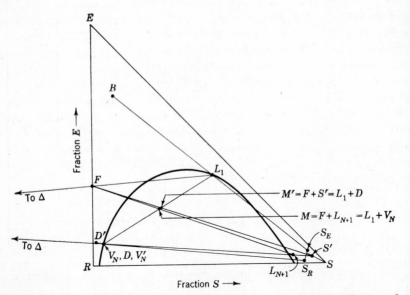

FIG. 7-34. Material-balance lines for the extraction process with raffinate reflux pictured in Fig. 7-33.

the pure make-up solvent. The difference point Δ is located as before by the intersection of lines through F and L_1 and through V_N and L_{N+1}.

$$\Delta_L = F - L_1 = V_N - L_{N+1}$$
or $$\Delta_R = L_1 - F = L_{N+1} - V_N$$

If the raffinate reflux had any effect on the stage-to-stage construction, the location of Δ would vary with the amount of reflux. Consider a case where F is completely specified and the position of L_1 has been fixed by the specification of one concentration. The solvent treat, S'/F in this case, has also been specified, so V_N can be located. A balance around the mixer (or presaturator) gives

$$V_N + S' = D + L_{N+1}$$

or
$$\Delta_L = V_N - L_{N+1} = D - S'$$

All four streams lie on the same line, and V_N and D are coincident. A change in the reflux does not change the position of the line. Only the position of L_{N+1} between V_N and S' changes with the reflux rate. Therefore, the point of intersection of the line through V_N and L_{N+1} (or D and S') with the line through F and L_1 is independent of the raffinate reflux ratio.

In view of the discussion above it can be seen that "minimum" raffinate reflux cannot have a meaning analogous to the concepts of minimum extract reflux and minimum solvent treat. Reduction of the raffinate reflux cannot in itself cause a tie line to become coincident with the operating lines. However, there will always be a minimum solvent treat (or a minimum combination of solvent treat and extract reflux) which will locate a $\Delta_{\min}$ regardless of what raffinate reflux rate is specified. At any specified V_N', when $\Delta = \Delta_{\min}$, the ratio V_N'/D will have a minimum value which is sometimes referred to as the minimum raffinate reflux ratio. For any given set of problem specifications (which locate F, L_1, and S') there is only one minimum solvent treat but there is an infinite number of minimum raffinate reflux ratios, since there can be an infinite number of specified raffinate reflux rates.

MASS-RATIO DESIGN DIAGRAMS

The construction lines may become overly crowded in portions of the triangular design diagrams described in the preceding sections. If so, it may be desirable to convert the equilibrium data points into mass ratios defined in Sec. 7-1 as

$$Y_S \text{ or } X_S = \frac{\text{amount of } S}{\text{combined amounts of } R \text{ and } E} \tag{7-18}$$

and
$$Y_E \text{ or } X_E = \frac{\text{amount of } E}{\text{combined amounts of } R \text{ and } E} \tag{7-19}$$

and plot these in the manner shown in Fig. 7-8. The part of the graphical construction which is congested on the triangular diagram will usually be spread on the mass-ratio diagram and vice versa. The mass ratios may be defined in terms of moles, weight, or volume as long as the ratios and the stream rates are in the same units. The expression of equilibrium data in volumetric units is discussed near the end of Sec. 7-1.

In the following descriptions of the mass-ratio design diagrams, it is assumed that the reader is familiar with the application of the lever-arm rule to triangular design diagrams. Therefore, the development of basic principles will not be discussed again at any great length. The purpose of the following sections is simply to show typical graphical constructions on the mass-ratio diagram for the various processes previously discussed and compare them to the analogous triangular solutions. No attempt will be made to make the mass-ratio diagrams

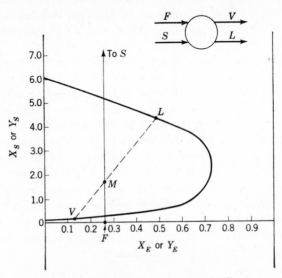

FIG. 7-35. Single-stage contact on a mass-ratio diagram.

correspond exactly to the triangular solutions because the lack of concentration grids on the illustrative figures makes the transfer of points uncertain. In the solution of real problems where the grid will be available, the transfer of points is easily accomplished. ⟨Any point in the triangular diagram is converted to a solvent-free basis by simply drawing a line from the S apex through the point to the R-E side. The value read from the ordinate can be plotted directly on the abscissa in the mass-ratio diagram. Values of X_S and Y_S must be calculated arithmetically.⟩

7-10. Single-stage Contact

This simplest of all extraction processes has been discussed previously in Sec. 7-4, and the graphical solution on the triangular diagram shown in Fig. 7-19. An analogous typical construction on the mass-ratio diagram is shown in Fig. 7-35. Note that a pure solvent stream falls at $X_S = \infty$, and therefore the addition point $M = F + S$ cannot be located from the lever-arm rule but must be calculated arithmetically from the specified S and F rates. Even when an impure solvent is used, the point which represents it will lie so high on the diagram that the graphical location of M is impractical. The line from F through M is vertical for a pure solvent but slants slightly to the left or right for an

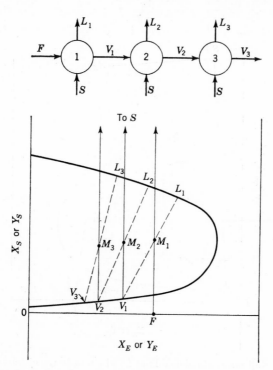

FIG. 7-36. Series of single-stage contacts with pure solvent.

impure solvent. The direction of the slant depends upon the composition of the solvent-free portion of the impure solvent.

The construction for a series of single-stage contacts with pure solvent is shown in Fig. 7-36, which is analogous to Fig. 7-20 except for the solvent purity.

7-11. Multistage Countercurrent Contact

The multistage countercurrent extraction process has been discussed in Sec. 7-5, and a flow diagram presented in Fig. 7-15. The over-all material balances for the process are shown in Fig. 7-22 on a triangular diagram.   The construction on the mass-ratio diagram analogous to that shown in Fig. 7-22 is illustrated in Fig. 7-37. Note the lack of congestion on the mass-ratio diagram in comparison with Fig. 7-22. The points S_E, S_R, and L_{N+1} which are jammed in the solvent corner on the triangular diagram are spread across the top of the mass-ratio diagram.

The location of that extract L_1 which would give the maximum concentration of E on a solvent-free basis can be found in Fig. 7-37 by

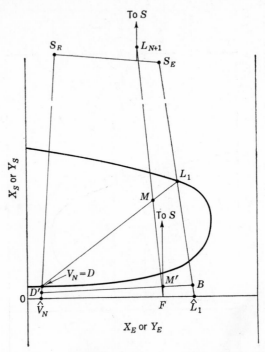

FIG. 7-37. Over-all material-balance lines for the multistage countercurrent process in Fig. 7-15. The analogous construction for a triangular diagram is in Fig. 7-22.

drawing a vertical line tangent to the solubility envelope. This construction corresponds to the tangent drawn from the solvent corner on the triangular diagram.

The stage-to-stage construction is shown in Fig. 7-38. As before, Δ is located by the intersection of the straight lines through V_N and L_{N+1} and through F and L_1.

7-12. Two Feeds

The reader is referred to Sec. 7-6 for a discussion of the two-feed process. The principles developed there apply directly to the mass-ratio diagram also. Material-balance lines analogous to those shown on the triangular diagram in Fig. 7-24 are shown on the mass-ratio diagram in Fig. 7-39. Figure 7-40 shows a typical stage calculation for the material balances in Fig. 7-39.

The subscripts R and L which were used to differentiate between the

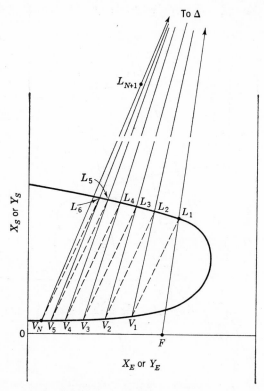

FIG. 7-38. Stage-to-stage construction for the material balances shown in Fig. 7-37.

two definitions of Δ and Δ' do not have the right- and left-hand significance which they have on the triangular diagram. The so-called right-hand difference lies above the mass-ratio diagram, while the so-called left-hand difference lies below.

7-13. Minimum Solvent Treat

The minimum solvent treat is determined on the mass-ratio diagram in exactly the same way as described for the triangular diagram in Sec. 7-7. The tie lines are extrapolated to their intersection with the line through V_N and L_{N+1}. The construction on a mass-ratio diagram is illustrated in Fig. 7-41.   At infinite solvent treat, Δ would be coincident with L_{N+1}. As the treat is decreased, Δ moves from L_{N+1} upward along the line through L_{N+1} and V_N.   The minimum treat occurs when Δ reaches the first intersection with an extrapolated tie line.   In Fig. 7-41 the tie lines are drawn in such a way as to cause this first intersection to occur above L_{N+1}.   This case is analogous to Fig. 7-26.

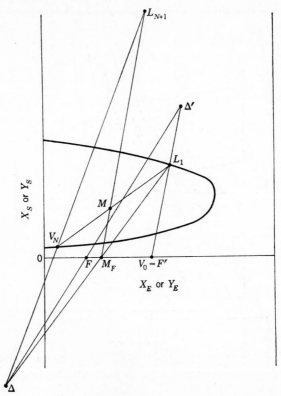

Fig. 7-39. Material-balance construction for the two-feed process in Fig. 7-23.

If the tie lines have large positive slopes, no intersections with the line through L_{N+1} and V_N will occur above L_{N+1}.   Therefore the treat can be reduced until $L_{N+1} = V_N$, at which time Δ is located at an infinite distance from L_{N+1}.   Further reduction causes Δ to reappear below V_N (now $\Delta = V_N - L_{N+1}$), and finally Δ will reach the first intersection which one of the extrapolated tie lines makes below V_N.   This situation is shown in Fig. 7-27 for the triangular diagram.   The analogous construction for the mass-ratio diagram is not shown but can easily be imagined from Fig. 7-41. For a more complete discussion of minimum solvent treat the reader is referred to Sec. 7-7.

7-14. Extract Reflux

Extract reflux may be useful in systems where the solvent shows no great preference for one component over the other, that is, in type II systems. A diagrammatic sketch of an extraction process with extract reflux is shown in Fig. 7-28. The corresponding material balances and the concepts of minimum and total extract reflux have been discussed in

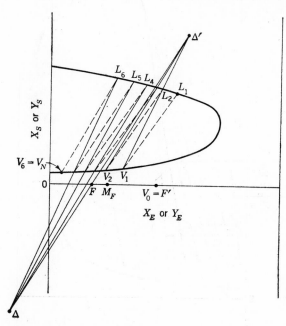

Fig. 7-40. A typical stage-to-stage construction for the material balances shown in Fig. 7-39.

Sec. 7-8. A typical graphical solution for the process in Fig. 7-28 at a reflux somewhere between minimum and total reflux is shown on triangular diagrams in Figs. 7-29 and 7-30. Figure 7-29 shows the material balances and location of the two difference points. Figure 7-42 shows the analogous construction on the mass-ratio diagram. As discussed previously in Sec. 7-8, the method used to locate the difference points depends upon the particular specifications with which the designer works. It is most convenient to specify the V_0/F ratio. The addition point $M_F = V_0 + F$ can then be located, and Eq. (7-11b) used to locate Δ.

$$\Delta = L_{N+1} - V_N = L_1 - M_F \qquad (7\text{-}11b)$$

Then Δ' is located by the application of

$$\Delta' = \Delta + F \qquad (7\text{-}8b)$$

If the degree of freedom utilized above to specify V_0/F is used instead to specify the external reflux ratio V_0/B, the location of Δ' and Δ is considerably more complicated. Equations (7-15) through (7-17) are then

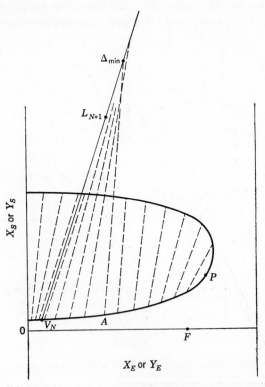

Fig. 7-41. Determination of the minimum solvent treat on a mass-ratio diagram. Tie line A gives the intersection closest to L_{N+1} and locates $\Delta_{\min}$. See Fig. 7-26 for the analogous triangular diagram.

used to locate Δ' on the line between V_0 and S_E. Then Δ is found from the simultaneous application of

$$\Delta = L_{N+1} - V_N$$
and
$$\Delta = \Delta' - F$$

The stage-to-stage construction on the mass-ratio diagram for the material balances shown in Fig. 7-42 is not shown. The principles of construction in a two-section column have been demonstrated amply in Fig. 7-40 for the mass-ratio diagram and in Figs. 7-25 and 7-30 for the triangular diagram.

Total Extract Reflux. A typical graphical construction for a column operating with total extract reflux but with continuous feed and continuous withdrawal of raffinate product is shown in Fig. 7-43. This

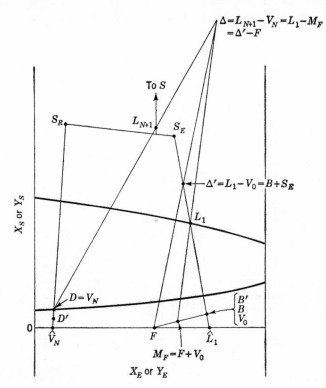

FIG. 7-42. Material-balance lines and location of Δ and Δ' for the process with extract reflux shown in Fig. 7-28.

figure is analogous to Fig. 7-31 except that Δ falls on the other side of the diagram. As described previously in Sec. 7-8, Δ' must coincide with S_E at total extract reflux, since $B = 0$ and

$$\Delta' = L_1 - V_0 = B + S_E \qquad (7\text{-}13)$$

Since $B = 0$, then

$$F + S = D'$$

where S is the make-up solvent. The points S_R and S_E are located by

244   DESIGN OF EQUILIBRIUM STAGE PROCESSES

specification (or rather from approximations of the performance of the solvent-recovery devices).

Minimum Extract Reflux. The concept of minimum extract reflux has been discussed thoroughly in Sec. 7-8 and illustrated on the triangular diagram in Fig. 7-32. The analogous construction for the mass-ratio diagram can be readily deduced from Figs. 7-42 and 7-43. At total extract reflux Δ' is coincident with S_E. To determine the minimum reflux, Δ' is moved toward L_1 until at some point in the column a tie line

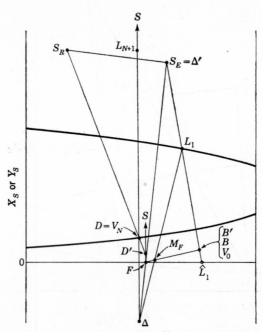

FIG. 7-43. Material balances for column operating with total extract reflux ($B = 0$) but with continuous addition of feed and continuous withdrawal of raffinate product.

and operating line become coincident. Usually this will occur with some tie line which lies between the lines $\overline{F\Delta'\Delta}$ and $\overline{M_F L_1 \Delta}$. Extrapolation of the tie lines in this region to locate that tie line which gives the highest intersection on the line $\overline{S_E L_1}$ will determine the minimum reflux in most cases. Usually the tie line which extrapolates through F will be the one to give the highest intersection, since it will be the last tie line in the extract end section. Tie lines in the raffinate end section should also be checked. Unusual distortion of the tie lines in that region can cause the coincidence between an operating and tie line to occur there before it happens in the extract end section.

7-15. Raffinate Reflux

An extraction process with raffinate reflux is shown in Fig. 7-33. As explained in Sec. 7-9, raffinate reflux cannot affect the number of equilibrium stages as extract reflux does. Also there is no minimum raffinate reflux in the sense that continued reduction of the reflux rate will bring

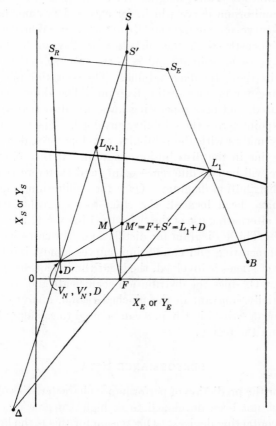

Fig. 7-44. Material-balance lines for the process with raffinate reflux pictured in Fig. 7-34.

about a coincidence between an operating line and a tie line at some point in the column. For a discussion of these characteristics of raffinate reflux the reader is referred to Sec. 7-9. The material balances for a column with raffinate reflux are illustrated on the mass-ratio diagram in Fig. 7-44. Figure 7-34 shows the analogous construction for the triangular diagram.

DISTRIBUTION DIAGRAMS

If a large number of stages are to be stepped off on a triangular or mass-ratio diagram, the accuracy can sometimes be improved through use of the distribution diagram. For this purpose, the distribution diagram is plotted in terms of the extract component E and takes on the general shape of the McCabe-Thiele diagram for distillation. The tie-line data furnish the equilibrium curve which, for a type I system, takes on the shape of the curve for component E in Fig. 7-5. When used for the stage-to-stage construction, the distribution diagram is usually constructed with x_E as the ordinate and y_E as the abscissa so that the equilibrium curve falls above the diagonal line. The operating line is obtained from the triangular or mass-ratio diagram. The difference point (or points) is located, and rays (operating lines) are drawn at random from the difference point across the solubility envelope. The two intersections which one ray makes with the solubility envelope provide one point on the operating line in the distribution diagram. If the column has two sections, there will be two difference points and therefore two operating lines on the distribution diagram. Once the equilibrium and operating-line curves have been located, the stage-to-stage calculation can be performed by alternating use of the two curves in exactly the same way as in the McCabe-Thiele diagram for distillation. In cases where a pinch point occurs (operating and equilibrium curves almost intersect) and a large number of stages is involved, the logarithmic plot described in Sec. 5-14 is useful. Or since the distribution coefficients and the phase rates will be essentially constant in the pinched region, analytical equations such as the one described in Chap. 8 can be used to calculate the number of stages within the region.

PERFORMANCE DATA

Methods for the prediction of performance characteristics of extraction processes have not been developed to so high a degree as methods for vapor-liquid contacting devices. One reason for this is the large number of mechanical arrangements utilized to mix and separate the two-liquid phases. Usually each device requires its own design methods. No generally accepted design procedure has been developed for any of the major contacting devices as yet. However, at the present rate of development it is likely that within a few years the treatment given the performance characteristics of vapor-liquid processes in the last three chapters of this book can be duplicated for some of the major liquid-liquid devices. At the present time the designer must depend upon the scattered reports in the literature concerning the particular mechanical device which he desires to use. Since many of these devices are patented, purchase of the equipment also provides information concerning the operating characteristics.

Perry (3) and Treybal (5) discuss the performance of extraction equipment. Treybal listed the stage efficiencies obtained in several discrete stage plants and found that the ratio of calculated equilibrium stages to actual stages fell between 0.71 and 1.0 in all cases. Efficiencies below 100 per cent in most industrial plants were attributed to incomplete settling rather than inadequate mixing. The effectiveness of mixing depends upon the degree of dispersion and the mass-transfer rates between phases. Settling time depends mainly upon the dispersion and the differences in gravity. High dispersion favors mass transfer but also favors entrainment of incompletely separated material from stage to stage. In a more recent paper (6), Treybal presents a method for the estimation of stage efficiencies in continuously operated, agitated, baffled vessels used in mixer-settler extractors. A complete up-to-date bibliography of papers dealing with the performance of all types of extraction devices is presented in the extraction section of the Unit Operation Review presented annually in *Industrial and Engineering Chemistry*.

NOMENCLATURE

B = final extract product rate after solvent recovery and after withdrawal of extract reflux (if any); moles, weight, or volume per unit time.

B' = extract product rate after solvent recovery but before withdrawal of extract reflux (if any); moles, weight, or volume per unit time.

C = number of components.

C.S.T. = critical solution temperature.

D = raffinate product rate after raffinate reflux (if any) has been withdrawn but before solvent recovery; moles, weight, or volume per unit time.

D' = final raffinate product rate after solvent recovery; moles, weight, or volume per unit time.

E = solute or extract component.

F = fresh feed rate; moles, weight, or volume per unit time. F' denotes feed at extract end when two feeds are used. $\hat{F}$ denotes solvent-free basis.

L = extract phase rate; moles, weight, or volume per unit time. $\hat{L}$ denotes solvent-free basis. In some of the figures, $\hat{L}_1$ is also used to denote the composition of L_1 if all the solvent were removed in the solvent recovery device. Since S_E is not, in general, pure solvent, $\hat{L}_1$ in these cases has a slightly different composition from L_1 on a solvent-free basis.

L_{N+1} = external stream entering stage N.

m = subscript referring to any stage in extract end of column.

M = number of equilibrium stages between feed stage and extract end. Also used to denote an addition point.

n = subscript referring to any stage in raffinate end of column.

N = total number of equilibrium stages in the column or train.

P = plait point.

$R = V'_N/D$ = raffinate reflux ratio. Also used to refer to major raffina
component.

R' = extract reflux ratio, amount refluxed to amount of B. $R' = V_0/B$
absence of F'.

S = fresh solvent rate; moles, weight, or volume per unit time.   Also used
denote solvent component.

$S' = S_E + S_R + S$ = recovered solvent plus make-up solvent. $S' = L_{N+1}$
V'_N.

S_E = solvent recovered from extract L_1; moles, weight, or volume per unit tim

S_R = solvent recovered from raffinate D; moles, weight, or volume per unit tim

t = temperature.

V = raffinate phase rate; moles, weight, or volume per unit time. $\hat{V}$ denot
solvent-free basis.   In some figures $\hat{V}_N$ is also used to denote the compos
tion of V_N if all the solvent were removed in the solvent-recovery devic
Since S_R is not, in general, pure solvent, $\hat{V}_N$ in these cases has a slightly di
ferent composition from V_N on a solvent-free basis.

x_i = concentration of i in extract phase.   Units consistent with units on phas
rates.

X_i = concentration on solvent-free basis; amount of i in extract phase/combine
amounts of R and E.

y_i = concentration of i in raffinate phase.   Units consistent with units on phas
rates.

Y_i = concentration on solvent-free basis; amount of i in raffinate phase/com
bined amounts of R and E.

z_i = concentration of i in feed stream.   Units consistent with units on phas
rates.

Z_i = feed concentration on solvent-free basis; amount of i in feed/combine
amounts of R and E.

Greek Symbols

Δ = hypothetical stream which represents the difference between two passin
streams. $\Delta_L = V_n - L_{n+1}$. $\Delta_R = L_{n+1} - V_n = -\Delta_L$. Δ refers to raf
finate end section when column has more than one section.

Δ' = same as Δ but for extract end section when column has more than on
section.

Δ_{min} = difference point corresponding to minimum solvent treat or minimum
extract reflux rate.

∞ = infinity.

PROBLEMS

The following problems will provide numerical examples for some of the extraction
processes discussed in this chapter. Either the triangular, mass-ratio, or distribution
diagram can be used to solve these examples. The student should apply at least two
of these three methods to each problem.

7-1. *One Feed—No Reflux.* Water (S) is to be used to separate a chloroform
(R)–acetone (E) mixture in a simple countercurrent extraction column such as the
one shown in Fig. 7-15. The feed ($F = V_0$) contains equal amounts of chloroform
and acetone on a weight basis. For the purposes of this problem, the solvent L_{N+1}
can be assumed to be pure water. The column will operate at essentially room
temperature (25°C). For a solvent/feed ratio of 1.565, calculate the product rates

nd compositions obtained with eight equilibrium stages. What is the minimum
ilvent treat? Brancker, Hunter, and Nash [*J. Phys. Chem.*, **44**, 683 (1940)] reported
ae data in Table 7-1 as weight fractions at 25°C.

TABLE 7-1. TIE-LINE AND SOLUBILITY DATA (WEIGHT FRACTIONS) FOR
THE ACETONE (*E*)–CHLOROFORM (*R*)–WATER (*S*) SYSTEM AT 25°C

Extract layer			Raffinate layer		
E	*R*	*S*	*E*	*R*	*S*
0.030	0.010	0.960	0.090	0.900	0.010
0.083	0.012	0.905	0.237	0.750	0.013
0.135	0.015	0.850	0.320	0.664	0.016
0.174	0.016	0.810	0.380	0.600	0.020
0.221	0.018	0.761	0.425	0.550	0.025
0.319	0.021	0.660	0.505	0.450	0.045
0.445	0.045	0.510	0.570	0.350	0.080

Solubility data outside the range of the above tie-line data are given below.

Acetone (*E*)	Chloroform (*R*)	Water (*S*)
0.573	0.354	0.073
0.605	0.285	0.110
0.600	0.220	0.180
0.592	0.178	0.230
0.585	0.145	0.270
0.566	0.110	0.324
0.556	0.100	0.344
0.540	0.086	0.374
0.532	0.080	0.388
0.516	0.070	0.414
0.490	0.056	0.454

7-2. Two Feeds–No Reflux. Monochlorobenzene (*S*) is to be used to recover
acetone (*E*) from water (*R*) in a countercurrent extraction at 25°C. Two acetone-
water mixtures must be processed. The feed-introduction points will be as shown in
Fig. 7-23 with $M = 2$. The feed compositions in weight fractions are

	F	*F'*
Acetone (*E*)	0.0600	0.200
Water (*R*)	0.9382	0.800
Monochlorobenzene (*S*)	0.0018	0

The solvent (L_{N+1}) can be assumed to be all monochlorobenzene and an L_{N+1}/F'
ratio of 1.208 will be used. The F/F' ratio is 0.10. How many equilibrium stages
are required to reduce the acetone concentration in the raffinate to 0.01 and what
will the product rates be?

Tie-line data (mass fractions) for this system have been presented by Othmer, White, and Trueger [*Ind. Eng. Chem.*, **33**, 1240 (1941)] and are listed in Table 7-2.

TABLE 7-2. TIE-LINE DATA FOR ACETONE (E)–WATER (R)–MONOCHLOROBENZENE (S) AS WEIGHT FRACTIONS AT 25°C

Raffinate layer			Extract layer		
E	R	S	E	R	S
0	0.9989	0.0011	0	0.0018	0.9982
0.05	0.9482	0.0018	0.0521	0.0032	0.9447
0.10	0.8979	0.0021	0.1079	0.0049	0.8872
0.15	0.8487	0.0024	0.1620	0.0063	0.8317
0.20	0.7969	0.0031	0.2223	0.0079	0.7698
0.25	0.7458	0.0042	0.2901	0.0117	0.6982
0.30	0.6942	0.0058	0.3748	0.0172	0.6080
0.35	0.6422	0.0078	0.4328	0.0233	0.5439
0.40	0.5864	0.0136	0.4944	0.0305	0.4751
0.45	0.5276	0.0224	0.5492	0.0428	0.4080
0.50	0.4628	0.0372	0.5919	0.0724	0.3357
0.55	0.3869	0.0631	0.6179	0.1383	0.2438
0.60	0.2741	0.1259	0.6107	0.2285	0.1508
0.6058	0.2566	0.1376	0.6058	0.2566	0.1376

7-3. *One Feed–Extract Reflux.* A 40 mole per cent methylcyclopentane (E)–60 mole per cent normal hexane (R) feed is to be contacted with aniline (S) as shown in Fig. 7-28. The solvent L_{N+1} contains 0.005 mole fraction methylcyclopentane as an impurity. The solvent recovered from the extract contains 0.10 mole fraction methylcyclopentane. The column will operate at 25°C. A reflux ratio V_0/B of 10 will be used. What is the solvent/feed ratio L_{N+1}/F and how many stages are required to produce a raffinate (D) with no more than 0.15 mole fraction methylcyclopentane and a saturated extract product (B) with 0.70 mole fraction methylcyclopentane? The equilibrium data for this system have been reported by Darwent and Winkler [*J. Phys. Chem.*, **47**, 442 (1943)] and are listed in Table 7-3.

TABLE 7-3. TIE-LINE DATA (MOLE PER CENT) FOR METHYLCYCLOPENTANE (E)–n-HEXANE (R)–ANILINE (S) SYSTEM AT 25°C

x_E	x_R	x_S	y_E	y_R	y_S
0	0.082	0.918	0	0.930	0.070
0.016	0.076	0.908	0.090	0.838	0.072
0.022	0.073	0.905	0.187	0.736	0.077
0.055	0.060	0.885	0.326	0.535	0.139
0.080	0.050	0.870	0.430	0.480	0.090
0.117	0.037	0.846	0.618	0.278	0.104
0.141	0.028	0.831	0.744	0.136	0.120
0.209	0.007	0.784	0.795	0.075	0.130
0.244	0	0.749	0.856	0	0.144

7-4. *One Feed–Raffinate Reflux.*   Ethyl ether (S) is to be used to recover ethanol (E) from a 30 per cent ethanol–70 per cent water feed at 25°C.   The solvent $(S'$ in Fig. 7-33) can be assumed to be pure ethyl ether.   If a raffinate reflux ratio V'_N/D of 0.05 is used, and if L_{N+1} is a saturated phase, how much solvent and how many stages are required to reduce the concentration of ethanol in the raffinate product to 2.5 weight per cent?   Equilibrium data in weight per cent for this system are tabulated in "International Critical Tables," vol. III, p. 405, and are also listed in Table 7-4 for convenience.

TABLE 7-4. TIE-LINE DATA (WEIGHT FRACTIONS) FOR ETHANOL (E)–WATER (R)–ETHYL ETHER (S) SYSTEM AT 25°C

x_E	x_R	x_S	y_E	y_R	y_S
0	0.013	0.987	0	0.940	0.060
0 029	0.021	0.950	0.067	0.871	0.062
0 067	0.033	0.900	0.125	0.806	0.069
0.102	0.048	0.850	0.159	0.763	0.078
0.136	0.064	0.800	0.186	0.726	0.088
0.168	0.082	0.750	0.204	0.700	0.096
0.196	0.104	0.700	0.219	0.675	0.106
0.220	0.130	0.650	0.231	0.650	0.119
0.241	0.159	0.600	0.242	0.625	0.133
0.257	0.193	0.550	0.256	0.590	0.154
0.269	0.231	0.500	0.265	0.552	0.183
0.278	0.272	0.450	0.274	0.515	0.211
0.282	0.318	0.400	0.280	0.470	0.250

REFERENCES

1. Gibbs, J. W.: *Trans. Conn. Acad. Arts. Sci.,* **3,** 152 (1876).
2. Janecke, E.: *Z. anorg. Chem.,* **51,** 132 (1906).
3. Perry, J. H.: "Chemical Engineers' Handbook," 3d ed., p. 747, McGraw-Hill Book Company, Inc., New York, 1950.
4. Randall, M., and B. Longtin: *Ind. Eng. Chem.,* **30,** 1063, 1188, 1311 (1938); **31,** 908, 1295 (1939); **32,** 125 (1940).
5. Treybal, R. E.: "Liquid Extraction," McGraw-Hill Book Company, Inc., New York, 1951.
6. Treybal, R. E.: *A.I.Ch.E. J.* **4,** 202 (1958).

CHAPTER 8

MULTICOMPONENT SEPARATIONS
General Short-cut Method

Short-cut methods for the approximate solution of multicomponent, multistage separation problems continue to serve useful purposes even though electronic computers are available to provide rigorous solutions. The available equilibrium data for some problems may not be sufficiently accurate to justify the longer rigorous methods, or in design studies, a large number of cases can be worked quickly by a short-cut method to pinpoint the optimum conditions and then the exact solution found by some longer rigorous method. The latter procedure saves expensive computer time, since the time required for a rigorous method usually exceeds that for a short-cut method by a factor of 10 or more. If a computer is not available, the designer usually must rely on the short-cut method alone, since the time required for the manual solution of a rigorous method is prohibitive.

Short-cut methods also have an academic usefulness in addition to the practical uses listed above. Most short-cut equations make use of average rates and distribution coefficients to predict the separation. The designer must exercise considerable judgment in the selection of meaningful averages if a good prediction is desired. The student, usually for the first time, finds that he must learn to think as the process acts. The effect of the K values on the individual component recoveries, the effect of temperature and pressure on the K values and the stream enthalpies, the probable behavior of the phase rate profiles, all these must be deduced beforehand if the short-cut method is to provide an easy, accurate answer. In the next chapter, other short-cut equations based on the conditions of total and minimum reflux will be presented, and these short-cut equations serve to explain the limiting conditions of total and minimum reflux for multicomponent systems just as the binary diagrams serve for binary systems.

This chapter presents the short-cut calculation method of Smith and Brinkley (12), which is applicable to all equilibrium-stage processes. The method is based upon an analytical equation which predicts the recovery of the individual components in the bottom-product streams. The equation was originally derived for the extraction process pictured in

Fig. 8-1. This process has three feed streams (including the solvent) and both extract and raffinate reflux. The equation takes into account the splits obtained on the individual components in the solvent-recovery device which removes S_E from L_1. Any component can enter in any one or all three of the feed streams, and any number of components and stages can be handled. The process in Fig. 8-1 is the most complicated process for which the equation is valid. For simpler processes the equation degenerates to correspondingly simpler forms. The forms of the equation applicable to the various processes are shown later in the sections which deal with the individual processes. Justification of the nomenclature used for the extraction processes is given at the beginning of Chap. 7.

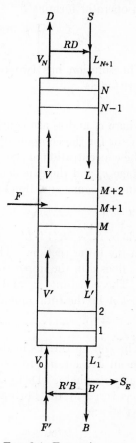

FIG. 8-1. Extraction process with two feeds and reflux at both ends.

8-1. Derivation of General Short-cut Equation

The calculus of finite differences can be used to relate the concentration of a given component to the stage number within any given section of an equilibrium-stage column. If the column contains two sections, the solutions of the difference equations for each of the two sections can be combined with a balance around the feed stage to provide an equation for the entire column. To permit a simple mathematical solution of the difference equations, it is necessary to assume constant flow rates and distribution coefficients within each column section. Even with this simplifying assumption, the derivation involves considerable algebraic manipulation and will be only outlined here. The complete derivation is shown in Appendix A. For background in the calculus of finite differences, the reader is referred to any standard text on the subject. Engineering applications of the calculus are illustrated in Marshall and Pigford (11), Tiller and Tour (16), and Tiller (15).

Writing a balance for any given component around stage $n + 1$, eliminating the y's, and rearranging provide the following difference equation for that component in the upper section of the column:

$$x_{n+2} - \left(\frac{K_{n+1}V_{n+1}}{L_{n+2}} + \frac{L_{n+1}}{L_{n+2}}\right) x_{n+1} + \frac{K_n V_n}{L_{n+2}} x_n = 0$$

The component subscript is omitted for the sake of simplicity.

The coefficients in the difference equation must be constant if a simple mathematical solution is desired. This requirement can be satisfied by assuming constant phase rates and distribution coefficients within the upper section of the column. The difference equation can then be written in operator form as

$$\left[E^2 - \left(\frac{KV}{L} + 1\right) E + \frac{KV}{L} \right] x_n = 0$$

The solution has two roots, KV/L and 1.0, and the solution can be written as

$$x_n = c_1(S_n)^n + c_2 \qquad (8\text{-}1)$$

where c_1 and c_2 are constants and $S_n = KV/L$ = the average stripping factor for the component in the upper section of the column. The x_n is the concentration of the component in the heavy or extract phase leaving stage n, and the exponent n on the stripping factor is the stage number. An analogous equation can be written for the lower section.

$$x_m = c_3(S_m)^m + c_4 \qquad (8\text{-}2)$$

where $S_m = K'V'/L'$ = the average stripping factor for the component in the lower section. The m has the same meaning in the lower section as does n in the upper section.

The constants c_1 and c_2 in Eq. (8-1) can be eliminated as follows: Let A be the total amount of the component entering the column.

$$A = F y_F + F' y_{F'} + S x_S$$

The concentrations in the feeds are represented by y's because the feeds in an extraction process (see Fig. 8-1) are usually more similar to the raffinate phase. Defining f as the fraction of the component which will be recovered at the bottom of the column, the following equality can be written:

$$D y_N = (1 - f)A$$

Substituting $(1 - f)A$ for $D y_N$ in

$$V_N y_N = D y_N + R D y_N = (1 + R) D y_N$$

gives

$$y_N = \frac{(1 + R)(1 - f)A}{V_N}$$

or

$$x_N = \frac{(1 + R)(1 - f)A}{K V_N}$$

Substitution for x_N in Eq. (8-1) gives

$$\frac{(1 + R)(1 - f)A}{KV} = c_1(S_n)^N + c_2$$

An analogous equation with S_n raised to the $N - 1$ power can be obtained from a balance around stage N. The two equations can then be combined to eliminate the two constants c_1 and c_2 and provide an equation for the upper section of the column. A similar procedure at the bottom end eliminates the constants c_3 and c_4 from Eq. (8-2) and provides an equation for the lower section of the column. The equations for the upper and lower sections can be combined with a component balance around the feed stage to provide the following equation for the entire process pictured in Fig. 8-1:

$$f = \frac{(1 - S_n{}^{N-M}) + q_S(S_n{}^{N-M} - S_n) + R(1 - S_n) + hq_{F'} S_n{}^{N-M}(1 - S_m{}^M)}{(1 - S_n{}^{N-M}) + hS_n{}^{N-M}(1 - S_m{}^M) + R(1 - S_n) + h[(1 + R')/(1 + gR')]S_m{}^M S_n{}^{N-M}(1 - S_m)}$$

$$(8\text{-}3)$$

The q_S and $q_{F'}$ are the fractions of the component which enter in the solvent and lower feed, respectively. The R and R' are the reflux ratios at the top and bottom ends of the column, and the g is the assumed recovery factor for the component in the solvent-recovery device. If a complete separation were obtained in the device between solvent and nonsolvent materials, the g for each of the solvent components would be 1.0 while the g's for all the other components would be zero. The value of h to be used depends upon the nature of the feed. For a feed more similar in nature to the light (vapor) or raffinate phase,

$$h = \frac{L}{L'} \frac{1 - S_n}{1 - S_m} \qquad (8\text{-}4)$$

For a feed more similar to the heavy (liquid) or extract phase,

$$h = \frac{K'}{K} \frac{L}{L'} \frac{1 - S_n}{1 - S_m} \qquad (8\text{-}5)$$

In distillation, Eq. (8-4) is used for a vapor feed while (8-5) is used for a liquid feed.

The quantity f in Eq. (8-3) represents the fraction of the component which will be recovered at the lower end of the column. Edmister (6) has defined similar recovery factors for absorption and distillation. The fractional recoveries of the various components will vary as their respective stripping factors vary. The use of Eq. (8-3) is straightforward once the average or effective stripping factors are known. The selection of the correct effective stripping factors is not so straightforward, and the methods of selection vary from process to process.

8-2. Simple Processes

Equation (8-3) will seldom be used as it is written in the preceding section because processes as complicated as that shown in Fig. 8-1 will seldom be encountered. For simpler processes, Eq. (8-3) degenerates to correspondingly simpler forms. The simplest multistage processes with which the designer must deal are represented by Fig. 8-2. Absorption,

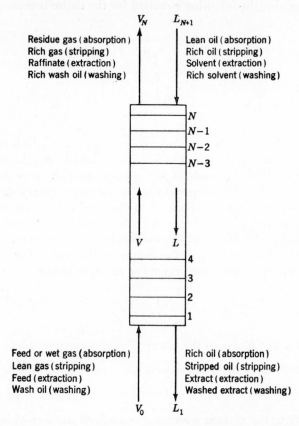

FIG. 8-2. Flow diagram for equilibrium-stage processes with one feed and no refluxes.

stripping, extraction, and washing all involve the countercurrent contact of a light and heavy phase, and all four processes can operate without reflux at either end. Figure 8-2 illustrates the analogy between the lean oil in absorption and the solvent in extraction. Both are used to absorb or extract certain components from a feed stream. The feed stream is a vapor in the case of absorption and a liquid in extraction, but in both cases the direction of net material transfer is from the feed to the solvent or absorber oil. If the concentrations of the entering streams are properly controlled, the net transfer of material will be in the opposite direction from the oil or solvent phase to the gas or raffinate phase. In this case the process is called stripping (vapor-liquid contact) or washing (liquid-liquid contact).

One thing detracts somewhat from the analogy between the four processes shown in Fig. 8-2. The extract or solvent-rich phase in extraction and washing is not always heavier than the raffinate or wash oil. When it is lighter, the solvent or solvent-rich phase must be introduced at the bottom of the column in a physical sense and will travel in a direction opposite to that traveled by the absorber oil in an actual column. For the purposes of calculation, this discrepancy in the analogy should be neglected and the diagram always drawn as shown in Fig. 8-2. The stages are then said to be numbered from the bottom or from the extract end, whichever phrase applies. Numbering the stages in this consistent manner and using the symbols L and V as shown in Fig. 8-2 permit identical equations to be written for all four processes.

The form of Eq. (8-3) which applies to the processes shown in Fig. 8-2 is obtained by setting $M = R = R' = F' = q_{F'} = 0$ and $S_n = S_m$. Equation (8-3) then reduces to the following form of the familiar Kremser (8) equation:

$$f = \frac{(1 - S^N) + q_S(S^N - S)}{1 - S^{N+1}} \tag{8-6}$$

where $S = KV/L =$ stripping factor and q_S is the fraction of a component which enters in the solvent or absorber oil stream, S or L_{N+1}. For a component which enters only in the feed, F or V_0, $q_S = 0$ and

$$f = \frac{1 - S^N}{1 - S^{N+1}} \tag{8-7}$$

For a component entering only in stream L_{N+1}, $q_S = 1.0$ and

$$f = \frac{1 - S}{1 - S^{N+1}} \tag{8-8}$$

When $S = 1.0$, these equations take on the indeterminate form $0/0$. The value of f at $S = 1.0$ is found by differentiating the numerator and denominator and evaluating the new fraction thus obtained. For example, for $S = 1.0$, Eq. (8-7) becomes $f = N/N + 1$.

The Kremser short-cut equation is often written (7, 14) in a form different from that shown in Eq. (8-6). For absorption it can be written for any component as

$$\frac{Y_0 - Y_N}{Y_0 - Y_{N+1}} = \frac{A^{N+1} - A}{A^{N+1} - 1} \tag{8-9}$$

where Y_{N+1} is the moles of the component per mole of entering gas which would exist in a vapor stream in equilibrium with the entering absorber oil. For stripping, the equation is written as

$$\frac{X_{N+1} - X_1}{X_{N+1} - X_0} = \frac{S^{N+1} - S}{S^{N+1} - 1} \tag{8-10}$$

where X_0 bears the same relation to the entering gas stream as Y_{N+1} does to the entering liquid stream. The stages are numbered from the bottom for both equations as they are written here. The left sides of Eqs. (8-9) and (8-10) are often called the efficiency of absorption or stripping, since they represent the ratio of the amount actually absorbed or stripped to the amount which would be absorbed or stripped if the exit stream were in equilibrium with the entering stream. Souders and Brown (14) have presented a graphical solution of Eqs. (8-9) and (8-10) which is convenient to use once the effective absorption or stripping factors are known.

If the stages are numbered from the bottom of the column for both absorption and stripping, it can be shown that both Eqs. (8-9) and (8-10) are identical with Eq. (8-6). For the sake of simplicity in nomenclature and uniformity in the calculation of both stripping and absorption problems as well as extraction and washing problems, the use of the Kremser short-cut equation in the form of Eq. (8-6) is recommended.

8-3. Absorption and Stripping

The mass transfer in absorption and stripping is almost completely in one direction. Consequently, the phase rates change rapidly through the column if an appreciable amount of material is transferred. In absorption, the wet gas which rises through the column loses material to the descending absorber oil but little of the relatively nonvolatile oil passes to the gas phase. As a result, the gas phase decreases in amount as it rises and the liquid phase increases a corresponding amount as it descends. The direction of transfer is reversed in stripping. The entering gas contains little or none of the components to be stripped from the rich oil, and these components pass from the oil to the gas phase to an extent described by their respective distribution coefficients. Stripping gases normally used are essentially insoluble in the oil phase, and little transfer of the gas to the liquid phase occurs. Consequently, the oil phase decreases and the gas phase increases in amount as they pass through the column. The assumptions of constant phase rates or a constant L/V ratio in absorption or stripping are valid only in the special case where the amount of material transferred is so small as to have a negligible effect on the phase rates.

Besides the effect on the phase rates, the unidirectional transfer of material creates thermal effects which must be considered. The material transferred undergoes a phase change, and the heats of vaporization or condensation are involved along with the smaller heats of solution. In absorption, most of the heat released as the gas components condense into the liquid phase must appear as sensible heat in the liquid, since insufficient material is being vaporized to utilize the released heat as heat of vaporization. Consequently, the temperature of the liquid phase rises as it passes down the column. The rise in liquid temperature occurs despite the fact that the liquid is at its bubble point (in an equilibrium stage). The vaporization of only a small amount of a light component (say methane) will cause a relatively large rise in the bubble point. The amount of sensible heat required to raise the liquid temperature is large compared with the heat required to vaporize the small amount of volatile material. The effect is opposite in stripping. The heat of vaporization of the light components as they pass from the liquid to the gas phase is supplied from the sensible-heat content of the liquid phase. Consequently, the temperature of the liquid phase decreases as it passes down the column. In the special case where only a very small amount of material is transferred, the temperature changes will be small and possibly negligible.

Because of the unidirectional nature of the transfer in absorption and stripping columns and the resulting variations in phase rates and temperatures, graphical solution methods are less useful than in the cases of binary distillation and ternary extraction systems. The change in the phase rates as they pass through the column causes the L/V ratio to change, and therefore the operating-line equation

$$y_n = \frac{L_{n+1}}{V_n} x_{n+1} + \frac{V_0 y_0 - L_1 x_1}{V_n}$$

does not plot as a straight line on an x-y diagram. Only in the special case of very small transfer will the L/V ratio remain essentially constant.

Absorption systems, like extraction systems, must contain at least three components—the two to be separated plus the absorber oil or solvent. Each component will have its own equilibrium curve as defined by its distribution coefficient. If the temperature changes from stage to stage, the distribution coefficient $K = y/x$ for each component changes and the equilibrium curves are not straight lines. Only in the special case of an isothermal column (and excluding liquid composition effects) will the equilibrium curves be straight. If the wet gas to an absorber is very dilute in the material being absorbed, the operation utilizes only those portions of the equilibrium curve very close to the origin, and these portions often can be drawn as straight lines (Henry's law holds).

To illustrate the above statements consider a three-stage absorber which operates at 100 psia and charges the following feed:

Component	y_0
C_1	0.70
C_2	0.15
C_3	0.10
C_4	0.05
	1.00

and has an L_{N+1}/V_0 ratio of 2.0. Figure 8-2 can be used as a schematic representation of this process. The wet gas and the absorber oil both enter at 80°F, and the oil is completely stripped of light hydrocarbons. To get a quick estimate of the amount absorbed, let us assume that $L/V = L_{N+1}/V_0$ and $t = 80$°F throughout the column. Both the operating lines and the equilibrium curves will be straight lines in this case and can be plotted as shown in Figs. 8-3a and b. The slope of the equilibrium curve (straight full line) for each component is the value of K at 100 psia and 80°F. The slope of all the operating lines (dashed lines) is the same and equals 2.0. The position of each operating line with respect to the corresponding equilibrium curve must be such as to satisfy the known concentrations in L_{N+1} and V_0 and also allow three stages to be stepped off between these concentrations. Consider the construction for propane in Fig. 8-3a. The point ($y_0 = 0.10$, x_1) represents passing streams V_0 and L_1 on the operating line (dotted), and it must lie somewhere along the horizontal line $y = 0.10$. Similarly, the point (y_3, $x_{N+1} = 0$) represents passing streams V_3 and L_{N+1} on the operating line, and since the entering absorber oil has been completely stripped, it must fall somewhere on the ordinate at $x = 0$. The slope of the operating line has been assumed to be 2.0, so by trial and error that position can be found where exactly three steps can be constructed between the intersections of the operating line with the lines $y = 0.10$ and $x = 0.0$. Similar trial-and-error constructions can be made for the other components. Actually the construction is not trial and error for the lightest and heaviest components. It can be seen that most of the methane is transferred in the top stage and by the time the oil leaves the second stage it is essentially saturated with methane. Therefore little change in methane concentration occurs across the bottom stage, and for all practical purposes the point (y_0, x_1) on the operating line coincides with the equilibrium curve. In the case of the butane (Fig. 8-3b) most of the transfer occurs at the bottom of the column, and to construct three stages it must be assumed that point y_3, x_{N+1} essentially coincides with the origin. This tendency of some components to accomplish the major portion of their transfer in some particular part of the column will

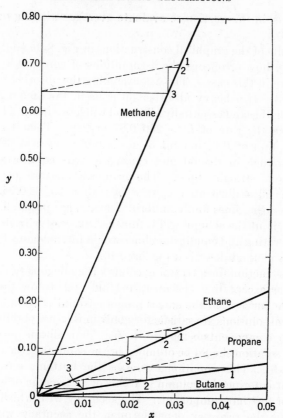

FIG. 8-3a. Graphical construction for absorber when constant temperature and L/V are assumed. The dashed lines are operating lines.

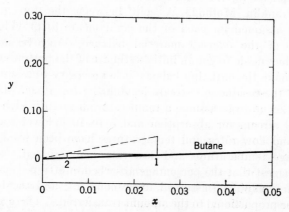

FIG. 8-3b. Graphical construction of stages for butane from Fig. 8-3a.

be used later in this section to evaluate average stripping factors for use in Eq. (8-6).

The results of the graphical constructions in Fig. 8-3a and b can be used to illustrate how erroneous the assumptions of constant L/V and temperature are in this case. The Σx_1 excluding the absorber oil is approximately 0.13. If a heavy absorber oil is used, the amount of oil in V_3 will be negligible and essentially all the oil will leave in L_1. For $V_0 = 100$, $L_{N+1} = 200$, the rate of $L_1 = 200/0.87$ or 229. Then $V_3 = 71$ moles. The absorption of this amount of gas would cause a rather large temperature change in the oil and cause the true equilibrium curves to deviate from straight lines. The erroneous nature of the answers obtained is also illustrated by the fact that $\Sigma y_3 \cong 0.75$. When the correct operating lines and equilibrium curves are used, all the Σy's and Σx's throughout the column will be unity. Successive trials to locate the correct operating and equilibrium lines would be analogous to the rigorous calculation method described in Chap. 13.

The construction for a stripping column is analogous to that shown for an absorber except that the operating lines fall below their respective equilibrium curves and the stages progress from left to right.

Graphical solutions are satisfactory only in the case of dilute concentrations of the components being transferred. For all other cases a rigorous numerical solution should be obtained or the solution estimated by means of Eq. (8-6). This equation expresses the bottoms recovery of each component entering the column in terms of an average or effective stripping factor for that component. The accuracy with which the equation can predict the recovery depends upon the accuracy with which the effective stripping factor can be estimated. The most widely known methods for predicting effective factors for hydrocarbon systems are those of Edmister (3, 4, 5) and Horton and Franklin (7).

Horton-Franklin Method. A split between the top and bottom products is assumed for each of the components to provide an initial assumption of the over-all material balance. An over-all enthalpy balance is then made to get an initial estimate of the bottom-oil temperature. To make the enthalpy balance it is necessary to assume the temperature of the residue gas stream leaving the top stage. It is usually sufficiently accurate to assume a temperature 5 to 10°F higher than the entering oil stream for absorption and 5 to 10°F lower for stripping. Once the end flow rates and temperatures have been fixed, the intermediate-stage temperatures and flow rates are assumed. Horton and Franklin suggest that the percentage absorbed or stripped on each plate through the column be assumed to be constant, and the temperature change to be proportional to the amount transferred. These assumptions can be expressed mathematically for absorption as

$$\left(\frac{V_N}{V_0}\right)^{1/N} = \frac{V_{n+1}}{V_n} \qquad (8\text{-}11a)$$

$$\frac{V_0 - V_n}{V_0 - V_N} = \frac{t_1 - t_{n+1}}{t_1 - t_{N+1}} \qquad (8\text{-}12a)$$

For stripping, the assumptions can be written as

$$\left(\frac{L_1}{L_{N+1}}\right)^{1/N} = \frac{L_n}{L_{n+1}} \qquad (8\text{-}11b)$$

$$\frac{L_{N+1} - L_n}{L_{N+1} - L_1} = \frac{t_{N+1} - t_n}{t_{N+1} - t_1} \qquad (8\text{-}12b)$$

The t_{N+1} is the temperature of the entering oil.

The stripping factor for any component on any stage can be calculated as soon as the stage temperatures and flow rates have been estimated from Eqs. (8-11) and (8-12). Judgment must now be exercised in the selection of that point in the column where the effective stripping factor for each component is to be evaluated. Horton and Franklin found that the effective factor should be evaluated at that point in the column where the transfer of that component becomes marginal, that is, at a point where the major portion of the transfer of the component has been accomplished. For the lightest components in absorption and the heaviest components in stripping, little transfer occurs and the effective stripping factors should be chosen at the bottom and at the top, respectively. For those components which experience the greatest transfer, the effective stripping factors should be chosen near the middle of the column. Horton and Franklin have suggested the following guides for components other than the stripping gas and absorber oil:

Absorption		Stripping	
S	$\dfrac{N - n + 1}{N}$	S	$\dfrac{n}{N}$
Above 10	1.0	0–0.1	1.0
2.5–10	0.9	0.1–0.4	0.9
1.0–2.5	0.8	0.4–1.0	0.8
0.25–1.0	0.7	1.0–4.0	0.7
Below 0.25	0.6	Above 4.0	0.6

The n represents that stage at which the effective stripping factor should be evaluated. If the reasoning is extended to the stripping gas and absorber oil, the effective stripping factors for these two materials should be evaluated at their point of entry, since little of either is transferred.

The estimated stripping factors can be used in Eq. (8-6) to calculate the fraction of each component which leaves the column in the bottom-product stream L_1. Then for each component

$$L_1 x_1 = f(V_0 y_0 + L_{N+1} x_{N+1}) \tag{8-13}$$
$$V_N y_N = (1 - f)(V_0 y_0 + L_{N+1} x_{N+1}) \tag{8-14}$$

where the subscripts refer to the stage number. The calculated recoveries must be checked against the splits originally assumed for each component, and if agreement is not sufficiently close, the calculations should be repeated with the calculated recoveries.

The procedure outlined above for the selection of effective stripping factors generally provides a good approximation of the various component recoveries even though the assumptions as to the amount of transfer on each stage may often deviate rather widely from the true values.

Edmister Method. The method of Edmister (3, 4, 5) for the estimation of effective stripping factors is easier to use and requires less judgment than the method of Horton and Franklin. The method as presented by Edmister predicts values of the absorption factors, $A = L/KV$. The derivation below is in terms of the stripping factor.

Horton and Franklin (7) have derived rigorous expressions for the efficiency of absorption or stripping in terms of the absorption and stripping factors. Their expression for an N-stage stripper when the stages are numbered from the bottom to the top is

$$
\begin{aligned}
\frac{X_{N+1} - X_1}{X_{N+1}} &= \frac{S_1 S_2 \cdots S_N + S_2 S_3 \cdots S_N + \cdots + S_N}{S_1 S_2 \cdots S_N + S_2 S_3 \cdots S_N + \cdots + S_N + 1} \\
&\quad - \frac{V_0 Y_0}{L_{N+1} X_{N+1}} \frac{S_2 S_3 \cdots S_N + S_3 S_4 \cdots S_N + \cdots + S_N + 1}{S_1 S_2 \cdots S_N + S_2 S_3 \cdots S_N + \cdots + S_N + 1}
\end{aligned} \tag{8-15}
$$

The effective stripping factor is that value which if substituted for each S_n in the above equation would leave the numerical value of the stripping efficiency $(X_{N+1} - X_1)/X_{N+1}$ unchanged. Letting S without a subscript denote the effective factor,

$$
\begin{aligned}
\frac{X_{N+1} - X_1}{X_{N+1}} &= \frac{S^N + S^{N-1} + \cdots + S}{S^N + S^{N-1} + \cdots + S + 1} \\
&\quad - \frac{V_0 Y_0}{L_{N+1} X_{N+1}} \frac{S^{N-1} + S^{N-2} + \cdots + S + 1}{S^N + S^{N-1} + \cdots + S + 1}
\end{aligned} \tag{8-16}
$$

Since
$$S^N + S^{N-1} + \cdots + S = \frac{S^{N+1} - S}{S - 1} \tag{8-17}$$

and
$$S^N + S^{N-1} + \cdots + S + 1 = \frac{S^{N+1} - 1}{S - 1} \tag{8-18}$$

the equation can be rewritten as

$$\frac{X_{N+1} - X_1}{X_{N+1}} = \left(1 - \frac{V_0 Y_0}{S' L_{N+1} X_{N+1}}\right) \frac{S^{N+1} - S}{S^{N+1} - 1} \tag{8-19}$$

where $S' = S$ when the equation is written in terms of the effective stripping factor.

A rigorous expression can be obtained for the effective factor S for a two-stage column. Comparison of Eqs. (8-19) and (8-15) shows that for two stages

$$\frac{S^3 - S}{S^3 - 1} = \frac{S_1 S_2 + S_2}{S_1 S_2 + S_2 + 1} \tag{8-20}$$

and

$$\frac{S^3 - S}{S'(S^3 - 1)} = \frac{S_2 + 1}{S_1 S_2 + S_2 + 1} \tag{8-21}$$

From Eq. (8-21) it can be seen that in terms of the stage stripping factors,

$$S' = \frac{S_2(S_1 + 1)}{S_2 + 1}$$

for two stages. The rigorous expression for S is obtained by clearing the fractions in (8-20) and factoring out the quantity $(S - 1)$ to give the quadratic equation

$$S^2 + S - S_2(S_1 + 1) = 0$$

which has the positive solution

$$S = \sqrt{S_2(S_1 + 1) + 0.25} - 0.5$$

Edmister suggests that the effective factor be considered to be independent of the number of stages and uses

$$S = \sqrt{S_N(S_1 + 1) + 0.25} - 0.5 \tag{8-22}$$

to calculate the effective factors in an N-stage column where N is the top stage and the bottom stage is numbered 1. An equation analogous to (8-22) but in terms of the absorption factors

$$A = \sqrt{A_1(A_N + 1) + 0.25} - 0.5 \tag{8-23}$$

was also derived by Edmister. Obviously, Eqs. (8-22) and (8-23) represent two different ways to obtain an effective factor. An A obtained by taking the reciprocal of an S calculated with Eq. (8-22) will not check numerically with the A obtained by the direct substitution of $1/S_N$ and $1/S_1$ into Eq. (8-23). Regardless of this fact, the equations provide "average" factors which generally give results comparable to those obtained by the method of Horton and Franklin. Edmister (5) has computed corrections based on arbitrarily assumed changes in the flow rates and temperatures which when applied to his effective factors make the results more consistent with those from a rigorous stage-to-stage calculation.

Except for the method for the estimation of the effective stripping factors, the Edmister approach is the same as that of Horton and Franklin. Equations (8-11) and (8-12) are used to estimate the stage flow rates and temperatures. Equation (8-22) or (8-23) can be used to provide an effective stripping factor for each component. Equation (8-6) provides the recovery in the bottom-product stream for each component, and Eqs. (8-13) and (8-14) are used to calculate the top and bottom product rates. If the calculated product rates do not check closely with the originally assumed splits, the calculations should be repeated with the calculated recoveries.

Irregular Profiles. Any method for the selection of effective factors is apt to be inadequate if unusual rate and temperature profiles exist in the column. The temperature profile in an absorber is often irregular. For example, the profile may exhibit a maximum if a large gas/liquid ratio is used and the gas enters at a temperature much below the temperature at which the oil leaves. This phenomenon is common in high-pressure absorbers where the gas/liquid ratio is high. The liquid initially increases in temperature in the top part of the column, but in the lower part the sensible heat transferred to the cold gas stream may overshadow the heat of condensation of the absorbed material, and as a result the liquid temperature begins to fall. If intercoolers are installed to remove heat from an absorber column and maintain a lower, more efficient temperature, the methods for the selection of an effective stripping factor will obviously not apply to the entire column and it would be necessary to apply Eq. (8-6) to each of the various column sections created by the coolers. Landes and Bell (9) describe more accurate approximation methods for towers with irregular temperature and flow-rate profiles. Rigorous stage-to-stage calculation methods are discussed in Chap. 13.

Example 8-1. An absorber which contains 20 trays and operates at 60 psia charges a wet gas at 90°F and of the following composition:

Component	Mole fraction
C_1	0.285
C_2	0.158
C_3	0.240
n-C_4	0.169
n-C_5	0.148
	1.000

The lean oil can be assumed to have the properties of normal octane and at present has a maximum circulation rate of 0.905 times the wet-gas rate

An inexpensive bottleneck-removal project would increase the oil-circulation rate to 1.104 times the wet-gas rate. At this expected rate the lean oil would enter the absorber at about 90°F and contain n-C_4 and n-C_5 to the extent of about 2 and 5 mole per cent, respectively. Estimate the recovery of each of the gas components at the new oil rate, and calculate the corresponding product rates and compositions. Assume that the absorber stage efficiency is 20 per cent for all components.

Solution. A basis of 100 moles of wet gas will be used. The corresponding oil rate is then 110.4 moles. The amounts of each component entering the column and the initial assumptions for the various recoveries are as follows:

Component	$V_0 y_0$	$L_{N+1} x_{N+1}$	Total moles entering	qs	Assumed		
					f	$L_1 x_1$	$V_N y_N$
C_1	28.5	0	28.5	0	0.1	2.85	25.65
C_2	15.8	0	15.8	0	0.2	3.16	12.64
C_3	24.0	0	24.0	0	0.6	14.40	9.60
n-C_4	16.9	2.21	19.11	0.1156	0.85	16.24	2.87
n-C_5	14.8	5.52	20.32	0.2717	0.95	19.30	1.02
Oil	0	102.67	102.67	1.0	1.0	102.67	0
	100.0	110.4	210.4			158.62	51.78

The residue gas will leave the tower at the temperature of the top stage. For the initial trial the top stage temperature will be assumed to be 5°F above that of the entering lean oil, or 95°F in this example. The temperature of the exit-rich oil (also the temperature of stage 1) is calculated by an over-all heat balance. Enthalpy data can be obtained from Figs. B-4 and B-5. Ideal-solution behavior will be assumed.

Component	$t_0 = 90°F$		$t_N = 95°F$		$t_{N+1} = 90°F$	
	H_V	$H_V(V_0 y_0)$	H_V	$H_V(V_N y_N)$	H_L	$H_L(L_{N+1} x_{N+1})$
C_1	5,600	159,600	5,600	144,500	4,300	
C_2	9,900	156,000	9,900	125,600	7,100	
C_3	13,700	328,100	13,700	132,100	7,200	
n-C_4	17,600	298,400	17,780	51,000	8,900	19,600
n-C_5	21,800	323,400	22,000	22,500	10,500	58,100
Oil	32,400		32,700		15,100	1,547,700
		1,265,500		475,700		1,624,700

Enthalpy of rich oil $= 1{,}265{,}500 + 1{,}624{,}700 - 475{,}700$
$$= 2{,}414{,}500 \text{ Btu/100 moles of wet gas}$$

The temperature of the rich oil stream is found by trial and error to be 135°F.

The phase-rate and temperature profiles are estimated by means of Eqs. (8-11a) and (8-12a).

$$\left(\frac{V_N}{V_0}\right)^{1/N} = \left(\frac{51.78}{100}\right)^{\frac14} = 0.8483 = \frac{V_{n+1}}{V_n} \qquad (8\text{-}11a)$$

This ratio, the balance $L_{n+1} = V_n + L_1 - V_0$, and the following rearrangement of Eq. (8-10)

$$t_{n+1} = t_1 - \frac{V_0 - V_n}{V_0 - V_N}(t_1 - t_{N+1}) = 135 - \frac{V_0 - V_n}{48.22} \quad (45) \quad (8\text{-}12a)$$

are used to calculate the following rate and temperature profiles:

	Wet gas	$n = 1$	$n = 2$	$n = 3$	$n = N = 4$	Lean oil
V_n	100.0	84.8	72.0	61.0	51.8	
L_n		158.6	143.4	130.6	119.6	110.4
V_n/L_n		0.535	0.502	0.467	0.433	
$V_0 - V_n$		15.2	28.0	39.0	48.2	
t_n	90	135	121	109	99	90

The assumed profiles can be used to calculate the effective stripping factors as shown below. The calculation is a trial and error, since the $(N - n + 1)/N$ ratios must be assumed to get the S's, which can then be used to check whether the assumed ratio agrees with the values suggested by Horton and Franklin. The values of V/L and t are read from plots of these variables vs n.

Component	$\dfrac{N - n + 1}{N}$	n	$\dfrac{V}{L}$	t	K	S
C_1	1.0	1.0	0.535	135	43.0	23.0
C_2	0.9	1.4	0.522	130	10.0	5.22
C_3	0.8	1.8	0.508	124	3.6	1.83
$n\text{-}C_4$	0.7	2.2	0.495	119	1.16	0.574
$n\text{-}C_5$	0.6	2.6	0.482	114	0.37	0.178
Oil		4.0	0.433	99	0.0155	0.0067

Substitution of the stripping factors in Eq. (8-6) gives the following component recoveries and product rates for trial 1:

Component	f	L_1x_1	V_Ny_N
C_1	0.0435	1.24	27.3
C_2	0.191	3.02	12.8
C_3	0.524	12.6	11.4
$n\text{-}C_4$	0.893	17.1	2.0
$n\text{-}C_5$	0.951	19.3	1.0
Oil	0.993	101.9	0.7
		155.2	55.2

The new over-all material balance can be used to calculate a new rich-oil temperature. The t_N calculated in the first trial should be used now for the temperature of the residue gas. The enthalpy of V_0 and L_{N+1} is unchanged from trial 1. The enthalpy of the new V_N at 99°F is 501,300 Btu/100 moles of wet gas. The corresponding temperature of the rich oil is 135°F, unchanged from trial 1. Equations (8-11a) and (8-12a) can now be written as

$$\left(\frac{V_N}{V_0}\right)^{1/N} = \left(\frac{55.23}{100}\right)^{\frac{1}{4}} = 0.862 = \frac{V_{n+1}}{V_n} \qquad (8\text{-}11a)$$

and $\qquad t_{n+1} = t_1 - \frac{V_0 - V_n}{V_0 - V_N}(t_1 - t_{N+1}) = 135 - \frac{V_0 - V_n}{44.78} \quad (45)$

$$(8\text{-}12a)$$

The use of these two expressions provides the following rate and temperature profiles:

	Wet gas	$n = 1$	$n = 2$	$n = 3$	$n = N = 4$	Lean oil
V_n	100.0	86.2	74.3	64.0	55.2	
L_n		155.2	141.4	129.5	119.2	110.4
V_n/L_n		0.555	0.526	0.495	0.463	
$V_0 - V_n$		13.8	25.7	35.9	44.8	
t_n	90	135	121	109	99	90

The profiles are so similar to those of trial 1 that the stripping factors will be little changed, and the effective stripping factors will be evaluated at the same values of n as used formerly. These values of n together with plots of the new profiles vs n provide effective stripping factors which when substituted in Eq. (8-6) give the following final component recoveries and product rates and compositions:

Component	S	f	$L_1 y_1$	$V_N y_N$	x_1	y_N
C_1	23.9	0.042	1.2	27.3	0.008	0.486
C_2	5.43	0.184	2.9	12.9	0.019	0.230
C_3	1.91	0.504	12.1	11.9	0.078	0.212
n-C_4	0.601	0.884	16.9	2.2	0.109	0.040
n-C_5	0.188	0.948	19.3	1.1	0.125	0.019
Oil	0.00718	0.993	101.9	0.7	0.661	0.013
			154.3	56.1	1.000	1.000

For comparison purposes, the effective stripping factors obtained by the Edmister equations

$$S = \sqrt{S_N(S_1 + 1) + 0.25} - 0.5 \qquad (8\text{-}22)$$

$$\frac{1}{S} = A = \sqrt{A_1(A_N + 1) + 0.25} - 0.5 \qquad (8\text{-}23)$$

are listed below beside the Horton-Franklin values.

Component	Horton-Franklin	Edmister	
		Eq. (8-22)	Eq. (8-23)
C_1	23.9	20.5	23.8
C_2	5.43	4.56	5.4
C_3	1.91	1.60	1.9
n-C_4	0.601	0.482	0.595
n-C_5	0.188	0.15	0.20
Oil	0.00718	0.007	0.0119

It can be seen that in this example Eq. (8-23) gives essentially the same effective stripping factors as the Horton-Franklin method whereas Eq. (8-22) gives values which are considerably lower. Since the profiles in an absorber or stripper can be quite irregular, the best method of averaging the stripping factors will vary from problem to problem.

The results of a rigorous stage-to-stage solution of this example problem are summarized in Table 13-1 in Chapter 13. The following tabulations compare the profiles from the rigorous solution with the second trial results in the short-cut method. The short-cut profiles below do not correspond exactly to the final calculated product rates. For example, the final values of V_4 and L_1 were 56.1 and 154.3, respectively. The other rates would change accordingly if a third trial were made.

Stage	Short-cut			Rigorous		
	t	V	L	t	V	L
1	135	86.2	155.2	130	88.8	152.6
2	121	74.3	141.4	126	82.6	141.3
3	109	64.0	129.5	118	76.1	135.1
4	99	55.2	119.2	109	57.8	128.7

As mentioned previously, the profiles calculated from the short-cut method often differ appreciably from the rigorous solution. Despite this the end or product rates usually agree much more closely. Also, the short-cut product compositions are usually quite close as shown below.

Component	x_1		y_N	
	Short-cut	Rigorous	Short-cut	Rigorous
C_1	0.008	0.008	0.486	0.473
C_2	0.019	0.018	0.230	0.226
C_3	0.078	0.072	0.212	0.225
n-C_4	0.109	0.109	0.040	0.043
n-C_5	0.125	0.127	0.019	0.017
Oil	0.661	0.666	0.013	0.016

Example 8-2. The rich oil produced in Example 8-1 is to be stripped with steam in a column which has 12 actual trays and operates at 45 psia. The entering steam is superheated and will have a temperature of 290°F at 45 psia. The rich oil will enter the column at 250°F. Assume a 20 per cent tray efficiency and neglect the solubility of water in the oil. Will a steam/rich oil ratio of 0.10 provide the lean-oil rate and composition assumed in Example 8-1?

Solution. The bubble point of the rich oil must always be checked to determine whether the rich oil will remain in the liquid phase when it enters the stripper. In this case the rich oil will be partially flashed and a flash calculation must be made to determine how much material passes directly out the overhead vapor line without entering into the stripping operation. The final trial in the equilibrium flash calculation is shown below:

Component	z	$K^{250°}_{45 \text{ psia}}$	$V = 0.325;\ \dfrac{L}{V} = 2.076$		$z - Vy$	y	x
			$1.0 + \dfrac{L}{VK}$	$\dfrac{z}{1.0 + L/VK}$			
C_1	0.008	74.5	1.028	0.0078	0.0002	0.0240	0.0003
C_2	0.019	23.8	1.087	0 0175	0.0015	0.0538	0.0022
C_3	0.078	11.1	1.187	0.0657	0.0123	0.2021	0.0182
$n\text{-}C_4$	0.109	5.4	1.384	0.0788	0.0302	0.2424	0.0447
$n\text{-}C_5$	0.125	2.4	1.865	0.0670	0.0580	0.2061	0.0859
Oil	0.661	0.32	7.488	0.0883	0.5727	0.2716	0.8487
	1.000			0.3251	0.6749	1.0000	1.0000

It can be seen that 32.5 per cent of the rich oil enters the stripper in the vapor phase and passes directly to the overhead gas line. On a basis of 100 moles of wet gas fed to the absorber, the actual feeds to the stripping operation are

$$L_{N+1} = (0.6749)(154.31) = 104.1 \text{ moles}$$
$$V_0 = (0.1)(104.1) = 10.41 \text{ moles}$$

The 104.1 moles of liquid feed to the top stripper tray has the composition shown in the last column of the flash calculation. The steam is free of hydrocarbons.

An assumption for each of the component splits provides an initial over-all material balance as follows:

Component	$V_0 y_0$	$L_{N+1} x_{N+1}$	Total moles entering	qs	Assumed		
					f	$L_1 x_1$	$V_N y_N$
C_1	0	0.03	0.03	1.0	0	0	0.03
C_2	0	0.23	0.23	1.0	0	0	0.23
C_3	0	1.89	1.89	1.0	0	0	1.89
$n\text{-}C_4$	0	4.65	4.65	1.0	0.30	1.40	3.26
$n\text{-}C_5$	0	8.94	8.94	1.0	0.60	5.36	3.58
Oil	0	88.34	88.34	1.0	0.95	83.92	4.42
Steam	10.41	0	10.41	0	0	0	10.41
	10.41	104.1	114.5			90.68	23.82

Since steam is involved, a datum of liquid at 32°F will be used in the enthalpy balance. Assuming $t_N = 240°F$, the enthalpy balance is as follows:

Enthalpy of $V_0 = (18)(1180.3)(10.41) = 221\ 160$ Btu

Component	$t_N = 240°F$		$t_{N+1} = 250°F$	
	H_V	$H_V(V_N y_N)$	H_L	$H_L(L_{N+1} x_{N+1})$
C_1	3,340	100	2,530	80
C_2	7,650	1,760	4,970	1,140
C_3	11,250	21,260	7,850	14,840
$n\text{-}C_4$	14,810	48,280	8,370	38,900
$n\text{-}C_5$	18,460	66,090	9,950	88,950
Oil	28,780	127,210	14,390	1,271,200
Steam	21,000	217,000		0
		481,700		1,415,100

Enthalpy of $L_1 = 221,160 + 1,415,100 - 481,700$
$$= 1,154,560 \text{ Btu}$$

The stripped-oil temperature corresponding to this enthalpy is found by trial and error to be 210°F. The phase-rate and temperature profiles can be estimated with Eqs. (8-11b) and (8-12b).

$$\left(\frac{L_1}{L_{N+1}}\right)^{1/N} = \left(\frac{90.68}{104.1}\right)^{1/2.4} = 0.9441 = \frac{L_n}{L_{n+1}} \qquad (8\text{-}11b)$$

Equation (8-12b) can be rearranged to

$$t_n = t_{N+1} - \frac{L_{N+1} - L_n}{L_{N+1} - L_1}(t_{N+1} - t_1) \qquad (8\text{-}12b)$$

$$= 250 - \frac{L_{N+1} - L_n}{13.42}(40)$$

The estimated profiles are as follows:

	Steam	$n = 1$	$n = 2$	$n = 2.4$	Rich oil
V_n	10.41	15.78	21.47	23.82	
L_n		90.68	96.05	98.33	104.1
V_n/L_n		0.1740	0.2235	0.2422	
$L_{n+1} - L_n$		13.42	8.05	5.77	0
t_n	290	210	226	233	250

The effective stripping factors can now be evaluated at the positions in the column suggested by Horton and Franklin. This is a trial-and-error procedure, since an n/N ratio must be assumed before an S can be calculated.

Component	n/N	n	V	L	t	K	S
C_1	0.6	1.44	18.2	93.0	217	70.0	13.7
C_2	0.6	1.44	18.2	93.0	217	21.0	4.11
C_3	0.7	1.68	19.7	94.3	221	9.5	1.98
$n\text{-}C_4$	0.8	1.92	21.1	95.7	225	4.4	0.970
$n\text{-}C_5$	0.9	2.16	22.5	97.0	229	1.96	0.455
Oil	1.0	2.4	23.8	98.3	233	0.20	0.0484
Steam	0	1.0	15.8	90.7	210	∞	∞

The infinite K values for the steam correspond to the assumption of complete insolubility of water in the oil made in the problem statement. The recoveries of the other components can be calculated by substituting the various effective stripping factors in Eq. (8-8).

Component	S	$1 - S$	$1 - S^{3.4}$	f	$L_1 x_1$	$V_N y_N$
C_1	13.7	−12.7	−7300	0.00174	0.000	0.030
C_2	4.11	− 3.11	−121	0.0257	0.006	0.224
C_3	1.98	− 0.98	−9.2	0.106	0.200	1.69
$n\text{-}C_4$	0.970	0.030	0.098	0.306	1.42	3.23
$n\text{-}C_5$	0.455	0.545	0.932	0.585	5.23	3.71
Oil	0.0484	0.9516	0.999	0.952	84.10	4.24
Steam						10.41
					91.0	23.5

The calculated values for L_1 and V_N are quite close to those assumed initially. Repetition of the calculations with the new over-all material balance would give essentially the same recoveries.

It can be seen that 4.24 moles of oil are carried overhead by the steam. This amount together with the $(0.3251)(154.31)(0.2716) = 13.62$ moles lost in the flash at the top of the stripper and the 0.72 lost overhead in the

Component	Mole fraction	
	Calculated	Example 8-1
C_1	0	0
C_2	0.000	0
C_3	0.002	0
$n\text{-}C_4$	0.013	0.02
$n\text{-}C_5$	0.048	0.05
Oil	0.937	0.93
	1.000	1.00

absorber (Example 8-1) sums to 18.58 moles of make-up oil required. Addition of the make-up to the L_1 calculated above would give 109.6 moles of lean oil compared with the 110.4 used in Example 8-1. The steam rate to the stripper could be lowered slightly if the lean-oil composition in Example 8-1 is to be produced. The calculated composition of the lean oil after the addition of the make-up oil is compared, in table at bottom of page 274, with the composition assumed in Example 8-1. It can be seen that the assumed lean-oil composition in Example 8-1 was sufficiently correct.

8-4. Distillation

Whereas mass transfer in absorption and stripping is essentially unidirectional, it is usually a two-directional process in distillation operations. The liquid fed to the top of the distillation column as reflux (or as a feed in a column which consists of only the stripping section) always contains components which are lower boiling than some of the components in the ascending vapor. The lighter components in the liquid tend to change places with the heavier components in the vapor. Whenever a heavier component in the vapor condenses, a lighter component in the liquid immediately utilizes the released heat of condensation as a heat of vaporization to escape to the vapor.

As in absorption and stripping, the liquid in an equilibrium distillation stage is at its bubble point, and this bubble-point temperature is determined by the stage pressure and liquid composition. In absorption and stripping, a relatively small amount of light components is "dissolved" in a large amount of essentially nonvolatile oil and a small change in the amount of light material radically changes the bubble-point temperature. This is a characteristic of "dumbbell" systems, where the term dumbbell is used to describe the volatility characteristics of the system. The light components comprise one end of the dumbbell, while the heavy absorber oil comprises the other end. Such feeds must sometimes be processed in distillation columns, but there the effect of reflux largely erases the dumbbell effect. Generally, distillation feeds are more "continuous" as far as the volatilities of the components are concerned. In the more continuous systems all components affect the bubble point. The release of heat due to the condensation of a heavier component cannot raise the liquid temperature greatly because only a slight increase in temperature immediately causes some components to move from the liquid to the vapor phase. The bulk of the heat of condensation therefore is absorbed as heat of vaporization rather than appearing as sensible heat. As the phases pass countercurrently from stage to stage, the various component concentrations continuously adjust themselves in conformity to their respective distribution coefficients and the energy required by a molecule to pass from the liquid to the vapor phase is supplied by the energy released by another molecule condensing from the vapor to the liquid. Mass transfer in both directions is thus maintained. The transfer in the two directions will not be exactly equal, since the latent heats of vaporization of the various components are not equal. For example, a mole of condensing n-butane at 1 atm will furnish energy sufficient to vaporize only 0.87 mole of n-pentane. Also, the column will not be completely adiabatic and loss or gain of heat may further unbalance the transfer from one phase to the other.

Despite unequal molar heats of vaporization and lack of adiabaticity, the fact that transfer in both directions does occur serves to maintain more constant phase rates than in absorption and stripping. Therefore, the averaging of phase rates to obtain an average or effective stripping factor for use in the short-cut equation is simpler than in the case of absorption or stripping.

The form of Eq. (8-3) applicable to a distillation column with a total condenser and one feed is obtained by setting

$$q_S = q_{F'} = R' = g = 0$$

and combining the second and fourth terms in the denominator to give

$$f = \frac{(1 - S_n^{N-M}) + R(1 - S_n)}{(1 - S_n^{N-M}) + R(1 - S_n) + hS_n^{N-M}(1 - S_m^{M+1})} \qquad (8\text{-}24)$$

Figure 8-4 shows the diagram to which Eq. (8-24) applies. Figure 8-4 was obtained from Fig. 8-1 by setting $S = F' = R' = S_E = 0$ and adding two arrows to represent the condenser and reboiler heat loads.

Equation (8-24) is applicable to distillation columns with either a partial reboiler or heat input to the bottom stage. The lowest equilibrium stage must always be numbered one regardless of whether it is a partial reboiler or the bottom equilibrium stage. Equation (8-24) is not strictly applicable to a column with a partial condenser, since it ignores any possible difference between the overhead and reflux compositions. However, the effect of a partial condenser can be closely approximated by increasing N by 1.0.

The number of design variables which must be specified to define a distillation process completely is discussed in Sec. 3-5. In computer work, particularly when an existing column is being studied, it is most convenient to include the following four items in the specifications:

1. N, the total number of stages
2. M, the number of stages below the feed stage
3. Maximum vapor rate at some point in the column
4. Total distillate rate D

The use of Eq. (8-24) is particularly convenient when these specifications are made. The specification of D and V or V' along with the specification of the feed stream variables (always specified) fixes the phase rates in both sections of the column. The determination of the stripping factors S_n and S_m then depends upon estimation of the individual K values. If ideal solutions are assumed, the K values are functions only of temperature at the specified tower pressures. Estimation of the K values in ideal-solution systems therefore reduces to the estimation of the temperature profile. Good initial estimates of the top and bottom temperatures are obtained by splitting the feed in a reasonable manner and performing dew- and bubble-point calculations on the resulting overhead and bottom streams. The feed-tray temperature can be estimated roughly by a bubble-point, dew-point, or flash calculation on the feed depending upon the thermal condition. However, it is easier, and often as accurate, to read the initially assumed feed-tray temperature from a straight line drawn between the assumed top and bottom temperatures.

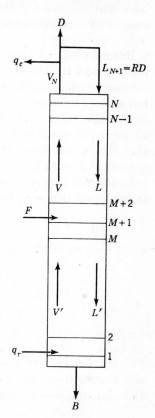

FIG. 8-4. Distillation column with a total condenser and either a partial or a total reboiler.

The liquid and vapor rates must be assumed constant within any column section, and their values are fixed by the specification of the feed, maximum vapor, and distillate rates and the thermal condition of the feed. Since the phase rates are held constant from trial to trial, the stripping factors can be varied only by changing the K values. This is done by adjusting the temperature profile. The top and bottom temperatures cannot be adjusted arbitrarily, since they must always correspond to the dew and bubble points of the overhead and bottom streams, respectively. The iteration variables, therefore, reduce to one, the feed-tray temperature. The calculator must adjust this temperature until the vapor- and distillate-rate specifications are met. This method of calculation is analogous to the technique of a column operator who adjusts the control-tray temperature until the column is loaded or until the product specifications are met.

The solution provided by Eq. (8-24) is based upon a close approximation to the operating conditions which will exist in the actual column. The results are, therefore, usually more reliable than estimates based upon fictitious conditions such as minimum or total reflux. Also, the use of Eq. (8-24) provides an estimate of the column temperature profile. This may be useful if the short-cut results are to be used as the first assumptions in a rigorous solution.

Example 8-3. A large butane-pentane splitter is to be shut down for repairs, and its $7500B /SD$ of feed diverted temporarily to a smaller column which is available for service as a spare. The smaller column has only 11 bubble-cap trays plus a partial reboiler. The feed enters on the middle tray. Past experience on similar feeds indicates that the 11 trays plus the reboiler are roughly equal to 10 theoretical stages and that the column has a maximum top vapor capacity of 1.75 times the volume of feed to be handled. The column will operate at a condenser pressure of 120 psia. The feed will be at its bubble point and has the following average composition:

Component	Fx_F
C_3	5
$i\text{-}C_4$	15
$n\text{-}C_4$	25
$i\text{-}C_5$	20
$n\text{-}C_5$	35
	100

For the purposes of economic evaluation, it is desired to plot curves which show the product compositions as functions of the overhead rate Calculate the product compositions which will be obtained when

$$\frac{D}{F} = 0.489$$

The regular splitter normally has less than 7 mole per cent $i\text{-}C_5$ in the overhead and less than 3 mole per cent $n\text{-}C_4$ in the bottoms. Will these product purities be met on the smaller column at $D/F = 0.489$?

Solution. The tray pressure drops will be neglected, and the entire column and reboiler will be assumed to be at 120 psia. At this pressure it is not necessary to obtain the K values from the convergence-pressure charts and the DePriester nomograph (Fig. B-3) can be used. If the constant molar overflow is assumed in each section, the rates in the uppe and lower sections are as follows:

Basis: 100 moles/hr of fresh feed.

Top Section

$D = (0.489)(100) = 48.9$

$V = (1.75)(100) = 175$

$L = 175 - 48.9 = 126.1$

$\dfrac{L}{V} = 1.388$

Bottom Section

$B = 100 - 48.9 = 51.1$

$V' = V = 175$

$L' = L + F = 226.1$

$\dfrac{V'}{L'} = 0.7739$

$$\frac{L}{L'} = \frac{126.1}{226.1} = 0.5577$$

$$R = \frac{126.1}{48.9} = 2.579$$

Since the feed enters in the middle of the column, $M = 5$ and $M + 1 = 6$.

The top- and bottom-stage temperatures can be estimated closely by assuming a split on the feed and making dew- and bubble-point calculations with the resulting overhead and bottoms compositions. The i-C$_5$ and n-C$_4$ concentrations obtained on the regular column can be used along with the specified D of 48.9 to estimate the split as follows:

Component	Fx_F	Dx_D	Bx_B	x_D	x_B
C$_3$	5	5.0	0.0	0.102	0.0
i-C$_4$	15	14.5	0.5	0.296	0.010
n-C$_4$	25	23.5	1.5	0.481	0.029
i-C$_5$	20	3.4	16.6	0.070	0.325
n-C$_5$	35	2.5	32.5	0.051	0.636
	100	48.9	51.1	1.000	1.000

Dew- and bubble-point calculations on the assumed overhead and bottoms compositions give top- and bottom-stage temperatures of 165 and 236°F, respectively. The bubble point of the feed is 185°F, and this will be used as the initial assumption of the feed-tray temperature. The average temperatures to be used to obtain K and K' values for the first trial are then

$$\frac{165 + 185}{2} = 175°F \quad \text{and} \quad \frac{185 + 236}{2} = 210°F$$

respectively, for the top and bottom sections.

The evaluation of Eq. (8-24) is best done in tabular form. Table 8-1 illustrates how the average temperatures and rates are used to make the first estimate of the product compositions. Equation (8-5) is used to calculate h, since a bubble-point (liquid) feed is assumed. The D calculated from the assumed temperature profile is considerably below the specification of $D = 48.9$. Also,

$$V_N = (1 + R)D = (3.579)(43.8) = 156.8$$

which is below the column capacity of 175. To load the column, the temperature profile must be raised. The calculated product compositions

TABLE 8-1. FIRST TRIAL IN EXAMPLE 8-3

Component	K_{175}	S_n	K_{210}	S_m	$\dfrac{K'}{K}$	$\dfrac{1-S_n}{1-S_m}$	h
C_3	3.03	4.21	3.83	2.96	1.26	1.63	1.15
i-C_4	1.53	2.12	2.07	1.60	1.35	1.87	1.41
n-C_4	1.17	1.62	1.62	1.25	1.38	2.46	1.90
i-C_5	0.575	0.798	0.825	0.638	1.43	0.558	0.447
n-C_5	0.485	0.673	0.720	0.557	1.48	0.738	0.611

Component	$S_n{}^5$	$1-S_n{}^5$	$R(1-S_n)$	Numerator	$S_m{}^6$	$hS_n{}^5(1-S_m{}^6)$
C_3	1290	-1289	-8.27	-1297	700	$-1,037,000$
i-C_4	43	-42	-2.90	-44.9	17.3	-987.7
n-C_4	11.6	-10.6	-1.61	-12.2	3.81	-61.7
i-C_5	0.323	0.677	0.521	1.20	0.069	0.1344
n-C_5	0.140	0.860	0.843	1.70	0.036	0.0826

Component	Denominator	f	Bx_B	Dx_D	x_B	x_D
C_3	$-1,038,000$	0.00125	0.006	4.994	0.000	0.114
i-C_4	$-1,032$	0.0435	0.625	14.375	0.011	0.328
n-C_4	-73.9	0.165	4.13	20.87	0.074	0.476
i-C_5	1.33	0.899	18.0	2.0	0.320	0.046
n-C_5	1.79	0.953	33.4	1.6	0.595	0.036
			56.16	43.83	1.000	1.000

should not be used to calculate new end temperatures. The amoun taken overhead was so far below the specification that the calculate compositions are not realistic in this trial. The calculated over-hea composition is too light, and a dew-point calculation based on this con position would be lower than the 165°F initially assumed. Similarl the calculated bottoms composition is also too light and its bubble poir is lower than the one initially assumed. It is better, therefore, not make new estimates of the overhead and bottoms temperatures at th

)int but rather retain the old ones and adjust only the feed-tray tem-
:rature for the second trial.

If 210°F is assumed for the feed-tray temperature, the average tem-
:ratures for the upper and lower sections become 187.5 and 223°F,
·spectively. The calculations when repeated with these temperatures
·ovide the following results:

Component	f	Bx_B	Dx_D	x_B	x_D
C_3	0.000789	0.004	4.996	0.000	0.101
$i\text{-}C_4$	0.0232	0.348	14.65	0.007	0.296
$n\text{-}C_4$	0.0900	2.25	22.75	0.045	0.459
$i\text{-}C_5$	0.8125	16.25	3.75	0.322	0.076
$n\text{-}C_5$	0.9031	31.61	3.39	0.627	0.067
		50.41	49.54	1.000	1.000

The calculated value of $D = 49.54$ is close enough to the specified
alue of 48.9 to permit the use of the calculated compositions in the
:timation of new end temperatures. Dew- and bubble-point calcula-
·ons provide temperatures of 163.5 and 234°F, respectively, for the top
:age and reboiler. Averaging these new end temperatures with the
·reviously assumed feed-tray temperature of 210°F gives 186.7 and
22°F, respectively, for the new average temperatures in the top and
·ottom sections. Repetition of the calculations with these temperatures
·rovides the final results shown below. The numbers in parentheses are
·e results of a rigorous computer solution (see Chap. 10) by the Thiele-
:eddes method as applied by Lyster et al. (10).

Component	f	Bx_B		Dx_D	
C_3	0.000825	0.004	(0.002)	4.996	(4.998)
$i\text{-}C_4$	0.0253	0.379	(0.330)	14.6	(14.6)
$n\text{-}C_4$	0.0964	2.41	(1.90)	22.6	(23.1)
$i\text{-}C_5$	0.8216	16.4	(16.4)	3.57	(3.56)
$n\text{-}C_5$	0.9098	31.8	(32.4)	3.16	(2.61)
		51.0		48.9	

The concentration of $n\text{-}C_4$ in the bottom product is 4.7 per cent, which
·xceeds the 3 per cent obtained on the larger column. Also, the $i\text{-}C_5$
·oncentration in the overhead product is 7.3 per cent, which exceeds the
· per cent previously obtained. The smaller column will not be able to
·eet those specifications at any overhead rate. Increasing D above
·8.9 would decrease the concentration of $n\text{-}C_4$ in the bottom product but
·ould also increase the concentration of $i\text{-}C_5$ in the overhead stream.

Likewise, decreasing D would remove $i\text{-}C_5$ from the overhead but would also increase the $n\text{-}C_4$ left in the bottom stream.

Since the product compositions predicted by the short-cut method were close to the rigorous solution, the estimated end temperatures must also be close. The short-cut method gave bottom- and top-stage temperatures of 234 and 163.5°F compared with the computer results of 234 and 165.7°F. The temperature profile in the computer solution sagged at the feed tray, and irregularities of this sort cannot be reproduced by the three-point profile obtained in the short-cut method. The feed-stage temperature obtained in the short-cut method was 210°F compared with the computer result of 202.3°F.

8-5. Extraction and Washing

Mass transfer in the liquid-liquid processes can range from unidirectional transfer to equimolar transfer in both directions. If the major raffinate and extract components were completely immiscible, only the solute would change phases, passing from the raffinate to the extract in extraction and from the extract to the raffinate in washing. If the major raffinate and extract components are miscible to a large extent, material besides the solute will pass in both directions and equimolar transfer may be approached. It can be seen that the ease with which good average phase rates can be obtained for use in the short-cut equation will vary from system to system.

Extraction and washing operations usually operate at a closely controlled temperature. The temperature is adjusted to take maximum advantage of the solubility relationships of the system. The temperature may be essentially constant throughout the column, or a temperature gradient may be maintained to induce internal reflux. In any case, the calculation of the separation to be obtained will be made at specified stage temperatures and the stage temperatures used in the calculations will not vary from trial to trial. Enthalpy balances must be made, of course, to design the facilities necessary to maintain the desired temperature, but these balances are not a part of the stage-to-stage or short-cut separation calculations.

The distribution coefficients for liquid-liquid systems cannot be calculated from only fugacities or obtained from such generalized correlations as the convergence-pressure charts. The coefficients must be measured experimentally and correlated in such a way that they can be predicted as a function of the stage composition. Ternary data can be represented by plotting the K values vs the concentration of any one of the three components. Quaternary data can be represented satisfactorily by plotting the K values vs the concentration of the major solvent component (lowest K values) with concentration parameters of the component which has the next lowest K values. Such a correlation is shown in Figs.

B-6 and B-7. Correlating equations for multicomponent systems are presented in Chap. 2.

The possible modifications of the extraction process are such that the flow pattern may vary from the simple one in Fig. 8-2 to the complicated one in Fig. 8-1. Likewise, the applicable equation may vary in complexity from Eq. (8-6) to Eq. (8-3). Some of the simpler forms of Eq. (8-3) have been presented for extraction by Alders (1), Tiller (15), and Tiller and Tour (16).

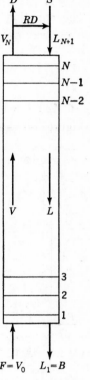

Equation (8-6) applies to simple extraction and washing just as it applies to absorption and stripping if the stream and stage designations of Fig. 8-2 are used. The next simplest extraction process is the one with one feed and raffinate reflux shown in Fig. 8-5. Figure 8-5 was obtained from Fig. 8-1 by setting

$$M = F' = R' = S_E = 0$$

In an analogous manner the applicable equation is obtained by setting

$$q_{F'} = R' = g = M = 0,$$

$S_n = S_m$, $V = V'$, and $L = L'$ to give

$$f = \frac{(1 - S^N) + q_S(S^N - S) + R(1 - S)}{(1 - S^{N+1}) + R(1 - S)} \qquad (8\text{-}25)$$

For $q_S = 0$,

$$f = \frac{1 - S^N + R(1 - S)}{1 - S^{N+1} + R(1 - S)} \qquad (8\text{-}26)$$

For $q_S = 1.0$,

$$f = \frac{1 - S + R(1 - S)}{1 - S^{N+1} + R(1 - S)} \qquad (8\text{-}27)$$

Fig. 8-5. Extraction process with one feed and raffinate reflux.

Casual inspection of Eqs. (8-26) and (8-27) might indicate that the use of raffinate reflux affects the recovery of a given component. However, as R changes, the stripping factor must also change and the net result is no change in f. This can be easily demonstrated for a one-stage process. The representative phase rates to be used in the calculation of the stripping factor for one stage are the equilibrium-phase rates leaving the stage.

Whenever two-column sections are involved, the form of Eq. (8-3) and its use both become more cumbersome. Effective stripping factors must be estimated for both sections, and this essentially doubles the calculation time relative to that required by Eq. (8-6) or (8-25). The equation for two feeds and no refluxes is obtained by setting

$$R = R' = g = 0$$

to give

$$f = \frac{(1 - S_n^{N-M}) + q_S(S_n^{N-M} - S_n) + hq_{F'}S_n^{N-M}(1 - S_m^{M})}{(1 - S_n^{N-M}) + hS_n^{N-M}(1 - S_m^{M+1})} \qquad (8\text{-}2\text{?}$$

The equation for one feed and extract reflux is obtained by settir $R = q_{F'} = 0$.

$$f = \frac{(1 - S_n^{N-M}) + q_S(S_n^{N-M} - S_n)}{(1 - S_n^{N-M}) + hS_n^{N-M}(1 - S_m^{M})}$$
$$+ h[(1 + R')/(1 + gR')]S_m^{M}S_n^{N-M}(1 - S_n$$
$$(8\text{-}2\text{?}$$

Figure 8-6 shows the flow diagram corresponding to Eq. (8-29). Th equations corresponding to other possible combinations of feeds and refluxes can be obtaine in a like manner.

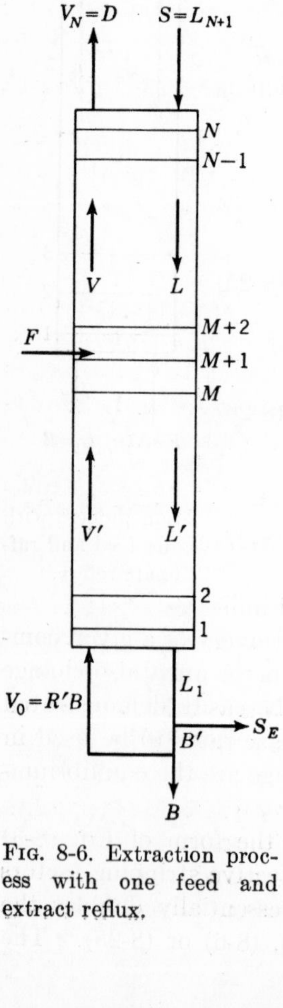

The use of the short-cut equation for ex traction and washing is more difficult than fc the vapor-liquid processes. It is not so eas to make a good first assumption for the variou component recoveries, since the K values ar not so predictable as those of vapor-liquid sys tems. Consequently, the initial estimation c the end compositions and the determination c the ranges of the respective K values throug the column or train are often difficult. It i often most convenient to obtain initial averag K values for each component by averaging a available tie-line data and using these averag values in the first trial. The results of the firs trial will then indicate the range of compositio in the column and permit the estimation o good average K values for the second trial Since the K values are usually not linear wit composition, the average values for the secon and subsequent trials must sometimes be ob tained by averaging several K values evaluate at various typical stage compositions along th section.

FIG. 8-6. Extraction process with one feed and extract reflux.

It is generally not convenient in extractio calculations to specify the product rates. (Th usual specifications for various extraction proc esses are shown in Sec. 3-5.) Consequently, the rates become iteration variables and vary from trial to trial. As mentioned previously, the initial estimate of the product rates to be used in the first trial is difficult. Fortunately, the short-cut equation predicts reasonable values for the product rates in the first trial even when the initial assumptions are greatly in error. If the initial assumptions are very bad, it is generally a good idea to repeat the first trial using the same K values with the calculated product rates.

The calculation procedure for extraction and washing problems can be summarized as follows:

1. Average all available tie-line data to obtain component K values for the first trial. Assume that $K = K'$ for each component if two column sections are involved.

2. Assume f's and calculate B and D with Eqs. (8-30) and (8-31). Use material balances to calculate any other end rates. Equations (8-32) and (8-34) are useful if extract reflux is used.

$$Bx_B + S_E x_{S_E} = f(Fy_F + F'y_{F'} + Sx_S) \qquad (8\text{-}30)$$

$$Dy_D = (1 - f)(Fy_F + F'y_{F'} + Sx_S) \qquad (8\text{-}31)$$

$$Bx_B = \frac{1 - g}{1 + gR'} (Bx_B + S_E x_{S_E}) \qquad (8\text{-}32)$$

$$L_1 x_1 = \frac{1 + R'}{1 - g} Bx_B \qquad (8\text{-}33)$$

3. Assume the extract rate to vary linearly from L_{N+1} to L_1. (In high solvent treat cases allowance should be made for a possible large change in L across stage N as the solvent becomes saturated with raffinate.) Calculate the raffinate rates by material balance. Average the rates in some manner to obtain realistic average extract and raffinate rates for each section. Arithmetic averages of the end values are often sufficient.

4. Use the average rates and K values to calculate average stripping factors for each component. Use Eq. (8-3) to predict the individual recoveries.

5. Calculate the amounts of each component leaving each end of the column with Eqs. (8-30) and (8-31). If extract reflux is used, calculate $L_1 x_1$ with Eqs. (8-32) and (8-33).

6. Use the calculated end compositions to estimate better K values. If the new K's differ appreciably from those used previously, repeat the calculations using the new K's.

For any given set of K values the calculations will converge to one unique set of end rates and compositions within three or four trials. However, it is not necessary to make more than one trial with each set of K values unless, as mentioned previously, the initial rate assumptions were widely in error, in which case the first trial may be repeated. It is seldom necessary to go through more than three sets of K values and usually two are sufficient.

Example 8-4. Liquid sulfur dioxide is used to extract toluene from paraffins. The toluene-rich extract from the extraction tower is then washed in a separate tower with a heavy paraffinic wash oil for further removal of the light paraffins which boil in the same temperature range as the toluene. Only limited distribution coefficient data are available. The following tabulation shows the approximate magnitude of the various component K values along with approximate rates and compositions of the four end streams from a plant test run on the washing tower.

Component	K	$L_{N+1}x_{N+1}$	V_0y_0	L_1x_1	V_Ny_N
Toluene	0.595	15.4	0	13.5	1.9
Paraffins	7.25	5.6	0	0.3	5.3
Olefins	1.71	1.3	0	0.6	0.7
Sulfur dioxide	0.222	77.7	0	75.4	2.3
Wash oil	29.2	0	10.3	2.4	7.9
		100.0	10.3	92.2	18.1

Estimate the number of theoretical stages in the column.

Solution. The individual splits and q_s values can be calculated from the plant material balance.

Component	f	q_s
Toluene	0.88	1.0
Paraffins	0.054	1.0
Olefins	0.46	1.0
Sulfur dioxide	0.97	1.0
Wash oil	0.23	0

In a manner analogous to that suggested by Horton and Franklin for absorption and stripping (Sec. 8-3), the effective stripping factor for each component will be evaluated at the point where the transfer of that component becomes marginal. Inspection of the K values indicates that little of the toluene and sulfur dioxide will change phases. One or two stages will probably be sufficient to effect the transfer that will occur, and therefore the stripping factors for these two components should be evaluated at the top of the column. In a similar manner, little of the wash oil will change phases, and its stripping factor should be evaluated at the bottom of the column. The paraffins will probably continue to transfer all the way through the column, and the paraffin stripping factor should be evaluated at the middle or slightly above. The olefin stripping factor should be calculated at some point in the upper half.

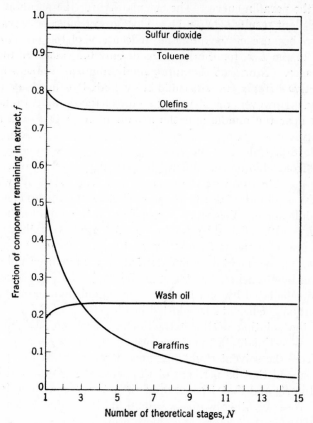

Fig. 8-7. Individual component recoveries as a function of N (Example 8-4).

Since the number of stages is not known, it is not feasible to select the optimum point for each component stripping factor until an approximate value of N has been obtained. For the first approximation the average phase rates can be estimated as follows:

$$\text{Average } V = \frac{10.3 + 18.1}{2} = 14.2$$

$$\text{Average } L = \frac{100 + 92.2}{2} = 96.1$$

The stripping factors calculated from these average rates when substituted in Eqs. (8-7) and (8-8) provide calculated recoveries as a function of the N assumed. Figure 8-7 shows a plot of the individual f's as a function of N. Note how all the individual recoveries except that of the paraffins become independent of the number of stages above $N = 3$. Since the paraffins are the *key* component, the estimate of N should be based on the paraffin curve. The calculated f for the paraffins approximates the plant results at $N \cong 11$. This answer must be accepted with reservation because of the probable inaccuracy of the plant test data. A change in the $L_1 x_1$ for the paraffins of only 0.1 causes N to vary by several stages. Now that the approximate number of stages is known, the respective stripping factors could be evaluated at the points indicated previously. However, the stripping factor of the paraffin would still be evaluated near the middle and the answer of $N = 11$ would not vary appreciably.

Example 8-5. Calculate the recovery at 25°C of acetone (20 weight per cent) from chloroform (80 weight per cent) with a mixed solvent composed of water (65 weight per cent) and acetic acid (35 weight per cent). Use a column with five theoretical stages and introduce the feed in the middle stage. Use a 1:1 solvent/feed ratio and an extract reflux ratio of $R' = 10$. The diagram in Fig. 8-6 applies to this process if N and M are set equal to 5 and 2, respectively.

Solution. Experimental tie-line data at 25°C have been published for this system by Brancker, Hunter, and Nash (2), and additional tie lines have been calculated by Smith (13). The equilibrium data in the form of distribution coefficients are plotted in Figs. B-6 and B-7.

Before the solution of the extraction problem can be started, it is necessary to estimate the performance of the solvent-recovery device that recovers the solvent from L_1. For the purposes of this problem it will be assumed that 98 per cent of the water and acetic acid in L_1 are recovered in S_E and that 1 per cent of the acetone and chloroform in L_1 appear in S_E. In other words $g = 0.98$ for the water and acetic acid and $g = 0.01$ for the acetone and chloroform.

The first step in the solution is the initial estimate of the over-all material balance. Inspection of the K values permits the estimation of reasonable recoveries for each of the components, and a good first assumption for the over-all material balance could be obtained in this manner. This procedure will not be followed in this example. Product rates which are not even of the correct order of magnitude will be assumed in order to illustrate the effect of poor initial assumptions on the speed of convergence.

A basis of $F = S = 100$ will be used. Assume that $D = B = 50$ and $S_E = S = 100$. Then $V_0 = (10)(50) = 500$ and

$$L_1 = S_E + B + V_0 = 650$$

Since the solvent rate is not high, the change in phase rates across stage N due to the entry of the unsaturated solvent will not be extraordinary.

In such cases it is usually satisfactory to assume the extract will vary linearly from L_{N+1} to L_1.

$$L_{M+1} \cong L_1 + \frac{M}{N}(S - L_1) = 650 + \frac{2}{5}(100 - 650) = 430$$

$$L_{M+2} \cong L_1 + \frac{M+1}{N}(S - L_1) = 650 + \frac{3}{5}(100 - 650) = 320$$

$$V_{M+1} \cong L_{M+2} + D - S = 320 + 50 - 100 = 270$$
$$V_M \cong L_{M+1} + V_0 - L_1 = 430 + 500 - 650 = 280$$

The change of only 10 between V_M and V_{M+1} indicates that the rates calculated above are unreasonable, since the difference should be about equal to the feed rate of 100. The discrepancy is caused by the 650 used for L_1 when, as shown by the final results, the final value is in the region of 110.

The estimated L's and V's at the end of each column section can now be averaged to provide average values for V, V', L, and L'. Initial average K values can be obtained by averaging all the available tie-line data. No distinction will be made between K and K' for the first trial. The average rates and K values can now be used to calculate S_n and S_m for each component. Substitution of the stripping factors in Eq. (8-29) provides the estimated recoveries. The K's and stripping factors used in the first trial are shown below with the calculated f's.

Component	$K = K'$	S_n	S_m	f
Acetone	2.12	1.49	1.50	0.144
Chloroform	8.0	5.60	5.66	0.00195
Water	0.04	0.0278	0.0281	0.972
Acetic acid	0.264	0.192	0.194	0.812

Equations (8-30) to (8-33) can now be used to calculate the various end rates. The calculated end rates are found to be so different from those assumed that trial 1 is repeated with the same K values but with the new rates. The calculated rates from trials 1 and 2 are shown below along with the initially assumed rates:

	B	D	L_1	V_0	S_E
Originally assumed	50	50	650	500	100
Results of trial 1	2.9	105	124	29	92
Results of trial 2	4.1	109	132	41	87

The calculated end rates would converge after three or four trials to a unique set of answers which correspond to the set of K values assumed. However, convergence with the initial set of K values is not desired, since the K values assumed are not likely to be the best possible estimates. Two trials were made with the first set of K values only because the initial rate assumptions were so erroneous. The end compositions calculated from such poor assumptions might not accurately define that portion of the two-phase region in which the process will operate. The results of the second trial should be more adequate for this purpose and should permit the selection of better average K values. Table 8-2 shows

TABLE 8-2. TRIAL 2 IN EXAMPLE 8-5

Component	S_n	S_m	f	$V_0y_0 + L_{N+1}x_{N+1}$	$Bx_B + S_Ex_{S_E}$
Acetone	2.32	0.425	0.197	20	3.93
Chloroform	8.74	1.60	0.00516	80	0.412
Water	0.0434	0.008	0.957	65	62.2
Acetic acid	0.300	0.055	0.707	35	24.8
				200	91.3

Component	Dy_D	$\dfrac{1-g}{1+gR'}$	Bx_B	$\dfrac{1+R'}{1-g}$	L_1x_1	x_1	y_D
Acetone	16.1	0.9	3.54	11.1	39.3	0.298	0.148
Chloroform	79.6	0.9	0.371	11.1	4.1	0.031	0.732
Water	2.82	0.00185	0.115	550	63.4	0.480	0.026
Acetic acid	10.2	0.00185	0.046	550	25.2	0.191	0 094
	108.7		4.07		132.0	1.000	1.000

the stripping factors used in trial 2 and how the recoveries resulting from these stripping factors are used with Eqs. (8-30) to (8-33) to calculate new end compositions. Figures B-7a and b furnish the K values required to calculate the x_5's for water and acetic acid from the calculated y_D's for these two components as follows:

Component	y_D	K_5	x_5
Water	0.026	0.044	0.588
Acetic acid	0.094	0.288	0.327

The concentration profiles of water and acetic acid in the extract phase are now assumed to be linear in order to approximate the plate compositions through the column. These approximate plate compositions can then be used with Figs. B-6a, b, c, and d to provide approximate K values for each component on each stage. In this example, the K values were averaged as follows:

Component	$K = (K_5 K_4 K_3)^{1/3}$	$K' = (K_1 K_2 K_3)^{1/3}$
Acetone	2.82	2.00
Chloroform	15.6	9.52
Water	0.0585	0.102
Acetic acid	0.294	0.300

This elaborate averaging procedure for the K values is usually necessary because the distribution coefficients seldom are linear functions of composition. When the K-value curves are not straight, an arithmetic average of two end values is inadequate.

The K values calculated from each trial are used in the succeeding trial. The trials are repeated until two successive sets of K values are in substantial agreement. It is necessary to make four trials in this example to obtain satisfactory convergence of the K values. At least one of the four trials was due to the poor initial rate assumptions.

The results of all four trials are shown in Table 8-3. Note how closely the results of the first trial check the final results despite poor initial assumptions for the rates and K values. In this problem the repetition of the first trial with the first set of K values appears to have hindered the convergence. The tabulation at the bottom of Table 8-3 shows the agreement obtained between the third and fourth set of K values.

The procedure described above will in most instances provide a good estimate of the separation. In some problems, however, it is difficult to estimate good average values for the K's and rates within each section. High solvent treat cases are difficult because of the large change from stage N to $N - 1$ which occurs as the unsaturated solvent dissolves a large part of V_{N-1}. The example above presents a similar difficulty at the other end of the column owing to the high extract reflux. With the g's assumed, the extract reflux is far removed from the solubility surface. When V_0 mixes with L_2, the addition point is still within the immiscible region but so close to the solubility surface that V_1 is only a small fraction of V_0. The solution of a large portion of V_0 in the extract phases causes the phase compositions in stage 1 to differ radically from stage 2, and as a result the K values in the two stages are widely different. As a result, the average V and K values used for the extract end section in the preceding example differ widely from the true values and the estimated separation is not accurate. The true separation is calculated in Example 12-1, where the above example is calculated rigorously. The following

TABLE 8-3. ORIGINAL ASSUMPTIONS AND SUMMARY OF TRIAL
RESULTS FOR EXAMPLE 8-5
(System: Acetone (A)–chloroform (B)–water (C)–acetic acid (D) at 25°C)

	End rates				
	B	D	S_E	L_1	V_0
Originally assumed	50	50	100	650	500
Results of trial 1	2.91	105.3	91.7	123.7	29.1
Results of trial 2	4.07	108.7	87.2	132.0	40.7
Results of trial 3	2.85	112.6	84.5	115.9	28.5
Results of trial 4	2.46	114.7	82.8	109.9	24.6

		Trial 1	Trial 2	Trial 3	Trial 4
x_1:					
	A	0.233	0.298	0.247	0.225
	B	0.013	0.031	0.012	0.007
	C	0.520	0.480	0.533	0.556
	D	0.234	0.191	0.208	0.211
x_B:					
	A	0.893	0.869	0.903	0.908
	B	0.048	0.091	0.042	0.030
	C	0.040	0.028	0.039	0.045
	D	0.018	0.011	0.015	0.017
x_D:					
	A	0.162	0.148	0.152	0.153
	B	0.758	0.732	0.709	0.697
	C	0.017	0.026	0.038	0.044
	D	0.062	0.094	0.101	0.107

	Upper section				Lower section			
	K_A	K_B	K_C	K_D	K'_A	K'_B	K'_C	K'_D
Originally assumed†	2.12	8.0	0.04	0.264	2.12	8.0	0.040	0.264
Results of trial 1†								
Results of trial 2	2.82	15.6	0.058	0.294	2.00	9.5	0.102	0.300
Results of trial 3	2.78	14.5	0.064	0.301	2.23	12.2	0.083	0.287
Results of trial 4	2.81	13.9	0.065	0.304	2.31	13.5	0.075	0.281

† K values used for trials 1 and 2 were arithmetic averages of all available tie-line data.

tabulation compares the rigorous results with some of the short-cut results in Table 8-3.

		Short-cut	Rigorous
B		2.46	5.32
L_1		109.9	146.6
	Acetone	0.225	0.325
x_1	Chloroform	0.007	0.066
	Water	0.556	0.436
	Acetic acid	0.211	0.173

It is obvious that the short-cut method should be used with caution whenever large solvent or extract reflux rates are used and the entering streams are unsaturated.

NOMENCLATURE

A = absorption factor, L/KV. Also used in derivation of Eq. (8-3) to denote the total amount of a component entering the column.

B = heavy or extract product rate; moles, weight, or volume per unit time.

c_1, c_2, c_3, c_4 = constant coefficients in solutions of difference equations.

D = light or raffinate product rate; moles, weight, or volume per unit time.

E = operator in difference equation. For example, $E(x) = x_{n+1}$, $E^2(x) = x_{n+2}$, etc.

f = fraction of a given component which is recovered in B (or $B + S_E$ in extraction).

F = upper feed rate.

F' = lower feed rate. If only one feed is used, $F' = 0$.

g = $S_E x_{S_E}/L_1 x_1$ = fraction of a given component in L_1 which is removed with S_E.

h = $(L/L')[(1 - S_n)/(1 - S_m)]$ if intermediate feed is similar to light (vapor) or raffinate phase ($K_{M+1} = K$). Use $h = (K'/K)(L/L')$ $[(1 - S_n/1 - S_m)]$ if intermediate feed is similar to heavy (liquid) or extract phase ($K_{M+1} = K'$). If the column contains only one section, $h = 1.0$.

H = enthalpy, Btu/lb mole. Subscripts V and L refer to vapor and liquid, respectively.

K = y/x = distribution coefficient. Primes denote lower or extract section of column.

L = heavy or extract phase rate; moles, weight, or volume per unit time.

m = any stage in lower or extract end of column (stages 1 to M).

M = number of theoretical stages below the feed stage. Feed stage is $M + 1$.

n = any stage in upper or raffinate end of column (stages $M + 2$ to N). If $M = 0$, n refers to any stage in column.

N = total number of theoretical stages including partial reboiler if any but excluding partial condenser if any.

q = heat stream. Subscripts c and r refer to condenser and reboiler.

q_S = fraction of a given component which enters in the fresh solvent stream S or in the stream L_{N+1} if raffinate reflux is not used.

$q_{F'}$ = fraction of a given component which enters in the lower feed F'.

$q_F = 1 - q_S - q_{F'}$ = fraction of a given component which enters in the upper feed F.

R = external reflux ratio at top or raffinate end of column; amount refluxed/amount of product D.

R' = external reflux ratio at bottom or extract end of column; amount refluxed/amount of product B

S = fresh solvent rate; moles, weight, or volume per unit time. Also, S is used for effective stripping factor when $S_n = S_m$.

S_E = solvent-rich material recovered in the extract solvent-recovery equipment; moles, weight, or volume per unit time.

$S_n = KV/L$ = effective component stripping factor in upper or raffinate end of column. The K, V, and L are representative averages for the section.

$S_m = K'V'/L'$ = effective component stripping factor in lower or extract end of column. The K', V', and L' are representative averages for the section.

t = temperature, °F.

V = light or raffinate phase rate; moles, weight, or volume per unit time.

x = concentration in heavy or extract phase. Units consistent with units on rates.

X = moles of any component in liquid per mole of liquid entering absorber or stripper.

y = concentration in light or raffinate phase. Units consistent with units on rates.

Y = moles of any component in vapor per mole of vapor entering absorber or stripper.

z = mole fraction in feed to flash separator.

PROBLEMS

8-1. A feed which contains 30 per cent C_3, 30 per cent $n\text{-}C_4$, and 40 per cent $n\text{-}C_5$ is to be charged to a column which has a total condenser and five equilibrium stages (including the partial reboiler). The feed must be introduced in the middle stage. The column will operate at a condenser pressure of 100 psia, and the interstage pressure drops can be neglected. At 100 psia the column has a vapor handling capacity on the top stage of twice the expected feed rate ($V_N/F = 2.0$). Make any other specifications necessary to calculate the product compositions obtained when $D/F = 0.3$ and the column is fully loaded. List the specifications used and estimate the product rates and compositions with Eq. (8-3).

8-2. How many equilibrium stages will be required to recover 80 per cent of the butane in a 95 per cent methane–5 per cent butane mixture by simple absorption with a nonvolatile absorber oil. Use a 1:1 ratio of oil to feed gas. Both the feed and oil enter at 99°F. The absorber pressure is 50 psia. Assume constant temperature and L and V rates through the column.

8-3. Paraxylene is recovered from a mixed xylenes stream by means of a low-temperature crystallization process. The filtrate leaving the crystallization plant contains 7 per cent paraxylene and the product paraxylene purity is 98 per cent. Recovery of paraxylene can be increased by prefractionation of the mixed xylenes stream to remove an orthoxylene concentrate which contains less than 7 per cent paraxylene. What would be the optimum D/F ratio on the prefractionator as far as 98 per cent paraxylene recovery is concerned if the following column is used?

Number of plates	50
Reboiler	Fired furnace
Feed location	26th plate
Maximum V_n/F	4.75
Plate efficiency	80%
Operating pressure	3 atm
Tray pressure drop	Neglect
Condenser	Total

The mixed xylenes stream has the following composition:

Component	Mole %
Toluene	1.0
Ethylbenzene	20.8
Paraxylene	18.0
Metaxylene	39.0
Orthoxylene	20.0
C_{9+}	1.2
	100.0

Hint: Try D/F ratios of 0.86, 0.90, and 0.94 in order to bracket the optimum.

8-4. Rework Example 8-5 when no reflux is used and the feed is introduced in stage 1. The equilibrium data for the acetone–chloroform–water–acetic acid system is given in Figs. B-6 and B-7.

REFERENCES

1. Alders, L.: "Liquid-Liquid Extraction," Elsevier Publishing Company, Amsterdam, 1955.
2. Brancker, A. V., T. G. Hunter, and A. W. Nash: *J. Phys. Chem.*, **44**, 683 (1940).
3. Edmister, W. C.: *Ind. Eng. Chem.*, **35**, 837 (1943).
4. Edmister, W. C.: *Trans. A.I.Ch.E.*, 42, 15, 403, 757 (1946).
5. Edmister, W. C.: *Chem. Eng. Progr.*, **44**, 615 (1948).
6. Edmister, W. C.: *A.I.Ch.E. J.*, **3**, 165 (1957).
7. Horton, G., and W. B. Franklin: *Ind. Eng. Chem.*, **32**, 1384 (1940).
8. Kremser, A.: *Natl. Petrol. News*, **22** (May 21, 1930).
9. Landes, S. H., and F. W. Bell: *Petrol. Refiner*, **39**(6) (1960).
10. Lyster, W. N., S. L. Sullivan, Jr., D. S. Billingsley, and C. D. Holland: *Petrol. Refiner*, **38**(6), 221 (1959); **38**(7), 151 (1959); **38**(10), 139 (1959).
11. Marshall, W. R., and R. L. Pigford: "The Application of Differential Equations to Chemical Engineering Problems," University of Delaware, Newark, Del., 1947.
12. Smith, B. D., and W. K. Brinkley: *A.I.Ch.E. J.*, **6**, 446 (1960).
13. Smith, J. C.: *Ind. Eng. Chem.*, **36**, 68 (1944).
14. Souders, M., Jr., and G. G. Brown: *Ind. Eng. Chem.*, **24**, 519 (1932).
15. Tiller, F. M.: *Chem. Eng. Progr.*, **45**, 391 (1949).
16. Tiller, F. M., and R. S. Tour: *Trans. A.I.Ch.E.*, **40**, 317 (1944).

CHAPTER 9

MULTICOMPONENT SEPARATIONS
Alternate Short-cut Distillation Methods

The general short-cut equation presented in Chap. 8 provides a good, practical solution method for most distillation problems. However, termination of the discussion of short-cut methods at this point would omit such famous and useful equations in the field of distillation as Fenske's total reflux equation† and Underwood's equations for minimum reflux. These equations can be used to solve practical problems, but their main value is in the light which they cast upon the workings of the distillation process.

9-1. Total Reflux—Fenske Equation

A column is said to be operating at total reflux when no feed is introduced and no products are withdrawn. The total reflux condition can be attained by simultaneously stopping the feed and product streams on a continuous column which is operating at steady state. Heat input to the reboiler and heat removal from the condenser are continued with one or the other being adjusted to maintain the over-all enthalpy balance when the feed and product streams are stopped. All the overhead vapor entering the condenser is condensed and returned to the column as liquid reflux. All the liquid overflow to the reboiler is vaporized and returned to the column as vapor. After steady state has been obtained under these conditions, the vapor and liquid streams which pass each other between any two stages are identical in rate and composition (but not in temperature and pressure) and the split obtained between V_N and L_1 is a maximum for any component at the given operating conditions.

Fenske (1) has presented an equation which relates the separation obtained between two components at total reflux to the number of equilibrium stages. The equation is derived for a binary but can be applied to any pair of components in a multicomponent mixture. The derivation of the Fenske equation depends upon the definition of the relative volatility α and upon the identity of composition between any

† This equation is often termed the Fenske-Underwood equation because both published the equation independently.

two passing streams. For any component i the definition of α can be written as

$$\alpha_{i-r} = \frac{y_i x_r}{y_r x_i} = \frac{K_i}{K_r} \tag{9-1}$$

where r denotes the component to which all the relative volatilities are referred. Equation (9-1) can be written for stage 1 as

$$\left(\frac{y_i}{y_r}\right)_1 = (\alpha_{i-r})_1 \left(\frac{x_i}{x_r}\right)_1 \tag{9-2}$$

Stage 1 is the lowest stage in the column and may be either the bottom contact stage or a partial reboiler. At total reflux, $(x_i)_2 = (y_i)_1$ and $(x_r)_2 = (y_r)_1$, so

$$\left(\frac{x_i}{x_r}\right)_2 = \alpha_1 \left(\frac{x_i}{x_r}\right)_1 \tag{9-3}$$

The subscript $i - r$ on the α was dropped for the sake of simplicity and replaced with the stage subscript. The definition of α provides

$$\left(\frac{y_i}{y_r}\right)_2 = \alpha_2 \left(\frac{x_i}{x_r}\right)_2$$

and substitution of this expression in Eq. (9-3) gives

$$\left(\frac{y_i}{y_r}\right)_2 = \alpha_2 \alpha_1 \left(\frac{x_i}{x_r}\right)_1 \tag{9-4}$$

Again, since $(x_i)_3 = (y_i)_2$ and $(x_r)_3 = (y_r)_2$,

$$\left(\frac{x_i}{x_r}\right)_3 = \alpha_2 \alpha_1 \left(\frac{x_i}{x_r}\right)_1 \tag{9-5}$$

The continued alternate use of the equilibrium and material-balance relationships provides the following relationship between the liquid entering stage N and the composition of the liquid in stage 1:

$$\left(\frac{x_i}{x_r}\right)_{N+1} = \alpha_N \alpha_{N-1} \cdots \alpha_1 \left(\frac{x_i}{x_r}\right)_1 \tag{9-6}$$

where the $(x_i/x_r)_{N+1}$ term refers to the reflux stream which is identical with the overhead vapor when a total condenser is used. For a total condenser Eq. (9-6) can be written as

$$\left(\frac{x_i}{x_r}\right)_D = \alpha_N \alpha_{N-1} \cdots \alpha_1 \left(\frac{x_i}{x_r}\right)_1 \tag{9-7}$$

If a partial condenser is used there will be an additional equilibrium stage above stage N. The vapor product available for withdrawal from the partial condenser is therefore related to the liquid in stage 1 by

$$\left(\frac{y_i}{y_r}\right)_D = \alpha_c \alpha_N \alpha_{N-1} \cdots \alpha_1 \left(\frac{x_i}{x_r}\right)_1 \qquad (9\text{-}8)$$

Equations (9-7) and (9-8) show that the maximum separation possible between components i and r is determined by the product of the α's associated with the various stages. The equations can be made more useful if the product of the α's is replaced by an average value α_a defined as

$$\alpha_a{}^N = \alpha_N \alpha_{N-1} \cdots \alpha_1 \qquad (9\text{-}9)$$

or
$$\alpha_a{}^{N+1} = \alpha_c \alpha_N \alpha_{N-1} \cdots \alpha_1 \qquad (9\text{-}10)$$

where N is the number of equilibrium stages including the partial reboiler if any but excluding the partial condenser if any. Equations (9-7) and (9-8) can then be written as

$$\left(\frac{x_i}{x_r}\right)_D = \alpha_a{}^N \left(\frac{x_i}{x_r}\right)_1 \qquad (9\text{-}11)$$

for a liquid overhead product and as

$$\left(\frac{y_i}{y_r}\right)_D = \alpha_a{}^{N+1} \left(\frac{x_i}{x_r}\right)_1 \qquad (9\text{-}12)$$

for a vapor overhead product from a partial condenser. Neither Eq. (9-11) nor (9-12) is applicable to a column with both a vapor and a liquid overhead product from a partial condenser.

Equations (9-11) and (9-12) are completely rigorous and involve no assumptions beyond the stipulation of total reflux. However, the equations cannot be used rigorously unless the values of α on each stage are known for each component, and the α's will not be known unless a complete stage-to-stage calculation has been performed. Therefore, in practical use of the Fenske equation, it is necessary to estimate the average α for each component. A convenient procedure is to estimate the α's at the top, middle, and bottom of the column and then assume α_a to be given by

$$\alpha_a = (\alpha_{\text{top}} \alpha_{\text{middle}} \alpha_{\text{bottom}})^{\frac{1}{3}} \qquad (9\text{-}13)$$

If the α's at the middle of the column cannot be estimated accurately, a geometric average of the top and bottom values only can be used.

It should be noted in Eqs. (9-11) and (9-12) that the mole fraction ratios can be replaced by mole, weight, or volume ratios, since the conversion factors cancel. The α's are always obtained from

$$\alpha_{i-r} = \frac{K_i}{K_r}$$

regardless of the units used in the ratios. For example, the most convenient forms of Eq. (9-11) for practical problems are

$$\frac{\text{Moles of } i \text{ in } D}{\text{Moles of } i \text{ in } B} = \alpha_a{}^N \frac{\text{moles of } r \text{ in } D}{\text{moles of } r \text{ in } B} \qquad (9\text{-}14)$$

and

$$\frac{\text{Barrels of } i \text{ in } D}{\text{Barrels of } i \text{ in } B} = \alpha_a{}^N \frac{\text{barrels of } r \text{ in } D}{\text{barrels of } r \text{ in } B} \qquad (9\text{-}15)$$

Similar forms can be written for Eq. (9-12). These two equations relate the splits on two components to the number of equilibrium stages required to effect this separation. Since the column is operating at total reflux, the number of stages is the minimum possible number N_m to accomplish the desired separation. If the splits on two components are specified, the required minimum of equilibrium stages N_m can be calculated. If the number of equilibrium stages and the split on one component are specified, the splits which will be obtained on all other components at total reflux can be calculated. A reasonably good estimate of the separation which will be accomplished in a plant column often can be obtained by specifying the split on one component, setting N_m equal to 60 per cent of the number of equilibrium stages (not actual stages), and using Eq. (9-14) or (9-15) to calculate the split on each of the other components. The number of equilibrium stages is found by multiplying the number of actual stages by an average stage efficiency.

Example 9-1. Use the final temperature profile and the splits on the $n\text{-}C_4$ and $i\text{-}C_5$ obtained in the solution of Example 8-3 to calculate the corresponding minimum number of equilibrium stages N_m necessary to accomplish the calculated separation at total reflux. Compare the calculated N_m to the number of equilibrium stages in the column. Use N_m to estimate the splits on the nonkey components, and compare the results with the solution of Example 8-3.

Solution. Normally, use of the Fenske equation to estimate the separation in a problem such as Example 8-3 would involve a trial-and-error procedure to establish the end temperatures. An initial separation of the feed between overhead and bottoms would be estimated, and dew and bubble points calculated to obtain first guesses as to the effective α's. The separation predicted by the Fenske equation would then be compared with the assumed separation, and new α's would be calculated if the agreement were not sufficiently close. In this example problem, the major point of interest is a direct comparison between the results obtained by the Fenske total reflux equation and those predicted by the general short-cut equation in Example 8-3. Therefore, the final temperature profile calculated in Example 8-3 will be used to evaluate the α's.

The K values are obtained from Fig. B-3. The tray pressure drops will be neglected, and all the K values read for a pressure of 120 psia.

Top temperature $= 163.5°F$
Middle temperature $= 210°F$
Bottom temperature $= 234°F$

Com- ponent	Top		Middle		Bottom		α_a
	K	α	K	α	K	α	
C_3	2.76	5.52	38.6	4.65	4.45	4.32	4.86
i-C_4	1.37	2.74	2.08	2.51	2.51	2.44	2.56
n-C_4	1.04	2.08	1.61	1.94	2.00	1.94	1.99
i-C_5	0.50	1.00	0.83	1.00	1.03	1.00	1.00
n-C_5	0.418	0.836	0.705	0.849	0.90	0.874	0.853

The α_a's were calculated with Eq. (9-13). Substitution of the mole fractions of n-C_4 and i-C_5 from the results of Example 8-3 into Eq. (9-14) gives

$$\frac{22.6}{2.41} = (1.99)^{N_m} \frac{3.57}{16.4}$$

$$N_m \ln 1.99 = \ln \frac{(16.4)(22.6)}{(3.57)(2.41)}$$

$$N_m = 5.5$$

This N_m is 55 per cent of the 10 equilibrium stages which the column was assumed to have. The N_m can be used with the split on either the n-C_4 or the i-C_5 to calculate the splits on the other components. The i-C_5 will be used, since the relative volatilities are already calculated with respect to that component. Equation (9-14) can be written as

$$\left(\frac{Dx_D}{Fx_F - Dx_D}\right)_i = \alpha_a{}^{N_m} \frac{Dx_{r,D}}{Bx_{r,B}} = \alpha_a{}^{5.5}(0.217)$$

or

$$Dx_{i,D} = \frac{0.217\alpha_a{}^{5.5}}{1 + 0.217\alpha_a{}^{5.5}} Fx_{i,F}$$

and the calculations performed as follows:

Component	α_a	$\alpha_a{}^{N_m}$	$0.217\alpha_a{}^{N_m}$	Fx_F	Dx_D	Bx_B
C_3	4.86	6000	1302	5	5.0	0.0
i-C_4	2.56	176	38.2	15	14.6	0.4
n-C_4	1.99	44	9.55	25	22.6	2.4
i-C_5	1.00	1.00	0.217	20	3.6	16.4
n-C_5	0.853	0.417	0.0905	35	2.9	32.1
				100	48.7	51.3

The Fenske equation gives a $D = 48.7$ compared with the 48.9 from the general short-cut method. The mole fractions in the overhead and bottoms from the two methods are compared in the following tabulation.

Component	Example 8-3		Fenske	
	x_D	x_B	x_D	x_B
C_3	0.102	0.000	0.102	0.000
i-C_4	0.299	0.007	0.300	0.008
n-C_4	0.462	0.047	0.464	0.047
i-C_5	0.073	0.322	0.074	0.320
n-C_5	0.064	0.624	0.060	0.625
	1.000	1.000	1.000	1.000

It can be seen that in this case the Fenske total reflux equation gives a good approximation of the separation if the temperature profile is known beforehand. It is somewhat less accurate if only the two end temperatures are used to estimate the average α's. Also, the close agreement illustrates the fact that the separation obtained at a reasonably large reflux does not differ much from the total reflux separation when the relative volatilities differ widely. The Fenske equation is less accurate when the reflux rate is low or the volatility differences are smaller. For very small volatility differences the Fenske equation can be widely in error.

9-2. Minimum Reflux—Underwood Equations

Minimum reflux is defined as that reflux rate which, if decreased an infinitesimal amount, would require an infinite number of stages to produce the specified separation between two components. (The two components for which the separation is specified are usually called the "key" components.) If an infinite or nearly infinite number of stages is involved, there must exist within the column at least one zone of constant composition where there is no change in composition of the liquid or vapor from stage to stage. Any specified separation of two components in a feed of given composition and thermal condition will have associated with it a unique minimum reflux rate and a unique set of limiting concentrations in the constant composition zones.

The concept of minimum reflux has no meaning unless the separation between *two* components is specified and the number of stages is *not* specified. Obviously, an existing column operating with a fixed number of stages can never produce a zone of constant composition even at zero reflux. Also, the specification of the split on a single component cannot give rise to a condition of minimum reflux, since such a specification can be met by taking more or less overhead product regardless of the reflux rate or number of stages.

Care must be taken to ensure that the true minimum reflux rate or ratio is obtained. Figure 9-1 utilizes the binary enthalpy-concentration diagram to illustrate that for a fixed feed the true minimum reflux ratio is the largest of all possible minimum reflux ratios. On the enthalpy-concentration diagram, the external reflux ratio L_{N+1}/D can be represented in terms of line segments on a vertical line through x_D as

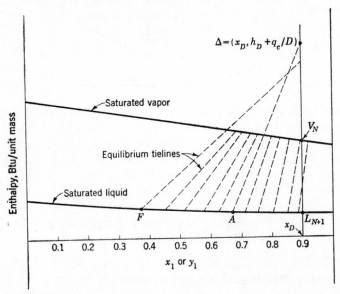

FIG. 9-1. Binary system with tie lines such that a zone of constant composition develops in the rectifying section at a higher reflux ratio than the ratio necessary to create such a zone at the feed plate.

$\overline{V_N\Delta}/\overline{V_N L_{N+1}}$. At minimum reflux there must be a coincidence between the operating lines and a tie line in order to produce a zone of constant composition. Or, in other words, the difference point Δ must coincide with the intersection of the extended tie line on the vertical through x_D. It is obvious that as the external reflux ratio is decreased, the zone of constant composition will develop around that tie line whose intersection on the x_D vertical is the highest. In Fig. 9-1 the tie line through point A gives the true minimum reflux ratio. Figure 9-2 illustrates the same principle on a y-x diagram in terms of the internal reflux ratio L/V. [The equilibrium curve is shaped in such a manner as to cause the development of a zone of constant composition in the rectifying section at an L/V higher than that L/V which would give such a zone around the feed stage.

Figures 9-1 and 9-2 show that for a given feed the zone of constant composition need not include the feed stage.   However, if the restriction of a given feed is dropped, it can be seen that the true minimum reflux depends somewhat upon the feed composition.   For example, a feed concentration of 80 per cent A would permit the production in a finite number of stages of the specified product purities in Fig. 9-2 without changing the L/V in the top section.

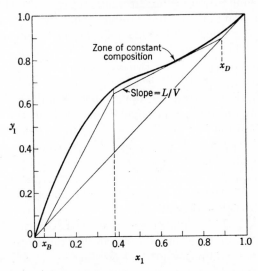

Fig. 9-2. Binary system with equilibrium curve such that a zone of constant composition develops in the rectifying section at a higher L/V ratio than the L/V ratio associated with the intersection of the q line with the equilibrium curve.

A multicomponent column is usually assumed to be at minimum reflux when a zone of constant composition appears in both the stripping and rectifying sections. [If a zone of constant composition appears in only one section as the reflux is slowly decreased, it is presumably possible to shift the feed-plate location (without changing the feed itself) to unload that section and cause the zone to disappear.] [The reflux could be reduced again until finally both sections are loaded equally and a zone of constant composition appears in both sections.] This reasoning for multicomponent systems presumes ideal behavior as far as the equilibrium data are concerned. The multicomponent mixture could behave similarly to the binary in Fig. 9-2 and produce a zone of constant composition which cannot be made to disappear by a slight shift in the feed location. [In this situation the feed composition would have to be moved all the way to the other side of the zone before the zone would disappear and permit the reflux to be reduced further.]

If certain components in a multicomponent mixture are assumed to be nondistributed and appear in only one of the product streams, the limiting composition calculated for the zone in the stripping section must differ from the composition calculated for the zone in the rectifying section. The limiting compositions would be calculated by making stage-to-stage calculations at the minimum reflux from the bottom up and from the top down until the zones of constant composition are reached. The limiting composition in the stripping section will include only those components present in the bottom product, while the limiting composition in the rectifying section will include only those components in the overhead stream. Therefore, in the case of nondistributed components the calculated compositions of the two zones cannot be identical and the zones cannot merge at the feed stages. There must be enough stages between the feed stage and the upper zone to reduce the heavy nondistributed components to a negligible concentration before the zone of constant composition is entered. Likewise, there must be some stages below the feed tray to reduce the light nondistributed components to a negligible concentration before the bottom zone is entered.

Modern computer solutions of multicomponent distillation problems adhere more closely to reality and assume that all components in the feed are distributed; that is, they all appear in both the overhead and bottom streams. Under these conditions (and with "normal" equilibrium relationships), the limiting compositions in the two zones will be identical and the two zones will merge at the feed stage.

It is important to know the minimum reflux ratio when only one case is to be solved, and this one solution must provide the number of stages required to obtain a specified separation between two components. In this type of solution the minimum reflux ratio is calculated and then multiplied by some factor (say 1.2 or 1.3) to give an operating reflux ratio L_{N+1}/D. This procedure ensures that the separation can be obtained in a finite number of stages but seldom provides an optimum solution. One reason for this is that both yield and purity of the products are important and specification of both is not convenient. The availability of electronic computers makes practical several solutions of the same problem. Usually the number of stages and one product rate are specified instead of the two product purities. For these specifications the concept of minimum reflux has no meaning. Any reflux ratio will provide a separation in the finite number of stages specified. The solution of a few cases pinpoints that N and R which provide the optimum product yields and purities. The rigorous solution methods described in Chap. 10 utilize these specifications and therefore are unconcerned with the minimum reflux ratio.

The equations of Underwood (5, 6) provide a convenient method for estimating the minimum reflux corresponding to any specified separation of two key components in a multicomponent feed. A small portion of Underwood's development is presented below as an aid in discussing the concept of minimum reflux and as background for the equations which will be used as part of the alternate short-cut method for distillation. Underwood's minimim reflux equations are restricted by the assumptions of constant molar overflow and constant α's. Their derivation will now be summarized.

The relation between the liquid compositions on any two adjacent plates in the rectifying section can be expressed in terms of the relative volatilities and reflux ratio as follows: For any component a balance around the top of the column gives

$$V_n y_n = L_{n+1} x_{n+1} + D x_D \tag{9-16}$$

If constant molar overflow is assumed, the subscripts on the rates can be dropped and the equalities

$$\frac{L}{V} = \frac{R}{R+1}$$
$$\frac{D}{V} = \frac{1}{R+1}$$

substituted in Eq. (9-16) to give

$$y_n = \frac{R}{R+1} x_{n+1} + \frac{x_D}{R+1} \tag{9-17}$$

It will be shown in Sec. 10-3 that for any given component the equilibrium relationship on stage n can be expressed as

$$y_n = \frac{\alpha x_n}{\displaystyle\sum_{i=1}^{i=C} \alpha_i x_{i,n}}$$

Substitution for y_n in Eq. (9-17) gives the following combination of the equilibrium and material-balance relationships:

$$\frac{\alpha x_n}{\displaystyle\sum_{i=1}^{C} \alpha_i x_{i,n}} = \frac{R}{R+1} x_{n+1} + \frac{x_D}{R+1} \tag{9-18}$$

A stage subscript on α is not necessary, since the relative volatilities are assumed to be constant over the column length.

Now define a function ϕ so that

$$\sum_{i=1}^{C} \frac{\alpha_i x_{i,D}}{\alpha_i - \phi} = R + 1 = \frac{L}{D} + 1 \tag{9-19}$$

and introduce this function into Eq. (9-18) by multiplying the equation by $\alpha/(\alpha - \phi)$.

$$\frac{\alpha^2 x_n/(\alpha - \phi)}{\displaystyle\sum_{i=1}^{C} \alpha_i x_{i,n}} = \frac{R}{R+1}\frac{\alpha x_{n+1}}{\alpha - \phi} + \frac{1}{R+1}\frac{\alpha x_D}{\alpha - \phi} \tag{9-20}$$

Writing Eq. (9-20) for each component and then summing the C equations so obtained give

$$\frac{\displaystyle\sum_{i=1}^{C} \frac{\alpha_i{}^2 x_{i,n}}{\alpha_i - \phi}}{\displaystyle\sum_{i=1}^{C} \alpha_i x_{i,n}} = \frac{R}{R+1}\sum_{i=1}^{C}\frac{\alpha_i x_{i,n+1}}{\alpha_i - \phi} + \frac{1}{R+1}\sum_{i=1}^{C}\frac{\alpha_i x_{i,D}}{\alpha_i - \phi} \tag{9-21}$$

By substitution of the definition of ϕ given by Eq. (9-19), Eq. (9-21) can be rewritten as

$$\frac{\displaystyle\sum_{i=1}^{C} \frac{\alpha_i{}^2 x_{i,n}}{\alpha_i - \phi} - \sum_{i=1}^{C} \alpha_i x_{i,n}}{\displaystyle\sum_{i=1}^{C} \alpha_i x_{i,n}} = \frac{R}{R+1}\sum_{i=1}^{C}\frac{\alpha_i x_{i,n+1}}{\alpha_i - \phi}$$

Expansion and rearrangement of the left-side numerator give

$$\frac{\phi \displaystyle\sum_{i=1}^{C} \frac{\alpha_i x_{i,n}}{\alpha_i - \phi}}{\dfrac{R}{R+1}\displaystyle\sum_{i=1}^{C} \alpha_i x_{i,n}} = \sum_{i=1}^{C}\frac{\alpha_i x_{i,n+1}}{\alpha_i - \phi} \tag{9-22}$$

If a condition of minimum reflux exists in the rectifying section, there must be a zone of constant composition in the section. Adjacent plates in the zone will have identical vapor and liquid compositions. If stages n and $n + 1$ are considered to be in the zone of constant composition, then $x_{i,n} = x_{i,n+1}$ and Eq. (9-22) reduces to

$$\phi = \frac{R_m}{R_m + 1}\sum_{i=1}^{C} \alpha_i x_{i,z} \tag{9-23}$$

where $x_{i,z}$ refers to the liquid concentration of component i in the zone of constant composition and R_m is the minimum reflux ratio.

Combination of Eq. (9-23) with a component material balance around the top of the column and through the zone of constant composition provides an expression explicit in $x_{i,z}$. Setting $x_{i,n} = x_{i,n+1} = x_{i,z}$ in

Eq. (9-18) and rearranging give for any given component

$$\sum_{i=1}^{C} \alpha_i x_{i,z} = \frac{\alpha x_z}{[R_m/(R_m + 1)]x_z + [(x_D)_m/(R_m + 1)]}$$

From Eq. (9-23)

$$\sum_{i=1}^{C} \alpha_i x_{i,z} = \phi \frac{R_m + 1}{R_m}$$

Combination of these two equations gives for any given component

$$\phi \frac{R_m + 1}{R_m} = \frac{\alpha x_z}{[R_m/(R_m + 1)]x_z + [(x_D)_m/(R_m + 1)]}$$

Denoting the given component by the subscript i and solving for $x_{i,z}$ give

$$x_{i,z} = \frac{\phi(x_{i,D})_m}{R_m(\alpha_i - \phi)} \tag{9-24}$$

where $(x_{i,D})_m$ refers to the concentration of i in D at minimum reflux.

Inspection of Eq. (9-19) shows that the number of roots (or possible values of ϕ) equals the number of components C. One root falls between $\phi = 0$ and $\phi = \alpha_C$, where α_C represents the relative volatility of the Cth or heaviest component (smallest α). The other roots fall between the α's for the various adjacent pairs of components. There is no root for $\phi > \alpha_1$, where α_1 represents the relative volatility of the lightest component (largest α). The preceding statements become obvious if the function

$$\sum_{i=1}^{C} \frac{\alpha_i x_{i,D}}{\alpha_i - \phi} - (R_m + 1)$$

is plotted vs ϕ. Whenever $\phi = \alpha_i$, the function becomes indeterminate. The values of ϕ when the function becomes zero are the root values.

Equations similar to Eqs. (9-19), (9-23), and (9-24) can be defined and derived for the stripping section.

$$\sum_{i=1}^{C} \frac{\alpha_i x_{i,B}}{\alpha_i - \psi} = -R' = -\frac{V'}{B} \tag{9-25}$$

$$\psi = \frac{R'_m + 1}{R'_m} \sum_{i=1}^{C} \alpha_i x_{i,z'} \tag{9-26}$$

$$x_{i,z'} = \frac{\psi(x_{i,B})_m}{(R'_m + 1)(\alpha_i - \psi)} \tag{9-27}$$

Equation (9-25) defines the function ψ for the stripping section, while Eqs. (9-26) and (9-27) relate ψ to the limiting zone compositions and to the composition of the bottoms at minimum reflux. As in the case of the rectifying section the number of possible roots in Eq. (9-25) equals the number of components.

The column is assumed to be operating at the true minimum reflux when both the stripping and rectifying sections contain zones of constant composition with the optimum feed location. (Possible abnormal volatilities in one section or the other are not considered.)   Underwood has shown that at the true minimum reflux, Eqs. (9-19) and (9-25) have one common root and that this root must lie between the α's for the light and heavy keys chosen.   Designating the common root as θ, Eqs. (9-19) and (9-25) can be rewritten as

$$\sum_{i=1}^{C} \frac{\alpha_i (x_{i,D})_m}{\alpha_i - \theta} = R_m + 1 \tag{9-28}$$

$$\sum_{i=1}^{C} \frac{\alpha_i (x_{i,B})_m}{\alpha_i - \theta} = -R_m' \tag{9-29}$$

where $\alpha_{hk} < \theta < \alpha_{lk}$. If Eqs. (9-28) and (9-29) are multiplied by D and by B, respectively, and then added, the resulting equation relates θ to the feed composition and thermal condition as follows:

$$\sum_{i=1}^{C} \frac{\alpha_i x_{i,F}}{\alpha_i - \theta} = 1 - q \tag{9-30}$$

Equation (9-30) is used to calculate the value of θ which lies between α_{hk} and α_{lk}.   Successive values of θ between α_{hk} and α_{lk} are assumed until the one value that satisfies Eq. (9-30) is found.   The value of θ so obtained is used with the assumed overhead composition to estimate R_m by means of Eq. (9-28).   As can be seen from Eq. (9-28), the overhead composition used in the calculation of R_m should be the $x_{i,D}$'s at minimum reflux.   However, the overhead composition at minimum reflux will never be known exactly.   It can be estimated by equations such as those of Shiras et al. (4), but the extra calculation is seldom worthwhile.   In practical calculations, the Fenske equation is normally used to provide a set of assumed $x_{i,D}$'s, which are then used in Eq. (9-28) to estimate the R_m corresponding to the assumed overhead composition.   As shown in the next section the R_m can be related to the actual R and an estimate of the column diameter thus obtained.

The preceding discussion has considered only adjacent key components.

n some cases, a component which would normally be chosen as one of the keys has only a small mole fraction in the feed. Rather than base a pecification on such a minor component, the next higher or lower omponent is chosen. The keys are then said to be split; that is, they ave a component between them. Underwood (6) has suggested a nethod for the calculation of the minimum reflux in such cases, and the eader is referred to his original article.

-3. Gilliland Correlation

A knowledge of N_m and R_m defines limiting conditions for a distillation olumn but otherwise serves little practical purpose unless these quanities can be related to the actual N and R. Gilliland (2) has presented a iseful correlation which relates N_m and R_m to the actual number of quilibrium stages and the actual reflux ratio necessary to obtain the lesired separation. The correlation is shown in Figs. B-8a and b, where $N - N_m)/(N + 1)$ is plotted vs $(R - R_m)/(R + 1)$. The ratios used or the ordinate and abscissa were chosen because they provide fixed nd points for the correlation curve (see Fig. B-8a). The right-hand end ioint represents total reflux conditions. At total reflux, R is infinite, $V = N_m$, and the ordinate and abscissa ratios reduce to zero and unity, espectively. The left-hand end point represents minimum reflux conditions. At minimum reflux, $R = R_m$, N becomes infinite, and the ibscissa and ordinate ratios reduce to zero and unity, respectively. lconomic design values of R usually range from 1.2 to 1.5 times the ninimum reflux ratio R_m. This range of R/R_m usually falls between 1.1 and 0.33 on the abscissa of Gilliland's correlation.

The values of N_m used in the correlation were calculated by Fenske's iquation. The values of R_m were calculated with Gilliland's (3) minimum eflux equation and not with the Underwood equations described in the irevious section. The exact effect of using Underwood's minimum eflux rather than one calculated by Gilliland's equations is not known iut is felt to be unimportant.

The values of R and N were obtained by stage-to-stage calculations on iight different systems. The thermal condition and composition of the eed for some of these systems were varied to give a total of 16 cases. lach case was calculated at various reflux ratios in order to obtain a elation between R and N for that particular case. Most of the calculated esults were taken from published examples in the literature. The ixamples covered systems containing from 2 to 11 components, a wide range of feed conditions, operating pressures from subatmospheric to 100 psig, relative volatilities between the key components ranging from 1.26 to 4.05, values of R_m from 0.53 to 7.0, and a range of equilibrium stages from 2.4 to 43.1. Despite the wide range of conditions covered, the correlation is good with the maximum deviation corresponding to 7 per cent of N. Gilliland points out that an error of a few tenths of a theoretical stage or a slight mislocation of the feed stage from the optimum would account for this deviation. He warns that the correlation be used with caution for systems exhibiting abnormal volatilities.

9-4. Alternate Short-cut Method

The Fenske and Underwood equations together with the Gilliland correlation provide a short-cut design method for distillation columns. The Fenske correlation is used to calculate N_m and estimate the overhead-product composition and rate. The Underwood equations are used to estimate the R_m which corresponds to the estimated overhead composition. The Gilliland correlation then relates N_m and R_m to the actual required N and R. The internal vapor rate which corresponds to the estimated R and overhead rate is calculated and translated into a required column diameter by means of a vapor loading equation such as the one described in Chap. 14.

The exact calculation procedure to be followed depends on whether a new column is being designed or the performance of an existing column is being studied. The former of these two problems is the easier, and it will be discussed first.

New Column. A column with one feed, a total condenser, and a partial reboiler will be used to discuss the alternate short-cut method. Analysis of this type of distillation column in Sec. 3-5 showed that $C + 2N + 9$ variables must be specified to define one unique operation of the column. These variables are specified as follows when using the alternate method.

Specifications	$N_i{}^u$	
Stage pressures (including reboiler)	N	
Condenser pressure		1
Stage heat leaks (excluding reboiler)	$N - 1$	
Heat leak and pressure in reflux divider		2
Feed stream	C	$+ 2$
N, total number of equilibrium stages		1
Feed location (optimum)		1
Bubble-point reflux		1
Split on two components or two purities		2
	$\overline{C + 2N + 9}$	

The assumption of optimum feed location is inherent in the Underwood minimum reflux conditions, since the concept of the true minimum reflux assumes that both the rectifying and stripping sections are loaded equally. Also, the assumption of saturated reflux is inherent in the Fenske and Underwood equations.

The column diameter required by the specified operation can be calculated by the following steps:

1. Assume a split for all components whose split is not specified. The over-all separation between B and D must satisfy the split or purity specifications on the key components.

2. Estimate the top, middle, and bottom temperatures with dew and bubble points as required on the overhead vapor, feed, and bottom product.

3. Calculate average α's with Eq. (9-13).

4. Use the Fenske equation to calculate the N_m which corresponds to the specified splits on the two key components. Once N_m is known, the splits on all the other components can be calculated.

5. If the overhead composition calculated in step 4 differs enough from that assumed in step 1 to affect the average α's repeat the previous calculations with the calculated composition.

6. Calculate the minimum reflux rate R_m which corresponds to the final values for the overhead composition. Equation (9-30) is used to obtain a value for θ which is then used along with the calculated overhead composition in Eq. (9-28) to estimate R_m.

7. The calculated values of N_m and R_m can now be used with the specified N to obtain R from the Gilliland correlation (Fig. B-8a or b).

8. The overhead rate D obtained in step 4 and the R from step 7 are related to the top vapor by $V = (1 + R)D$. Once V and V' are known approximately, the vapor loading calculation method from Chap. 14 can be used to estimate the required column diameter.

As mentioned previously, there is a basic inconsistency in the use of the total reflux overhead composition in the Underwood minimum reflux equations. If desired, an additional step can be inserted between steps 5 and 6 for the estimation of the minimum reflux overhead composition. The equations of Shiras, Hanson, and Gibson (4) are well suited for this purpose. This additional calculation makes the method more consistent but usually has little effect on the final answer.

No example is presented for the design of a new column. However, the numerical calculations involved in the various steps are illustrated in Example 9-2, which shows the application of the alternate short-cut method to an existing column.

Existing Column. The prediction of the performance of an existing column in a new service is a more common problem than the design of a new column. It is also a more difficult problem when the alternate short-cut method is used, since N, N_m, R, and the vapor rate must all conform to the physical realities. A column with a total condenser and partial reboiler will again be used as an example. The problem specifications when an existing column is considered are, with the exception of the feed location and the split specifications, simply a listing of the physical limitations of the unit.

Specifications		N_i^u
Stage pressures (including reboiler)		N
Condenser pressure		1
Stage heat leaks (excluding reboiler)		$N - 1$
Heat leak and pressure in reflux divider		2
Feed stream	C	$+ 2$
N, total number of equilibrium stages		1
Feed location (optimum)		1
Bubble-point reflux		1
Maximum vapor rate		1
Split on one component or one purity		1
		$C + 2N + 9$

Note that one of the split specifications used previously has been replaced with a specification of the maximum vapor rate. Also, note that regardless of where the feed plate is actually located, it must be assumed to be located in the optimum location if the Underwood equations are to be used. In other words, the results of the alternate short-cut method will be based on the optimum rather than the actual feed-plate location.

The major iteration variable when the alternate method is applied to an existing column is the value of N_m used to predict the separation. The assumed value of N_m must be adjusted until the required vapor rate matches the known column capacity. The individual calculation steps are described below.

1-3. Same as shown previously for a new column.

4. Assume that $N_m = 0.6N$.

5. Use the assumed N_m along with the one specified split or purity to calculate the splits on all the other components with the Fenske equation.

6. If the overhead composition calculated in step 5 differs enough from that assumed in step 1 to affect the average α's, repeat the previous calculations with the calculated composition.

7. Calculate the minimum reflux rate R_m which corresponds to the values for the overhead composition. Equations (9-30) and (9-28) are required.

8. Use the specified V and the value of D from step 5 to calculate the corresponding R.

9. Use the calculated values of R and R_m along with the specified N to get a new value of N_m from the Gilliland correlation. If this new N_m differs appreciably from the value assumed in step 4, the calculations should be repeated with the new N_m.

The calculation procedure can be lengthened by using the results obtained above with the Shiras, Hanson, and Gibson (4) minimum reflux equations to estimate the minimum reflux overhead composition. This composition can be used to estimate a new N_m in the manner described above. As mentioned previously the additional calculations are seldom worth the trouble.

Example 9-2. Rework Example 8-3 using the alternate short-cut method. Specify the i-C_5 concentration in the overhead product as 7.0 per cent. The solution of Example 8-3 by the general short-cut equation presented in Chap. 8 indicated that the spare column could not provide an n-C_4 concentration in the bottom stream of 3 per cent or less without exceeding the 7 per cent i-C_5 specification in the overhead stream. Does the alternate short-cut method concur in this prediction?

Solution. The split for each component between the overhead and bottom products must be assumed in order to obtain average α's for use in the Fenske equation. Since the overhead rate is not specified, any reasonable split on the feed can be assumed. The only restriction is that the concentration of the i-C_5 in the overhead be 7 per cent. A perfect separation between the n-C_4 and i-C_5 would give $D = 45$ and $B = 55$. These values will be used for the initial assumptions. The Dx_D for i-C_5 is then $(0.07)(45) = 3.15$. The other component splits are arbitrarily assumed as follows:

Component	Fx_F	Dx_D	Bx_B
C_3	5	5	
i-C_4	15	15	
n-C_4	25	21.85	3.15
i-C_5	20	3.15	16.85
n-C_5	35		35.00
	100	45.00	55.00

The column pressure has been specified as 120 psia, and the pressure drop through the column will be neglected. Dew and bubble points on the assumed overhead and bottom streams, respectively, give 151.5 and 235°F for the top and bottom temperatures. The bubble point for the feed at 120 psia is 185°F, and this will be used for the middle temperature. The calculation of average α's is as follows:

Component	151.5°F		185°F		235°F		α_a
	K	α	K	α	K	α	
C_3	2.52	5.66	3.25	5.16	4.45	4.28	5.00
i-C_4	1.23	2.76	1.70	2.70	2.52	2.42	2.63
n-C_4	0.915	2.06	1.30	2.06	2.00	1.92	2.01
i-C_5	0.445	1.00	0.63	1.00	1.04	1.00	1.00
n-C_5	0.36	0.809	0.54	0.857	0.90	0.865	0.843

Assume that $N_m = 0.6N = 0.6(10) = 6.0$. Use this number together with the assumed split on the i-C_5 to calculate the corresponding splits on the other components. Equation (9-14) can be written for this purpose as follows:

$$\left(\frac{Dx_D}{Fx_F - Dx_D}\right)_i = \alpha_a{}^{6.0}\frac{3.15}{16.85} = \alpha_a{}^{6.0}(0.1869)$$

or

$$Dx_{i,D} = \frac{0.1869\alpha_a{}^{6.0}}{1 + 0.1869\alpha_a{}^{6.0}}Fx_{i,F}$$

The evaluation of this equation for each component is shown below.

Component	α_a	$\alpha_a{}^{6.0}$	$0.1869\alpha_a{}^{6.0}$	Fx_F	Dx_D	Bx_B
C_3	5.00			5	5.0	0.0
i-C_4	2.63	330	61.7	15	14.8	0.2
n-C_4	2.01	66	12.3	25	23.1	1.9
i-C_5	1.00	1.00	0.187	20	3.15	16.85
n-C_5	0.843	0.36	0.0672	35	2.20	32.80
				100	48.25	51.75

The new overhead and bottom compositions differ enough from the initial assumptions to require the calculation of new average α's. Dew and bubble points on the overhead and bottoms streams give new top and bottoms temperatures of 159 and 236°F, respectively. These temperatures together with the 185°F bubble point of the feed provide a new set of average α's. The new average α's vary only slightly from those used previously, and the D and B will not vary appreciably from the 48.25 and 51.75 calculated above with the original α's. The split on the i-C_5 necessary to meet the 7 per cent specification will be approximately $(0.07)(48.25) = 3.4$ moles in D and 16.6 moles in B. Equation (9-14) can now be written as

$$\frac{Dx_D}{Fx_F - Dx_D} = \alpha_a{}^{6.0}\frac{3.4}{16.6} = \alpha_a{}^{6.0}(0.2048)$$

The new product rates and compositions are as follows:

Component	$\alpha_a{}^{6.0}$	$0.2048\alpha_a{}^{6.0}$	Fx_F	Dx_D	Bx_B	x_D	x_B
C_3			5	5.0	0.0	0.102	0.000
i-C_4	322	65.9	15	14.8	0.2	0.301	0.004
n-C_4	68	13.9	25	23.3	1.7	0.473	0.033
i-C_5	1.00	0.205	20	3.4	16.6	0.069	0.327
n-C_5	0.415	0.085	35	2.7	32.3	0.055	0.636
			100	49.2	50.8	1.000	1.000

The i-C$_5$ concentration in the new overhead composition is 6.91 per cent. The split on the i-C$_5$ could be adjusted to meet the 7 per cent specification exactly; however, the 6.91 per cent is close enough for the present.

The Underwood minimum reflux equations can now be used to estimate the minimum reflux ratio R_m which corresponds to the calculated separation between the two key components n-C$_4$ and i-C$_5$. There is, of course, a basic inconsistency in using a total reflux overhead composition for the overhead composition at minimum reflux. However, as mentioned previously, the methods available for estimating the x_D's at minimum reflux are cumbersome to use and the increased accuracy, if any, is not worth the additional calculation time.

A bubble-point feed ($q = 1.0$) was specified in Example 8-3, so Eq. (9-30) can be written as

$$\sum_{i=1}^{i=C} \frac{\alpha_i x_{i,F}}{\alpha_i - \theta} = 0 \qquad (9\text{-}30)$$

The calculation of the root for θ that lies between the α's for the keys proceeds as follows:

Component	x_F	α	αx_F	$\theta = 1.36$		$\theta = 1.365$	
				$\alpha - \theta$	$\dfrac{\alpha x_F}{\alpha - \theta}$	$\alpha - \theta$	$\dfrac{\alpha x_F}{\alpha - \theta}$
C$_3$	0.05	4.99	0.2495	3.63	0.0687	3.625	0.0688
i-C$_4$	0.15	2.62	0.3930	1.26	0.3119	1.255	0.3131
n-C$_4$	0.25	2.02	0.5050	0.66	0.7651	0.655	0.7710
i-C$_5$	0.20	1.00	0.2000	−0.36	−0.5556	−0.365	−0.5479
n-C$_5$	0.35	0.864	0.3024	−0.496	−0.6097	−0.501	−0.6036
	1.00				−0.0196		+0.0014

Interpolation gives $\theta = 1.3647$. This value is used with Eq. (9-28) to give R_m.

Component	x_D	αx_D	$\alpha - \theta$	$\dfrac{\alpha x_D}{\alpha - \theta}$
C$_3$	0.102	0.5090	3.6253	0.1404
i-C$_4$	0.301	0.7886	1.2553	0.6282
n-C$_4$	0.473	0.9555	0.6553	1.4581
i-C$_5$	0.069	0.0690	−0.3647	−0.1892
n-C$_5$	0.055	0.0475	−0.5007	−0.0949
	1.000			$1.9426 = R_m + 1$

The calculated $R_m = 1.9426 - 1.0 = 0.9426$. Since the difference $(\alpha - \theta)$ is involved in the calculations, slight changes in the value of θ used can have an appreciable effect on the R_m obtained.

The actual reflux ratio R is obtained from the maximum allowable top vapor rate of 175 and the calculated D of 49.2.

$$L_{N+1} = V_N - D$$

$$R = \frac{V_N}{D} - 1 = \frac{175}{49.2} - 1 = 2.557$$

The values of R_m, R, and N can now be used with the Gilliland correlation to check the value of $N_m = 6.0$ which was assumed at the beginning of the solution.

$$\frac{R - R_m}{R + 1} = \frac{2.557 - 0.943}{2.557 + 1.0} = 0.454$$

From Fig. B-8a, $(N - N_m)/(N + 1) = 0.268$. Since $N = 10$, the corresponding value of N_m is 7.05, which differs appreciably from the 6.0 assumed.

Repetition of the calculations with $N_m = 7.0$ gave $R = 2.519$,

$$R_m = 0.9782$$

and a calculated check value of $N_m = 6.94$. The calculated D and B rates and compositions and the average α's were as follows:

Component	α_a	Dx_D	Bx_B	x_D	x_B
C_3	4.98	5.00	0	0.1004	0.0
$i\text{-}C_4$	2.61	14.91	0.09	0.2996	0.0017
$n\text{-}C_4$	2.02	24.16	0.84	0.4852	0.0168
$i\text{-}C_5$	1.00	3.48	16.52	0.0700	0.3283
$n\text{-}C_5$	0.851	2.23	32.87	0.0448	0.6532
		49.78	50.32	1.000	1.0000

The results indicate that the 7 per cent $i\text{-}C_5$ in D and the 3 per cent $n\text{-}C_4$ in B specifications can be met easily. The general short-cut equation solution and the rigorous computer results listed in Example 8-3 are more conservative and indicate that the specifications could not be met. The general short-cut equation predicted 7.3 per cent $i\text{-}C_5$ in D and 4.7 per cent $n\text{-}C_4$ in B, while the rigorous computer results were 7.3 per cent $i\text{-}C_5$ in D and 3.7 per cent $n\text{-}C_4$ in B. The discrepancy is understandable, since the results above are based on total reflux while the results from the general short-cut and rigorous solutions are based on an R of 2.579.

NOMENCLATURE

B = heavy product rate; moles, weight, or volume per unit time.

C = number of components in the system.

D = light product rate; moles, weight, or volume per unit time.

F = feed rate; moles, weight, or volume per unit time.

i = subscript referring to component i.

$K = y/x$ = distribution coefficient. Prime denotes lower section of column.

L = heavy phase rate; moles, weight, or volume per unit time. Prime denotes lower section of column.

m = subscript denoting value at minimum condition.

n = subscript denoting any stage in column.

N = total number of equilibrium stages in column including partial reboiler if any but excluding partial condenser if any.

N_m = total number of equilibrium stages required to effect the specified separation at total reflux. N_m includes the partial reboiler if any but excludes the partial condenser if any. With a partial condenser, the total number of equilibrium stages at total reflux is $N_m + 1$.

q = fraction of feed which is liquid at the feed tray temperature and pressure. See Sec. 5-6 for more complete definition.

r = subscript referring to component on which relative volatilities are based.

$R = L_{N+1}/D$ = external reflux ratio at top of column.

R_m = minimum external reflux ratio, L_{N+1}/D, at top of column.

$R' = V_1/B$ = reboil ratio at bottom of column.

V = light phase rate; moles, weight, or volume per unit time.

x = concentration in heavy phase. Units consistent with units on rates.

y = concentration in light phase. Units consistent with units on rates.

z = subscript which denotes a concentration in the zone of constant composition in the upper section of the column. z' has the same meaning for the lower section.

Greek Symbols

α = relative volatility. $\alpha_{i-r} = K_i/K_r$, where r refers to reference. α_a = average for column for component under consideration. α_{hk} and α_{lk} denote α for heavy key and light key, respectively.

θ = common root between Eqs. (9-19) and (9-25).

ϕ = a function defined by Eq. (9-19).

ψ = a function defined by Eq. (9-25).

PROBLEMS

9-1. The fractionation efficiencies of two types of trays were compared by operating both under identical column conditions and then comparing the overhead and bottoms purities obtained with each. A binary feed of C_3 and $n\text{-}C_4$ was used. The overhead and bottoms purities obtained were as follows:

	C_3 in bottoms	C_4 in distillate
Type A, mole %	3.2	0.5
Type B, mole %	1.5	2.4

Use the Fenske equation to select the best tray.   Assume that $\alpha_a = 2.0$.

9-2. Estimate the solution to Prob. 8-1 with the Fenske equation.   How do the specifications for a column under total reflux differ from those used in the solution to Prob. 8-1?

9-3. What would be the calculated optimum split on the prefractionator in Prob. 8-3 if the Fenske equation is used to estimate the split?

REFERENCES

1. Fenske, M. R.: *Ind. Eng. Chem.*, **24,** 482 (1932).
2. Gilliland, E. R.: *Ind. Eng. Chem.*, **32,** 1220 (1940).
3. *Ibid.*, **32,** 1101 (1940).
4. Shiras, R. N., D. N. Hanson, and C. H. Gibson: *Ind. Eng. Chem.*, **42,** 871 (1950).
5. Underwood, A. J. V.: *J. Inst. Petrol.*, **31,** 111 (1945); **32,** 598 (1946); **32,** 614 (1946).
6. Underwood, A. J. V.: *Chem. Eng. Progr.*, **44,** 603 (1948).

CHAPTER 10

MULTICOMPONENT SEPARATIONS
Rigorous Methods for Distillation

The specification of a relatively small number of variables will cause a distillation column to produce a unique set of stream rates, compositions, temperatures, and pressures. The purpose of any rigorous calculation method is the prediction of the unique set of conditions which will exist in the operating column under any desired set of specified variables. The number of unspecified (but fixed) variables is large compared with the number of specified variables. In any engineering problem where this is true, it is usually necessary to assume values for some of the unspecified variables in order to begin the calculations which will predict values of these same variables. The calculated values are then compared with the assumed values in the classic trial-and-error procedure, and the calculations are repeated or stopped according to the degree of agreement between the two sets of values. All the multicomponent calculation methods described in this book are of this general nature.

In the discussion to follow, the total number of variables in the distillation process will be divided into the following three groups:

1. Specified variables
2. Assumed iteration variables
3. All other iteration variables

The *specified variables* are those which are specified to define a unique operation of the process. The necessary number of specified variables for any process can be determined by the principles set forth in Chap. 3. These variables do not change from trial to trial but remain constant throughout the calculation of any given problem.

The *assumed iteration variables* are those variables which are assumed at the beginning of each trial in order that the trial calculations can be made. The values of all the iteration variables (assumed or otherwise) change from trial to trial. The values of the assumed iteration variables which are calculated in each trial are used as the initial assumptions for the next trial.

The third group includes all the variables which are not included in groups 1 and 2. The true, final values of all group 2 and group 3 variables are fixed once the specified variables are enumerated. The purpose of the trial-and-error calculation is to determine those true values.

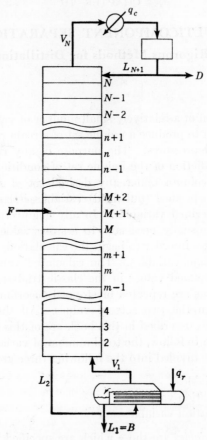

FIG. 10-1. Flow diagram and nomenclature for the distillation process considered in Chap. 10.

The purpose of this chapter is to describe the basic computational methods which have been proposed for the multicomponent, multistage distillation process. The similarities and differences among the various methods will be more evident if all are applied to the same process and the same variables are specified throughout. The process which will be used is a single-feed column with a total condenser and a partial reboiler. See Fig. 10-1 for a schematic flow diagram.

This process has $C + 2N + 9$ independent design variables (see Sec. 3-5), and these will be specified as follows:

Specifications		$N_i{}^u$
Pressure in each stage (including reboiler)		N
Pressure in condenser		1
Pressure and heat leak in reflux divider		2
Heat leak in each stage (excluding reboiler)		$N - 1$
Feed stream	C	$+ 2$
N, total number of stages		1
M, number of stages below the feed stage		1
Condenser load or reflux temperature		1
Distillate rate		1
Vapor rate at some point in the column (usually on top stage)		1
		$\overline{C + 2N + 9}$

If an existing column is being studied, and if the column is operated at capacity, all the items above with the exception of the feed stream and the distillate rate are fixed by the physical characteristics of the column. If a new column is being designed, a greater choice of variables is possible, but for the sake of simplicity in the ensuing discussion only the set above will be used.

10-1. Major Steps

All calculation methods for multistage separation processes can be divided into the following basic parts:

1. *Specification of the process.* This is accomplished by listing a sufficient number of *specified variables* as was done at the beginning of the chapter for the process to be discussed in this chapter.

2. *Preliminary assumptions.* These assumptions provide initial values for the *assumed iteration* variables which must be assumed to start the trial-and-error procedure. All the computational methods to be discussed in this chapter include the following among these variables:

a. Temperature profile
b. L and V profiles

The term profile refers to the curve obtained in a plot of the variable in question vs the stage number. Assumption of the temperature profile, for example, fixes the temperature in each stage for that trial.

In addition to the temperature, L, and V profiles, some calculation methods require the assumption of the end-product compositions. The initial assumptions for the product compositions can be just a guess or obtained from some short-cut method such as the Fenske equation presented in Chap. 9 or the general short-cut equation described in Chap. 8.

The latter method also provides an initial estimate of the temperature profile.

The trial-and-error calculations are greatly facilitated if all components are assumed to be *distributed* components; that is, they appear to some extent in both the overhead and bottom products. Often, the concentrations of very light components in the bottom product and very heavy components in the overhead product will be so small as to require the use of negative exponents of 10 to denote the decimal location.

3. *Stage-to-stage calculations.*

4. *Convergence procedure.* All or some of the stage concentrations obtained in step 3 are corrected to provide better values of the assumed iteration variables for the next trial.

5. *Calculation of new temperature and rate profiles.* This step precedes step 4 in some of the methods to be described.

6. *Convergence check.* The decision as to whether to stop or repeat the calculations must be made.

The manner in which the various computational methods handle these six parts is presented in the following sections.

10-2. Nomenclature

It is convenient to simplify the writing of the distillation equations by substitution of the following definitions.

$$v_{i,n} = V_n y_{i,n} \qquad v_{i,m} = V_m y_{i,m}$$
$$l_{i,n} = L_n x_{i,n} \qquad l_{i,m} = L_m x_{i,m}$$
$$d_i = D x_{i,D} \qquad b_i = B x_{i,B}$$

The v_i, l_i, d_i, and b_i represent the moles of component i in the given stream. Another convenient grouping of variables is accomplished by use of the absorption and stripping factors, which are defined as

$$A_{i,n} = \frac{L_n}{K_{i,n} V_n} \qquad A_{i,m} = \frac{L_m}{K_{i,m} V_m}$$
$$S_{i,n} = \frac{1.0}{A_{i,n}} \qquad S_{i,m} = \frac{1.0}{A_{i,m}}$$

10-3. Lewis-Matheson Method

The Lewis-Matheson method (10) will be described first because it performs the stage-to-stage calculations algebraically in a manner analogous to the way they were performed graphically in the binary McCabe-Thiele and Ponchon-Savarit methods described in Chaps. 5 and 6. The modification of the Lewis-Matheson method presented below is essentially that described by Bonner (3). Bonner's procedure differs from the original Lewis-Matheson method in the specifications made. Instead of specifying the split between two components and then calculating from both ends toward the feed stage to determine the number of stages required, Bonner simplified the convergence problem by specifying the number of stages in each section and then calculating the product compositions necessary to give a feed-stage match with the specified number of stages. However, Bonner's basic calculation procedure and his use of the feed-stage match as a convergence device are the same as that of Lewis and Matheson.

Preliminary Assumptions. The only iteration variables which must be assumed in the Lewis-Matheson method are the end-product compositions. The corresponding stage temperatures and phase rates can be obtained by trial-and-error calculations around each stage as the calculations proceed through the column. However, in computer practice it has been found to be more convenient to assume initial temperature and rate profiles and use these in the first stage-to-stage calculation through the column. The resulting stage compositions are then used to calculate new temperature and rate profiles for the next trial. This procedure eliminates the trial-and-error calculation of temperature in each stage. Bonner therefore found it convenient to assume the following iteration variables:

1. End-product compositions
2. Temperature profile
3. L and V profiles

The assumed end-product compositions are, of course, related by the over-all material balance and may be selected by any one of the three methods described in Sec. 10-1.

Stage-to-stage Calculations. The assumed composition of the vapor leaving the top stage can be calculated from the assumed distillate composition for any given type of condenser. When a total condenser is used, $y_{i,N} = x_{i,D} = x_{i,N+1}$. A dew-point calculation at the assumed top stage temperature provides the corresponding equilibrium liquid composition for that stage. If the dew-point calculation is performed by the method described in Chap. 4, a trial-and-error search for the correct temperature is required. However, as mentioned previously, the trial-and-error calculation of the dew-point temperature corresponding to the assumed overhead composition is not desirable. Instead, the assumed temperature for stage N is used to calculate a liquid composition for stage N. This is most conveniently done through use of the relative volatilities instead of K values in the equilibrium relationships. The equilibrium equations to be used on each stage in the rectifying section are derived by writing the definition of α for component i on any stage,

$$(\alpha_{i-r})_n = \left(\frac{y_i x_r}{x_i y_r}\right)_n$$

and rearranging it to

$$x_{i,n} = \left(\frac{y_i}{\alpha_{i-r}}\right)_n \left(\frac{1}{K_r}\right)_n$$

At the dew point, the Σx_i's $= 1.0$ and

$$(K_r)_n = \sum_{i=1}^{i=C} \left(\frac{y_i}{\alpha_{i-r}}\right)_n \tag{10-1}$$

This equation will be used later to provide a new assumed temperature for each stage. For the present, however, it will be rearranged to provide the desired expression for $x_{i,n}$. Taking the reciprocal of both sides, multiplying by $(y_i/\alpha_{i-r})_n$, and canceling like terms give the following general equation:

$$x_{i,n} = \frac{(y_i/\alpha_{i-r})_n}{\Sigma(y_i/\alpha_{i-r})_n} \tag{10-2}$$

Multiplication of both the numerator and denominator by V_n permits the equation to be written as

$$x_{i,n} = \frac{(v_i/\alpha_{i-r})_n}{\Sigma(v_i/\alpha_{i-r})_n} \tag{10-3}$$

Actually this equation should be written as

$$\frac{x_{i,n}}{\Sigma x_{i,n}} = \frac{(v_i/\alpha_{i-r})_n}{\Sigma(v_i/\alpha_{i-r})_n}$$

This form emphasizes the fact that the liquid composition calculated is a normalized one. In general, the stage temperature used to evaluate the relative volatilities will not be the dew point of the calculated vapor rising from the stage and therefore Σx_i will not be 1.0. Hence the need to normalize the calculated $x_{i,n}$'s.

Equation (10-3) is used to calculate the L_N composition which corresponds to the assumed overhead composition and stage N temperature. The overflow rate L_N in the assumed L profile can be used to calculate all the $l_{i,N}$'s which are used in component balances around the top of the column,

$$v_{i,N-1} = l_{i,N} + d_i$$

to provide the composition of V_{N-1}. Equation (10-3) is used again to calculate the composition of L_{N-1}, and component material balances around the top furnish the composition of V_{N-2}. This alternating use of the equilibrium and material-balance equations is continued until the calculated composition of the liquid overflow from the feed stage L_{M+1} is obtained.

The correctness of the feed-stage composition calculated by working down from the top must now be checked by analogous calculations from the bottom. An equilibrium relation analogous to Eq. (10-3) and suitable for calculating up the column is derived as follows: The definition of α for component i on any stage in the stripping section is

$$(\alpha_{i-r})_m = \left(\frac{y_i x_r}{x_i y_r}\right)_m$$

This can be rearranged to give

$$y_{i,m} = (K_r \alpha_{i-r} x_i)_m$$

Since $\Sigma y_i = 1.0$ at the bubble point,

$$\left(\frac{1}{K_r}\right)_m = \sum_{i=1}^{i=C} (\alpha_{i-r} x_i)_m \qquad (10\text{-}4)$$

Taking the reciprocal of both sides, multiplication by $(\alpha_{i-r} x_i)_m$ and cancellation of terms give the following general equation:

$$y_{i,m} = \frac{(\alpha_{i-r} x_i)_m}{\Sigma(\alpha_{i-r} x_i)_m} \qquad (10\text{-}5)$$

If both the numerator and denominator are multiplied by L_m,

$$y_{i,m} = \frac{(\alpha_{i-r} l_i)_m}{\Sigma(\alpha_{i-r} l_i)_m} \qquad (10\text{-}6)$$

The assumed bottoms-product composition and the assumed stage 1 temperature are used with Eq. (10-6) to calculate the composition of V_1. The assumed V profile is used to make component material balances around the bottom of the column,

$$l_{i,2} = v_{i,1} + b_i$$

to furnish the composition of L_2. Equation (10-6) is used to calculate the composition of V_2, and component balances around the bottom provide the composition of L_3. The calculations are continued up the column until the composition of the feed-stage overflow stream L_{M+1} is obtained. The composition of L_{M+1} thus obtained is compared with the composition obtained from the top-down calculations. The mismatch is then used to correct the assumed end compositions. The manner in which this is done is discussed later under Convergence Procedure.

New Temperature and Rate Profiles. Bonner found it convenient to combine the calculation of the new stage temperatures and phase rates for the next trial with the stage-to-stage calculations. Therefore, these steps will be discussed before the convergence procedure.

Equation (10-1) provides the K value of the reference component at the dew point. The temperature which corresponds to this K value is the new stage temperature for the next trial. If the α's in Eq. (10-1) are evaluated at the true dew-point temperature, the K_r will, of course, correspond to this temperature. In all trials but the last one, the α's will be evaluated at some temperature other than the correct dew point of the vapor, but owing to the small variation of α with temperature the K_r obtained will not vary widely from the true dew-point value in any trial. Equation (10-1) provides new stage temperatures for the rectifying section, and Eq. (10-4) performs the same service for the stripping section. As the new stage temperatures are obtained in the stage-to-stage calculations, they are set aside for use in the next trial.

Enthalpy balances around the ends of the column are used to calculate new L and V profiles. The newly calculated stage compositions can be used with either the old or new temperature profiles to evaluate molar enthalpies for the streams in the enthalpy balance. If the assumption of ideal vapor and liquid solutions is made, the molar enthalpies are obtained from

$$H_n = \sum_{i=1}^{C} y_{i,n} H_{i,n}$$

$$h_n = \sum_{i=1}^{C} x_{i,n} h_{i,n}$$

In the rectifying section,

$$V_n H_n = L_{n+1} h_{n+1} + D h_D + q_c$$

Since $L_{n+1} = V_n - D$, it can be eliminated and the equation solved for V_n to give

$$V_n = \frac{D(h_D - h_{n+1}) + q_c}{H_n - h_{n+1}} \tag{10-7}$$

The balances in the stripping section are made around the bottom of the column.

$$L_{m+1} h_{m+1} + q_r = V_m H_m + B h_B$$

Eliminating L_{m+1} with $L_{m+1} = V_m + B$ gives

$$V_m = \frac{B(h_{m+1} - h_B) + q_r}{H_m - h_{m+1}} \tag{10-8}$$

The new rates calculated in this manner are set aside to be used as the assumed L and V profiles for the next trial.

Convergence Procedure. The mismatch in the two calculated feed-stage compositions is used to correct the assumed end compositions. An equation which relates a calculated feed-stage liquid concentration to the assumed bottoms concentration for that component can be derived as follows: The derivation will be for component i, but the subscript i will be omitted for the sake of simplicity.

The equilibrium expression for component i in stage 1 can be expressed in terms of the stripping factor $S_m = K_m V_m / L_m$ as follows:

$$y_1 = K_1 x_1$$
$$V_1 y_1 = \frac{K_1 V_1}{L_1} L_1 x_1 = S_1 L_1 x_1 = S_1 B x_B$$
$$v_1 = S_1 l_1 = S_1 b$$

This equilibrium relationship can be written for any stage in the stripping section as

$$v_m = S_m l_m$$

The component i material balance around stage 1 is

$$l_2 = v_1 + b$$

Substitution for v_1 from the equilibrium relationship gives

$$l_2 = S_1 b + b = b(S_1 + 1)$$

Substitution of this expression in

$$v_2 = S_2 l_2$$

gives
$$v_2 = b(S_1 S_2 + S_2)$$

Substitution of this equation in the component i balance around the bottom two stages,

$$l_3 = v_2 + b$$

gives
$$l_3 = b(S_1 S_2 + S_2 + 1)$$

Continuation of this procedure to stage M gives

$$l_{M+1} = b(S_1 S_2 \cdots S_M + S_2 S_3 \cdots S_M + \cdots + S_{M-1} S_M + S_M + 1) \tag{10-9}$$

Equation (10-9) relates the calculated moles of component i in L_{m+1} to the assumed moles of i in the bottoms product in terms of the stripping factors which were used on each stage. The stripping factors on each stage are determined by the assumed temperature and phase rate profiles.

Equation (10-9) can be written in terms of mole fractions for component i as

$$x_{M+1} = x_B \left[\frac{B}{L_{M+1}} (S_1 S_2 \cdots S_M + S_2 S_3 \cdots S_M + \cdots \right.$$
$$\left. + S_{M-1} S_M + S_M + 1) \right] \qquad (10\text{-}10)$$

or as
$$x_{M+1} = x_B f(S) \qquad (10\text{-}11)$$

where $f(S)$ represents the expression in the brackets in (10-10). For any set of assumed temperature and rate profiles, $f(S)$ is a constant which equals the ratio x_{M+1}/x_B. The change in x_B required to produce a desired change in x_{M+1} for any given $f(S)$ is given by

$$x_{M+1} + \Delta x_{M+1} = (x_B + \Delta x_B) f(S) \qquad (10\text{-}12)$$

Subtraction of (10-11) from (10-12) gives

$$\Delta x_{M+1} = \Delta x_B f(S) = \Delta x_B \frac{x_{M+1}}{x_B} \qquad (10\text{-}13)$$

where the x_{M+1} is that concentration of i in L_{M+1} which was obtained in the bottom-up calculations. The x_{M+1} obtained in the top-down calculations can be related to the assumed overhead product composition in a similar manner to give

$$\Delta x_{M+1} = \frac{x_{M+1}}{x_D} \Delta x_D \qquad (10\text{-}14)$$

The mismatch or error between the two calculated $x_{i, M+1}$'s for the jth trial is denoted as

$$e_j = (x_{M+1,B} - x_{M+1,D})_j \qquad (10\text{-}15)$$

where the subscript B is used with the $x_{i, M+1}$ obtained from the stripping section stage-to-stage calculation and the D denotes the value obtained in the top-down calculations. The mismatch for the $j + 1$ trial can be expressed as

$$e_{j+1} = (x_{M+1,B} - x_{M+1,D})_{j+1} \qquad (10\text{-}16)$$

Since
$$(x_{M+1,B})_{j+1} = (x_{M+1,B} + \Delta x_{M+1,B})_j$$
and
$$(x_{M+1,D})_{j+1} = (x_{M+1,D} + \Delta x_{M+1,D})_j$$

where $\Delta x_{M+1,B}$ and $\Delta x_{M+1,D}$ represent the changes in the two calculated feed-stage concentrations between trials j and $j + 1$, Eq. (10-16) can be rewritten as

$$e_{j+1} = (x_{M+1,B} + \Delta x_{M+1,B})_j - (x_{M+1,D} + \Delta x_{M+1,D})_j$$

Substitution from Eq. (10-15) gives

$$e_{j+1} = e_j + (\Delta x_{M+1,B} - \Delta x_{M+1,D})_j$$

The purpose of the convergence method is to make $e_{j+1} = 0$. The Δx's must then be such as to make

$$(\Delta x_{M+1,D} - \Delta x_{M+1,B})_j = e_j$$

The trial subscripts can now be dropped. Substitutions from Eqs. (10-13), (10-14), and (10-15) give

$$\frac{x_{M+1,D}}{x_D} \Delta x_D - \frac{x_{M+1,B}}{x_B} \Delta x_B = x_{M+1,B} - x_{M+1,D}$$

Replacing Δx_D with $\Delta d/D$ and Δx_B with $\Delta b/B$, using the relation

$$\Delta d = -\Delta b$$

and adding the component subscript give

$$\Delta d_i = \frac{(x_{M+1,B} - x_{M+1,D})_i}{(x_{M+1,D}/d)_i + (x_{M+1,B}/b)_i} \tag{10-17}$$

Equation (10-17) provides the correction to d_i which would reduce the mismatch at the feed stage for component i to zero in the next trial if the temperature and rate profiles were unchanged. Unfortunately the profiles change for several trials and the trials must be continued until the profiles become constant.

If the product rates are fixed by specification, then $\Sigma \Delta d_i$ and $\Sigma \Delta b_i$ must be zero. The sum of the corrections from Eq. (10-17) will usually not sum to zero. The sum of the positive Δd_i's will exceed the negative Δd_i's or vice versa. In order that $\Delta D = 0$, the sum of the positive corrections made must equal the sum of the negative corrections made. This means that the corrections to some components must be ignored. Bonner has suggested that the lightest and heaviest components be corrected first and the middle components be allowed to go partially or completely uncorrected if necessary. The lightest and heaviest components exert a greater influence on the temperature profile than do the middle components.

Convergence Check. Convergence is checked by noting the largest component mismatch which occurs at the feed stage. A maximum error is specified, and as soon as all the component errors fall below this specification, the calculations are stopped.

Example 10-1. Rework Example 8-3 by the Lewis-Matheson method. Plot the three temperatures obtained in the solution by the general short-cut method in Example 8-3 vs the stage number to provide the first assumption for the temperature profile. Specify a bubble-point reflux (q_c is the latent heat of V_n) and a bubble-point feed. Assume constant molar overflow for the first trial. Use the product compositions obtained in Example 8-3 as the first assumption for the end-product compositions.

Solution. One complete trial with enthalpy balances and convergence corrections will be made. The results of this trial are shown in Table 10-6 and Fig. 10-2, where they are compared with the final converged

TABLE 10-1. INITIAL ASSUMPTIONS FOR EXAMPLE 10-1

Component	d	$v_N = d + l_{N+1}$	$l_1 = b$
C_3	4.996	17.9	0.00412
$i\text{-}C_4$	14.62	52.3	0.379
$n\text{-}C_4$	22.59	80.7	2.410
$i\text{-}C_5$	3.57	12.8	16.43
$n\text{-}C_5$	3.16	11.3	31.84
	48.9	175.0	51.1

Stage	V	L	t	α				
				C_3	$i\text{-}C_4$	$n\text{-}C_4$	$i\text{-}C_5$	$n\text{-}C_5$
10	175	126.1	163.5	5.65	2.87	2.15	1.0	0.830
9			178.5	5.25	2.72	2.07		0.839
8			191.3	4.96	2.60	2.00		0.847
7			202.0	4.71	2.52	1.97		0.854
6		226.1	210.0	4.60	2.50	1.95		0.858
5			216.4	4.50	2.47	1.95		0.862
4			221.7	4.42	2.46	1.94		0.865
3			226.3	4.38	2.45	1.94		0.867
2			230.3	4.36	2.44	1.94		0.869
1		51.1	234.0	4.31	2.43	1.93		0.871

solution obtained on a computer with the method of Lyster et al. (11), which is described in the next section.

The initially assumed end compositions, rate profiles, and temperature profiles are shown in Table 10-1. The temperature assumptions fix the relative volatilities on each stage, and the calculated α_i's are shown beside the assumed temperature profile. Note that the $i\text{-}C_5$ (heavy key) has been used as the reference component.

Table 10-2 shows the top-down stage-to-stage calculations which involve alternating use of the equilibrium relationship

$$x_{i,n} = \frac{(v_i/\alpha_{i-r})_n}{\Sigma(v_i/\alpha_{i-r})_n} \qquad (10\text{-}3)$$

and the component material balance around the top end of the column,

$$v_{i,n-1} = l_{i,n} + d_i$$

The top-down calculations begin with the assumed top vapor composition and are continued until the composition of the liquid overflow from the

TABLE 10-2. TOP-DOWN CALCULATIONS FOR EXAMPLE 10-1

Component	v_{10}	$\dfrac{v_{10}}{\alpha_{10}}$	x_{10}	l_{10}	v_9	$\dfrac{v_9}{\alpha_9}$
C_3	17.9	3.17	0.0371	4.68	9.68	1.84
$i\text{-}C_4$	52.3	18.22	0.2135	26.92	41.54	15.27
$n\text{-}C_4$	80.7	37.54	0.4398	55.46	78.05	37.70
$i\text{-}C_5$	12.8	12.80	0.1500	18.91	22.48	22.48
$n\text{-}C_5$	11.3	13.62	0.1596	20.12	23.28	27.74
	175.0	85.35	1.0000	126.1	175.0	105.03

Component	x_9	l_9	v_8	$\dfrac{v_8}{\alpha_8}$	x_8	l_8	v_7
C_3	0.0175	2.21	7.21	1.45	0.0119	1.50	6.50
$i\text{-}C_4$	0.1454	18.33	32.95	12.67	0.1041	13.13	27.75
$n\text{-}C_4$	0.3590	45.27	67.86	33.93	0.2789	35.17	57.76
$i\text{-}C_5$	0.2140	27.00	30.57	30.57	0.2513	31.69	35.26
$n\text{-}C_5$	0.2641	33.30	36.46	43.04	0.3538	44.61	47.77
	1.0000	126.1	175.0	121.66	1.0000	126.1	175.0

Component	$\dfrac{v_7}{\alpha_7}$	x_7	l_7	v_6	$\dfrac{v_6}{\alpha_6}$	x_6
C_3	1.38	0.0104	1.31	6.31	1.37	0.0098
$i\text{-}C_4$	11.01	0.0828	10.44	25.06	10.02	0.0717
$n\text{-}C_4$	29.32	0.2206	27.82	50.41	25.85	0.1849
$i\text{-}C_5$	35.26	0.2653	33.45	37.02	37.02	0.2648
$n\text{-}C_5$	55.93	0.4209	53.08	56.24	65.54	0.4689
	132.90	1.000	126.1	175.0	139.80	1.000

feed stage (stage 6) is obtained. Table 10-3 shows the analogous calculations for the bottom section. The bottom-up calculations begin with the assumed bottoms-product composition and involve the alternating use of the equilibrium relationship

$$y_{i,m} = \frac{(\alpha_{i-r}l_i)_m}{\Sigma(\alpha_{i-r}l_i)_m} \qquad (10\text{-}6)$$

TABLE 10-3. BOTTOM-UP CALCULATIONS FOR EXAMPLE 10-1

Component	l_1	$\alpha_1 l_1$	y_1	v_1	l_2	$\alpha_2 l_2$
C_3	0.00412	0.01776	0.0003	0.0525	0.0566	0.247
$i\text{-}C_4$	0.379	0.921	0.0185	3.24	3.62	8.83
$n\text{-}C_4$	2.41	4.65	0.0935	16.36	18.77	36.41
$i\text{-}C_5$	16.43	16.43	0.3303	57.80	74.23	74.23
$n\text{-}C_5$	31.84	27.73	0.5574	97.55	129.39	112.44
	51.1	49.75	1.0000	175.00	226.1	232.16

Component	y_2	v_2	l_3	$\alpha_3 l_3$	y_3	v_3
C_3	0.0011	0.192	0.194	0.850	0.0034	0.595
$i\text{-}C_4$	0.0380	6.65	7.029	17.22	0.0690	12.07
$n\text{-}C_4$	0.1568	27.44	29.85	57.91	0.2321	40.62
$i\text{-}C_5$	0.3197	55.95	72.38	72.38	0.2901	50.77
$n\text{-}C_5$	0.4844	84.77	116.61	101.10	0.4054	70.94
	1.0000	175.0	226.1	249.46	1.0000	175.0

Component	l_4	$\alpha_4 l_4$	y_4	v_4	l_5	$\alpha_5 l_5$	y_5
C_3	0.599	2.65	0.0097	1.70	1.704	7.67	0.0254
$i\text{-}C_4$	12.45	30.63	0.1123	19.65	20.03	49.47	0.1636
$n\text{-}C_4$	43.03	83.48	0.3059	53.53	55.94	109.1	0.3608
$i\text{-}C_5$	67.20	67.20	0.2463	43.10	59.53	59.53	0.1969
$n\text{-}C_5$	102.78	88.90	0.3258	57.01	88.85	76.59	0.2533
	226.1	272.86	1.0000	175.0	226.1	302.36	1.0000

Component	v_5	l_6	x_6
C_3	4.44	4.444	0.0196
$i\text{-}C_4$	28.63	29.01	0.1283
$n\text{-}C_4$	63.14	65.55	0.2900
$i\text{-}C_5$	34.46	50.89	0.2251
$n\text{-}C_5$	44.33	76.17	0.3370
	175.0	226.1	1.0000

and the component balance equation

$$l_{i,m+1} = v_{i,m} + b_i$$

The bottom calculations are continued until the composition of the liqui‹
overflow from the feed stage is obtained. The assumed V's and L's are
used in the material balances in both sections of the column.

The $\Sigma v/\alpha$'s and $\Sigma \alpha l$'s in Tables 10-2 and 10-3 can be divided by the appropriate V and L, respectively, and then used in Eqs. (10-1) and (10-4) to provide new temperatures for each stage. The calculation of these new temperatures is shown in Table 10-4. Note that the feed-stage

TABLE 10-4. NEW TEMPERATURE PROFILE FOR EXAMPLE 10-1

Stage	$\sum \dfrac{v}{\alpha}$	$\sum \dfrac{y}{\alpha} = K_{i\text{-}C_5}$	$\Sigma \alpha l$	$\Sigma \alpha x$	$K_{i\text{-}C_5}$	New t
10	85.3	0.488				160.0
9	105.0	0.600				180.1
8	121.6	0.695				192.0
7	132.9	0.759				200.0
6	139.8	0.799				205.0
6			337.0	1.49	0.671	189.5
5			302.4	1.34	0.748	200.0
4			272.9	1.21	0.828	210.0
3			249.5	1.10	0.907	219.0
2			232.2	1.03	0.974	228.0
1			49.7	0.974	1.03	234.0

TABLE 10-5. NEW V AND L PROFILES FOR EXAMPLE 10-1

n	$h_D - h_{n+1}$	$D(h_D - h_{n+1}) + q_c$	$H_n - h_{n+1}$	V	L
10	0	1,431,000	8,180	175.0	117.9
9	-1240	1,371,000	8,220	166.8	116.0
8	-2500	1,309,000	7,940	164.9	113.2
7	-3320	1,269,000	7,830	162.1	115.1
6	-4080	1,232,000	7,510	164.0	

m	$h_{m+1} - h_B$	$B(h_{m+1} - h_B) + q_r$	$H_m - h_{m+1}$	V	L
6					215.8
5	-3080	1,317,000	8,000	164.7	216.2
4	-2360	1,354,000	8,200	165.1	217.6
3	-1660	1,390,000	8,350	166.5	217.3
2	-1020	1,423,000	8,560	166.2	222.3
1	- 420	1,453,000	8,490	171.2	51.1

temperature obtained from the top-down calculations does not coincide with that from the bottom-up calculations.

It has been pointed out previously that stage temperatures obtained as shown in Table 10-4 will not coincide exactly with the bubble points of the calculated stage liquid compositions. The true bubble points can be found by an iterative procedure in which the temperatures shown in

Table 10-4 are used to calculate new α's which are used to obtain new values of $\Sigma y/\alpha$ or $\Sigma \alpha x$ for each stage. Equations (10-1) and (10-4) are used again to find the temperatures which correspond to these summations. If these temperatures match those used to obtain the α's, the iteration is stopped. This procedure would provide a better temperature profile for the next trial and would undoubtedly reduce the total number

TABLE 10-6. CORRECTION OF END COMPOSITIONS IN EXAMPLE 10-1

Component	$x_{M+1,B}$	$x_{M+1,D}$	$(x_{M+1,B} - x_{M+1,D})$	$\left(\dfrac{x_{M+1,D}}{d}\right)$	$\left(\dfrac{x_{M+1,B}}{b}\right)$	Δd
C_3	0.0196	0.0098	0.0098	0.00196	4.76	0.00206
$i\text{-}C_4$	0.1283	0.0717	0.0566	0.00490	0.339	0.165
$n\text{-}C_4$	0.2900	0.1849	0.1051	0.00819	0.120	0.819
$i\text{-}C_5$	0.2251	0.2648	−0.0397	0.0742	0.0137	−0.452
$n\text{-}C_5$	0.3370	0.4689	−0.1319	0.1484	0.0106	−0.830
	1.0000	1.0000				−0.296

Component	Actual correction	d		
		Assumed	Trial 1 result	Final computer solution
C_3	+0.002	4.996	4.998	4.998
$i\text{-}C_4$	+0.165	14.62	14.78	14.67
$n\text{-}C_4$	+0.819	22.49	23.41	23.10
$i\text{-}C_5$	−0.156	3.57	3.41	3.56
$n\text{-}C_5$	−0.830	3.16	2.33	2.61
	0	48.9	48.9	48.9

of trials required for convergence. However, the calculation time per trial would increase.

The liquid composition for each stage in the rectifying section is shown in Table 10-2. The corresponding vapor compositions are obtained by dividing the $v_{i,n}$'s by 175, which is the assumed V in this first trial. Likewise, Table 10-3 lists the $y_{i,m}$'s, and the $x_{i,m}$'s can be obtained by dividing the $l_{i,m}$'s by the assumed L of 226.1. The phase compositions can now be used with the stage temperatures from Table 10-4 to obtain molar stream enthalpies by means of the following equations, which assume ideal solutions:

$$H_n = \sum_{i=1}^{C} y_{i,n} H_{i,n}$$

$$h_n = \sum_{i=1}^{c} x_{i,n} h_{i,n}$$

An enthalpy balance around the total condenser provides q_c as follows:

$$q_c = V_N(H_N - h_D)$$
$$= 175(18,900 - 10,720)$$
$$= 1.431 \times 10^6 \text{ Btu/100 moles of feed}$$

An enthalpy balance around the entire column then provides q_r.

$$q_r = Dh_D + Bh_B + q_c - Fh_F$$
$$= 48.9(10,720) + 51.1(16,910) + 1,431,500 - 100(13,450)$$
$$= 1,474,870 \text{ Btu/100 moles of feed}$$

Once q_c and q_r are available, Eqs. (10-7) and (10-8) can be used to calculate new V and L profiles. Those calculations are shown in Table 10-5. It should be noted that the enthalpy balances as shown in Eqs. (10-7) and (10-8) involve differences between large numbers of the same order of magnitude. An insufficient number of digits in the calculations or irregularities in the new stage temperatures can cause considerable scatter in the calculated rates and even erratic oscillation of the V and L profiles during the first few trials. Ways of damping these oscillations are described briefly in the next section.

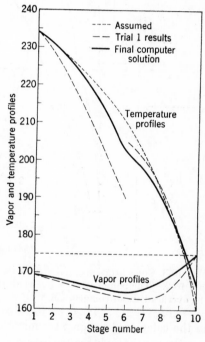

FIG. 10-2. Comparison of the assumed and calculated profiles for the first trial in Example 10-1 with the final computer solution.

The final calculation to be performed in the first trial is the correction of the assumed end compositions by means of Eq. (10-17). This is done in Table 10-6. Note that the correction actually made on the i-C_5 was only -0.156 instead of the -0.452 calculated from Eq. (10-17). The reduction was necessary to make the $\Sigma \, \Delta d_i$'s $= 0$. The last three columns of Table 10-6 compare the d_i's obtained in Example 8-3 (assumed values) and those obtained in the first trial in this example with the results from a computer solution based on the Thiele-Geddes method (described in the next section). The computer program calculated its own K values, and they undoubtedly varied somewhat from the nomograph values. Therefore, the final solution results given do not correspond exactly to the solution which would be obtained if the manual solution of this example were completed. Figure 10-2 compares the assumed and calculated temperature and vapor profiles with the computer solution results. Note the mismatch at the feed stage in the calculated temperature profile. A similar mismatch occurred in the vapor profile. However, the scatter

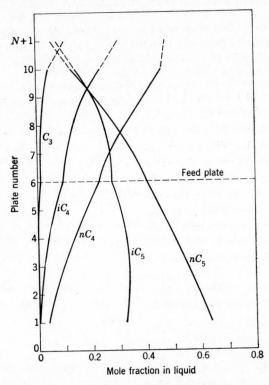

FIG. 10-3. Concentration profiles from the final computer solution of Examples 10-1 and 10-2.   The points at $N + 1$ refer to the reflux composition which is the same as the overhead vapor composition.

in the calculated vapor rates masked the mismatch, and a smooth curve was drawn through the feed stage.

It is instructive to plot the concentration profiles for the various components as shown in Fig. 10-3, where the stage liquid concentrations are plotted vs stage number.   Note the discontinuities at the feed stage and the fact that the feed-stage liquid composition is considerably different from that of the feed.   It can be seen from the $n\text{-}C_4$ and $i\text{-}C_5$ profiles that the separation between these two key components is improving rapidly with stage number and that additional stages in the top section would be worthwhile.

Plots such as Fig. 10-3 illustrate graphically any "pinched" regions which might occur in the calculations. If a zone of constant composition exists, each component profile becomes asymptotic to a vertical line which represents the concentration of that component in the zone. Such a zone can be produced at any reflux ratio in a stage-to-stage calculation by withholding the introduction of the feed. At the true minimum reflux the zone will appear in both the top-down and bottom-up calculations regardless of the feed location. The last statement assumes "normal" volatility behavior. The subject of minimum reflux is discussed more thoroughly in Sec. 9-2.

10-4. Thiele-Geddes Method

The Thiele-Geddes method (13) as applied by Lyster, Sullivan, Billingsley, and Holland (11) will be described in this section. The Thiele-Geddes method together with the convergence method of Lyster et al. is generally faster than the Lewis-Matheson method and the convergence procedure described in the preceding section. It is not known whether this difference in calculation time would exist if the more efficient convergence procedure of Lyster et al. were coupled with the Lewis-Matheson method.

Lyster et al. have applied the Thiele-Geddes method to multifeed columns with or without side streams. For the applications to processes other than the one discussed in this chapter, the reader is referred to their original paper.

Preliminary Assumptions. The end compositions need not be assumed. The trial calculations can be made if the following are assumed:

1. Temperature profile
2. L and V profiles

Stage-to-stage Calculations. The stage compositions in the Thiele-Geddes method are obtained by stage-to-stage calculations from both ends toward the feed stage as in the Lewis-Matheson method. The two methods differ in that the Thiele-Geddes method initially calculates the ratios $v_{i,n}/d_i$, $l_{i,n}/d_i$, $v_{i,m}/b_i$, and $l_{i,m}/b_i$ rather than v_i or l_i.

The working equations are derived as follows: In the rectifying section the equilibrium relationship for component i on any stage n can be expressed in terms of the absorption factor and d_i.

$$x_n = \frac{y_n}{K_n}$$

$$L_n x_n = \frac{L_n}{K_n V_n} V_n y_n$$

$$l_n = A_n v_n$$

$$\frac{l_n}{d} = \frac{v_n}{d} A_n \qquad (10\text{-}18)$$

The general component i material balance around the top of the column can be written as

$$v_n = l_{n+1} + d$$

or
$$\frac{v_n}{d} = \frac{l_{n+1}}{d} + 1 \qquad (10\text{-}19)$$

Increasing the subscripts in Eq. (10-18) by one and substituting for l_{n+1}/d in Eq. (10-19) give the following combined equilibrium and material-balance relationship for component i.

$$\frac{v_n}{d} = \frac{v_{n+1}}{d} A_{n+1} + 1 \qquad (10\text{-}20)$$

Or if v_n/d is eliminated in Eq. (10-19), the following relationship between the liquid ratios results:

$$\frac{l_n}{d} = A_n \left(\frac{l_{n+1}}{d} + 1 \right) \qquad (10\text{-}21)$$

The absorption factors on each stage in the rectifying section are fixed by the temperature- and rate-profile assumptions.

Equation (10-21) is used to calculate the l/d ratio on each stage in the rectifying section. The calculations are started by writing the equation for stage N.

$$\frac{l_N}{d} = A_N \left(\frac{l_{N+1}}{d} + 1 \right)$$

With a total condenser where $x_{i,D} = x_{i,N+1}$, the ratio

$$\frac{l_{N+1}}{d} = \frac{L_{N+1}}{D} = R$$

A knowledge of the reflux ratio (obtained from specified distillate and top vapor rates) permits the calculation of l_N/d from which l_{N-1}/d is obtained, etc. Equation (10-21) is applied to each stage in succession until the ratio l_{M+2}/d in the overflow from the stage above the feed stage is obtained. The calculations are then switched to the stripping section.

The equilibrium relationship for component i in the stripping section can be expressed in terms of the stripping factor and b_i.

$$y_m = K_m x_m$$
$$V_m y_m = \frac{K_m V_m}{L_m} L_m x_m$$
$$v_m = S_m l_m$$
$$\frac{v_m}{b} = \frac{l_m}{b} S_m \qquad (10\text{-}22)$$

The general component i material balance around the bottom end of the column can be written as

$$\frac{l_{m+1}}{b} = \frac{v_m}{b} + 1 \qquad (10\text{-}23)$$

Combination of the equilibrium and material-balance equations gives

$$\frac{l_{m+1}}{b} = \frac{l_m}{b} S_m + 1 \qquad (10\text{-}24)$$

The bottom-up calculations are started by writing Eq. (10-24) for stage 1 as

$$\frac{l_2}{b} = \frac{V_1 K_1}{B} + 1 = S_1 + 1$$

The assumed temperature and rate profiles provide S_1 and all the other stripping factors. Equation (10-24) is applied to each of the stripping stages in sequence until the ratio l/b in the liquid entering the feed stage is obtained.

The manner in which the rectifying and stripping section calculations are meshed at the feed stage depends upon the thermal condition of the feed. Figure 10-4 shows three possible ways in which the fresh feed can affect the L and V rates between the feed stage and stage $M + 2$. The bar superscript denotes the stream rate when the stream *enters* a stage, while no bar denotes the rate when the stream leaves a stage.

The top-down and bottom-up calculations have provided values of $l_{i,M+2}/d_i$ and $\bar{l}_{i,M+2}/b_i$, respectively. For a bubble-point feed,

$$v_{i,M+1} = \bar{v}_{i,M+1}$$

and a combination of Eqs. (10-19) and (10-23) provides

$$\frac{b_i}{d_i} = \frac{v_{i,M+1}/d_i}{\bar{v}_{i,M+1}/b_i} = \frac{l_{i,M+2}/d_i + 1}{\bar{l}_{i,M+2}/b_i - 1}$$

$$(10\text{-}25)$$

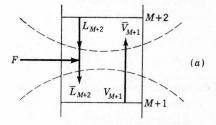

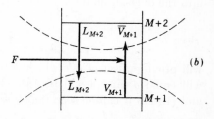

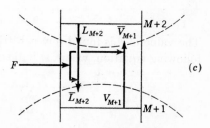

FIG. 10-4. Effect of feed on stream rates just above feed stage $M + 1$. (*a*) Subcooled or bubble-point feed. (*b*) Superheated or dew-point feed. (*c*) Partially flashed feed.

The reader is referred to the original paper by Lyster et al. for the equations which apply to other feed conditions.

The over-all component i material balance can be used to calculate b_i and d_i individually.

$$d_i + b_i = Fx_{i,F}$$

$$d_i = \frac{Fx_{i,F}}{1 + b_i/d_i} \tag{10-26}$$

$$b_i = (b_i/d_i)d_i \tag{10-27}$$

The product compositions calculated in this manner are those which would give a perfect feed-stage match if the same temperature and rate profiles were retained and the stage-to-stage calculations were made from each end of the column by the Lewis-Matheson method.

The calculated d_i's will not sum to the specified D in any trial until convergence is obtained. Therefore, in each trial the calculated d_i's must be corrected in such a manner as to make

$$\sum_{i=1}^{i=C} d_i' = D_{\text{specified}} \tag{10-28}$$

where the prime denotes the corrected values of d_i. Lyster et al. have suggested a simple but effective way of doing this.

Convergence Procedure. Define a quantity θ which when introduced into the over-all material balance [Eq. (10-26)]

$$d_i' = \frac{Fx_{i,F}}{1 + (b_i/d_i)\theta} \tag{10-29}$$

will provide a set of d_i'''s which satisfies Eq. (10-28). At convergence d_i will equal d_i' and θ will be 1.0. By comparison between Eqs. (10-26) and (10-29) it can be seen that

$$b_i' = \theta \frac{b_i}{d_i} d_i' \tag{10-30}$$

The value of θ is found for each trial by a trial-and-error solution of the following equation, where D is the specified distillate rate:

$$D - \sum_{i=1}^{i=C} \frac{Fx_{i,F}}{1 + (b_i/d_i)\theta} = 0 \tag{10-31}$$

The reader is referred to the original paper by Lyster et al. for convenient methods of making this trial-and-error calculation on a computer.

New Temperature and Rate Profiles. The stage-to-stage calculations provided values of l_i/d_i and l_i/b_i for each stage in the rectifying and stripping sections, respectively. These are used in the following equations to calculate normalized liquid concentrations in each stage:

$$x_{i,n} = \frac{(l_{i,n}/d_i)d_i'}{\displaystyle\sum_{i=1}^{i=C} (l_{i,n}/d_i)d_i'} \tag{10-32}$$

$$x_{i,m} = \frac{(l_{i,m}/b_i)b_i'}{\displaystyle\sum_{i=1}^{i=C} (l_{i,m}/b_i)b_i'} \tag{10-33}$$

Note that the calculated ratios are multiplied by the corrected d_i's and b_i's. For this reason the calculation of the stage compositions must be done after the convergence procedure has been applied.

The new temperature profile is obtained by bubble-point calculations in each stage. Lyster et al. describe three computer methods for bubble- and dew-point calculations. One of these is a direct iteration with Eq. (10-4). The stage temperature from the old profile is used to evaluate the α_{i-r}'s. Equation (10-4) is used to calculate K_r, which provides the temperature for the next evaluation of the α_{i-r}'s. The cycle is continued until the temperature of the calculated K_r is the same temperature used to evaluate the α_{i-r}'s. Another method described is the use of Newton's approximation. Newton's method is presented later in Sec. 10-6.

The new stage compositions and temperatures and the assumption of ideal solutions are used to calculate the new rate profiles in exactly the same way as described in the previous section [see Eqs. (10-7) and (10-8)].

Lyster et al. found that the successive temperature and rate profiles tended to oscillate around the correct profiles. This was particularly true for wide boiling feeds. It was found that the number of trials could be materially reduced if the oscillations were damped. The swing in the temperature profile was reduced by averaging the old and new temperature profiles to provide the profile actually used in the next trial. The stability of the rate profiles was increased by the use of the same L and V profiles for the first two trials. In later trials the change in the vapor rates was restricted to a certain percentage of the old V's. The new V's calculated within this restriction were then averaged with the old V's to provide the new vapor-rate profile. The specified D was used in all material and enthalpy balances. These "forcing procedures" together with the θ method of convergence usually provided satisfactory convergence within seven or eight trials.

Convergence Check. Convergence is obtained when

$$D_{\text{specified}} - D_{\text{calculated}} = \text{specified error}$$

Seven or eight trials are usually sufficient to satisfy this convergence check when the specified error is no less than 5 in the fourth significant digit of D. After this many trials the temperature and rate profiles are usually known to well within 1 per cent. Direct iteration (no forcing procedures and the θ method not used) usually required three to four times as many trials to attain this degree of convergence.

Example 10-2. Rework Example 8-3 by the Thiele-Geddes method. Plot the three temperatures obtained in the solution by the general short-cut method vs the stage number to provide the first assumption for the

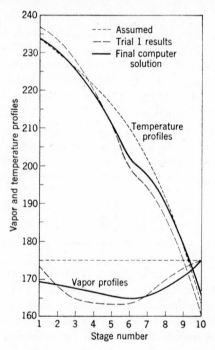

Fig. 10-5. Comparison of the assumed and calculated profiles for the first trial in Example 10-2 with final computer solution.

temperature profile. Specify a bubble-point reflux (q_c is the latent heat of V_N) and a bubble-point feed, and assume constant molar overflow for the first trial.

Solution. One complete trial with enthalpy balances and convergence corrections will be made. The results of this trial are shown in Tables 10-10 and 10-11 and Fig. 10-5 and compared with the converged computer solution obtained by the method of Lyster et al. The concentration profiles from the final solution are shown in Fig. 10-3.

The initial assumptions for the temperature and rate profiles along with the corresponding K values are shown in Table 10-7. The top-down calculations are accomplished with the following equations:

$$\frac{l_N}{d} = A_N(R + 1)$$

$$\frac{l_n}{d} = A_n\left(\frac{l_{n+1}}{d} + 1\right)$$

The calculation of the $l_{i,M+2}/d_i$ ratios from the top down is shown in Table 10-8. The calculation of the $l_{i,M+2}/b_i$ ratios from the bottom up is

TABLE 10-7. INITIAL ASSUMPTIONS FOR EXAMPLE 10-2

Stage	V	L	t	K				
				C_3	C_4	$n\text{-}C_4$	$i\text{-}C_5$	$n\text{-}C_5$
10	175	126.1	163.5	2.77	1.38	1.04	0.500	0.420
9			178.5	3.10	1.60	1.22	0.590	0.495
8			191.3	3.40	1.78	1.37	0.685	0.585
7			202.0	3.63	1.94	1.49	0.770	0.660
6		226.1	210.0	3.84	2.06	1.60	0.825	0.702
5			216.4	4.00	2.21	1.73	0.895	0.765
4			221.7	4.15	2.28	1.80	0.925	0.800
3			226.3	4.28	2.36	1.88	0.965	0.835
2			230.3	4.36	2.43	1.94	1.000	0.870
1			234.0	4.42	2.50	1.99	1.030	0.890

TABLE 10-8. TOP-DOWN CALCULATIONS FOR EXAMPLE 10-2

Component	$R + 1$	A_{10}	$\dfrac{l_{10}}{d}$	$\dfrac{l_{10}}{d} + 1$	A_9	$\dfrac{l_9}{d}$
C_3	3.58	0.260	0.931	1.931	0.232	0.448
$i\text{-}C_4$	3.58	0.522	1.87	2.87	0.450	1.29
$n\text{-}C_4$	3.58	0.693	2.48	3.48	0.590	2.05
$i\text{-}C_5$	3.58	1.44	5.16	6.16	1.22	7.52
$n\text{-}C_5$	3.58	1.72	6.16	7.17	1.46	10.5

Component	$\dfrac{l_9}{d} + 1$	A_8	$\dfrac{l_8}{d}$	$\dfrac{l_8}{d} + 1$	A_7	$\dfrac{l_7}{d}$
C_3	1.448	0.212	0.307	1.307	0.198	0.259
$i\text{-}C_4$	2.29	0.405	0.927	1.927	0.371	0.715
$n\text{-}C_4$	3.05	0.526	1.60	2.60	0.484	1.26
$i\text{-}C_5$	8.52	1.05	8.95	9.95	0.936	9.31
$n\text{-}C_5$	11.5	1.23	14.1	15.1	1.09	16.5

accomplished with

$$\frac{l_{m+1}}{b} = \frac{l_m}{b} S_m + 1$$

and the results are shown in Table 10-9.

Since a bubble-point feed was specified, Eq. (10-25) is used to calculate the b_i/d_i ratios.

$$\frac{b_i}{d_i} = \frac{l_{i,7}/d_i + 1}{l_{i,7}/b_i - 1} \qquad (10\text{-}25)$$

Obtaining the calculated ratios from Tables 10-8 and 10-9 and performing the operations indicated in Eqs. (10-25) and (10-26) provide the calculated d_i's as follows:

Component	$\dfrac{l_7}{d} + 1$	$\dfrac{\bar{l}_7}{b} - 1$	$\dfrac{b}{d}$	$1 + \dfrac{b}{d}$	$F x_F$	d
C_3	1.26	5450	0.000231	1.000	5	5.00
$i\text{-}C_4$	1.71	175	0.00977	1.01	15	14.85
$n\text{-}C_4$	2.26	47.7	0.0474	1.047	25	23.88
$i\text{-}C_5$	10.3	2.46	4.19	5.19	20	3.85
$n\text{-}C_5$	17.5	1.54	11.4	12.4	35	2.82
					100	50.4

The calculated D is 50.4 compared with the specified value of 48.9. Equation (10-31) is now used to find that value of θ which satisfies Eq. (10-28). The value of θ (obtained by trial and error) which gives $\Sigma d_i' = 48.9$ is 1.25. The corrected d_i's can now be used with Eq. (10-30) to calculate the corrected b_i's. These calculations are shown below.

Component	$\dfrac{b}{d}$	d'	$d'\theta$	b'	x_D	x_1
C_3	0.000231	5.00	6.25	0.00144	0.102	0
$i\text{-}C_4$	0.00977	14.82	18.53	0.181	0.303	0.004
$n\text{-}C_4$	0.0474	23.60	29.50	1.40	0.482	0.027
$i\text{-}C_5$	4.19	3.21	4.01	16.79	0.066	0.329
$n\text{-}C_5$	11.4	2.30	2.88	32.7	0.047	0.640
		48.9		51.1	1.000	1.000

Table 10-10 shows the calculation of the new corrected liquid compositions for each stage by means of Eqs. (10-32) and (10-33). Bubble-point calculations (not shown) on the corrected liquid compositions provide new stage temperatures. The bubble-point calculations also provide new vapor compositions for each stage.   Ideal solutions are now assumed,

TABLE 10-9. BOTTOM-UP CALCULATIONS FOR EXAMPLE 10-2

Component	S_1	$\dfrac{l_2}{b}$	S_2	$\dfrac{l_2}{b} S_2$	$\dfrac{l_3}{b}$	S_3
C_3	15.1	16.1	3.37	54.3	55.3	3.31
$i\text{-}C_4$	8.56	9.56	1.88	18.0	19.0	1.83
$n\text{-}C_4$	6.81	7.81	1.50	11.7	12.7	1.45
$i\text{-}C_5$	3.53	4.53	0.774	3.51	4.51	0.747
$n\text{-}C_5$	3.05	4.05	0.673	2.73	3.73	0.646

Component	$\dfrac{l_3}{b} S_3$	$\dfrac{l_4}{b}$	S_4	$\dfrac{l_4}{b} S_4$	$\dfrac{l_5}{b}$
C_3	183.0	184.0	3.21	590.6	591.6
$i\text{-}C_4$	34.8	35.8	1.76	63.0	64.0
$n\text{-}C_4$	18.4	19.4	1.39	27.0	28.0
$i\text{-}C_5$	3.37	4.37	0.716	3.13	4.13
$n\text{-}C_5$	2.41	3.41	0.619	2.11	3.11

Component	S_5	$\dfrac{l_5}{b} S_5$	$\dfrac{l_6}{b}$	S_6	$\dfrac{l_6}{b} S_6$	$\dfrac{l_7}{b}$
C_3	3.10	1834	1835	2.97	5450	5451
$i\text{-}C_4$	1.71	109.4	110.4	1.59	175	176
$n\text{-}C_4$	1.34	37.5	38.5	1.24	47.7	48.7
$i\text{-}C_5$	0.693	2.86	3.86	0.638	2.46	3.46
$n\text{-}C_5$	0.592	1.84	2.84	0.543	1.54	2.54

and the molar enthalpies of the vapor and liquid streams in the column are calculated with

$$H_n = \sum_{i=1}^{C} y_{i,n} H_{i,n}$$

$$h_n = \sum_{i=1}^{C} x_{i,n} h_{i,n}$$

An enthalpy balance around the total condenser provides q_c as follows:

$$q_c = V_N(H_N - h_D)$$
$$= 175(18,900 - 10,720) = 1,426,250 \text{ Btu/100 moles of feed}$$

An enthalpy balance around the entire column then provides q_r.

$$q_r = Dh_D + Bh_B + q_c - Fh_F$$
$$= (48.9)(10,630) + (51.1)(17,080) + 1,426,250 - (100)(13,540)$$
$$= 1,464,850 \text{ Btu/100 moles of feed}$$

These values of q_c and q_r are now used along with the stream enthalpies

TABLE 10-10. STAGE COMPOSITIONS FROM FIRST TRIAL IN EXAMPLE 10-2

Component	x_1	$\dfrac{l_2}{b}b'$	x_2	$\dfrac{l_3}{b}b'$	x_3	$\dfrac{l_4}{b}b'$	x_4
C_3	0.000	0.0232	0.000	0.0796	0.000	0.279	0.001
i-C_4	0.004	1.73	0.008	3.44	0.016	6.48	0.030
n-C_4	0.027	10.9	0.049	17.8	0.081	27.2	0.124
i-C_5	0.329	76.1	0.344	75.7	0.346	73.4	0.335
n-C_5	0.640	132.4	0.599	122.0	0.557	111.5	0.510
	1.000	221.1	1.000	219.0	1.000	218.9	1.000

Component	$\dfrac{l_5}{b}b'$	x_5	$\dfrac{l_6}{b}b'$	x_6	$\dfrac{l_7}{d}d'$	x_7
C_3	0.852	0.004	2.65	0.011	1.295	0.012
i-C_4	11.6	0.052	20.0	0.085	10.6	0.097
n-C_4	39.2	0.176	53.9	0.230	29.7	0.271
i-C_5	69.3	0.311	64.8	0.277	29.9	0.273
n-C_5	101.7	0.447	92.9	0.397	37.9	0.347
	226.6	1.000	234.2	1.000	109.4	1.000

Component	$\dfrac{l_8}{d}d'$	x_8	$\dfrac{l_9}{d}d'$	x_9	$\dfrac{l_{10}}{d}d'$	x_{10}
C_3	1.535	0.013	2.240	0.019	4.66	0.038
i-C_4	13.7	0.120	19.1	0.162	27.7	0.228
n-C_4	37.8	0.331	48.4	0.410	58.5	0.481
i-C_5	28.7	0.252	24.1	0.204	16.6	0.136
n-C_5	32.4	0.284	24.1	0.204	14.2	0.117
	114.1	1.000	118.0	1.000	121.7	1.000

to calculate new V and L profiles as shown in Table 10-11. The new temperature profile from the previous bubble-point calculations is also shown in Table 10-11. Note that it is not necessary to list two values of V, L, and temperature for the feed stage (stage 6) because the Thiele-Geddes method gives a perfect match at the feed stage in each trial.

The new temperature and rate profiles (which would be used as the assumptions for trial 2) are compared in Fig. 10-5 with the final solution obtained on a computer with the Thiele-Geddes method. The computer solution is not based on K values from the monograph in the Appendix and therefore does not represent exactly the final solution which would be obtained if the manual solution were completed. However, this discrepancy does not detract materially from the value of the comparison. Note that the V profile improved considerably in the first trial. Also,

TABLE 10-11. NEW TEMPERATURE AND RATE PROFILES FOR EXAMPLE 10-2

n	New t	$h_D - h_{n+1}$	$D(h_D - h_{n+1}) + q_c$	$H_n - h_{n+1}$	V	L
10	160.0	0	1,426,000	8150	175.0	124.0
9	175.0	−1220	1,367,000	7900	172.9	119.6
8	186.0	−2190	1,319,000	7830	168.5	117.6
7	194.0	−3010	1,279,000	7680	166.5	114.2
6	200.0	−3490	1,256,000	7700	163.1	214.3

m	New t	$h_{m+1} - h_B$	$B(h_{m+1} - h_B) + q_r$	$H_m - h_{m+1}$	V	L
5	211.0	−2480	1,338,000	8200	163.2	214.7
4	220.0	−1780	1,374,000	8400	163.6	215.5
3	228.0	−1130	1,407,000	8560	164.4	221.1
2	233.5	− 560	1,438,000	8460	170.0	224.0
1	237.5	− 210	1,454,000	8410	172.9	51.1

the temperature profile began to take on the shape at the feed stage shown by the final solution, although the new stage temperatures at the column ends are more incorrect than those based on the short-cut results from Example 8-3. Averaging of the new and old temperature profiles · would provide a better profile for trial 2, but this would not be the case for the vapor profile. The reason the averaging technique is not so important in this example is that the initially assumed temperature profile was so good. Use of the general short-cut method (Chap. 8) to provide the initial temperature profile should largely eliminate the erratic oscillation of the profiles in the early trials, which occurs when the initial assumptions are not so good.

The concentration profiles from the final trial in the computer solution are plotted in Fig. 10-3 and discussed at the end of the previous section.

10-5. Absorption or Stripping-factor Method

The absorption or stripping-factor method relates the concentration of component i in a stage to its concentration in any other stage in terms of the absorption or stripping factors for the intervening stages. The basic derivation was used in the development of the convergence method described in Sec. 10-3. See Eq. (10-9). The absorption and stripping-factor method of calculation has been applied to distillation by Edmister (6) and by Donnell and Turbin (5) among others. The method was originally developed for absorber and stripper calculations by Kremser (9), Brown and Souder (4), and Horton and Franklin (8). Presentation of the equations will be deferred until Chaps. 12 and 13, where they will be applied to multicomponent extraction and absorption calculations.

10-6. Amundson-Pontinen Method

Amundson and Pontinen (1, 2) have aptly described the multicomponent distillation problem as a system of simultaneous algebraic equations in the compositions in which a set of parameters, the temperatures, must be fixed in order that the over-all material and heat balances will not be violated. Whereas the Lewis-Matheson, Thiele-Geddes, and absorption factor methods solve the simultaneous equations in the stage concentrations by means of some sort of stage-to-stage calculations, the method of Amundson and Pontinen solves the equations directly. This direct line of attack is made practical by the availability of very fast computers with large memory capacity. The organization of the equations is such that process variations like multiple feeds and side streams are readily handled. The solution of such problems has been described (1, 2).

Preliminary Assumptions. The iteration variables which must be assumed in the Amundson-Pontinen method are as follows:

1. Temperature profile
2. V profile

The assumption of the V profile along with the specification of D fixes the L profile, of course, but the assumed L's need not be evaluated, since they do not appear in the equations used by Amundson and Pontinen. As in the Thiele-Geddes method, it is not necessary to assume any concentrations to start the calculations. This last statement does not apply, of course, to any approximate product compositions which are assumed to aid in the selection of a realistic temperature profile.

Stage-to-stage Calculations. The phrase "stage-to-stage calculations" does not describe correctly this part of the over-all computation as far as the Amundson-Pontinen method is concerned. Instead of the stage compositions being calculated by some sort of stagewise calculation, the systems of simultaneous equations which relate the component stage concentrations to the temperature and vapor profiles are solved simultaneously. Amundson and Pontinen suggest the use of matrices in the solution of the simultaneous equations. Any method of solving large sets of simultaneous equations is, of course, equally applicable. The only practical requirement is that the method used to obtain the stage compositions by simultaneous solution of the equations be faster than a stagewise calculation. Since many of the arithmetical operations involved in the stagewise calculations are also involved in setting up the simultaneous equations, the method used to solve the equations must be fast in order to be competitive.

The distillation equations can be expressed in matrix form as follows: A component i balance around stage 1 (the reboiler if it is an equilibrium stage) can be written as

$$-(B + V_1K_1)x_1 + (V_1 + B)x_2 = 0$$

where the component i subscripts are omitted to simplify the nomenclature. The component i balance around stage 2 is

$$(V_1K_1)x_1 - (V_1 + B + V_2K_2)x_2 + (V_2 + B)x_3 = 0$$

This equation can be generalized and written for stage m in the stripping section as

$$(V_{m-1}K_{m-1})x_{m-1} - (V_{m-1} + B + V_mK_m)x_m + (V_m + B)x_{m+1} = 0$$

Moving now to the other end of the column and assuming for the sake of generality that the column has a partial condenser, the component i balance around the condenser can be written as

$$(V_NK_N)x_N - [V_N + (K_{N+1} - 1)D]x_{N+1} = 0$$

If the condenser is a total one, a 1.0 is substituted for K_{N+1}. A component i balance around stage N gives

$$(V_{N-1}K_{N-1})x_{N-1} - (V_{N-1} - D + V_NK_N)x_N + (V_N - D)x_{N+1} = 0$$

This equation can be generalized and written for any stage n in the rectifying section as

$$(V_{n-1}K_{n-1})x_{n-1} - (V_{n-1} - D + V_nK_n)x_n + (V_n - D)x_{n+1} = 0$$

The component i balance around the one-feed stage in the process under consideration is

$$(V_MK_M)x_M - (V_M + B + V_{M+1}K_{M+1})x_{M+1} + (V_{M+1} - D)x_{M+2} = -Fx_F$$

It is only in the component i balance around the feed stage that the right-hand side of the equation has a value other than zero.

If the component i balances around the individual stages are written in order from stage 1 to stage $N + 1$ (partial condenser), and if the equal signs and like values of x_t are aligned vertically, it can be seen that the left-hand sides of the equations form a square array which has

$N + 1$ rows and $N + 1$ columns of x_i's. The coefficients of the x's are zero throughout the square array except on the main diagonal and the two adjacent diagonals. The laws of multiplication of conformable matrices permit the square array for component i to be represented as the product of the two conformable matrices $\overline{\overline{M}}_i$ and $\bar{x}_i$. The set of simultaneous equations for component i can then be represented by the matrix equation

$$\overline{\overline{M}}_i \bar{x}_i = \bar{F}_i$$

where $\bar{x}_i$ and $\bar{F}_i$ are single-column matrices with $N + 1$ rows as shown below.

$$\bar{x}_i = \begin{matrix} x_{i,1} \\ x_{i,2} \\ \cdot \\ \cdot \\ \cdot \\ x_{i,M+1} \\ \cdot \\ \cdot \\ \cdot \\ x_{i,N} \\ x_{i,N+1} \end{matrix} \qquad \bar{F}_i = \begin{matrix} 0 \\ 0 \\ \cdot \\ \cdot \\ \cdot \\ -Fx_{i,F} \\ \cdot \\ \cdot \\ \cdot \\ 0 \\ 0 \end{matrix}$$

The elements in the matrix $\overline{\overline{M}}_i$ are the coefficients of the x_i's in the set of simultaneous equations. The elements are all zero except on the main diagonal and the two adjacent diagonals. Since there are $N + 1$ simultaneous equations for each component, the matrix is square and of order $N + 1$. The nonzero elements can be evaluated from the assumed temperature and V profiles. The matrix can then be inverted and multiplied by $\bar{F}_i$ to provide the concentrations of component i in each stage as follows:

$$\bar{x}_i = \overline{\overline{M}}_i^{-1} \bar{F}_i \qquad (10\text{-}34)$$

If the assumed temperature and V profiles are not correct, the stage concentrations calculated from (10-34) will not be the correct final values and will not in general sum to 1.0. Amundson and Pontinen found that the unnormalized mole fractions calculated from (10-34) ranged from values greater than 1.0 to negative values. The significance of the negative values of the stage concentrations is made more evident by an inspection of the following equation derived in Sec. 10-3:

$$l_{m+1} = b(S_1 S_2 \cdots S_m + S_2 S_3 \cdots S_m \\ + \cdots + S_{m-1} S_m + S_m + 1) \quad (10\text{-}9)$$

A similar equation can be written for the top section to relate the stage concentrations to d in terms of the stage absorption factors. These equations show that negative stage concentrations are possible only if the corresponding b_i or d_i is negative. Also, they show that the sign on the stage concentrations for any component must be constant throughout any one section of the column. For example, if b_i is a negative number, all stage concentrations for component i below the lowest feed or side stream must be negative because of the direct proportionality shown in Eq. (10-9). If b_i is a positive number, then all the stage concentrations for i in the bottom section of the column must be positive. It is, of course, possible to start with a positive b_i and after a stage-to-stage calculation (without normalization of the calculated mole fractions) up through the first feed tray find that the stage concentrations become negative because

$$L_{M+2}x_{m+2} = V_{M+1}y_{M+1} + Bx_B - Fx_F$$

and too small a value was chosen for $b_i = Bx_{i,B}$. The method of Amundson and Pontinen, however, does not assume b_i's (or d_i's) nor does it make a stage-to-stage calculation which passes through a feed stage. Instead, the method solves directly for those stage concentrations (including the b_i's and d_i's) which correspond to the assumed rate and temperature profiles. These same stage concentrations are obtained by a different procedure in the method of Lyster et al. which cannot produce negative values. The negative values obtained by Amundson and Pontinen must therefore be due to the numerical difficulties involved in the solution of large sets of simultaneous equations.

Convergence Procedure. The calculation method proposed by Amundson and Pontinen does not provide for the correction of the calculated stage concentrations to better values before they are used to calculate a new temperature profile. The only operations performed on the calculated concentrations are to change any negative numbers to zero and then normalize the resultant stage compositions so that $\Sigma x_i = 1.0$ in each stage. The iterations are continued without any special forcing or convergence procedures until satisfactory convergence is obtained.

New Temperature and Rate Profiles. The new temperature profile is obtained in the same manner as in all the methods described previously. Bubble-point calculations based upon the normalized liquid compositions are performed in each stage. Lyster et al. (11) describe three methods of performing bubble-point calculations on a computer. One of these is the Newton approximation† which utilizes the linear terms in a Taylor series expansion of some temperature function. Amundson and Pontinen utilize Newton's approximation method and the normalized liquid con-

---

† Greenstadt, Bard, and Morse (7) have used Newton's approximation to estimate new values for all the other trial values as well as temperature.

centrations to provide new stage temperatures as follows: Let

$$\phi_n = \sum_i^C K_{i,n} x_{i,n}$$

where ϕ_n is a function of temperature only for any given set of $x_{i,n}$'s. The expansion in the neighborhood of t'_n of the temperature function ϕ_n as a Taylor series through only the linear terms gives

$$\phi_n(t_n) = \phi_n(t'_n) + \frac{\partial \phi_n}{\partial t_n}(t_n - t'_n)$$

where t'_n is the previously assumed temperature on stage n. The desired value of the new temperature t_n is, of course, the one where $\phi_n = 1.0$. Setting $\phi_n(t) = 1.0$ and solving for t_n give

$$t_n = t'_n + [1 - \phi_n(t'_n)]\left(\frac{\partial \phi_n}{\partial t_n}\right)^{-1} \tag{10-35}$$

where $\phi_n(t'_n)$ is the value of $\Sigma K_{i,n} x_{i,n}$ obtained at the old stage n temperature. The partial derivative is obtained by

$$\frac{\partial \phi_n}{\partial t} = \sum_i^C x_i \frac{\partial K_{i,n}}{\partial t}$$

where $K_{i,n}$ is represented in the computer as a polynomial function of temperature,

$$K_{i,n} = a_i + b_i t_n + c_i t_n{}^2 + \cdots$$

The derivatives of the polynomials of each of the components are evaluated at t'_n to provide a value for $\partial \phi_n / \partial t_n$.

Once the new temperature profile has been obtained, it can be used with the calculated stage compositions to obtain molar enthalpy values for all the vapor and liquid streams. The component enthalpies like the K values are represented in the computer as polynomial functions of temperatures. The assumption of ideal solutions permits the calculation of the molar stream enthalpies from

$$H_n = \sum_i^C y_{i,n} H_{i,n}$$

and

$$h_n = \sum_i^C x_{i,n} h_{i,n}$$

The individual V's can be expressed by a series of enthalpy balances around the individual stages as follows. Around stage 1,

$$(h_2 - H_1)V_1 = B(h_1 - h_2) - q_r$$

Around stage 2,

$$(H_1 - H_2)V_1 + (h_3 - H_2)V_2 = B(h_2 - h_3)$$

Around stage m in the stripping section,

$$(H_{m-1} - h_m)V_{m-1} + (h_{m+1} - H_m)V_m = B(H_m - h_{m+1})$$

Around the feed stage,

$$(H_M - h_{M+1})V_M + (h_{M+2} - H_{M+1})V_{M+1} = Dh_{M+2} + Bh_{M+1} - Fh_F$$

Around stage n in the rectifying section,

$$(H_{n-1} - h_n)V_{n-1} + (h_{n+1} - H_n)V_n = D(h_{n+1} - h_n)$$

Around stage N,

$$(H_{N-1} - h_N)V_{N-1} + (h_{N+1} - H_N)V_N = D(h_{N+1} - h_N)$$

Since the last equation includes V_N, it is not necessary to include a balance around the condenser.

The simultaneous equations in the V's if written in order from stage 1 to stage N form a square array similar to the one formed by the composition equations except that it is of order N. The set of simultaneous equations can be represented by the matrix equation

$$\overline{\overline{H}}\,\bar{V} = \bar{G}$$

where $\bar{V}$ and $\bar{G}$ are single-column matrices.

$$
\bar{V} =
\begin{array}{l}
V_1 \\
V_2 \\
\cdot \\
\cdot \\
\cdot \\
\cdot \\
\cdot \\
\cdot \\
V_{N-1} \\
V_N
\end{array}
\qquad
\bar{G} =
\begin{array}{l}
B(h_1 - h_2) - q_r \\
B(h_2 - h_3) \\
\cdot \\
\cdot \\
Dh_{M+2} + Bh_{M+1} - Fh_F \\
\cdot \\
\cdot \\
\cdot \\
-D(h_{N-1} - h_N) \\
-D(h_N - h_{N+1})
\end{array}
$$

The matrix $\overline{\overline{H}}$ is of order N and has as elements the coefficients of the V's in the simultaneous equations. The coefficients contain only enthalpy terms which are evaluated from the stage compositions and temperatures. The $\bar{G}$ matrix elements can also be evaluated. The individual V's for the next trial are then provided by

$$\bar{V} = \overline{\overline{H}}^{-1}\bar{G} \tag{10-36}$$

Once the new V profile is obtained, the next trial can be started if the convergence is not satisfactory.

Convergence Check. The trials are repeated until all the $\Sigma x_{i,n}$'s throughout the column differ from unity by less than a specified amount.

10-7. Relaxation Method

Rose, Sweeny, and Schrodt (12) have applied the familiar relaxation method to the solution of the continuous, steady-state multicomponent distillation problem. The relaxation method is not competitive with those methods described earlier in this chapter when applied to the steady-state problem, but it does have a unique value in that it permits the approximation of the transient conditions which will exist between the time when a disturbance occurs in the column and the time when steady state again prevails. The reader is referred to the original paper by Rose et al. for the steady-state application. In this section, the application of the method to the transient problem will be briefly discussed.

Consider the process described at the beginning of the chapter to be operating at steady state under the specifications set forth. Now assume that the composition of the feed changes while the feed rate and all the other specifications remain constant. The change in feed composition will cause a change in all the stage compositions, temperatures, and rates. The new steady-state conditions can be predicted by one of the methods in the previous four sections, but none of these methods can describe the transient conditions as the column moves from the original steady-state conditions to the new ones. The application of the relaxation method to the transient problem will be described briefly. The method can be divided into the same major parts as the calculation methods described previously.

Preliminary Assumptions. If the true transient conditions are to be approximated in the problem under consideration, the steady-state conditions which exist at the time when the disturbance occurs must be taken as the initial values of the assumed iteration variables. These variables include

1. Temperature profile
2. L and V profiles
3. All stage compositions

The relaxation method utilizes the imbalance in the material entering and leaving a stage to predict the change in the composition of the stage liquid which will occur within a short time interval. Therefore, it is also necessary to assume values for the following:

4. Liquid holdup in each stage (vapor holdup is neglected)
5. Time interval considered

The values used for the holdup and the time interval do not affect the final answers obtained. However, there are certain arithmetical restrictions on the magnitude of the values assumed. The holdup assumed should be large with respect to the feed rate. Rose et al. suggest a value of five to ten times the amount of feed which enters the column during the time interval. The time interval considered and the holdup used should be such that the change in the stage liquid composition during the time interval is moderate. It will be seen that distorted new values of the stage concentrations would be obtained if the holdup were small with respect to V and L and the time interval long.

Stage-to-stage Calculations. Denote the total moles of liquid holdup on each stage as P. All stages including the partial reboiler will be assumed to have the same holdup. Denote the number of the time interval under consideration as j. Then $j + 1$ refers to the succeeding time interval. The change in the number of moles of component i in stage n during the jth time interval is given by

$$P(x_{i,n,j+1} - x_{i,n,j}) = \Delta t(v_{n-1} + l_{n+1} - v_n - l_n)_{i,j}$$

where the Δt is the length chosen for the time intervals. Rearranging provides the concentrations of component i on stage n at the end of the jth time interval.

$$x_{i,n,j+1} = x_{i,n,j} + \frac{\Delta t}{P} (v_{n-1} + l_{n+1} - v_n - l_n)_{i,j} \qquad (10\text{-}37)$$

The equation for the stripping section is obtained by replacing n with m. The equation for the new feed-stage concentration of component i at the end of the jth interval is

$$x_{i,M+1,j+1} = x_{i,M+1,j} + \frac{\Delta t}{P} (Fx_{i,F})$$

$$+ \frac{\Delta t}{P} (v_M + l_{M+2} - v_{M+1} - l_{M+1})_{i,j} \qquad (10\text{-}38)$$

At time zero all the stage compositions, temperatures, and flow rates are those of the original steady-state condition with the exception of the $Fx_{i,F}$ term, which has just changed instantaneously to its new value. Application of Eq. (10-38) to the feed stage gives a new value for $x_{i,M+1}$ at the end of the first time interval Δt. None of the other stage liquids will reflect any change during the first time interval, since streams entering and leaving every stage except the feed stage at time zero are in material balance. During the second time interval, the changed vapor and liquid streams from the feed stage will affect the stage above and the one below the feed stage. The disturbance will move up and down from the feed stage one stage per time interval until the ends of the column are reached. The vapor in equilibrium with the liquid in each stage at the end of each time interval is obtained with $y_i = K_i x_i$. However, the calculation of the new vapor concentrations should be postponed until the new temperature and rate profiles are obtained.

New Temperature and Rate Profiles. At the end of each time interval, the new stage liquid compositions can be used in bubble-point calculations to provide a new temperature profile and new K values for each component in each stage. The vapor concentrations are then obtained from

$$y_{i,n,j+1} = (Kx)_{i,n,j+1}$$

The new temperature and phase compositions permit the calculation of new molar enthalpies for each vapor and liquid stream. The new L and V profiles are then obtained from enthalpy balances [see Eqs. (10-7) and (10-8)]. The calculations for the next time interval can now be started.

Convergence Procedure. Convergence is obtained by simple iteration through a sufficient number of time intervals. As mentioned previously the time intervals must be small if convergence is to be achieved. A large number of intervals must be considered before the column conditions achieve the new steady-state values. At first, the transient conditions approach the steady-state values rapidly, but the approach then becomes asymptotic and convergence is slow. Since the calculations involved in each time interval correspond to a trial in the methods described in previous sections, it can be seen that the relaxation method is comparatively slow.

Convergence Check. Convergence is achieved when additional time intervals produce no further change in the column conditions; that is, a steady-state condition has been achieved.

NOMENCLATURE

$A_{i,n} = L_n/K_{i,n}V_n$ = absorption factor for component i.

$b_i = Bx_{i,B}$ = moles of component i in bottom product. Primes denote corrected values.

B = heavy product rate; moles per unit time.

C = number of components.

d = differential operator.

$d_i = Dx_{i,D}$ = moles of component i in overhead product. Primes denote corrected values.

D = light product rate; moles per unit time.

$e_{i,j}$ = mismatch at feed stage for component i in jth trial in Lewis-Matheson method. See Eq. (10-15).

f = function.

F = feed rate; moles per unit time.

h = liquid enthalpy, Btu/lb mole. Subscripts n and m refer to stage number.

H = vapor enthalpy, Btu/lb mole. Subscripts n and m refer to stage number.

i = subscript referring to component i.

j = subscript referring to trial number.

$K_i = y_i/x_i$ = equilibrium-distribution coefficient for component i.

$l_{i,n} = L_n x_{i,n}$ = moles of component i in the liquid overflow from stage n.

L = heavy phase rate; moles per unit time.

m = any stage in stripping section (stages 1 to M).

M = number of equilibrium stages below the feed stage. Feed stage is $M + 1$.

n = any stage in rectifying section.

N = total number of equilibrium stages including the partial reboiler if any, but excluding the partial condenser if any.

$N_i{}^u$ = degrees of freedom in the process. See Chap. 3.

P = moles of liquid holdup in an equilibrium stage.

q = heat stream. Subscripts c and r refer to condenser and reboiler heat loads, respectively.

r = subscript referring to reference component on which relative volatiles are based.

R = external reflux ratio at top of column; amount refluxed per amount of product.

$S_{i,n} = K_{i,n}V_n/L_n$ = stripping factor for component i.

t = time. Δt refers to length selected for each time interval in relaxation method.

$v_{i,n} = V_n y_{i,n}$ = moles of component i in the vapor rising from stage n.

V = light phase rate; moles per unit time.

x_i = mole fraction of component i in heavy phase.

y_i = mole fraction of component i in light phase.

Greek Symbols

$\alpha_{i-r} = K_i/K_r$ = relative volatility of component i with respect to component r.

∂ = partial differential operator.

Δ = increment in variable.

θ = correction factor defined by Eq. (10-29)

REFERENCES

1. Amundson, N. R., and A. J. Pontinen: *Ind. Eng. Chem.*, **50**, 730 (1958).
2. Amundson, N. R., A. J. Pontinen, and J. W. Tierney: *A.I.Ch.E. J.*, **5**, 295 (1959).
3. Bonner, J. S.: *Proc. Am. Petrol. Inst.*, **36**, sec. III, 238 (1956).
4. Brown, G. G., and M. Souder, Jr.: *Ind. Eng. Chem.*, **24**, 519 (1932).
5. Donnell, J. W., and K. Turbin: *Chem. Eng.*, **58**, 112 (July, 1951).
6. Edmister, W. C.: *A.I.Ch.E. J.*, **3**, 165 (1957).
7. Greenstadt, J., Y. Bard, and B. Morse: *Ind. Eng. Chem.*, **50**, 1644 (1958).
8. Horton, G., and W. B. Franklin: *Ind. Eng. Chem.*, **32**, 1384 (1940).
9. Kremser, A.: *Natl. Petrol. News*, **22** (May 21, 1930).
10. Lewis, W. K., and G. L. Matheson: *Ind. Eng. Chem.*, **24**, 494 (1932).
11. Lyster, W. N., S. L. Sullivan, Jr., D. S. Billingsley, and C. D. Holland: *Petrol. Refiner*, **38**(6), 221 (1959); **38**(7), 151 (1959); **38**(10), 139 (1959).
12. Rose, A., R. F. Sweeny, and V. N. Schrodt: *Ind. Eng. Chem.*, **50**, 737 (1958).
13. Thiele, E. W., and R. L. Geddes: *Ind. Eng. Chem.*, **25**, 289 (1933).

CHAPTER 11

MULTICOMPONENT SEPARATIONS
Azeotropic and Extractive Distillations

The design engineer must often effect the separation of components which boil so closely together that their separation by simple distillation is impracticable or impossible. However, the very fact that the components boil at the same temperature provides a basis for their eventual separation. Compounds of similar chemical structure (members of homologous series) cannot boil at the same temperature because of the molecular weight difference. Compounds which do boil at the same temperature must therefore be from different compound classes or series and differ from one another in some aspect of their molecular structure. The molecular dissimilarities will cause each of the close-boiling components to react differently when the liquid-phase environment is changed by the addition of a new component commonly called a "solvent" or "entrainer." A successful solvent or entrainer must enhance the volatility of one key component more than the other. In order to do this, it must make the escape of one type of molecule from the solution more difficult, or conversely, it must make easier the escape of the other type present.

Distillation processes which effect the separation of close-boiling key components by the addition of a new component are termed azeotropic or extractive distillations. Azeotropic distillation refers to those processes where the added component forms an azeotrope (constant-boiling mixture) with one of the keys and the azeotrope becomes either the overhead or bottom product. If the azeotrope formed boils at a lower temperature than the original feed, it will be taken overhead. If the azeotrope formed by the addition of the solvent boils higher than the original feed, the azeotrope will be the bottom product. The former case is more common. In many cases, the azeotrope formed will not be a binary one but will contain certain amounts of both components; i.e., the point representing the azeotrope will fall within a triangular composition diagram rather than on one side. Such an azeotrope is termed a *ternary* azeotrope, since it contains amounts of both keys plus the solvent. If the ratio of the original two components in the ternary azeotrope is different from that in the feed, removal of the ternary azeotrope as the overhead (or bottom) product achieves a separation between them.

Extractive distillation is similar to the azeotropic process where the solvent leaves at the bottom except that no azeotrope is formed between the added component and either of the keys. Also, the added component is commonly referred to as the "solvent" rather than the "entrainer." The use of the word solvent is natural because of the strong analogy between extractive distillation and liquid-liquid extraction. In both cases a solvent is added which has a different selectivity for each of the components to be separated. The same analogy holds for azeotropic distillation except that there is nothing in liquid-liquid systems analogous to an azeotrope.

Both azeotropic and extractive distillation depend upon a controlled modification of the deviations from Raoult's law which each of the key components exhibits. The modification is obtained (a) by the choice of the solvent and (b) by control of the solvent concentration in the various stages. The second item is a design factor and can be handled by introduction of the solvent or entrainer at the appropriate rate and place in the column. The choice of a solvent is a more difficult subject and generally cannot be made without resorting to experimental verification of the predicted results. The present knowledge of the liquid state is not sufficient to permit exact quantitative prediction of the phase relationships which will result upon the addition of the new component to the mixture. However, qualitative predictions of the direction of deviations from Raoult's law can be made from the classifications of Ewell et al. (17), and quantitative predictions of the activity coefficients of the various components at infinite dilution in the solvent are possible by the method of Pierotti et al. (36). These two papers are discussed in the next section. If further information on the system is desired, experimental measurements of the activity coefficients of the keys at the desired solvent concentration must be made. If vapor-liquid equilibrium data are available for all three of the binary systems formed by the two keys and the solvent, the ternary equilibria can be predicted quantitatively by the methods of Chap. 2.

11-1. Prediction of Deviations

Hydrogen Bonding. Ewell, Harrison, and Berg (17) have classified all liquids into five groups according to the ability of their molecules to form intermolecular bonds with other like or unlike molecules. The ability to enter into intermolecular bonds depends upon the degree of polarization within the molecule. If the molecule contains "negative" atoms such as oxygen, nitrogen, fluorine, or chlorine, the electrons in the molecule tend to lie closer to these large atoms than to the hydrogen atoms to which they are attached. Consequently, the hydrogen atoms have a deficiency of negative charge and become positive with respect to the negative atoms. A hydrogen atom attached to a negative atom or to a carbon atom which in turn is attached to a sufficiently negative atom or group of atoms will attempt to satisfy its electron deficiency by the formation of an intermolecular bond with a negative (or donor) atom in another molecule. The intermolecular attachment through the hydrogen atom is called a hydrogen bond. Its strength depends upon the nature of the negative atoms between which the hydrogen is coordinating. Ewell et al. give the following examples of strong and weak bonds:

Strong	Weak
$O \rightarrow HO$	$N \rightarrow HN$
$N \rightarrow HO$	
$O \rightarrow HN$	

$$O \rightarrow \atop N \rightarrow \left\{ \begin{array}{l} HCCl_3 \\ HCCl-CCl \\ HCNO_2 \\ HCCN \end{array} \right.$$

A schematic representation of the concept of the hydrogen bond is shown in Fig. 11-1. The bonds may cause the formation of bimolecular complexes, or they may form chainlike or three-dimensional aggregates among large numbers of molecules. The energy involved in the bonds may be of the order of several kilocalories per mole (37).

It can be seen from Fig. 11-1 that a hydrogen bond is formed only when an active hydrogen atom (deficient in electrons) comes into contact with a donor atom. Some pure liquids contain both active hydrogens and donor atoms, some contain only donor atoms, some contain only active hydrogens, and some contain neither active hydrogens nor donor atoms. Ewell et al. utilized this natural classification to divide all liquids into five groups (the liquids with both active hydrogen and donor atoms were divided into two groups according to the strength of the bonds formed) which are described below.

Class I. Liquids capable of forming three-dimensional networks of strong hydrogen bonds, e.g., water, glycol, glycerol, amino alcohols, hydroxylamine, hydroxy acids, polyphenols, amides, etc. Compounds such as nitromethane and acetonitrile also form three-dimensional networks of hydrogen bonds, but the bonds are much weaker than those involving OH and NH groups. Therefore, these types of compounds are placed in class II.

Class II. Other liquids composed of molecules containing both active hydrogen atoms and donor atoms (oxygen, nitrogen, and fluorine), e.g., alcohols, acids, phenols, primary and secondary amines, oximes, nitro compounds with α-hydrogen atoms, nitriles with α-hydrogen atoms, ammonia, hydrazine, hydrogen fluoride, hydrogen cyanide, etc.

Class III. Liquids composed of molecules containing donor atoms but not active hydrogen atoms, e.g., ether, ketones, aldehydes, esters,

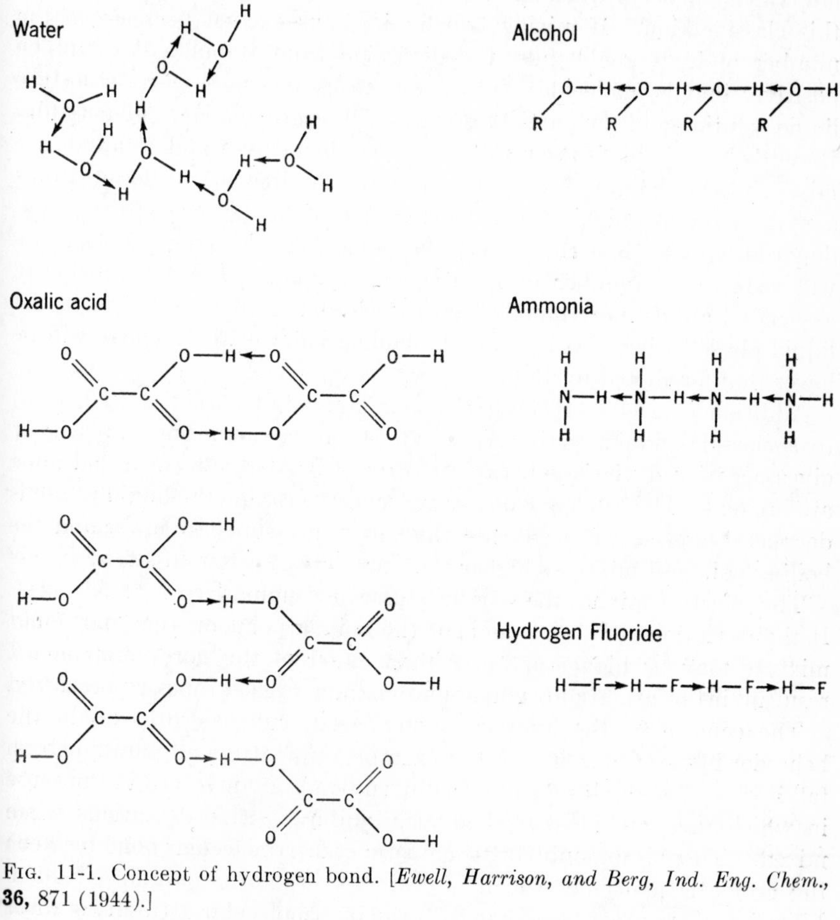

FIG. 11-1. Concept of hydrogen bond. [*Ewell, Harrison, and Berg, Ind. Eng. Chem.,* **36,** 871 (1944).]

tertiary amines (including pyridine type), nitrocompounds and nitriles without α-hydrogen atoms, etc.

Class IV. Liquids composed of molecules containing active hydrogen atoms but no donor atoms. These are molecules having two or three chlorine atoms on the same carbon as a hydrogen atom or one chlorine on the same carbon atom and one or more chlorine atoms on adjacent carbon atoms, e.g., $CHCl_3$, CH_2Cl_2, CH_3CHCl_2, CH_2Cl-CH_2Cl, $CH_2Cl-CHCl-CH_2Cl$, $CH_2Cl-CHCl_2$, etc.

Class V. All other liquids, i.e., liquids having no hydrogen bond-forming capabilities, e.g., hydrocarbons, carbon disulfide, sulfides, mercaptans, halohydrocarbons not in class IV, and nonmetallic elements such as iodine, phosphorus, sulfur, etc.

The direction and approximate extent of the deviations from Raoult's law which will occur when any two liquids are mixed can be deduced from this classification. If mixing two liquids causes a net decrease in the number of hydrogen bonds, the deviation from Raoult's law will be positive; that is, molecules will be able to escape more easily from the liquid solution and the partial pressures will be greater than predicted by Raoult's law. This is the result obtained when class I and V liquids are mixed. Class I liquids contain both active hydrogen and donor atoms and therefore are highly bonded. Dilution of the active hydrogen and donor atoms with molecules which possess no bond-forming potentialities will reduce the number of intermolecular bonds, make the formerly associated atoms more mobile, and thereby make their escape from the liquid phase easier. As a result the boiling point of the mixture will be lower than predicted by Raoult's law.

Addition of a class III liquid (donor atoms but no active hydrogen) to a class IV liquid (active hydrogen but no donor atoms) invariably gives negative deviations because hydrogen bonds are formed and none are broken. The tendency of the molecules to escape the liquid phase is decreased, and as a consequence the partial pressures are lower and the boiling points of mixtures higher than predicted by Raoult's law.

The effect of mixing the various classes is summarized in Table 11-1. If the deviations are large enough, the boiling point of some particular mixture may be higher or lower than either of the pure components, resulting in the formation of maximum or minimum azeotropes, respectively.

The concept of the hydrogen bond is not sufficient to explain the behavior in many mixtures. For example, aromatics and paraffins both fall in class V and, although no hydrogen bonds are involved in the sense in which they are illustrated above, exhibit positive deviations when mixed. There is undoubtedly some form of intermolecular bond between the aromatic molecules which is destroyed when the paraffin is added. Likewise, there must be a change in the intermolecular attractions when olefins and saturated paraffins are mixed, which produces the slight deviations observed in such mixtures. Hydrogen bonds are probably just one example of several bonding mechanisms which are operative in a liquid mixture.

Activity Coefficients at Infinite Dilution. Pierotti, Deal, and Derr (36) have presented a method whereby a more quantitative prediction of the extent of the deviations from Raoult's law can be made than is possible through the classification system of Ewell et al. Their method is based upon an empirical equation which relates the excess partial molar free energy of a solute at infinite dilution in a solvent to a series of interaction constants. The excess partial molar free energy is related to the activity coefficient as follows: If 1 mole of the solute in the standard state (Pierotti et al. defined their standard state as the pure liquid at its vapor pressure

TABLE 11-1. SUMMARY OF DEVIATIONS FROM RAOULT'S LAW†

Classes	Deviations	Hydrogen bonding
I + V ⎱ II + V ⎰	Always + deviations; I + V, frequently limited solubility	H bonds broken only
III + IV	Always − deviations	H bonds formed only
I + IV ⎱ II + IV ⎰	Always + deviations; I + IV, frequently limited solubility	H bonds both broken and formed, but dissociation of class I or II liquids is more important effect
I + I I + II I + III II + II II + III	Usually + deviations, very complicated groups. Some − deviations give some max azeotropes	H bonds both broken and formed
III + III III + V IV + IV IV + V V + V	Quasi-ideal systems, always + deviations or ideal; azeotropes, if any, will be minima	No H bonds involved

† From R. H. Ewell, J. M. Harrison, and L. Berg, *Ind. Eng. Chem.*, **36,** 871 (1944).

at the temperature of the solution) is put into a given amount of solvent, the change in free energy of the solute is related to its activity in the mixture by

$$\bar{F}_1 - F_1^\circ = RT \ln \frac{\bar{f}_1}{f_1^\circ} = RT \ln a_1$$

For a nonideal solution, $a_1 = \gamma_1 x_1$ and

$$\bar{F}_{1,\text{nonideal}} - F_1^\circ = RT \ln \gamma_1 + RT \ln x_1 \qquad (11\text{-}1)$$

If the solution formed had been ideal, $\gamma_1 = 1.0$, $a_1 = x_1$, and

$$\bar{F}_{1,\text{ideal}} - F_1^\circ = RT \ln x_1 \qquad (11\text{-}2)$$

The excess partial molar free energy is defined by the difference between (11-1) and (11-2).

$$F_1^E = \bar{F}_{1,\text{nonideal}} - \bar{F}_{1,\text{ideal}} = RT \ln \gamma_1 \qquad (11\text{-}3)$$

If the amount of solvent used were infinite with respect to the amount of solute, the $F_1{}^E$ and γ_1 obtained would be the values at infinite dilution. The γ_1 at infinite dilution corresponds to the end point ($x_2 = 1.0$) of the γ_1 vs solvent plot.   Figures 11-2 and 11-3 illustrate the fact that these end points vary in a rather regular manner with a systematic change

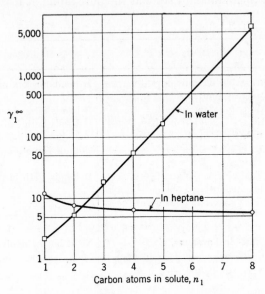

FIG. 11-2. Activity coefficients at 100°C for the straight-chain normal alcohols at infinite dilution in water and heptane.

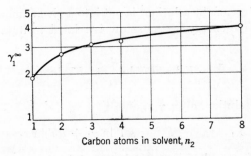

FIG. 11-3. Activity coefficients for water at 100°C and infinite dilution in the straight-chain normal alcohols vs the solvent carbon number.

in the molecular structure of the solvent or solute.   In Fig. 11-2 the number of carbon atoms in the solute is varied while the molecular structure of the solvents remains constant.   In Fig. 11-3, the carbon number of the solvent is varied.   Pierotti et al. predict the change in γ^∞ with molecular structure by means of a correlation which treats log

γ^∞ as a sum of contributions from individual interactions between pairs of structural groups in the solvent and solute molecules. Let R and X represent the hydrocarbon radical and the functional group, respectively, in a molecule. Consider a solution of R_1X_1 at infinite dilution in R_2X_2. The interactions which can occur between two R_1X_1 molecules are $X_1 - X_1$, $X_1 - R_1$, and $R_1 - R_1$. The analogous interactions between two R_2X_2 molecules are $X_2 - X_2$, $X_2 - R_2$, and $R_2 - R_2$. The interactions between solute and solvent molecules are $X_1 - X_2$, $X_2 - R_1$, $X_1 - R_2$, and $R_1 - R_2$. The number of the various interactions which can occur will depend upon the number of carbon atoms in the solute and solvent molecules because the carbon atoms tend to dilute the effect of the functional groups. The manner in which chain length is taken into account is shown in Eq. (11-4), which is the general correlation for methylene homologs R_1X_1 at infinite dilution in methylene homologs R_2X_2.

$$\frac{F_1{}^E}{2.303RT} = \log \gamma_1{}^\infty = A_{1,2} + \frac{B_2 n_1}{n_2} + \frac{C_1}{n_1} + D(n_1 - n_2)^2 + \frac{F_2}{n_2} \quad (11\text{-}4)$$

where $F_1{}^E$ = excess partial molar free energy of solute 1 at infinite dilution in solvent 2.

$\gamma_1{}^\infty$ = activity coefficient of R_1X_1 at infinite dilution in R_2X_2.

n_1, n_2 = number of carbon atoms in hydrocarbon radicals R_1 and R_2, respectively.

$A_{1,2}$ = coefficient which depends on the nature of the solute and solvent functional groups X_1 and X_2. It represents interactions $X_1 - X_1$, $X_1 - X_2$, and $X_2 - X_2$ and can be considered to be the sum of interactions for zero carbon-numbered members of the homologous solute and solvent series.

B_2 = coefficient which depends only on the nature of the solvent functional group X_2. The term $B_2 n_1/n_2$ accounts for the increase in the number of $X_2 - X_2$ interactions broken and the increase in the number of $R_1 - X_2$ interactions which occur as the solute molecule increases in carbon number. These interactions are inversely proportional to n_2, the solvent carbon number, because the X_2 groups are diluted as the number of methylene groups in R_2 increases.

C_1 = coefficient which depends only on the solute functional group X_1. The C_1/n_1 term accounts for the change in the number of $X_1 - X_1$ interactions as the solute carbon number increases and the polar groups are diluted by the methylene groups of R_1.

D = coefficient independent of both X_1 and X_2. The $D(n_1 - n_2)^2$ accounts for changes in $R_1 - R_1$, $R_1 - R_2$, and $R_2 - R_2$ interactions with carbon number.

F_2 = coefficient which essentially depends only on the nature of the solvent functional group X_2. The F_2/n_2 accounts for the change in the numbers of $X_2 - X_2$ interactions as they are diluted by an increase in the methylene groups in R_2.

Values of the required constants for several homologous series with various functional groups are shown in Table 11-2.† The constants are given at three temperatures. Either the constants or the calculated γ^∞'s can be plotted vs temperature for the purposes of interpolation or extrapolation. A brief discussion of the systems shown will illustrate most of the characteristics of the general correlation and how it varies to fit the particular system under consideration.

Whenever the solvent contains no carbon atoms (water in Table 11-2), those terms in the correlation which depend on n_2 are not included. The correlation for any homologous series in water involves only A, B, and C types of terms, where B represents a combination of the B_2/n_2 and D constants.

Representation of solutes in which the functional group does not fall at the end of the molecule requires special rules for the evaluation of n_1. As shown in Table 11-2, each branch of the hydrocarbon radical is counted separately starting at the carbon atom to which the functional group is attached. For example, in tertiary butyl alcohol, the central carbon is counted in each branch; that is, $n_1' = n_2'' = n_3''' = 2$ even though the total carbon atom is four. Each branch is considered to be a separate radical and to make its own contribution to the A and C coefficients. A more general grouping which satisfactorily represents all alcohols with one set of constants is also shown in Table 11-2. Other modifications of the C term are used for acetals and alkyl-substituted cyclic compounds.

The alcohols in paraffin series illustrate those systems in which the solvent contains no functional group and the interactions represented by the B and F terms do not occur. The C_1 constant is the same as in the alcohol-water series. Note also that the D constant is the same for the paraffin solvent regardless of the solute (compare the alcohol-paraffin and the ketone-paraffin constants in Table 11-2). This carry-over of constants from system to system often reduces the experimental work required for a new system. The carry-over of constants breaks down for the C_1 constant when the ketone-water and ketone-paraffin systems are compared.

† Additional data are available in Pierotti, Deal, and Derr, Document 5782, American Document Institute, Washington, D.C. See the original article for details.

Mixtures of water in homologs of alcohols (and ketones) require only $A_{1,2}$ and F_2 constants. Effects of branching in the solvent molecule are included in the F term.

When ketones or aldehydes are dissolved in alcohols, both solute and solvent contain functional and alkyl groups and all five terms are included in Eq. (11-4). All the constants except $A_{1,2}$ and B_2 are available from systems listed previously.

Paraffins in ketones represent systems in which no A or C terms are required because the solute contains no functional group. The usual B, D, and F terms are included.

Table 11-3 presents the correlation constants for various homologous series of hydrocarbons in specific solvents. Whenever the solvent structure remains constant, F_2/n_2 and B_2/n_2 remain constant and can be combined with the A term. Also, a distinction must be made between paraffinic carbons and carbons included in a ring. It was found that the cyclic nucleus was best treated as a functional group with a number of carbon atoms equal to $(n_r - 4)$ or 2 for a six-carbon ring. Letting n_p represent the number of paraffinic carbons in the alkyl side chains, the carbon number of the solute becomes $n_p + 2$ for a six-carbon ring and Eq. (11-4) can be rewritten for paraffin and alkyl cyclic solutes as

$$\log \gamma_1^\infty = K_1 + Bn_p + \frac{C_1}{n_p + 2} + D(n_1 - n_2)^2 \qquad (11\text{-}5)$$

where n_p is the number of paraffinic carbons in the solute and n_1 and n_2 are total carbon numbers in the solute and solvent, respectively. When paraffins are the solute, the C term is dropped because the solute contains no functional group. Likewise for heptane as a solvent, the B term becomes zero because no functional groups exist in a mixture of a paraffin in a paraffin.

The accuracy of the correlation by Pierotti et al. is quite good, particularly in view of the large number of systems covered. The over-all average deviation in γ_1^∞ was about 8 per cent in the 44 sets of systems correlated.

The correlations can be used to estimate γ values for binary systems over the entire composition range. The two γ^∞'s provide the two constants in the Van Laar or three-suffix Margules equations (see Chap. 2), and these equations can be used to predict γ's as a function of composition. The prediction is often very rough, and the different equations give different answers. If the binary has an azeotrope and the azeotropic conditions are known, another point is available which, together with the infinite dilution values, permits a three-constant equation to be used to predict more complicated binaries. Such predictions are subject to the usual pitfalls of curve fitting. The predicted curve will pass

TABLE 11-2. CORRELATION CONSTANTS FOR ACTIVITY COEFFICIENTS†

(Infinite dilution of various combinations of homologous series of solutes and solvents‡)

Solute series	Solvent series	Temp, °C	A Constant	B Term	B Constant	C Term	C Constant	D Term	D Constant	F Term	F Constant
n-Acids	Water	25	−1.00	Bn_1	0.622	$C(1/n_1)$	0.490	None		None	
		50	−0.80	Bn_1	0.590		0.290				
		100	−0.620	Bn_1	0.517		0.140				
n-Primary alcohols	Water	25	−0.995	Bn_1	0.622	$C(1/n_1)$	0.558	None		None	
		60	−0.755	Bn_1	0.583		0.460				
		100	−0.420	Bn_1	0.517		0.230				
n-sec-alcohols	Water	25	−1.220	Bn_1	0.622	$C(1/n_1' + 1/n_1'')$	0.170	None		None	
		60	−1.023	Bn_1	0.583		0.252				
		100	−0.870	Bn_1	0.517		0.400				
n-tert-alcohols	Water	25	−1.740	Bn_1	0.622	$C\left(\dfrac{1/n_1' +}{1/n_1'' +}\;1/n_1'''\right)$	0.170	None		None	
		60	−1.477	Bn_1	0.583		0.252				
		100	−1.291	Bn_1	0.517		0.400				
Alcohols, general	Water	25	−0.525	Bn_1	0.622	$C\left[\begin{array}{c}(1/n_1'-1)+\\(1/n_1''-1)+\\(1/n_1'''-1)\end{array}\right]$	0.475	None		None	
		60	−0.33	Bn_1	0.583		0.39				
		100	−0.15	Bn_1	0.517		0.34				
n-Allyl alcohols	Water	25	−1.180	Bn_1	0.622	$C(1/n_1)$	0.558	None		None	
		60	−0.929	Bn_1	0.583		0.460				
		100	−0.650	Bn_1	0.517		0.230				
n-Aldehydes	Water	25	−0.780	Bn_1	0.622		0.320	None		None	
		60	−0.400	Bn_1	0.583		0.210				
		100	−0.03	Bn_1	0.517		0.0				
n-Alkene aldehydes	Water	25	−0.720	Bn_1	0.622	$C(1/n_1)$	0.320	None		None	
		60	−0.540	Bn_1	0.583		0.210				
		100	−0.298	Bn_1	0.517		0.0				

Solute	Solvent	t	A	B-term	B	C-term	C	D-term	D	F-term	F
n-Ketones	Water	25	-1.475	Bn_1	0.622	$C(1/n'_1 + 1/n''_1)$	0.500	None		None	
		60	-1.040	Bn_1	0.583		0.330				
		100	-0.621	Bn_1	0.517		0.200				
n-Acetals	Water	25	-2.556	Bn_1	0.622	$C\left[\dfrac{1}{n'_1} + \dfrac{1}{n''_1} + \dfrac{2}{n''_1}\right]$	0.486	None		None	
		60	-2.184	Bn_1	0.583		0.451				
		100	-1.780	Bn_1	0.517		0.426				
n-Ethers	Water	20	-0.770	Bn_1	0.640	$C[1/n'_1 + 1/n''_1]$	0.195	None		None	
n-Nitriles	Water	25	-0.587	Bn_1	0.622	$C(1/n_1)$	0.760	None		None	
		60	-0.368	Bn_1	0.583		0.413				
		100	-0.095	Bn_1	0.517		0.00				
n-Alkene nitriles	Water	25	-0.520	Bn_1	0.622	$C(1/n_1)$	0.760	None		None	
		60	-0.323	Bn_1	0.583		0.413				
		100	-0.074	Bn_1	0.517		0.00				
n-Esters	Water	20	-0.930	Bn_1	0.640	$C(1/n'_1 + 1/n''_1)$	0.260	None		None	
n-Formates	Water	20	-0.585	Bn_1	0.640	$C(1/n_1)$	0.260	None		None	
n-Mono-alkyl chlorides	Water	20	1.265	Bn_1	0.640	$C(1/n_1)$	0.073	None		None	
n-Paraffins	Water	16	0.688	Bn_1	0.642	None	None	None		None	
n-Alkyl benzenes	Water	25	3.554	Bn_1	0.622	$C[(1/n_1 - 4)]$	-0.466	None		None	
n-Alcohols	Paraffins	25	1.960	None		$C\left[(1/n'_1 - 1) + (1/n''_1 - 1) + (1/n'''_1 - 1)\right]$	0.475	$D(n_1 - n_2)^2$	-0.00049	None	
		60	1.460	None			0.390		-0.00057		
		100	1.070	None			0.340		-0.00061		
n-Ketones	Paraffins	25	0.0877	None		$C(1/n'_1 + 1/n''_1)$	0.757	$D(n_1 - n_2)^2$	-0.00049	None	
		60	0.016	None			0.680		-0.00057		
		100	-0.067	None			0.605		-0.00061		
Water	n-Alcohols	25	0.760	None		None	None	None		$F'(1/n_2)$	-0.630
		60	0.680	None							-0.440
		100	0.617	None							-0.280

TABLE 11-2. CORRELATION CONSTANTS FOR ACTIVITY COEFFICIENTS (Continued)

Solute series	Solvent series	Temp, °C	A Constant	B Term	B Constant	C Term	C Constant	D Term	D Constant	F Term	F Constant
	sec-Alcohols	80	1.208	None		None		None		$F(1/n_2' + 1/n_2'')$	-0.690
	n-Ketones	25	1.857	None		None		None		$F(1/n_2' + 1/n_2'')$	-1.019
		60	1.493								-0.73
		100	1.231								-0.557
Ketones	n-Alcohols	25	-0.088	$B(n_1/n_2)$	0.176	$C(1/n_1' + 1/n_1'')$	0.50	$D(n_1 - n_2)^2$	-0.00049	$F(1/n_2)$	-0.630
		60	-0.035	$B(n_1/n_2)$	0.138		0.33		-0.00057		-0.440
		100	-0.035	$B(n_1/n_2)$	0.112		0.20		-0.00061		-0.280
Aldehydes	n-Alcohols	25	-0.701	$B(n_1/n_2)$	0.176	$C(1/n_1)$	0.320		-0.00049	$F(1/n_2)$	-0.630
		60	-0.239	$B(n_1/n_2)$	0.138		0.210		-0.00057		-0.440
Esters	n-Alcohols	25	0.212	$B(n_1/n_2)$	0.176	$C(1/n_1' + 1/n_1'')$	0.260	$D(n_1 - n_2)^2$	-0.00049	$F(1/n_2)$	-0.630
		60	0.055	$B(n_1/n_2)$	0.138		0.240		-0.00057		-0.440
		100	0.0	$B(n_1/n_2)$	0.112		0.220		-0.00061		-0.280
Acetals	n-Alcohols	60	-1.10	$B(n_1/n_2)$	0.138	$C(1/n_1' + 1/n_1'' + 2/n_1''')$	0.451		-0.00057		-0.440
Paraffins	Ketones	25	None	$B(n_1/n_2)$	0.1821	None		$D(n_1 - n_2)^2$	-0.00049	$F(1/n_2' + 1/n_2'')$	0.402
		60		$B(n_1/n_2)$	0.1145				-0.00057		0.402
		90		$B(n_1/n_2)$	0.0746				-0.00061		0.401

† From G. J. Pierotti, C. H. Deal, and E. L. Derr, Ind. Eng. Chem., **51**, 95 (1959).

‡ General expression, $\log \gamma^\circ = A_{1,2} + B_2(n_1/n_2) + C_1(1/n_1) + D(n_1 - n_2)^2 + F_2(1/n_2)$.

TABLE 11-3. CORRELATION† CONSTANTS FOR ACTIVITY COEFFICIENTS‡
(Various homologous series of hydrocarbons in specific solvents)

Temp, °C	Solute series	Solute-dependent C's (Term = $\dfrac{C}{n_p+2}$)	Heptane	Methyl ethyl ketone	Furfural	Phenol	Ethyl alcohol	Triethylene glycol	Diethylene glycol	Ethylene glycol
					Solvent-dependent B's (term = $B_p n_p$)					
25			0.0	0.0455	0.0937	0.0625	0.088		0.191	0.275
50			0.0	0.033	0.0878	0.0590	0.073	0.161	0.179	0.249
70			0.0	0.025	0.0810	0.0586	0.065		0.173	0.236
90			0.0	0.019	0.0686	0.0581	0.059	0.134	0.158	0.226
					Solute-solvent dependent K's (term = K)					
25	Paraffins	0.0	0.0	0.335	0.916	0.870	0.580		0.875	
50		0.0	0.0	0.332	0.756	0.755	0.570	0.72	0.815	1.208
70		0.0	0.0	0.331	0.737	0.690	0.590		0.725	1.154
90		0.0	0.0	0.330	0.771	0.620	0.610	0.68	0.72	1.089
25	Alkyl cyclohexanes	−0.260	0.18	0.70	1.26	1.20	1.06		1.675	
50		−0.220		0.650	1.120	1.040	1.01	1.46	1.61	2.36
70		−0.195	0.131	0.581	1.020	0.935	0.972		1.550	2.22
90		−0.180	0.09	0.480	0.930	0.843	0.925	1.25	1.505	2.08
25	Alkyl benzenes	−0.466	0.328	0.277	0.67	0.694	1.011		1.08	
50		−0.390	0.243		0.55	0.580	0.938	0.80	1.00	1.595
70		−0.362	0.225	0.240	0.45	0.500	0.900		0.96	1.51
90		−0.350	0.202	0.239	0.44	0.420	0.862	0.74	0.935	1.43

TABLE 11-3. CORRELATION† CONSTANTS FOR ACTIVITY COEFFICIENTS‡ (Continued)

Temp, °C	Solute series	Solute-dependent C's ($\text{Term} = \dfrac{C}{n_p + 2}$)	Solvents							
			Heptane	Methyl ethyl ketone	Furfural	Phenol	Ethyl alcohol	Triethylene glycol	Diethylene glycol	Ethylene glycol
25	Alkyl naphthalenes	−0.10	0.53	0.169	0.46	0.595	1.06	0.75	1.00	1.92
50		−0.14	0.53	0.141	0.40	0.54	1.03		1.00	1.82
70		−0.173	0.53	0.215	0.39	0.497	1.02	0.83	0.991	1.765
90		−0.204	0.53	0.232		0.445			1.01	
25	Alkyl Tetralins	+0.28	0.244	0.179	0.652	0.378		1.00	1.43	
50		+0.24			0.528	0.364			1.38	
70		+0.21	0.220	0.217	0.447	0.371		0.893	1.33	
90		+0.19			0.373	0.348			1.28	
25	Alkyl Decalins	−0.43	0.356	0.871	1.54	1.411		1.906	2.46	
50		−0.368			1.367	1.285			2.25	
70		−0.355		0.80	1.253	1.161		1.68	2.07	
90		−0.320			1.166	1.078			2.06	

† Expression, $\log \gamma^° = K + B_p n_p + (C/n_p + 2) + D(n_1 - n_2)^2$ where for all systems:

$$D = -49 \times 10^{-5} \quad -55 \times 10^{-5} \quad -58 \times 10^{-5} \quad -61 \times 10^{-5}$$
$$t°C = 25 \qquad\quad 50 \qquad\qquad 70 \qquad\qquad 90$$

‡ From G. J. Pierotti, C. H. Deal, and E. L. Derr, *Ind. Eng. Chem.*, **51**, 95 (1959).

through the three known points but may deviate widely between these points.

The main value of the correlation lies in the selection of the solvents for liquid-liquid extraction and extractive distillation processes. Figure 11-4 compares the values of the product $p_1^* \gamma_1^\infty$ for several homologous series at infinite dilution in water. Such predictions provide a measure of the ease of separation of the various types of compounds in an extractive distillation with water as the solvent. Similar plots for other solvents

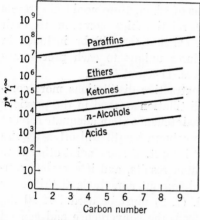

FIG. 11-4. Predicted volatilites of various homologous series at infinite dilution in water. [*Pierotti, Deal, and Derr, Ind. Eng. Chem.*, **51**, 95 (1959).]

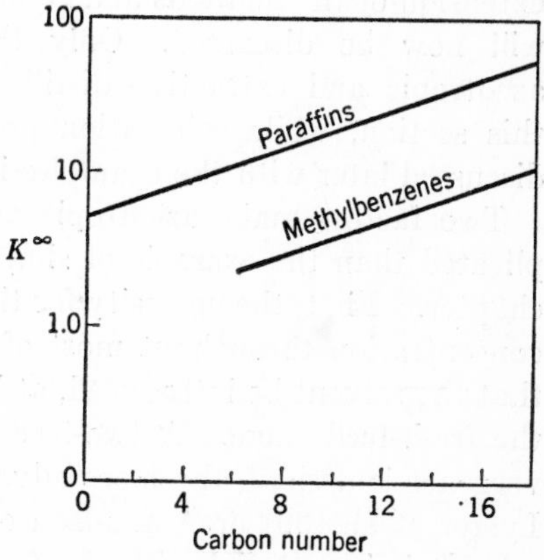

FIG. 11-5. Limiting distribution coefficients K^∞ of paraffins and aromatics between heptane and phenol at 70°C. [*Pierotti, Deal, and Derr, Ind. Eng. Chem.*, **51**, 95 (1959).]

provide a basis for choosing between solvents in so far as selectivity is concerned.

The limiting liquid-liquid distribution coefficient K^∞ for a given solute between two solvents can be obtained from the ratio of the two γ_1^∞'s. Figure 11-5 shows predicted values of K^∞ between heptane and phenol (assumed to be immiscible) for paraffins and the homologous series of methylbenzenes. The K^∞'s for each homologous series are a straight-line function of carbon number, and the straight lines representing the two homologous series are parallel. The relative volatility $\alpha_{1-2}^\infty = K_1^\infty / K_2^\infty$ of component 1 with respect to component 2, where 1 and 2 are compounds with the same carbon number but from different homologous series, is the antilog of the vertical distance between the two parallel lines. The relative volatility (or relative distribution ratio) calculated in this manner is a measure of the inherent ability of the solvent pair in question to separate the two components. In some cases, hydrocarbon systems, for instance, one of the solvents considered may be the major raffinate component in the feed. This is the case in Fig. 11-5, which illustrates the inherent ability of phenol to recover methylbenzenes from paraffins in a liquid-liquid extraction.

11-2. Basic Design Relationships

The same basic relationships (equilibrium relations, material and enthalpy balances) described in Chap. 10 for ideal distillations are used in the design of azeotropic and extractive distillation columns. The extension of the methods and equations of Chap. 10 to nonideal systems will now be discussed. Only those relationships common to both azeotropic and extractive distillation calculations will be included in this section, while calculation procedures unique to each process are discussed later with the examples in Secs. 11-8 and 11-10.

Two factors make azeotropic and extractive distillations more complicated than the example of simple distillation worked in the previous chapter. First, the necessity for the maintenance of an adequate solvent concentration throughout most of the column length will often require that the solvent be introduced, at least in part, at some point other than the fresh-feed stage. A two-feed column results, and if complete convergence is desired, the more advanced computer methods described by Lyster et al. (30) or Amundson et al. (1, 2) must be used. If all the solvent is introduced with the fresh feed, the convergence and general calculation methods described in Chap. 10 are directly applicable.

The second factor which complicates azeotropic and extractive distillation calculations is the nonideal phase behavior. The assumption of ideal liquid solutions is not valid, and activity coefficients are not unity. Also, partial enthalpies instead of pure component enthalpies must be used in the calculation of total phase enthalpies.

Equilibrium Relationships. Distribution coefficients obtained from any source which assumes ideal liquid-phase behavior at any pressure will be grossly inaccurate. Experimental determination of the activity coefficients is required, and these activity coefficients must be correlated in such a way as to permit their calculation as a function of liquid-phase composition. The equations of Wohl and of Redlich and Kister which were described in Chap. 2 provide multicomponent correlation forms suitable for computer use. In some cases ternary systems are conveniently correlated by the γ vs x plots suggested by Carlson and Colburn (6) and used by Colburn and Schoenborn (9) and Scheibel and Friedland (40). Plots of log γ_i vs x_i with parameters of $x_j/(x_j + x_k)$ are made for each of the three components. The two extreme curves on each plot are for the binaries $i - j$ and $i - k$, and all the ternary data fall within these boundaries. Colburn and Schoenborn (9) point out that this correlation scheme fails completely for many systems. Often the curves at intermediate concentrations fall outside the limiting binary curves.

The distribution coefficient $K_i = y_i/x_i$ is obtained from the activity coefficient by Eq. (1-13) or (1-14). If ideal gas solutions are assumed and the departure from ideal gas behavior can be represented adequately by the second virial coefficient, Eq. (1-14) reduces to

$$K_i = \frac{y_i}{x_i} = \frac{\gamma_i p_i^*}{p} \exp \frac{(V_i^L - B_i)(p - p_i^*)}{RT} \tag{11-6}$$

For a given liquid-phase composition and stage temperature all the quantities on the right-hand side of (11-6) are known and the K_i can be calculated. If the equilibrium relation is to be used in the form of relative volatilities,

$$\alpha_{i-r} = \frac{\gamma_i p_i^*}{\gamma_r p_r^*} \exp \left[\frac{(V_i^L - B_i)(p - p_i^*)}{RT} - \frac{(V_r^L - B_r)(p - p_r^*)}{RT} \right] \tag{11-7}$$

The exponential terms in (11-6) and (11-7) should be included in the calculation of K's and α's only if they were included in the calculation of the activity coefficients from the original experimental data. The experimental data must be quite accurate to justify inclusion of the terms at 1 atm pressure. At elevated pressures the correction terms are sizable enough to include, even though the data are of only ordinary accuracy.

The use of (11-6) or (11-7) presumes a knowledge of the stage temperature. In computer practice, the stage temperatures are fixed by the assumption of a temperature profile at the beginning of the trial. These previously assumed stage temperatures can then be used with the calculated liquid compositions to evaluate (11-6) or (11-7) and provide equilibrium data for the stage-to-stage calculations. After the convergence procedure has supplied corrected stage compositions, it is necessary to calculate a new temperature profile by means of bubble-point calculations on each stage. Methods for bubble-point calculations were described in Sec. 10-4. However, in nonideal systems it may be more convenient to rearrange (11-6) to give

$$p = \sum_{i=1}^{c} y_i p = \sum_{i=1}^{c} \gamma_i x_i p_i^* \exp \frac{(V_i^L - B_i)(p - p_i^*)}{RT} \tag{11-8}$$

and then obtain the bubble point by successive assumptions of temperature until the calculated pressure equals the specified stage pressure.

Enthalpy Balances. The enthalpies of the vapor and liquid phases leaving stage n can be calculated from

$$H_n = \sum_{i=1}^{C} y_{i,n}\bar{H}_{i,n} \tag{11-9}$$

$$h_n = \sum_{i=1}^{C} x_{i,n}\bar{h}_{i,n} \tag{11-10}$$

Negligible error is involved in the assumption that $\bar{H}_{i,n} = H_{i,n}$, and therefore the vapor-phase enthalpy can be calculated from a knowledge of the pure component enthalpies only. This is generally not true for the liquid phase, and values of $\bar{h}_{i,n}$ are required. These are easily calculated if partial molar heats of solution L_i are available as a function of composition. The partial molar heat of solution for the liquid phase is defined as

$$L_{i,n} = \bar{h}_{i,n} - h_{i,n} \tag{11-11}$$

If heat must be added to the mixture to maintain a constant temperature as a mole of i is dissolved at constant composition, then $\bar{h}_i > h_i$ and L_i is positive. If heat is evolved, $\bar{h}_i < h_i$ and L_i is negative.

The experimental determination of the L_i's over the entire range of composition is a task which rivals in magnitude the measurement of the activity coefficients, and consequently such data are not often available to the designer. Fortunately, the partial molar heats of solution can be predicted with sufficient accuracy from the activity coefficient data (31) by means of Eq. (2-61),

$$\left(\frac{d \log \gamma_i}{d(1/T)}\right)_x = \frac{\bar{h}_i - h_i}{2.3R} = \frac{L_i}{2.3R} \tag{2-61}$$

Figure 2-9 shows a typical binary plot of $\ln \gamma_1$ vs reciprocal absolute temperature with parameters of composition. The average L_1 at any composition is obtained from the slope of the curve. In most systems the change in L_i with temperature is negligible over the range in question, and a plot of $\log \gamma_i$ vs $1/T$ is essentially a straight line.

The Margules equations can be used to represent analytically the information provided by a plot such as Fig. 2-9. Colburn (10) has shown the following derivation for the Margules equations as written by Wohl. The parallel derivation for the Redlich-Kister equation is as follows: The binary form of the two-constant Redlich-Kister equation is

$$\log \gamma_1 = x_2^2[B_{12} + C_{12}(4x_1 - 1)] \tag{11-12}$$

Differentiating with respect to $1/T$ at constant composition gives

$$\left(\frac{\partial \log \gamma_1}{\partial(1/T)}\right)_x = x_2{}^2\left[\frac{\partial B_{12}}{\partial(1/T)} + (4x_1 - 1)\frac{\partial C_{12}}{\partial(1/T)}\right] \tag{11-13}$$

Since
$$\lim_{x_1 \to 0} \log \gamma_1 = B_{12} - C_{12} \tag{11-14}$$

$$\lim_{x_2 \to 0} \log \gamma_2 = B_{12} + C_{12} \tag{11-15}$$

Eq. (2-61) can be written as

$$\left[\frac{\partial(B_{12} - C_{12})}{\partial(1/T)}\right]_{x_1=0} = \frac{L_{12}{}^\infty}{2.3R} \tag{11-16}$$

or
$$\left[\frac{\partial(B_{12} + C_{12})}{\partial(1/T)}\right]_{x_2=0} = \frac{L_{21}{}^\infty}{2.3R} \tag{11-17}$$

The last two equations can be solved for $\partial C_{12}/\partial(1/T)$ and added to give

$$2\left(\frac{\partial C_{12}}{\partial(1/T)}\right)_x = \frac{L_{21}{}^\infty - L_{12}{}^\infty}{2.3R} \tag{11-18}$$

Substitution in (11-13) gives the following expression for the partial molar heat of solution as a function of the infinite dilution heats:

$$L_1 = x_2{}^2[L_{12}{}^\infty + 2x_1(L_{21}{}^\infty - L_{12}{}^\infty)] \tag{11-19}$$

The expression for L_2 is obtained by substituting 1 for 2 and vice versa. The $L_1{}^\infty$ and $L_2{}^\infty$ are obtained from the $x_1 = 0$ and $x_2 = 0$ curves in plots such as Fig. 2-9 or, in other words, from a plot of the quantities $(B_{12} - C_{12})$ and $(B_{12} + C_{12})$ vs $1/T$.

Often, the experimental data will not be extensive enough to plot log γ's vs reciprocal temperature at constant composition. In such cases the correlation of Pierotti et al. described in Sec. 11-1 can be used to predict $\gamma_i{}^\infty$ at three temperatures. The logarithms of these values can then be plotted to give $L_i{}^\infty$. An illustration of this type of calculation is included in the example on extractive distillation, Example 11-2.

Equation (11-19) can easily be extended to multicomponent mixtures. From Eqs. (11-16) to (11-18) it is seen that the binary constants for the $i - j$ binary are related to the partial molar heats at infinite dilution as follows.

$$\left(\frac{\partial B_{ij}}{\partial(1/T)}\right)_x = \frac{1}{2.3R}\frac{L_{ji}{}^\infty + L_{ij}{}^\infty}{2} \tag{11-20}$$

$$\left(\frac{\partial C_{ij}}{\partial(1/T)}\right)_x = \frac{1}{2.3R}\frac{L_{ji}{}^\infty - L_{ij}{}^\infty}{2} \tag{11-21}$$

Differentiating the two-constant multicomponent form of the Redlich-Kister equation with respect to $1/T$, substituting for $\partial \log \gamma_r/\partial(1/T)$ from (2-61), and eliminating the binary constants with Eqs. (11-20) and (11-21) provide the following expression for L_r in terms of the liquid composition and the binary heats of solution at infinite dilution:

$$L_r = \tfrac{1}{2}(1 - x_r) \sum_i x_i(L_{ir}{}^\infty + L_{ri}{}^\infty) - \tfrac{1}{2} \sum_{\substack{i \neq j \\ i \neq r}} (L_{ji}{}^\infty + L_{ij}{}^\infty)x_i x_j$$

$$+ \tfrac{1}{2} \sum_i (L_{ir}{}^\infty - L_{ri}{}^\infty)x_i[2x_r(1 - x_r + x_i) - x_i]$$

$$- \sum_{\substack{i \neq j \\ i \neq r}} (L_{ji}{}^\infty - L_{ij}{}^\infty)x_i x_j(x_i - x_j) \quad (11\text{-}22)$$

This equation should predict the partial molar heats of solution as well as the two-constant equation correlates the activity coefficient data. Once the L_i's are known, the total phase enthalpy is obtained easily from Eqs. (11-11) and (11-10).

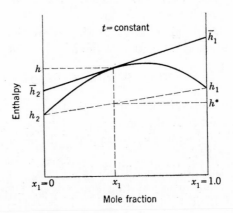

FIG. 11-6. Graphical representation of liquid enthalpies and heats of solution in a binary system at a composition of x_1.

The physical significance of the heat quantities under discussion is illustrated graphically in Fig. 11-6. The system depicted requires that enthalpy be added to the mixture to maintain the temperature at a constant value. Such behavior is typical of systems which exhibit positive deviations from Raoult's law. The total mixture enthalpy h is therefore greater than the ideal solution enthalpy h^*, where

$$h^* = x_1 h_1 + x_2 h_2$$

and $h = x_1 \bar{h}_1 + x_2 \bar{h}_2$. It follows from the general thermodynamic equations for any extensive property (12) that

$$\bar{h}_1 = h + (1 - x_1)\left(\frac{\partial h}{\partial x_1}\right)_T \qquad (11\text{-}23)$$

$$\bar{h}_2 = h - x_1\left(\frac{\partial h}{\partial x_1}\right)_T \qquad (11\text{-}24)$$

Therefore a tangent to the mixture enthalpy curve at x_1 will give as its intercepts at $x_1 = 0$ and $x_1 = 1.0$ the values of $\bar{h}_2$ and $\bar{h}_1$, respectively. The partial molar heats of solution L_1 and L_2 can be read from the figure as the differences $(\bar{h}_1 - h_1)$ and $(\bar{h}_2 - h_2)$, respectively.

The difference $(h - h^*)$ is defined as the integral heat of solution q per mole of mixture at a composition of x_1. It is the amount of heat which must be added to the mixture to maintain isothermal conditions and is related to the partial molar heats of solution by

$$q = h - h^* = x_1 L_1 + x_2 L_2 \qquad (11\text{-}25)$$

For a multicomponent mixture on stage n,

$$q = h_n - h_n^* = \sum_{i=1}^{C} x_{i,n} L_{i,n} \qquad (11\text{-}26)$$

Since
$$h_n^* = \sum_{i=1}^{C} x_{i,n} h_{i,n}$$

and
$$\bar{h}_{i,n} = L_{i,n} + h_{i,n}$$

it can be seen that (11-26) and (11-10) are identical. The use of either equation depends upon the prior prediction of the $L_{i,n}$'s.

The integral heat of solution can be expressed in terms of 1 mole of solution as done in (11-25), or it can be expressed on the basis of 1 mole of any component. Equation (11-25) can then be written as

$$q = x_1 q_1 = x_2 q_2 = h - h^* = x_1 L_1 + x_2 L_2 \qquad (11\text{-}27)$$

At dilute mixtures of 1 in 2 the relation between h and x_1 approaches a straight line. As a rough approximation it can be assumed to be straight, and under this assumption $L_2 = 0$ and (11-27) can be written as

$$x_1 q_1 = h - h^* \cong x_1 L_1 \qquad (11\text{-}28)$$

and $q_1 \cong L_1$. This approximation is sometimes useful in the estimation of the temperature rise in the liquid phase in an extractive distillation column between the solvent entry point and some lower point in the column. The reader is referred to the example problem on extractive distillation in Perry (35) for an application of this approximation.

Mertes and Colburn (31) calculated so-called integral heats of solution for butane, isobutane, and 1-butene in furfural by graphical integration of the binary partial molar heats of solution from $x_1 = 0$ to $x_1 = x_1$. The result is not the integral heat of solution as defined by (11-27) but rather the average partial molar heat of solution of component 1 as x_1 increases from zero to x_1.

$$x_1 \text{ (average } L_1) = \int_0^{x_1} L_1 \, dx_1 \tag{11-29}$$

Multiplication of this average L_1 by the number of moles of component 1 present in the final solution gives the contribution which 1 makes to the total heat effect but neglects the effect on the partial molar enthalpy of component 2.

The magnitude of the error involved in the n-butane–furfural system if the heats of solution are assumed to be zero is illustrated by an example presented by Colburn (10). The partial molar heat of solution of n-butane in a 20 per cent butane–80 per cent furfural mixture at 150°F was estimated to be 2420 Btu/lb mole of butane by means of Eq. (11-19). The molar heat of vaporization of pure n-butane at 150°F is 7820. The net heat required to vaporize 1 mole of n-butane from this mixture is therefore $7820 - 2420 = 5400$ Btu. Omission of the heat of solution would cause a 45 per cent error in the actual enthalpy required for vaporization in this case.

AZEOTROPIC DISTILLATION

An azeotrope is a liquid mixture which exhibits a maximum or a minimum boiling point. The occurrence of a maximum or minimum on the temperature vs composition surface is caused by positive or negative deviations from Raoult's law. However, deviation from Raoult's law is not in itself sufficient to cause the occurrence of an azeotrope. The boiling points of the pure components must be sufficiently close together to permit a maximum or minimum to occur. Close-boiling compounds with small deviations may produce an azeotrope, whereas other compounds which form very nonideal mixtures cannot exhibit an azeotrope because of a wide difference in the pure component boiling points. Azeotropes occur infrequently among compounds whose boiling points differ by more than 20 to 30°C.

11-3. Binary Azeotropes

Diagrams for typical homogeneous binary azeotropes are shown in Figs. 11-7 to 11-10. An azeotrope is termed homogeneous if only one liquid phase is present in the azeotropic mixture. If two liquid phases

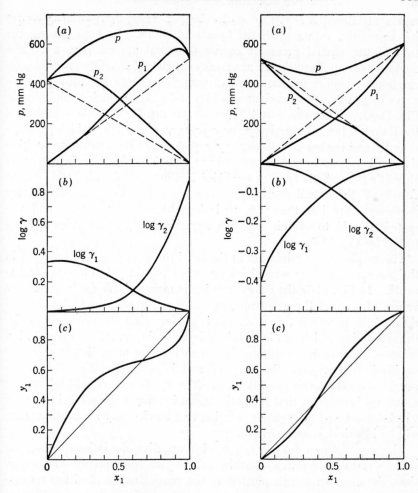

Fig. 11-7. Chloroform (1)–methanol (2) system at 50°C. Azeotrope formed by positive deviations from Raoult's law (dashed lines). (*Data of Sesonske, Dissertation, University of Delaware.*)

Fig. 11-8. Acetone (1)–chloroform (2) system at 50°C. Azeotrope formed by negative deviations from Raoult's law (dashed lines). (*Data of Sesonske, Dissertation, University of Delaware.*)

are present, the azeotrope is heterogeneous. The t-x diagram for a typical heterogeneous azeotrope is shown in Fig. 11-11.

The data in Figs. 11-7 and 11-8 were taken under isothermal conditions. The variations in the component partial pressures with composition are shown in the top diagram (a) in each case. In Fig. 11-7a the partial pressure of each component falls above the straight dashed line which represents Raoult's law, $p_i = x_i p_i^*$. In Fig. 11-8a the partial pressures fall below the values predicted by Raoult's law. In both figures, the sum of the partial pressures equals the total system pressure. The deviations from Raoult's law are great enough and the differences in pure component boiling points are small enough to cause the total pressure curve to go through a maximum in Fig. 11-7a and through a minimum in Fig. 11-8a. When the partial pressures are greater than those predicted by Raoult's law, the activity coefficients $\gamma_i = y_i p / x_i p_i^*$ will be greater than 1.0 and the log γ's will have a positive sign. Because of the positive sign on log γ, such systems are said to exhibit positive deviations from Raoult's law. This type of behavior is illustrated in Fig. 11-7. Conversely, if the actual partial pressures are less than $x_i p_i^*$, the activity coefficients are less than 1.0, their logarithms are negative, and the system is said to exhibit negative deviations. This behavior is illustrated by Fig. 11-8.

The maximum or minimum in the total pressure curves is reflected in the y-x curves as a reversal in the relative volatility of the two components. In Fig. 11-7c the chloroform is the more volatile $(y_1 > x_1)$ at low values of x_1. At the azeotrope, $y_1 = x_1$, and at concentrations to the right of the azeotrope, $y_1 < x_1$. It is obvious that the azeotrope limits the separation which can be obtained in an ordinary, refluxed distillation column. For such a column, the operating line must lie between the equilibrium curve and the diagonal and intersect the diagonal at the overhead-product concentration x_D. The maximum x_D will fall at the azeotrope because at that point the operating line and equilibrium curve will intersect and all streams entering and leaving any succeeding stage will be identical in composition.

The y-x diagram for a system with negative deviations shows the same reversal in relative volatility of the two components. However, since the diagram can be plotted vs the concentration of either component, the shape of the diagram does not differ basically from the diagram for a positive deviation system. Figure 11-8c would appear quite similar to Fig. 11-7c if plotted vs the concentration of x_2.

Azeotropes are usually classified as *minimum boiling* or *maximum boiling*. These terms are not very meaningful for isothermal data and are less basic than the terms *maximum pressure* and *minimum pressure*. An azeotrope which results from positive deviations from Raoult's law will boil at a lower temperature than either pure component and is therefore termed a minimum-boiling azeotrope. A maximum-boiling azeotrope results from negative deviations from Raoult's law. The systems in Figs. 11-7 and 11-8 are classified as minimum boiling and maximum boiling, respectively, even though the t vs x curve was held on a horizontal line throughout the composition ranges. The terms minimum boiling

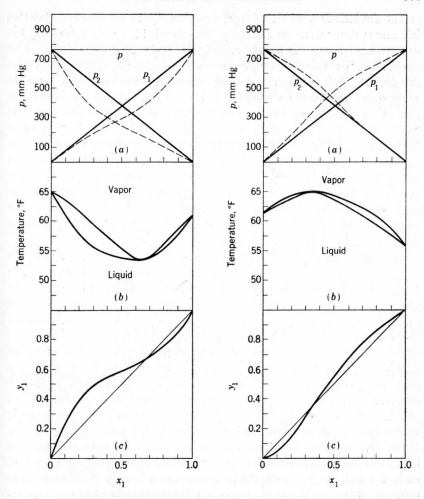

FIG. 11-9. Chloroform (1)–methanol (2) system at 757 mm Hg. Minimum boiling azeotrope formed by positive deviations from Raoult's law (dashed lines). (*Data from Perry, "Chemical Engineers' Handbook," 3d ed., McGraw-Hill Book Company, Inc., New York, 1950.*)

FIG. 11-10. Acetone (1)–chloroform (2) system at 760 mm Hg. Maximum boiling azeotrope formed by negative deviations from Raoult's law (dashed lines). (*Data from Perry, "Chemical Engineers' Handbook," 3d ed., McGraw-Hill Book Company, Inc., New York, 1950.*)

and maximum boiling are better illustrated by Figs. 11-9 and 11-10, respectively. The data in those systems are isobaric, and t vs x curves can be plotted to show the variation in boiling points with composition. The partial pressure vs x curves show the nature of the deviations, and the signs of the omitted log γ curves can be deduced from them.

The deviations from Raoult's law are so large in the positive direction in some systems that immiscibility results and this immiscibility persists up to the boiling temperature. Figure 11-11 illustrates the typical behavior of the majority of such systems (35). Most binary systems with immiscible regions at the boiling temperature form heterogeneous azeotropes; that is, the constant-boiling mixture contains two liquid phases. Such an azeotrope furnishes the same obstacle to complete

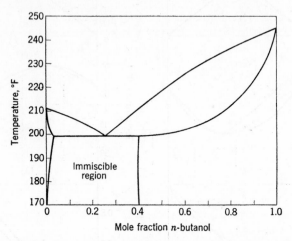

FIG. 11-11. *n*-Butanol-water system at 760 mm Hg forms a heterogeneous azeotrope. (*From Perry, "Chemical Engineers' Handbook," 3d ed., McGraw-Hill Book Company, Inc., New York, 1950.*)

separation by simple distillation as does a homogeneous azeotrope. However, the immiscibility can often be used to split the original feed into two feeds, which can then be fed to two simple distillation columns each of which produces the minimum-boiling azeotrope as the overhead product and one of the pure components as the bottom product. Such a process for the *n*-butanol–water system is described briefly in Sec. 11-7. All heterogeneous azeotropes are minimum boiling because they are the result of positive deviations from Raoult's law.

11-4. Prediction of Binary Azeotropes

The azeotrope composition will shift as the pressure (or the temperature) of the system is changed. The azeotrope will move to the right or left on the composition scale and will often disappear altogether. It is possible in some cases to separate two components which exhibit an azeotrope at 1 atm by carrying out the distillation at subatmospheric or superatmospheric pressures. Several rules and methods have been suggested for the prediction of azeotropic compositions as a function of pressure. The method used depends upon the data available, and in the following discussion the various methods will be grouped according to this criterion.

No Azeotropic Data. Horsley (24) has presented a set of charts which permit the estimation of the azeotropic composition and temperature for a very large number of systems when only the pure component boiling points are known. If vapor-pressure data are available for the two

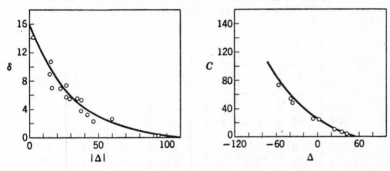

FIG. 11-12. Azeotropic compositions and boiling points for methanol-hydrocarbon systems as a function of boiling-point differences between the pure components. (*L. H. Horsley, "Azeotropic Data," Advances in Chemistry Series, American Chemical Society, Washington, D.C.*)

C = azeotropic composition in weight per cent first component.
δ = boiling point of lower boiling component minus azeotropic boiling point.
$|\Delta|$ = absolute difference in boiling points of components.
Δ = boiling point of first component minus boiling point of second component.

components, the difference in boiling points can be obtained at any pressure and therefore the Horsley charts can be used to predict the shift in azeotropic temperature and composition with pressure. Figure 11-12 shows curves for one of the 45 sets of systems covered by the Horsley charts. Table 11-4 compares the predicted with the experimental azeotropic conditions as a function of pressure for the methanol-benzene system.

Integrated forms of the Gibbs-Duhem equation are useful in the prediction of azeotrope behavior. The azeotropic composition in mixtures where the two log γ vs x_1 curves are symmetrical (or almost symmetrical) can be predicted reasonably well by the following equation (35):

$$\left[\frac{\log (p/p_1^*)}{\log (p/p_2^*)}\right]^{\frac{1}{2}} + 1 = \frac{1}{x_1} \tag{11-30}$$

This equation is derived by writing the symmetrical two-suffix Margules equations,

$$\log \gamma_1 = A(1 - x_1)^2$$
$$\log \gamma_2 = A x_1^2$$

and dividing one into the other. The x_1 in Eq. (11-30) is the mole fraction of component 1 in the azeotrope at the total pressure p. Since most systems are not symmetrical, Eq. (11-30) should be used with caution.

TABLE 11-4. EFFECT OF PRESSURE ON AZEOTROPE IN
METHANOL(1)–BENZENE(2) SYSTEM†

p, mm Hg	B.P., °C		Δ, °C	Azeotropic B.P., °C		Azeotrope wt. %	
	1	2		Calcd.	Exp.	Calcd.	Exp.
200	35	43	− 8	23	26	30	34
400	50	61	−11	39	42	33	36
760	65	80	−15	55	57	39	40
6,000	130	162	−32	125	124	54	55
11,000	153	193	−40	150	149	64	63

† Horsley, "Azeotropic Data," Advances in Chemistry Series, American Chemical Society, Washington, D.C.

Sometimes t-y-x data will be available over the entire range of composition but the azeotropic composition will not be known exactly. The azeotrope under isobaric conditions can be pinpointed by plotting both p_2^*/p_1^* and γ_1/γ_2 vs temperature. The two curves will intersect at the azeotrope where $y = x$ and $\gamma_1/\gamma_2 = p_2^*/p_1^*$. The azeotropic composition can then be read from a plot of γ_1/γ_2 vs x. For isothermal data, only the γ_1/γ_2 vs x plot is required, since p_2^*/p_1^* is a constant.

Carlson and Colburn (6) have shown how the azeotropic relation $\gamma_1/\gamma_2 = p_2^*/p_1^*$ can be used to approximate the azeotropic conditions at other pressures from the t-y-x data at one pressure. The ratio of activity coefficients γ_1/γ_2 is assumed to be independent of temperature, and the γ_1/γ_2 vs x plot at the conditions of the experimental data is assumed to be valid at the new pressure. Subject to this assumption (which involves the ratio of the heats of mixing of the two components and their respective changes with temperature) the azeotropic temperature and composition at the new pressure can be found by a trial-and-error procedure. A temperature is assumed, the ratio p_2^*/p_1^* calculated, and the corresponding composition obtained from the γ_1/γ_2 vs x plot. The assumption is verified by substitution in the expression

$$p = y_1 p_1 + y_2 p_2 = \gamma_1 x_1 p_1^* + \gamma_2 x_2 p_2^*$$

which assumes ideal gas behavior.

Azeotrope Data at One Pressure. The azeotropic conditions at one pressure are known for a large number of systems (24). Methods which permit the prediction of the azeotropic conditions at other pressures from one experimentally determined point are therefore of particular interest. Wrewski (47) offered the generalization that as the boiling point of a positive azeotrope (minimum boiling) rises, the concentration of the component with the higher molar latent heat of vaporization will increase. Conversely the composition of a negative azeotrope (maximum boiling) will shift toward the component with the lower molar heat of vaporization. Even in cases where Wrewski's rule holds, it is often desired to have more quantitative information on the effect of pressure.

Coulson and Herington (11) extended the use of the two-suffix Margules equations to this problem. They assumed the Margules constant A to be inversely proportional to temperature, $A = a/T$. The two-suffix equations can then be written as

$$\log \gamma_1 = \frac{a}{T} x_2{}^2$$

$$\log \gamma_2 = \frac{a}{T} (1 - x_2)^2$$

Subtracting one equation from the other gives at the azeotrope

$$\log \frac{\gamma_1}{\gamma_2} = \log \frac{p_2^*}{p_1^*} = \frac{a}{T} (2x_2 - 1)$$

Writing the same equation at another temperature T' and then dividing the equation at T' into the equation at T gives

$$\frac{(2x_2 - 1)_T}{(2x_2 - 1)_{T'}} = \frac{T \log (p_1^*/p_2^*)_T}{T' \log (p_1^*/p_2^*)_{T'}} \tag{11-31}$$

Equation (11-31) relates the azeotropic composition and temperature at one pressure to the azeotropic conditions at another pressure. Since few systems have symmetrical log γ vs x curves, Eq. (11-31) should be used with caution.

Gilliland (21) has made similar use of the three-suffix Margules equation to predict the effect of temperature on the azeotrope. He assumed that the constants were inversely proportional to the one-fourth power of temperature and then rewrote the Margules equations at the azeotrope as follows:

$$T^{0.25} \log \frac{p}{p_1^*} = x_2{}^2(b + cx_2) \tag{11-32}$$

$$T^{0.25} \log \frac{p}{p_2^*} = x_1{}^2(b + cx_2 + 0.5c) \tag{11-33}$$

Azeotropic data at one pressure are used to obtain the proportionality constants b and c by simultaneous solution of Eqs. (11-32) and (11-33). These constants along with the vapor pressures at the new temperature are then used in Eq. (11-34) to obtain the azeotropic composition at the new temperature.

$$T^{0.25} \log \frac{p_1^*}{p_2^*} = b(2x_1 - 1) - c(1 - 3x_1 + 1.5x_1^2) \qquad (11\text{-}34)$$

Equation (11-34) is obtained by the subtraction of (11-32) from (11-33). Either Eq. (11-32) or (11-33) provides the total pressure at the new temperature. The method is time consuming because it involves the simultaneous solution of two equations plus the solution of a quadratic.

Joffe (25) has presented an approach similar to that of Gilliland except that the Van Laar equations are used. The temperature dependence of the constants is taken into account by the following modifications of the Van Laar equations.

$$\left(\frac{T}{T_0}\right)^n \log \gamma_1 = \frac{A_{12}x_2^2}{[(A_{12}/A_{21})x_1 + x_2]^2} \qquad (11\text{-}35)$$

$$\left(\frac{T}{T_0}\right)^n \log \gamma_2 = \frac{A_{21}x_1^2}{[x_1 + (A_{21}/A_{12})x_2]^2} \qquad (11\text{-}36)$$

where T_0 is the absolute temperature at which azeotrope composition is known and T is the temperature corresponding to the new azeotropic composition x_1. Equations (11-35) and (11-36) reduce to the Van Laar equations used by White (48) when the exponent n is taken as 1.0. Subtraction of (11-35) from (11-36) gives

$$\left(\frac{T}{T_0}\right)^n \log \frac{\gamma_2}{\gamma_1} = A_{12}A_{21} \frac{A_{12}x_1^2 - A_{21}x_2^2}{(A_{12}x_1 + A_{21}x_2)^2} \qquad (11\text{-}37)$$

Simplification results if a quantity R is defined as $R = A_{12}x_1 + Bx_2$ from which is obtained

$$x_1 = \frac{R - A_{21}}{A_{12} - A_{21}} \qquad (11\text{-}38)$$

and

$$x_2 = \frac{A_{12} - R}{A_{12} - A_{21}} \qquad (11\text{-}39)$$

Substitution of (11-38) and (11-39) into (11-37) along with the azeotropic condition $\log (\gamma_2/\gamma_1) = \log (p_1^*/p_2^*)$ provides

$$R^2 = \frac{A_{12}A_{21}}{1 - [(A_{12} - A_{21})/A_{12}A_{21}](T/T_0)^n \log (p_1^*/p_2^*)} \qquad (11\text{-}40)$$

The Van Laar constants are obtained by the simultaneous solution of Eqs. (11-35) and (11-36) at the known azeotropic composition where $T = T_0$. Equations (11-40) and (11-38) then provide the azeotropic composition at any other temperature T. Some assumption must be made for the exponent n. Joffe found that $n = 0$ (no effect of temperature on Van Laar constants) gave good results in two aqueous systems (water–ethyl acetate and water–ethyl alcohol). In mixtures of organic liquids (methanol-benzene, carbon tetrachloride–ethyl acetate, and ethyl alcohol–ethyl acetate) the best results were obtained with $n = 1.0$. The total pressure of the azeotrope at temperature T is found from

$$\log \gamma_2 = \left(\frac{T_0}{T}\right)^n \frac{A_{21}A_{12}{}^2 x_1{}^2}{R^2} \tag{11-41}$$

and
$$p = \gamma_2 p_2^*$$

Azeotrope Data at Two or More Pressures. The range of pressure over which a binary azeotrope can exist is well illustrated by the method

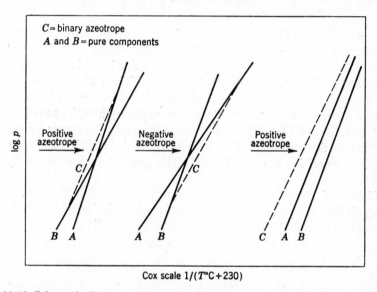

C = binary azeotrope
A and B = pure components

log p

Positive azeotrope

Negative azeotrope

Positive azeotrope

C

C

B   A   A   B   C   A   B

Cox scale $1/(T°C + 230)$

FIG. 11-13. Schematic diagram of vapor-pressure curves of binary azeotropes. (*L. H. Horsley, "Azeotropic Data," Advances in Chemistry Series, American Chemical Society, Washington, D.C.*)

of plotting described by Nutting and Horsley (24). The plot of the azeotropic temperature vs pressure (this can be called the vapor-pressure curve for the azeotrope) is essentially a straight line on a Cox chart as are the vapor-pressure curves of the pure components. Since an azeotrope by definition has either a higher or a lower vapor pressure than either of the pure components, the "vapor-pressure curve" for the azeotrope will always lie above or below the curves for the pure components. This is illustrated schematically in Fig. 11-13, where lines A and B are the vapor-pressure curves for the pure components and C is the "vapor-pressure curve" for the azeotrope. If curve C crosses either A or B, the azeotropic vapor pressure is no longer greater or less than both of the components, and the system will become nonazeotropic at the point of intersection. If no intersection occurs, the azeotrope will persist up to the critical pressure. In cases where the azeotropic temperature and pressure are known only at one pressure, a rough estimate of the effect of pressure can be obtained if the azeotrope curve is drawn with a slope equal to the average of the slopes of the two component curves.

Skolnik (44) has suggested that the intersections in Fig. 11-13 be found algebraically. The vapor-pressure lines on a Cox chart are represented by the Antoine equation

$$\log p_i^* = A_i - \frac{B_i}{t + 230} \qquad (11\text{-}42)$$

the constants of which can be determined from two vapor-pressure data points. Simultaneous solution of the Antoine equation for the azeotrope with each of the pure component Antoine equations gives the points of intersection, if any. If the azeotropic conditions at only one pressure are known, the "vapor pressure" at any other pressure can be estimated from a vapor-pressure–temperature nomograph such as the one of Lippincott and Lyman (29). The reader is referred to the paper by Skolnik for further details. Skolnik also shows that the logarithm of the azeotropic composition is essentially a straight-line function of the boiling point.

Licht and Denzler (28) have shown by a thorough thermodynamic analysis that the pressure-temperature relationship for all types of azeotropes can be correlated by a Clapeyron-type equation

$$\frac{dp}{dT} = \frac{\lambda}{T \, \Delta V} \qquad (11\text{-}43)$$

The λ is the specific latent heat of vaporization. It is the enthalpy required to form one unit weight of vapor from the appropriate amounts of the pure liquids. It therefore includes both the latent heat and the heat of mixing at the azeotropic composition. The ΔV is the specific volume increase in the formation of a unit weight of vapor from appropriate amounts of the pure liquids. The applicability of the Clapeyron-type equations justifies the consideration of the azeotrope as a "single component" with its own unique vapor-pressure curve. The pressure-temperature relationship is therefore subject to plotting in any of the various ways in which vapor-pressure data are handled (p vs T in ordinary vapor-pressure curve, $\log p$ vs $1/T$, Duhring's rule plots, Othmer-type charts, etc.). Othmer and Ten Eyck (34) have extended by analogy the Clapeyron-type equation of Licht and Denzler to show that both the azeotropic composition and the partial pressures of the components in the azeotrope are essentially straight-line functions of the total system pressures on a log-log plot. The location of all these straight lines depends upon a knowledge of the azeotropic conditions at two pressures. If the azeotrope is known at only one pressure, some knowledge of the heat effects is necessary to establish the slope.

11-5. Ternary Azeotropes

The total pressure surface and the temperature surfaces for the liquid and vapor phases in ternary systems often exhibit maxima and minima analogous to those found in binary systems. At the high and low points the vapor and liquid surfaces are tangent, and at the point of tangency the vapor and liquid phases in equilibrium are identical in composition. These ternary azeotropes offer the same obstacle to complete separation as do azeotropes in binary systems.

The contours of the pressure and temperature surfaces over the range of ternary compositions are influenced strongly but not completely determined by the characteristics of the three limiting binaries. In other words, the surfaces which connect the three binary systems are not flat planes and may exhibit valleys and ridges which are lower or higher than the maximum and minimum points on the periphery of the diagram. Quite often a ternary system which contains two or three minimum-boiling azeotropes will exhibit a depression in the temperature surfaces which reaches a point lower than any of the binary azeotropes, but the existence of the binary azeotropes does not guarantee the presence of a ternary azeotrope. Extensive experimental data within the ternary diagram are required to define the surfaces accurately. Once the surfaces are known, however, the general direction which a distillation process will take becomes obvious. Figure 11-14 shows the liquid-temperature surface for the system methyl ethyl ketone (1)–n-heptane (2)–toluene (3). The liquid-temperature surface is described by contour lines of constant temperature (full lines). The 88 and 104°C liquid-surface contour lines are connected by a series of arrows (tie lines) to their corresponding vapor-surface contour lines. The arrows connect liquid and vapor compositions which are in equilibrium and therefore represent the separation obtained in one equilibrium stage. The liquid isotherms show the high point of the diagram to be at the toluene corner while the lowest point is the MEK-heptane binary azeotrope. All points within the diagram fall between these high and low points, and no ternary azeotrope exists. In general, the path of a distillation can be described by saying that the composition of the liquid running down the column tends to go "uphill" on the temperature surface while the vapor composition moves "downhill." The direction of the tie lines in Fig. 11-14 illustrates this tendency. The actual path followed by a continuous distillation is, of course, affected by the material-balance restrictions imposed by the designer. Continuous distillation paths will be discussed further in the following sections in connection with example problems.

A ternary system may contain any combination of minimum- and maximum-boiling binary azeotropes which may or may not be accompanied

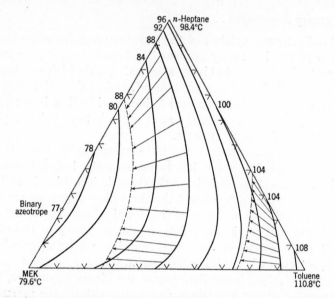

FIG. 11-14. Liquid-temperature contour lines (dark) in methyl ethyl ketone (1)–n-heptane (2)–toluene (3) system. Vapor-temperature contour lines (dashed) shown at 88 and 104°C. Arrows connect equilibrium-phase compositions. [*Data of Steinhauser and White, Ind. Eng. Chem.,* **41**, 2912 (1949).]

by a ternary azeotrope. Ewell and Welch (18) discuss several of the possible combinations and show how the contours of the system can be investigated by batch distillations. In a batch distillation, the still-pot composition must move on a straight line away from the overhead-product composition. (The preceding statement is exact only if the liquid holdup in the column is negligible.) The overhead product will always be some low point on the temperature surfaces which can be reached from the still-pot composition without passing over any ridges; that is, the temperature profile in the column cannot have any maxima or minima. (The last statement assumes that the distillation apparatus has a sufficient number of equilibrium stages to "reach" from the still-pot composition to the low point.) The foregoing requirements permit a ternary diagram to be split into triangular regions where the vertices of each region represent the overhead products which will be obtained in succession during the batch rectification of a feed whose composition falls within the region. The methanol–methylene chloride–acetone system shown in Fig. 11-15a has three such regions, ADE, DEB, and EBC. Point 1 represents a feed in region ADE. The lowest point on the temperature surfaces for this ternary lies at the methanol–methylene chloride binary azeotrope (point E), and this point can be reached from

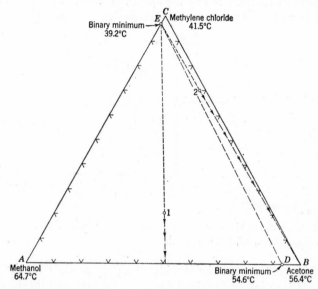

Fig. 11-15a. Binary azeotropes and batch-distillation regions in the methanol–methylene chloride–acetone system. The batch distillation curves for points 1 and 2 are shown in Fig. 11-15b.

point 1 as it can, in fact, be reached from any point in the entire system. The first overhead product will therefore be the methanol–methylene chloride azeotrope as shown on the batch-distillation curve for point 1 in Fig. 11-15b. The still-pot composition will move away from the overhead-product composition as shown by arrows until the methylene chloride is exhausted and only the methanol-acetone binary remains. When the still-pot composition reaches the base of the diagram, the lowest point which can be reached becomes the methanol-acetone binary azeotrope and this mixture becomes the second overhead product. The still-pot composition now moves toward the methanol corner, and after the acetone is exhausted, pure methanol is taken overhead. In an analogous manner, a feed located within the region DEB will furnish overhead products in the following order: (1) methanol–methylene chloride binary azeotrope, (2) methanol-acetone binary azeotrope, and (3) pure acetone.

A feed in region *EBC* will give the methanol–methylene chloride azeotrope until the methanol is exhausted, after which pure methylene chloride and pure acetone come over in order. If the feed lies on a boundary between two regions, only two products will be obtained, since two components are exhausted simultaneously. As shown by the distillation curve for point 2 in Fig. 11-15*b*, the methanol–methylene chloride azeotrope is followed only by pure acetone.

Continuous distillation is less flexible than a batch process in so far as the number of pure products is concerned. No more than two pure

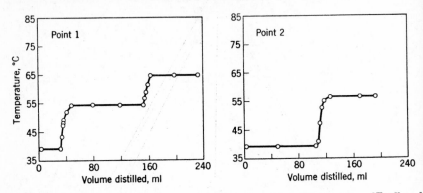

Fig. 11-15*b*. Batch-distillation curves for points 1 and 2 in Fig. 11-15*a*. [*Ewell and Welch, Ind. Eng. Chem.*, **37**, 1224 (1945).]

products (an azeotrope will be considered a pure product for the purpose of this discussion) can be obtained from one column, and these two products must lie at the ends of a straight material-balance line which passes through the total feed composition. For example, if sufficient stages were available, a feed corresponding to point 1 in Fig. 11-15*a* could be split into an overhead product corresponding to the methanol–methylene chloride azeotrope and a bottom product corresponding to the still-pot composition after the methylene chloride has been exhausted in the batch rectification. Whereas continued operation of the batch rectification separated the resulting methanol-acetone binary, the bottoms product from the continuous process must be charged to another column for separation.

It should be understood that a plot of the stage vapor and liquid compositions in either a continuous or batch distillation will not, in general, coincide with the straight material-balance lines shown in Fig. 11-15*a*. The separation made in each stage depends upon the tie line, and the direction of the tie line depends upon the contours of the vapor and liquid surfaces at the particular point. Figure 11-16 shows two of the infinite number of distillation paths which would be followed by a batch distillation of a methyl ethyl ketone–n-heptane–toluene mixture under total reflux and with negligible holdup in the column. Point 1 represents the initial still-pot composition, and path A represents the equilibrium-stage compositions at total reflux along the column before the still-pot composition has been changed by the withdrawal of any overhead product. Each arrow connects a liquid to its equilibrium vapor and therefore represents an equilibrium stage. The arrows were located from the extensive data of Steinhauser and White (46). Since the column is at total reflux,

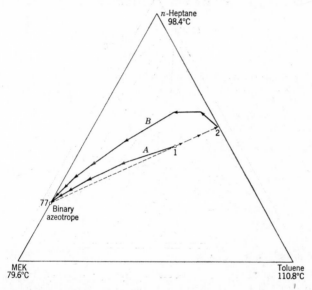

FIG. 11-16. Two total reflux distillation paths which occur in the batch distillation of the MEK-heptane-toluene mixture represented by point 1.

vapor and liquid streams which pass each other between stages are identical in composition and, therefore, the liquid end of each tie line falls upon the vapor end of the preceding tie line. Point 2 represents the still-pot composition at the point where the methyl ethyl ketone is almost exhausted, and path B represents the stage compositions which correspond to this pot composition. As long as any methyl ethyl ketone remains in the still pot, the distillation path will bend around to the binary azeotrope. When the last trace of ketone disappears from the still, the path will become coincident with the side of the diagram and end at the heptane corner.

The total reflux distillation path in a continuous column is identical with the total reflux path in a batch column if the end compositions are the same. The ends of the arrows are, however, not coincident at any reflux other than total reflux. As will be brought out later in the example problems, the path of a continuous distillation can be controlled somewhat by the proper choice of reflux ratio and feed entry points.

Separation by batch distillation is somewhat more complicated when the ternary diagram has a ridge across it. A ridge will always occur when the ternary includes one maximum-boiling binary azeotrope and two minimum-boiling ones. Such a system is illustrated in Fig. 11-17a.

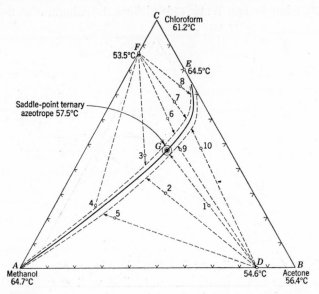

FIG. 11-17a. Binary azeotropes and ternary saddle-point azeotrope in acetone-chloroform-methanol system. Material-balance lines are shown for 10 feeds. The corresponding distillation curves are shown in Fig. 11-17b.

The peak of the ridge runs from the methanol corner to the maximum-boiling chloroform-acetone binary azeotrope (point E) as shown by the heavy curved line. The ridge sags in the middle of the diagram owing to the influence of the two minimum-boiling binary azeotropes (points D and F). At the low point in this sag, the vapor and liquid surfaces become tangent (point G) and a ternary azeotrope exists which is neither the highest nor the lowest point on the temperature surfaces. A ternary azeotrope which occurs at a low point on a ridge was termed a "saddle-point" azeotrope by Ostwald (33) in 1901. The top of the ridge is evidently quite flat. The ends of the dashed lines (with arrows) indicate the still-pot compositions when the overhead product switches from one of the minimum-boiling binary azeotropes to the ternary azeotrope. This switch seemingly occurs before the still-pot composition reaches the peak of the ridge, indicating that the top of the ridge is quite flat.

The batch-distillation curves for the various feeds (points 1 to 10) in g. 11-17a are shown in Fig. 11.17b. For feed 1, the overhead product is e methanol-acetone azeotrope until the still-pot composition has mbed the ridge and reached the constant-boiling mixture at point G, iich becomes the second and last product obtained. For the rest of the eds on the acetone side of the ridge (points 2, 5, 9, and 10), the initial

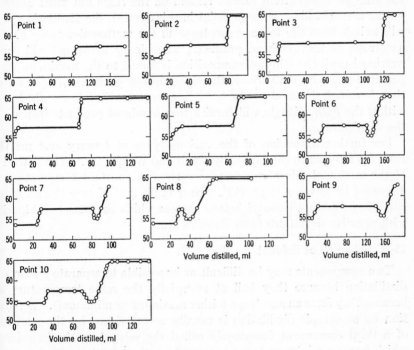

FIG. 11-17b. Batch-distillation curves for points 1 to 10 in Fig. 11-17a. [*Ewell and Welch, Ind. Eng. Chem.*, **37**, 1224 (1945).]

verhead product is always the methanol-acetone binary, but as the still-ot composition climbs the ridge, the distillation paths swing out farther nd farther from the straight material-balance line toward point G until iey finally break away from point D and run down to G. The same eneral comments apply to the first two overhead products obtained rom points on the chloroform side of the ridge with the exception that ie methanol-chloroform azeotrope is always the first product instead of ie methanol-acetone one.

The third overhead product obtained in the distillations shown in Fig. 1-17b depends upon where the still-pot composition intersects the ridge. or points 2 to 5 the intersection occurs on the methanol side of the iddle-point azeotrope, and since the ridge is fairly straight on that side, the still-pot composition moves along the ridge away from the sad point until the methanol corner is reached. Methanol then becomes t third and last product for feeds 2 to 5. The situation is different points 6 to 10 where the ridge is intersected between the saddle po and the chloroform-acetone binary azeotrope. The ridge is not straig in this region. As the second product (ternary azeotrope) is withdraw the still-pot composition cannot remain on the ridge but must desce on the acetone side. At some point in this descent, the distillation pa will switch from the ternary azeotrope to the methanol-acetone bina azeotrope as the overhead product. The binary azeotrope will co overhead until the still-pot composition is forced to the maximum te perature point on the surface, the chloroform-acetone azeotrope, whi then becomes the final product. It can be seen that feeds which f within the *FGE* triangle will furnish four overhead products instead the usual three.

For further discussion of the various types of ternary systems t reader is referred to the article by Ewell and Welch (18). The behavi of the methanol-chloroform-acetone system at constant temperature discussed by Severns et al. (43). Karr et al. (27) discuss the aceton chloroform–methyl isobutyl ketone system which contains two binari with negative deviations from Raoult's law.

11-6. Selection of Solvent

Two components may be difficult or impossible to separate by simp distillation because they boil at essentially the same temperature because they form an azeotrope (either maximum or minimum). Separ tion by azeotropic distillation is usually accomplished by the additi of a third component (commonly called the solvent or the entraine which forms a minimum-boiling azeotrope with at least one of the clos boiling pair. One of the azeotropes formed (if more than one is forme by the solvent must boil considerably below (or above) either of the clos boiling pair or the azeotrope which they form. As an example, consid the separation of cyclohexane (80.0°C) and benzene (80.2°C). The two components form a minimum-boiling azeotrope (77.4°C) at 50.2 mo per cent benzene (39). Perry (35) lists a large number of compoun which form azeotropes with one or both of these components. How ever, the number of suitable solvents is small, since one of the azeotrop formed with the solvent must boil lower (or higher) than the benzen cyclohexane azeotrope (77.4°C), and if the separation is to be an eas one, it should boil considerably lower. Of all the possible solvents list by Perry only methanol and acetone form azeotropes which boil mo than 10°C below 77.4°C. The pure component and azeotrope boilin points are shown in Figs. 11-18 and 11-19. For both solvents, the so

nt-cyclohexane binary azeotrope is the lowest point on the temperature
s composition surfaces (no ternary azeotropes). Without any specific
nowledge of the surface contours within the diagram, it seems safe to
ssume that no ridges occur across any part of the acetone diagram in
ig. 11-18; that is, the surfaces can be assumed to slope toward the ace-
ne-cyclohexane azeotrope from all parts of the diagram. It should,
erefore, be possible to withdraw pure benzene as the bottom product

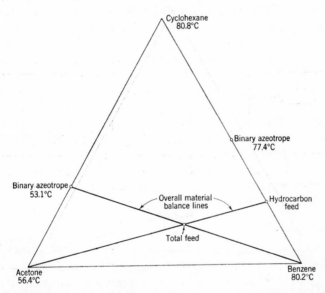

FIG. 11-18. Pure component and azeotrope boiling points for acetone-cyclohexane-
enzene system at 1 atm. Material-balance lines show the solvent treat required
produce pure benzene as the bottoms product.

hile the acetone-cyclohexane azeotrope is taken overhead. The mate-
ial-balance lines for such an operation are shown in Fig. 11-18. Enough
cetone is added to the cyclohexane-benzene feed to give a total feed
hich lies on a straight material-balance line between pure benzene and
he acetone-cyclohexane azeotrope. The total reflux distillation path
ould probably lie below the over-all material-balance line because the
lope toward the acetone corner is greater than toward the cyclohexane-
enzene azeotrope. Enough stages must be provided, of course, to
each from the benzene corner to the acetone-cyclohexane azeotrope if
oth are to be withdrawn as products. If too few stages are provided,
he operator through control of the column temperatures can withdraw
ne or the other as a product but not both. If the temperatures are
aised to permit withdrawal of pure benzene as the bottom product, the

overhead product which lies at the other end of the material-balance li
and distillation path will fall somewhere in the middle of the diagram.
It seems probable from Fig. 11-19 that methanol cannot provide t)
complete separation between benzene and cyclohexane envisioned wi
acetone. Methanol forms a minimum-boiling azeotrope with bo
cyclohexane and benzene, and the presence of these two low spots ma
well give rise to a ridge across the diagram from the methanol corn

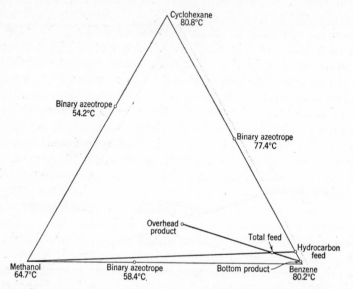

FIG. 11-19. Pure component and azeotrope boiling points for methanol-cyclohexan
benzene system at 1 atm. The material-balance lines represent the plant operatio
reported by Ratliff and Strogel. [*Petrol. Refiner,* **33**(5), 151 (1954).]

toward the opposite side. This ridge would obstruct a distillation pat
from the methanol-benzene azeotrope to the methanol-cyclohexane azec
trope, and it may extend far enough into the diagram to prevent a pat
between benzene and the methanol-cyclohexane azeotrope. Any distilla
tion path which originates from the pure benzene corner would probabl
swing down toward the methanol-benzene azeotrope and then woul
find its way to the lower methanol-cyclohexane azeotrope blocked b
the ridge. Material-balance restrictions eliminate the methanol-benzen
azeotrope as the overhead product, and the distillation path would en
somewhere inside the diagram. The correctness of these supposition
is supported by Ratliff and Strogel (38), who describe briefly the purifica
tion of benzene by azeotropic distillation with methanol. A hydrocarbo
feed containing 95 per cent benzene and 5 per cent close-boiling paraffins
naphthenes, and olefins was mixed with methanol to give a total feed o

.1 per cent methanol, 86.3 per cent benzene, and 4.6 per cent impurities. The bottoms product from the azeotropic distillation tower contained 99 per cent benzene and essentially no methanol. The overhead product was not a methanol-impurities azeotrope. Instead, the distillation ended at an overhead composition of 36 per cent methanol, 48.6 per cent benzene, and 15.4 per cent impurities. If the impurities may be represented by cyclohexane, the material-balance lines for this operation can be drawn on Fig. 11-19 as shown. Since the bottom product falls on the benzene-cyclohexane side, the first tie line above the bottom product must coincide with that side of the diagram and therefore the distillation path must lie above the material-balance line for at least part of the way to the overhead composition. Ratliff and Strogel do not tell how many stages the column contained, so it is not known whether too few stages or the presumed ridge is responsible for the failure of the column to produce a methanol-impurity mixture as the overhead product.

The principles illustrated in the previous discussion are not altered if the pair to be separated form a maximum-boiling azeotrope instead of a minimum. The objective in the choice of the solvent is still the formation of a constant-boiling mixture on one of the opposite sides which will differ appreciably in boiling temperature from the original azeotrope. The azeotrope (or azeotropes) formed by the solvent may be either maximum boiling or minimum boiling. The magnitude rather than the direction of the difference in boiling point between the solvent azeotrope and the original one is the important factor. It should be noted that the nature of the azeotropes formed by the solvent determines which of the two original components will be recovered as a bottoms product. In Fig. 11-18, the minimum-boiling acetone-cyclohexane azeotrope permits benzene to be withdrawn as the bottoms product. If a solvent which formed a lower boiling azeotrope with benzene than with cyclohexane were used, cyclohexane would be the bottoms product. If the original pair has an appreciable difference in boiling points, a solvent is more apt to be successful if it increases the volatility difference rather than reverse it as is done in Fig. 11-18.

One of the original pair need not be removed as a pure product for the separation to be successful. The solvent may form minimum azeotropes with both components, and these two azeotropes may differ appreciably in boiling point. If so, the two solvent azeotropes might be removed as the overhead and bottoms products. For example, the methanol and benzene azeotrope in Fig. 11-19 could be separated completely by the addition of cyclohexane to give a total feed which falls on a straight line between the methanol-cyclohexane and benzene-cyclohexane azeotropes, which could then be withdrawn as overhead and bottom products, respectively. The same sort of separation between cyclohexane and

benzene could be accomplished with the two methanol azeotropes exce
for the probable existence of a ridge between the two azeotropes.

In many cases of practical interest the lowest point on the temperatu
surface does not coincide with any of the binary azeotropes on the sid
of the composition triangle. In these systems, the valleys formed l
the minimum-boiling binary azeotropes deepen as they run into t
diagram and their point of intersection defines a ternary azeotrope
the lowest point in the triangle. The distillation path will run down
this point rather than to one of the higher binary azeotropes, and t
overhead product will have the composition of the low point if enou
stages are available to reach it. Systems with one binary azeotro
seldom form a ternary azeotrope. On the other hand, three minimun
boiling binary azeotropes do not guarantee that a ternary azeotro
will exist.

The presence of a ternary azeotrope does not make separation of tl
original two components impossible. If the ratio of the two in tl
ternary azeotrope is sufficiently different from their ratio in the fresh fee
a practical separation can be achieved. In fact, this sort of operatic
is quite common. The best known example is probably the separatic
of the water–ethyl alcohol azeotrope with benzene or trichloroethylene
the solvent. Both solvents form ternary azeotropes which are remove
as overhead products. A simplified diagram of the benzene process
shown in Fig. 11-24. The process is somewhat more complicated tha
when a simple binary azeotrope is taken overhead.

It can be seen that the number of possible solvents for any azeotrop
process is severely limited. Only those compounds which boil within
limited range (0 to 40°C) of the original pair and which exhibit sufficient
large positive or negative deviations to form an azeotrope with one
them have any chance of being successful solvents. Added to the
volatility restrictions are such requirements as low cost, stability, safet
and ease of recovery from the product stream (or streams) in which th
solvent appears.

In the above discussion, the azeotropic distillation systems were alwa
assumed to be ternaries. In actual practice the feeds will often contai
more than two components. For example, benzene will be separate
from a mixture of paraffins, naphthenes, and olefins rather than fro
pure cyclohexane. However, a study of the benzene-cyclohexane syste
will provide most of the answers for the more complicated system.

11-7. Azeotropic Processes

The distillation column (or columns) in which the azeotropic separatio
occurs often does not represent the major part of the capital expenditur
necessary to construct the entire process. If a solvent is used, facilitie

must be provided for the recovery of the solvent from the overhead- and bottom-product streams. Often these facilities are quite extensive. Also, feed-preparation facilities are often required. The feed to be separated must be either a pair of pure components or a pair of component classes which have been cut to a very narrow boiling range by previous distillations if the separation is to be made in one azeotropic distillation. As an illustration of this last requirement consider the benzene-purification process reported by Ratliff and Strogel (38) which was discussed in the previous section. Obviously pure benzene could not be removed as a bottom product if the original feed had not been cut previously to eliminate toluene and heavier aromatics. Also, if the impurities had not been cut to a very narrow boiling range around the boiling point of benzene, the feed would have contained impurities whose large boiling-point differences from methanol would have prevented the formation of an azeotrope with the solvent.

To illustrate the total facilities which may be involved in an azeotropic distillation process, several processes will be discussed more or less in the order of increasing complexity.

Heterogeneous Azeotropes. Separation of heterogeneous azeotropes can be accomplished without the addition of a solvent. A typical heterogeneous azeotrope is the normal butanol-water system shown in Fig. 11-11. The process by which the butanol and water are separated is depicted in Fig. 11-20. It can be seen from the phase diagram that a vapor in the region of the azeotrope will, when condensed, form two liquid phases, one which is rich in water and one which is rich in butanol. These two phases can be separated in a decanter to furnish two feeds. The butanol-rich feed is sent to the butanol column where, since water is the more volatile component in this concentration range, high-purity n-butanol is removed as the bottoms product while a vapor which approaches the azeotropic composition is taken overhead. The water-rich phase is fed to the water column. At high water concentrations, butanol is the more volatile component and therefore water is the bottom product. A vapor which approaches the azeotrope is taken overhead, where it is mixed with the vapor from the butanol column, condensed, mixed with the fresh feed, and sent to the decanter for phase separation. Live steam can be used for heat input to the water column, since the bottom product from that column is water. The introduction of fresh feed into the decanter is advisable when the feed contains concentrations of butanol lower than that obtained in the butanol-rich phase from the decanter. Otherwise the fresh feed should be introduced directly to the butanol column. The process is described more thoroughly by Perry (35). The calculation of the required stages in each column can be made with binary diagrams and the methods described in Chap. 5. Extremely high purity butanol and water are produced as products, and the log-log extension of the binary diagram as described in Sec. 5-14 must be used.

One Azeotrope Product. The acetone-cyclohexane-benzene system used as an example in the previous section comes under this heading. The acetone-cyclohexane binary azeotrope would be taken overhead. The acetone could then be recovered from the cyclohexane by water

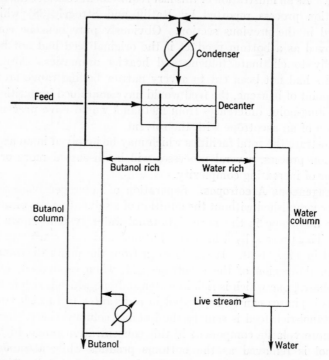

Fig. 11-20. Separation of a heterogeneous azeotrope by means of a decanter and two distillation columns. No solvent is involved. (*Perry, "Chemical Engineers' Handbook," 3d ed., McGraw-Hill Book Company, Inc., New York, 1950.*)

washing followed by separation of the acetone-water mixture by simple distillation. A simplified flow diagram for this type of process is shown in Fig. 11-21. The process described by Ratliff and Strogel (38) for the purification of benzene with methanol corresponded roughly to Fig. 11-21. Pure benzene was withdrawn as the bottom product, while the methanol was recovered from the overhead mixture of methanol, benzene and impurities by water washing. The methanol and water were then separated by simple distillation and recycled.

Two Binary Azeotropes. The over-all process may become more complicated if the system contains two binary azeotropes. Benedic and Rubin (4) describe two such processes, one where the solvent is miscible with the overhead at all temperatures and one where the solvent becomes only partially miscible when the overhead is condensed. Figure 11-22 illustrates the separation of toluene from a close-boiling paraffin with methanol as the solvent. After condensation of the overhead product (methanol-paraffin azeotrope) the methanol is still miscible with

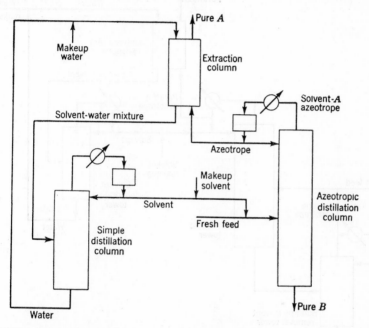

FIG. 11-21. Schematic diagram for separation of A and B with a solvent that forms a minimum boiling azeotrope with A. The solvent is miscible with both A and water while A is immiscible with water.

the paraffin and can be recovered by extraction with water followed by a simple distillation to separate the water and methanol. The recovered solvent is returned to the azeotropic tower via the reflux. To ensure a very low concentration of paraffin in the bottom product a slight excess of solvent is used to ensure that the bottom-product point will fall on the solvent side of the toluene corner. The methanol is then separated from the toluene in a binary distillation tower which takes overhead the methanol-toluene azeotrope. The small amount of this material taken overhead is added to the feed stream before entry to the big azeotropic tower. Figure 11-23 illustrates the major facilities required when the solvent becomes partially immiscible with the paraffin after condensation of the overhead from the big azeotropic tower. Nitromethane exhibits such behavior when used as the solvent in the toluene-paraffin system. The overhead product is split into two liquid phases after condensation. Part of the upper paraffin-rich layer is returned as reflux to the azeotropic tower, while the remainder is separated from the solvent by simple distillation to provide an essentially pure paraffin overhead

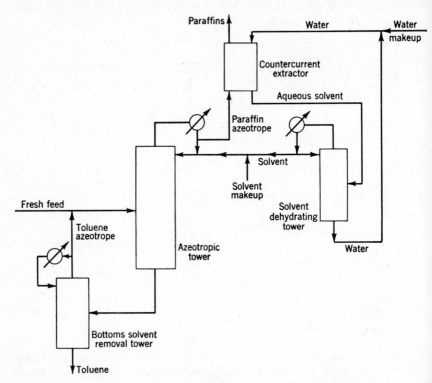

Fig. 11-22. Schematic flow diagram for separation of toluene-paraffin mixture with methanol. The methanol forms miscible azeotropes with both components. [*Benedict and Rubin, Trans. A.I.Ch.E.*, **41**, 353 (1945).]

product. The solvent-rich layer from the separator and the solvent recovered from the paraffin are returned to the azeotropic tower with the reflux. The same situation exists with the bottom toluene product as was described above for the methanol solvent.

Ternary Azeotrope Overhead. The addition of the solvent to the close-boiling pair may cause the formation of a ternary azeotrope which can then be withdrawn as a product from the azeotropic column. An example is the dehydration of ethanol-water mixtures with benzene (35). Benzene forms a minimum-boiling ternary azeotrope with ethanol and water which boils at a lower temperature than the ethanol-water binary azeotrope and also contains a higher ratio of water to ethanol. Figure 11-24 shows a schematic flow diagram for a process which produces pure alcohol from the ethanol-water azeotrope. The benzene can be added to the feed stream but is returned to the primary column via the reflux in Fig. 11-24. The ternary azeotrope is taken overhead, while pure

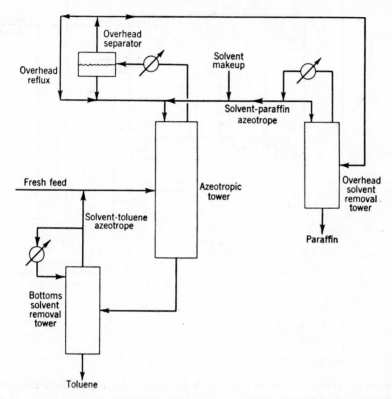

FIG. 11-23. Schematic flow diagram for separation of toluene-paraffin mixture with nitromethane. The nitromethane-paraffin azeotrope is immiscible after condensation. [*Benedict and Rubin, Trans. A.I.Ch.E.*, **41**, 353 (1945).]

alcohol is withdrawn from the bottom. The overhead vapor forms two layers upon condensation. The upper benzene-rich layer is returned to the primary column via the reflux. The lower water-rich layer is fed to a secondary column which also produces the ternary azeotrope as the overhead product. The bottom product from the secondary column is a mixture of alcohol and water which is split in a third tower into a bottoms product of pure water and an overhead product which is the ethanol-water binary azeotrope. This overhead stream is recycled to the feed to the primary tower. Trichloroethylene is another solvent which will separate ethanol and water in essentially the same process as shown in Fig. 11-24. Coates (8) also discusses both processes and presents a numerical example for the estimation of the stages required in the primary column when trichloroethylene is used as the solvent.

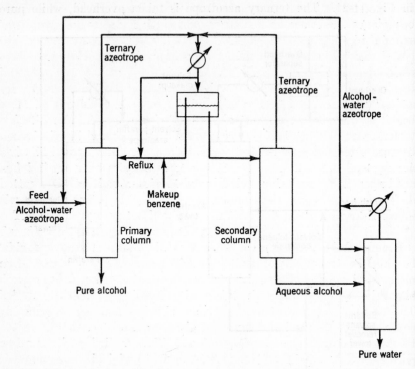

FIG. 11-24. Schematic flow diagram for separation of the alcohol-water azeotrope with benzene. Benzene forms a ternary azeotrope which boils lower than the binary azeotrope. (*Perry, "Chemical Engineers' Handbook," 3d ed., McGraw-Hill Book Company, Inc., New York,* 1950.)

11-8. Example Calculation

The short-cut methods in Chaps. 8 and 9 are less reliable for azeotropic separations than for more ideal systems. Nevertheless, the general short-cut method of Chap. 8 can be used profitably for orientation studies if the flow diagram is one for which the short-cut equation is applicable. If all the feed is inserted at one stage, or if the solvent is introduced with the reflux while the feed enters some intermediate stage, the equation is applicable and its basic assumptions will be strained no more for an azeotropic distillation than in the extraction and washing xamples worked in Sec. 8-5. Care must be exercised in the selection f average K values and average flow rates, since both may change ppreciably through the column.

Rigorous computer calculations are made as described for ideal distilla-ion systems in Chap. 10 with the appropriate extensions for nonideal •ehavior as described in Sec. 11-2. Such methods are too time con-uming for manual calculations. Approximate answers which are)etter than those from the short-cut equation and which require only a easonable time for calculation can be obtained with a simplified stage-o-stage calculation in which enthalpy balances are omitted but the ssumption of constant volatilities is not made. The example problem)elow is worked by this method. The separation of n-heptane and oluene with methyl ethyl ketone as the entrainer is used as an example)ecause extensive equilibrium data are available (46). This system is iot the most practical example of azeotropic distillation because heptane nd toluene differ enough in volatility to permit their separation by simple listillation. However, the example does illustrate the principles involved n azeotropic calculations.

Example 11-1. A 45 per cent toluene–55 per cent n-heptane feed is to)e split into essentially pure toluene and heptane. It is known that nethyl ethyl ketone (MEK) forms a minimum-boiling azeotrope with ieptane. No other binary or ternary azeotropes are formed with MEK. The azeotrope and pure component boiling points along with part of the ?quilibrium isotherms and tie lines for this system are shown in Fig. 11-14. Perform orientation studies to determine reasonable values for product purities, reflux rate, feed locations and compositions, and total number of stages.

Solution. Obviously a large number of cases must be worked to optimize the various parameters. Complete optimization is impractical without the services of a computer. However, intelligent application of the procedure described below will establish a good design basis with a reasonable amount of manual calculation.

It should be noted that the approach followed in this example is the opposite to that used in Chaps. 8, 9, and 10. In the example problems in those chapters the number of stages, feed locations, diameter, etc., were assumed to be fixed a priori and the corresponding separation was calculated. In this example and the one in the next section, a desired separation is specified and the required column parameters calculated.

Solvent-introduction Point. The first decision to be made concerns the solvent-introduction point. This subject is discussed by Benedict and Rubin (4), Colburn (10), Hachmuth (22), and others. In most cases, it seems best to introduce the solvent near the top of the column. However, azeotropic systems can vary so much that generalizations are dangerous. The optimum introduction point will depend upon the volatility of the solvent with respect to the nonsolvent components and possibly upon whether the solvent forms azeotropes with one or both keys and whether or not the bottom product is an azeotrope. A solvent which forms an azeotrope with only the overhead key and which has a considerably lower boiling point than either key should probably be introduced at least in part below the fresh feed if an appreciable solvent concentration is to be maintained below the feed stage. Enough stages must be provided, of course, below the solvent-introduction point to strip the solvent from the bottom product. If the solvent is relatively nonvolatile with respect to the keys, it can be introduced at the top and still be present in reasonable concentrations at the fresh-feed stage and below. If a solvent azeotrope is removed at both the top and bottom, the optimum solvent-introduction point may fall anywhere along the column.

The solution presented here assumes that half of the solvent is introduced with the fresh feed and the other half at a point below the feed. This case seems reasonable because the MEK boils 18.8°C below heptane and 31.2°C below toluene and therefore the solvent concentration would decrease rapidly below the fresh-feed stage if no solvent were introduced below. Other solvent-introduction points should, of course, be investigated before the final design is made.

Specifications. The column will be assumed to have a total condenser and a partial reboiler. With two feeds, the degrees of freedom can be shown by the principles of Chap. 3 to be $2C + 2N + 12$. The specifications tabulated below will be made to define one unique set of column conditions.

Specifications	$N_i{}^u$
Pressure in N stages	N
Pressure in condenser	1
Pressure and heat leak in reflux divider	2
Heat leak in $N - 1$ stages	$N - 1$
Reflux temperature	1
F_1 and F_2	$2C + 4$
Feed locations	2
One overhead purity	1
One bottom purity	1
One reflux ratio L/V	1
	$\overline{2C + 2N + 12}$

Atmospheric pressure in the condenser and negligible pressure drop through the column will be assumed. Heat leaks at all points will be neglected. The reflux temperature will be the bubble point or saturation temperature. The feeds are assumed to be saturated liquids. These specifications fix the temperature and pressure of each feed for any given compositions. The specifications of the feed compositions and rates are made below under Material Balances. The feed locations will be the optimum ones corresponding to the other specifications. The remaining specifications are made below.

Material Balances. Toluene concentrations of 0.005 and 0.99 in the overhead and bottoms will be specified. A solvent/feed ratio which will

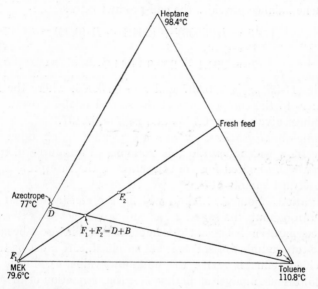

Fig. 11-25. Material-balance lines for a solvent/fresh-feed ratio of 1.941. A line from the toluene corner through $F_1 + F_2$ locates D.

give an overhead composition slightly below the MEK-heptane azeotrope will be used. As will be discussed later, the overhead composition must lie on that side of the azeotrope from which the distillation path approaches the azeotrope. The slope of the temperature surfaces from the toluene corner is greater in the direction of the MEK corner, and consequently the distillation path from the toluene corner will swing down toward the MEK corner and then approach the azeotrope from below. Therefore D must be below the azeotrope. The material-balance line construction for a solvent/feed ratio of 1.941 is shown in Fig. 11-25. This ratio plus the previous decision to split the solvent evenly between the fresh feed stage and the lower solvent feed stage provides the following feed rates and compositions per 100 moles of fresh feed. F_1

refers to the lower (pure MEK) and F_2 to the upper feed, which is a mixture of MEK and the fresh feed.

Component	$F_1 x_{F_1}$	x_{F_1}	$F_2 x_{F_2}$	x_{F_2}
MEK	97.05	1.00	97.05	0.4925
n-C$_7$	0	0	55.00	0.2791
Toluene	0	0	45.00	0.2284
	97.05	1.00	197.05	1.0000

A toluene balance provides the end-product rates.

$$45 = D(0.005) + (294.1 - D)(0.99)$$
$$D = 249.9$$
$$B = 294.1 - 249.9 = 44.2$$

From Fig. 11-25, $x_{M,B} \cong 0.007$ and $x_{H,B} \cong 0.003$, where the first subscript refers to the component and the second to the stream. Component balances give $x_{M,D} = 0.7755$ and $x_{H,D} = 0.2197$.

Operating-line Equations. Phase-rate profiles based upon constant molar overflow will be used in this example. Procedures whereby heat effects can be estimated for the correction of the rate profiles are illustrated later in Example 11-2.

Specification of one L/V ratio will fix the constant molar overflow profiles throughout the column for any given set of external rates. Since the separation between two components has in effect been specified while the total number of stages is not specified, care must be taken that the specified reflux is above the minimum value. Colburn (10) and Perry (35) discuss analytical methods for the estimation of the minimum reflux. However, such methods are of doubtful value in systems which deviate widely from their inherent assumptions. Minimum reflux ratios in various parts of the column can be checked accurately by the method illustrated in Fig. 11-26. A tie line (dashed) is located somewhere along the probable path of the distillation, and the length of the tie line compared with that segment of the material-balance line which represents the difference point. For example, in the bottom section of the column, B is the difference point and $L_{n+1} - V_n = B$. Therefore L_{n+1} can be located on the straight line which connects the vapor end of the tie line for stage n to the difference point B. The internal reflux ratio L/V is given by the ratio of the line segment $\overline{BV_n}$ to the segment $\overline{BL_{n+1}}$. If the ratio is too small, L_{n+1} will lie closer to B than L_n and the tie line for stage $n + 1$ will make less progress toward D than did the tie line for stage n. Eventually the tie lines and material-balance lines will become coincident and a zone of constant composition is produced. The L'_{n+1} represents a reflux below the minimum, while the L_{n+1} represents a workable reflux.

The same check should be made with a tie line close to D. The reflux ratio represented on the diagram is obviously below the minimum, and the calculated stage compositions will swing upward around D until the tie lines and material-balance lines become coincident and no further progress is made. It is not so important to check a tie line in the middle of the column. If a pinch begins to develop there, it can be eliminated by the introduction of a feed.

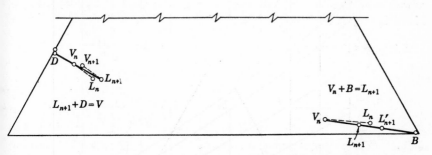

FIG. 11-26. Estimation of minimum reflux ratio in top and bottom sections of column. Dashed lines represent tie lines while heavy lines are material-balance lines.

An internal reflux L/V of 0.6 was found to be above the minimum reflux. The corresponding constant molar overflow rates throughout the column are as follows: The single primes denote the middle section while the double primes refer to the bottom section of the column.

$$D = 249.9$$
$$V = V' = V'' = 624.7$$
$$L = 374.8$$
$$L' = L + F_2 = 374.8 + 197.0 = 571.8$$
$$L'' = L' + F_1 = 571.8 + 97.0 = 668.8$$

The operating-line equations for the various sections are

$$\text{Top: } x_{i,n+1} = 1.6667 y_{i,n} - 0.6667 x_{i,D}$$
$$\text{Middle: } x_{i,n+1} = 1.0925 y_{i,n} + 0.3446 x_{i,F_2} - 0.437 x_{i,D}$$
$$\text{Bottom: } x_{i,n+1} = 0.9341 y_{i,n} + 0.06609 x_{i,B}$$

A summary of the equation constants is given in the following tabulation:

Component	x_D	x_B	Bottom, $0.06609x_B$	Middle, $0.3446x_{F_2} - 0.437x_D$	Top, $0.6667x_D$
MEK	0.7754	0.007	0.0005	−0.1691	0.5170
n-C_7	0.2196	0.003	0.0002	0.0002	0.1464
Toluene	0.0050	0.990	0.0654	0.0765	0.0033
	1.0000	1.000			

Equilibrium Data. Extensive data for the MEK-heptane-toluene system have been published by Steinhauser and White (46). The system was correlated adequately with the two-constant Redlich-Kister equations, and the authors listed values for the binary constants. However, for manual calculations, evaluation of the equations to predict γ's and

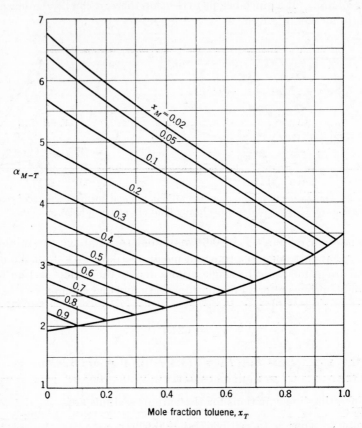

FIG. 11-27. Relative volatility of MEK vs x_T with parameters of x_M.

the subsequent bubble-point calculations required to obtain K's or α's from the γ's are too time consuming. It is more convenient to plot α's directly vs two concentrations as shown in Figs. 11-27 and 11-28. Data at low-toluene and high-MEK concentrations are not so complete as desirable, but utilization of the binary data and the fact that α_M must equal α_H at the azeotrope (zero toluene) provide checks on the location of the curves in these areas.

Enthalpy Data. The effect of temperature on the activity coefficients at constant composition was not measured. Therefore values of L^∞ cannot be obtained from activity data. However, it would be possible to predict infinite dilution activity coefficients as a function of temperature by the correlation of Pierotti et al. (36) described in Sec. 11-1 and from these γ^∞'s obtain L^∞'s. An illustration of this calculation is included in Example 11-2 for extractive distillation. No enthalpy balances will be made in this example.

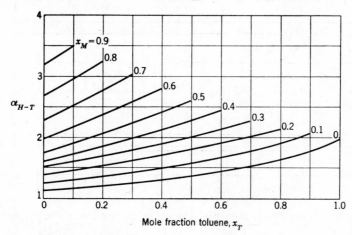

FIG. 11-28. Relative volatility of heptane vs x_T with parameters of x_M.

Stage-to-Stage Calculations. The stage-to-stage calculations are performed by the alternate use of the equilibrium relationship as expressed by

$$y_{i,n} = \frac{(\alpha_{i-r}x_i)_n}{\Sigma(\alpha_{i-r}x_i)_n} \qquad (10\text{-}5)$$

and the appropriate operating-line equation as written above. The calculations are started at the bottom stage and proceed upward until a vapor stream which approximates the specified overhead composition is produced. The feeds are inserted at their optimum locations. The optimum location is found by a trial-and-error procedure in which the composition of V_{n+1} obtained when the feed is not inserted on stage n is compared with the V_{n+1} obtained when the feed is inserted on stage n. The lowest stage at which the feed can be inserted and produce a V_{n+1} closer to D than would otherwise be obtained is the optimum feed stage. This calculation utilizes the ultimate criterion for feed-plate location; namely, the optimum location is the one which minimizes the number of stages.

The stage-to-stage calculations are shown in Table 11-5,† and the stage

† The reader should not be concerned about the number of digits carried in these calculations. It is usually customary to carry some "insignificant" figures in repetitive calculations of this type and then round off the final answers.

TABLE 11-5. STAGE-TO-STAGE CALCULATIONS FOR EXAMPLE 11-1

Component	x_1	α_1	$\alpha_1 x_1$	y_1	$0.9341y_1$	x_2
MEK	0.007	3.48	0.0244	0.0239	0.0223	0.0228
n-C_7	0.003	1.98	0.0059	0.0058	0.00542	0.00562
Toluene	0.990	1.00	0.9900	0.9703	0.9064	0.9718
	1.000		1.0203	1.0000		1.0000

α_2	$\alpha_2 x_2$	y_2	$0.9341y_2$	x_3	α_3	$\alpha_3 x_3$
3.42	0.0780	0.0735	0.0687	0.0692	3.3	0.2289
1.97	0.0111	0.0105	0.00981	0.0100	2.0	0.0200
1.00	0.9718	0.9160	0.8556	0.921	1.0	0.9210
	1.0609	1.0000		1.0002		1.1694

y_3	$0.9341y_3$	x_4	α_4	$\alpha_4 x_4$	y_4	$0.9341y_4$
0.1953	0.1824	0.1829	3.0	0.5487	0.3965	0.3703
0.0171	0.0160	0.0162	2.12	0.0343	0.0248	0.0232
0.7876	0.7357	0.8010	1.00	0.8010	0.5787	0.5405
1.0000		1.0001		1.3840	1.0000	

x_5	α_5	$\alpha_5 x_5$	y_5	$0.9341y_5$	x_6
0.3708	2.67	0.9900	0.5997	0.5601	0.5606
0.0234	2.35	0.0550	0.0333	0.0311	0.0313
0.6059	1.00	0.6059	0.3670	0.3428	0.4082
1.0001		1.6509	1.0000		1.0001

α_6	$\alpha_6 x_6$	y_6	$0.9341y_6$	x_7
2.40	1.3454	0.7327	0.6844	0.6849
2.64	0.0826	0.0450	0.0420	0.0422
1.00	0.4082	0.2223	0.2076	0.2730
	1.8362	1.0000		1.0001

α_7	$\alpha_7 x_7$	y_7	$0.9341y_7$	x_8	α_8
2.28	1.5616	0.7983	0.7456	0.7460	2.25
2.88	0.1215	0.0621	0.0580	0.0582	2.96
1.00	0.2730	0.1396	0.1304	0.1958	1.00
	1.9561	1.0000		1.0000	

TABLE 11-5. STAGE-TO-STAGE CALCULATIONS FOR EXAMPLE 11-1 (*Continued*)

$\alpha_8 x_8$	y_8	
		This V_8 composition is not as close to D as the V_8 obtained
1.6785	0.8201	below where the feed is introduced on stage 7. Therefore,
0.1723	0.0842	the calculations are continued below starting with y_7 and
0.1958	0.0957	using the operating line for the middle section of the column.
2.0466	1.0000	

y_7	$1.0925y_7$	x_8	α_8	$\alpha_8 x_8$	y_8
0.7983	0.8721	0.7030	2.30	1.6169	0.7930
0.0621	0.0678	0.0680	2.84	0.1931	0.0947
0.1396	0.1525	0.2290	1.00	0.2290	0.1123
1.0000		1.0000		2.0390	1.0000

$1.0925y_8$	x_9	α_9	$\alpha_9 x_9$	y_9	$1.0925y_9$
0.8664	0.6971	2.39	1.6661	0.7751	0.8468
0.1035	0.1037	2.74	0.2841	0.1322	0.1444
0.1227	0.1992	1.00	0.1992	0.0927	0.1013
	1.0000		2.1494	1.0000	

x_{10}	α_{10}	$\alpha_{10} x_{10}$	y_{10}	$1.0925y_{10}$	x_{11}
0.6777	2.47	1.6739	0.7495	0.8188	0.6497
0.1446	2.64	0.3817	0.1709	0.1867	0.1869
0.1778	1.00	0.1778	0.0796	0.0870	0.1635
1.0001		2.2334	1.0000		1.0001

α_{11}	$\alpha_{11} x_{11}$	y_{11}	$1.0925y_{11}$	x_{12}	α_{12}
2.59	1.6827	0.7285	0.7959	0.6268	2.67
2.48	0.4635	0.2007	0.2193	0.2195	2.37
1.00	0.1635	0.0708	0.0773	0.1538	1.00
	2.3097	1.0000		1.0001	

$\alpha_{12} x_{12}$	y_{12}	
1.6736	0.7129	Comparison of this V_{12} with the one below indicates that the
0.5202	0.2216	feed should be introduced on stage 11.
0.1538	0.0655	
2.3476	1.0000	

TABLE 11-5. STAGE-TO-STAGE CALCULATIONS FOR EXAMPLE 11-1 (*Continued*)

y_{11}	$1.6667y_{11}$	x_{12}	α_{12}	$\alpha_{13}x_{12}$	y_{12}
0.7285	1.2142	0.6972	2.54	1.7709	0.7493
0.2207	0.3345	0.1881	2.54	0.4778	0.2022
0.0708	0.1180	0.1147	1.00	0.1147	0.0485
1.0000		1.0000		2.3634	1.0000

$1.6667y_{12}$	x_{13}	α_{13}	$\alpha_{13}x_{13}$	y_{13}	$1.6667y_{13}$
1.2489	0.7319	2.52	1.8444	0.7629	1.2715
0.3370	0.1906	2.60	0.4956	0.2050	0.3417
0.0808	0.0775	1.00	0.0775	0.0321	0.0535
	1.0000		2.4175	1.0000	

x_{14}	α_{14}	$\alpha_{14}x_{14}$	y_{14}	$1.6667y_{14}$	x_{15}
0.7545	2.50	1.8862	0.7717	1.2862	0.7692
0.1953	2.60	0.5078	0.2078	0.3463	0.1999
0.0502	1.00	0.0502	0.0205	0.0342	0.0309
1.0000		2.4442	1.0000		1.0000

α_{15}	$\alpha_{15}x_{15}$	y_{15}	$1.6667y_{15}$	x_{16}	α_{16}
2.5	1.9230	0.7743	1.2905	0.7735	2.51
2.64	0.5297	0.2133	0.3555	0.2091	2.60
1.00	0.0309	0.0124	0.0207	0.0174	1.00
	2.4836	1.0000			

$\alpha_{16}x_{16}$	y_{16}	$1.6667y_{16}$	x_{17}	α_{17}	$\alpha_{17}x_{17}$
1.9415	0.7759	1.2932	0.7762	2.52	1.9560
0.5437	0.2172	0.3620	0.2156	2.58	0.5562
0.0174	0.0069	0.0115	0.0082	1.00	0.0082
2.5026	1.0000				2.5204

y_{17}	
0.7761	The toluene concentration in V_{17} is below the specified value of 0.005.
0.2207	
0.0032	
1.0000	

compositions are plotted in Fig. 11-29.   Most of the operating lines and
tie lines in the figure have been omitted for the sake of clarity.   Those
V's and L's marked with dotted circles are those which would have been
obtained if a feed had not been introduced on the preceding stage.

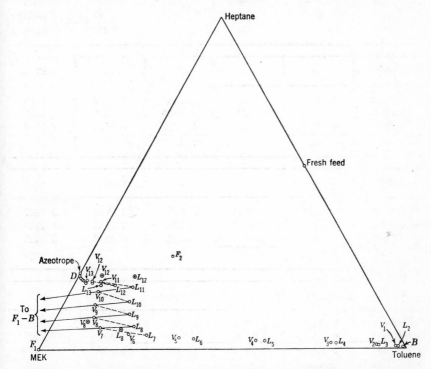

FIG. 11-29. Distillation path for Example 11-1.   Stream compositions are those from
Table 11-5.   The tie lines and material-balance lines for the top and bottom sections
are omitted for clarity.   The circled points are those which would exist if the feed
had not been introduced on the previous plate.

The effect of a slight mislocation of the overhead product composition
is shown in Fig. 11-30.   In the calculations represented there, D was
assumed to lie slightly above the azeotrope.   If the upper feed is introduced on stage 11, the angles which the tie lines and operating lines make
with each other are such as to prevent the calculated stage compositions
from approaching D.   As a consequence the distillation path veers away
to the left and would eventually reach the MEK-heptane side where the
tie lines and operating lines would become coincident.   This is not a
minimum reflux condition but rather illustrates the impossibility of
approaching D from the other side of the azeotrope.   If the feed is not
introduced on stage 11, the distillation path will continue upward.

However, it is not possible to circle the azeotrope and approach D from above because a zone of constant composition is reached in a few stages. In summary, the assumed D must lie on the side of the azeotrope from which the distillation path approaches the azeotrope.

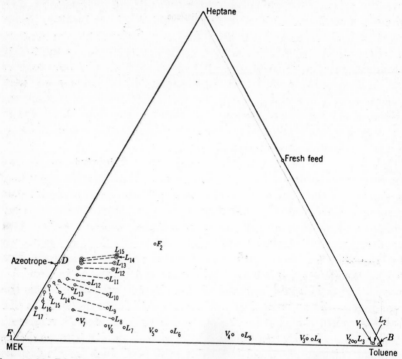

Fig. 11-30. Distillation paths obtained when D is located slightly above the azeotrope. Circled points are obtained when feed is not introduced on stage 11. All material-balance lines and some tie lines are omitted for clarity.

Other cases should be worked for other solvent-introduction points and other reflux rates. Fortunately the solvent/feed ratio need not be varied. Its value is fixed by the need to remove high-purity toluene from the bottom and the desirability of a maximum heptane content in the overhead stream. Once the best solvent-introduction point and reflux ratio are known approximately, enthalpy balances should be used to correct the phase rates. Enthalpy balances in nonideal systems are illustrated in Example 11-2.

EXTRACTIVE DISTILLATION

Extractive distillation is a simpler process than azeotropic distillation. The solvent always boils so far above the fresh-feed components that the formation of new azeotropes is impossible. Because of its relative nonvolatility, the solvent always leaves the column at the bottom and, in order to maintain a high concentration throughout most of the column, must always be introduced above the fresh-feed stage. The solvent cannot be put in on the top stage because a few stages must be provided above the solvent entry point to reduce the solvent concentration to a negligible amount before the overhead product is withdrawn. Since the volatility of the solvent is so low, the overflow of solvent from stage to stage is relatively constant from the solvent-introduction point to the bottom stage. The solvent is in this respect analogous to the absorber oil in an absorber column. Like the absorber oil, the solvent will tend to increase in temperature owing to the condensation of nonsolvent material as the solvent flows down the column, but since the liquid in each stage is at its bubble point, the stage temperature is controlled by the stage pressure and composition.

High solvent concentrations are usually desirable to maximize the difference in volatilities between the keys. However, the solubility relationships of the system must be known and care taken to ensure that the solvent concentration is maintained within a miscible range. Also, it should be kept in mind that although a higher solvent concentration may decrease the required number of stages owing to an increased volatility difference between the keys, it also increases the liquid overflow rate and therefore requires larger heat loads and column diameters. A vapor feed may be more desirable because a liquid fresh feed dilutes the descending solvent and results in a lower solvent concentration in the bottom section. Reflux at the top of the column also dilutes the solvent, and increased reflux is therefore not always synonomous with increased separation. For a given solvent/fresh feed ratio there is usually an optimum reflux rate which strikes a balance between the inherent benefits of reflux and the effect of solvent concentration on the volatilities. The solvent concentration is controlled by manipulation of the solvent, fresh feed, and reflux rates.

The recovery of the solvent is also simpler for extractive than for azeotropic distillation. The solvent does not form an azeotrope with the nonsolvent material in the bottom product, and the separation can be made by simple distillation. Quite often the calculation of the solvent-recovery column can be accomplished with a binary diagram.

The characteristics and design of extractive distillation columns have been widely discussed in the literature (3, 4, 7, 8, 10, 22, 35, 45), and several papers which report plant operations have appeared (5, 14, 15, 16, 23).

11-9. Selection of Solvent

The number of possible solvents available for a separation by extractive distillation is usually much larger than for an azeotropic separation, since the volatility restrictions are less severe. The only volatility restriction is that the solvent boil sufficiently higher than the nonsolvent material to prevent the formation of an azeotrope. The general approach in the selection of a solvent is to select a compound which is more similar to the higher boiling key and then go up the homologous series for that compound until a homolog is found which boils high enough to make a solvent-nonsolvent azeotrope impossible. Molecular similarity to the higher boiling key will cause the solvent to behave more ideally with that key than with the lighter key. The mixture becomes more unlike the light key, which then exhibits greater positive deviations.

TABLE 11-6. HOMOLOGS OF ACETONE (56.4°C) AND METHANOL (64.7°C)

Solvent	B.P., °C	Solvent	B.P., °C
Methyl ethyl ketone	79.6	Ethanol	78.3
Methyl n-propyl ketone	102.0	Propanol	97.2
Methyl i-butyl ketone	115.9	Water	100.0
Methyl n-amyl ketone	150.6	Butanol	117.8
		Amyl alcohol	137.8
		Ethylene glycol	197.2

The effect on the light key of adding the solvent is essentially the same as increasing the concentration of the heavy key on a binary γ vs x plot. The deviation of the light key from ideality increases as its concentration decreases. The separation between the two keys could easily be made if a high concentration of the heavy key could be maintained in the liquid on each stage. This is, of course, impossible, since the ratio of light key to heavy key must increase near the top of the column, hence the need for a solvent which has the same effect as a high heavy-key concentration but which can be separated easily from the light key.

The solvent may, of course, be more similar to the light key and cause it to be removed at the bottom while the heavy key goes overhead. However, in this case the natural volatility difference must be overcome and reversed, and this usually decreases the effectiveness of the solvent.

Scheibel (41) illustrates the homologous series method of solvent selection with the acetone-methanol azeotrope. Table 11-6 lists several of the homologs for both acetone and methanol along with their normal boiling points. Of the acetone homologs listed, only methyl ethyl ketone forms an azeotrope with methanol and is therefore unsuitable for use as a solvent. None of the methanol homologs form an azeotrope with acetone, and supposedly all would function as a solvent. Use of an alcohol as the solvent would take advantage of the natural volatility of acetone compared with methanol and would probably give larger relative volatilities than if one of the ketones were used. The lowest boiling homolog which does not form an azeotrope is usually the best solvent because the increase in size of the methylene radical tends to dilute the similarity of the homologs with one of the keys. The boiling point above which the homologs no longer form an azeotrope can be found from plots similar to Fig. 11-31.

Molecular similarity will usually give a qualitative indication of how the keys will react to the proposed solvent. For example, ring compounds such as phenol and aniline are more similar to aromatics such as

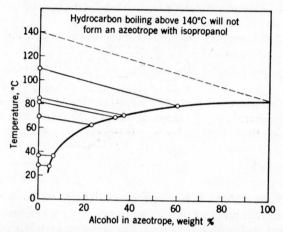

FIG. 11-31. Relationship between boiling points and compositions of azeotropes of isopropanol and hydrocarbons and boiling points of pure hydrocarbons. [*Scheibel*, *Chem. Eng. Progr.* **44**, 927 (1948).]

benzene and toluene than to paraffins and naphthenes. Unsaturated ring compounds such as furfural will have a higher affinity for olefins than for saturated paraffins. If the molecular similarity concept is insufficient, the classifications of Ewell et al. (17) based on hydrogen-bonding potentialities will often predict qualitatively the key for which the solvent will have the greater affinity. The correlation of Pierotti et al. (36) is valuable for a more quantitative prediction of the solvent effect. The use of these methods is described in Sec. 11-1. Newman et al. (32) present an example of the application of the Ewell et al. method to the selection of a solvent.

11-10. Example Calculation

The solvent liquid concentration in an extractive distillation column normally decreases rapidly from some high value in the reboiler to some lower value in the third or fourth stage which then persists to the fresh-feed stage. If a vapor feed is used, this essentially constant solvent concentration obtains until the solvent-introduction point is reached. The constancy of the solvent concentration makes extractive distillation more susceptible to treatment by shortcut methods than azeotropic distillation. The most commonly used short-cut approach utilizes the McCabe-Thiele diagram with the equilibrium curve drawn on a solvent-free basis at the desired value of solvent concentration. If a vapor feed is used, one diagram is sufficient to reach from the bottom stage to the solvent-introduction point. If the feed is a liquid, the solvent liquid concentration will change across the feed stage and two diagrams will be necessary. The change in solvent concentration across the bottom few plates is usually ignored. This method of attack has been illustrated by Perry (35), Atkins and Boyer (3), Smith and Dresser (45), and others. Chambers (7) shows how two x-y diagrams can be used concurrently to step off stages without the assumption that the solvent concentration is constant.

Rigorous computer solutions of extractive distillation problems can be made by the general calculation methods described in Chap. 10 and Sec. 11-2. Such methods require too much time for manual solution. Good approximate answers can be obtained from a simplified stage-to-stage calculation which does not assume a constant solvent concentration but does assume constant molar overflow rates. The overflow rates can subsequently be corrected by enthalpy balances, and the calculations repeated to obtain a better approximation.

The example below is solved by a simplified stage-to-stage calculation with a subsequent correction of the phase rates. The separation of methylcyclohexane (MCH) from toluene with phenol as a solvent is accomplished. Dunn et al. (15) studied the effect of several solvents on the volatilities of toluene and nonaromatic material boiling between 210 and 235°F. Several solvents give satisfactory volatility ratios with furfural, acetonyl acetone, nitrobenzene, nitrotoluene, and aniline giving ratios as large or larger than phenol. Furfural boils low enough to form azeotropes with saturated hydrocarbons boiling close to toluene and therefore would be unsuitable as a solvent. Phenol was selected above the other compounds because it is stable, relatively nontoxic, and readily available. A study of the methylcyclohexane-toluene system should be representative of the toluene-nontoluene system studied by Dunn et al. Figure 11-32 shows a simplified flow diagram for the entire process.

Example 11-2. Approximately how many stages will be required in the extractive distillation column in Fig. 11-32 to recover 95 per cent of the toluene as a 99 per cent pure product (solvent free) from a 50-50 mixture of toluene and methylcyclohexane when phenol is used as the solvent?

Solution. The general approach used in this example is similar to that of Example 11-1 and opposite to the approach used in Chaps. 8, 9, and 10.

In the example problems in those chapters the number of stages, feed locations, diameter, etc., were assumed to be fixed a priori and the corresponding separation calculated. In this example, as in Example

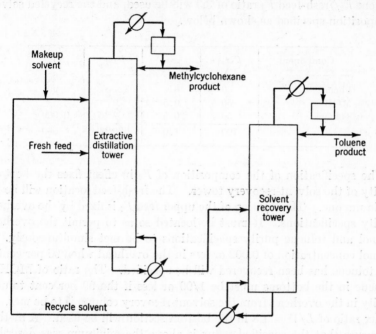

FIG. 11-32. Simplified flow diagram for an extractive distillation process to separate toluene and methylcyclohexane.

11-1, a desired separation is specified and the required column parameters calculated.

Specifications. A column with a total condenser and partial reboiler will be used. With two feeds, the column will have $2C + 2N + 12$ degrees of freedom. The items listed below will be specified.

Specifications	$N_i{}^u$
Pressure in N stages	N
Pressure in condenser	1
Pressure and heat leak in reflux divider	2
Heat leak in $N - 1$ stages	$N - 1$
Reflux temperature	1
F_1 and F_2	$2C + 4$
Fresh-feed location	1
Phenol concentration in overhead	1
Ratio of MCH to toluene in bottoms	1
Recovery of toluene	1
One reflux ratio L/V	1
	$2C + 2N + 12$

Atmospheric pressure in the condenser and a 0.2-psi drop per actual stage will be specified to fix all the stage pressures. Heat leaks at all points will be neglected. Bubble-point reflux and feeds will be used. A solvent F_2/fresh-feed F_1 ratio of 3.3 will be used, and the recycled solvent composition specified as shown below.

Component	$F_1 x_{F_1}$	x_F	$F_2 x_{F_2}$	x_{F_2}
Phenol	0	0	327.03	0.991
MCH	50	0.5	0	0
Toluene	50	0.5	2.97	0.009
	100	1.0	330.00	1.000

The specification of the composition of F_2 in effect fixes the bottoms purity of the solvent-recovery tower. The fresh-feed location will be the optimum one. The location of the upper feed F_2 is fixed by the overhead purity specifications. It must be located so as to permit the overhead phenol and toluene purity specifications to be met simultaneously. A phenol concentration of 0.002 or less in the overhead when 95 per cent of the toluene has been recovered will be specified. The ratio of MCH to toluene in the bottoms must be 1/99 or less if the 99 per cent toluene purity in the overhead from the solvent recovery column is to be met. A reflux ratio of $L/V = 1.8$ in the bottom section will be used. A method to ensure that the specified reflux is above the minimum was described in Example 11-1.

Material Balances

Toluene in $B = 0.95(100)(0.5) + (330)(0.009) = 50.47$
MCH in $B = (1/99)(50.47) = 0.51$
Phenol in $B = (330)(0.991) - (0.002)D = 327.03 - 0.002D$

Therefore

$$B = 50.47 + 0.51 + 327.03 - 0.002D = 378.01 - 0.002D$$
Toluene in $D = (0.05)(100)(0.05) = 2.5$
MCH in $D = (0.5)(100) - 0.51 = 49.49$
Phenol in $D = 0.002D$

Therefore

$$D = 2.5 + 49.49 + 0.002D$$
$$D = 51.99 + 0.002D$$
$$D = \frac{51.99}{0.998} = 52.09$$

The over-all balances are summarized below.

Component	Dx_D	x_D	Bx_B	x_B
Phenol	0.10	0.002	326.90	0.8651
MCH	49.49	0.950	0.51	0.0013
Toluene	2.50	0.048	50.50	0.1336
	52.09	1.000	377.91	1.0000

Operating-line Equations. Constant molar overflow will be assumed for the first pass through the column. The L/V in the bottom section has been specified as 1.8. The rates throughout the column are as follows, where the single primes refer to the middle section and the double primes to the bottom section of the column:

$$B = 337.91$$
$$L'' = B + V'' = B + 0.5555L'' = 850.3$$
$$L' = L'' - F_1 = 850.3 - 100 = 750.3$$
$$L = L' - F_2 = 750.3 - 330 = 420.3$$
$$V = V' = V'' = L'' - B = 472.4$$

The operating-line equations for the various sections are as follows:

$$\text{Top: } x_{i,n+1} = 1.124y_{i,n} - 0.1239x_{i,D}$$
$$\text{Middle: } x_{i,n+1} = 0.6296y_{i,n} + 0.4398x_{i,F_2} - 0.06942x_{i,D}$$
$$\text{Bottom: } x_{i,n+1} = 0.5555y_{i,n} + 0.4444x_{i,B}$$

The operating-line constants are summarized in the following tabulation.

Component	Bottom, $0.4444x_{i,B}$	Middle, $0.4398x_{i,F_2} - 0.06942x_{i,D}$	Top, $0.1239x_{i,D}$
Phenol	0.3845	0.4357	0.0002
MCH	0.0006	-0.0659	0.1177
Toluene	0.0594	0.0007	0.0059

It is not known as yet whether the feed and reflux rates listed above will furnish a satisfactory solvent-concentration profile in the column. Benedict and Rubin (4), Colburn (10), and Perry (35) show how the solvent concentrations above and below the feeds can be approximated. However, the method requires an estimate of the average volatility of solvent to nonsolvent material, and since this volatility is a function of the solvent concentration, a trial-and-error procedure is required. The same information can be obtained more simply by an inspection of the operating-line equation for the bottom section. Owing to the low volatility of phenol, its vapor concentration will be small and $0.5555y_n$ will be small compared with $0.4444x_B$. Since $0.4444x_B = 0.3845$ for phenol, it can be seen that the phenol will drop to about 40 per cent in the liquid on the third or fourth stage from the bottom. This concentration will persist until the feed is introduced, at which time it will increase slightly. If this solvent concentration were not satisfactory, the solvent/fresh-feed ratio or the reflux ratio would be adjusted to give a more satisfactory value. The rates listed above provide solvent concentrations in the 40 to 50 per cent range, which is satisfactory.

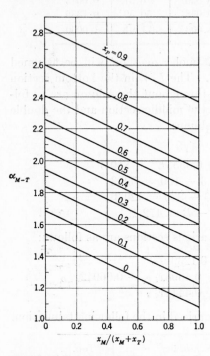

FIG. 11-33. Relative volatility of methylcyclohexane to toluene.

Equilibrium Data. For the purposes of manual calculations where enthalpy balances are not required, the equilibrium data are most conveniently used in the form of relative volatilities. Figures 11-33 and 11-34 plot the volatilities of MCH and phenol relative to toluene vs the fraction MCH in the liquid on a solvent-free basis. The experimental data were reported by Drickamer et al. (13).

The use of relative volatilities avoids the necessity for bubble-point calculations during the stage-to-stage calculations. However, if enthalpy balances are to be made, it is necessary to predict activity coefficients as a function of liquid composition in order to estimate heats of solution.

The activity coefficients can be predicted arithmetically with some form of the Margules-type equations or, in some cases, graphically from the log γ vs x plots suggested by Carlson and Colburn (6) and used by Colburn and Schoenborn (9) and Scheibel and Friedland (40). The latter correlation does not work for many systems, since the curves at intermediate concentrations do not always fall between the limiting binary curves.

Stage-to-Stage Calculations. The stage-to-stage calculations are tabulated in Table 11-7.† The equilibrium relationships are used in the form of Eq. (10-5).

$$y_{i,n} = \frac{(\alpha_{i-r}x_i)_n}{\Sigma(\alpha_{i-r}x_i)_n} \qquad (10\text{-}5)$$

† The reader should not be concerned about the large number of digits carried in these calculations. It is customary to carry some "insignificant" figures in repetitive calculations of this type and then round off the final answers.

The calculations start at the bottom stage and proceed upward through the column until the overhead purity specifications are met. The fresh feed is inserted on stage 8, which is the optimum location. The optimum location is determined easily from Fig. 11-35, which shows the stage compositions calculated in Table 11-7. The L_9 composition obtained is

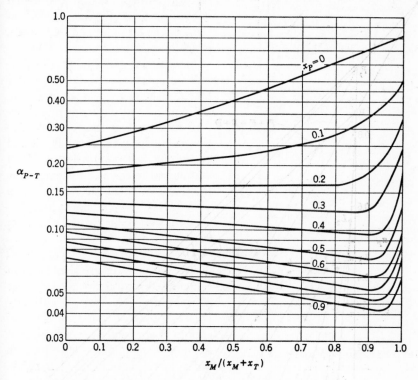

FIG. 11-34. Relative volatility of phenol to toluene.

farther to the right when $\Delta' = B - F_1$ rather than B is used as the difference point.

The location of F_2 must be approached somewhat differently. It must be located in such a manner as to cause the phenol and toluene purity specifications to be met as nearly simultaneously as possible. Insertion of F_2 on stage 13 approaches this requirement as closely as possible. The use of stage 12 as the solvent entry point causes the phenol concentration to drop below its specified value long before the toluene specification is reached.

Enthalpy Balances. The calculations described above give a good approximate answer and can be performed in a reasonable amount of time. However, the answers obtained will not be exact owing to the assumption of constant molar overflow. The error caused by this assumption may be negligible, or it may be appreciable. Therefore, it is usually worth-while to estimate the error involved by making enthalpy balances to calculate more correct phase-rate profiles. Such calculations are quite

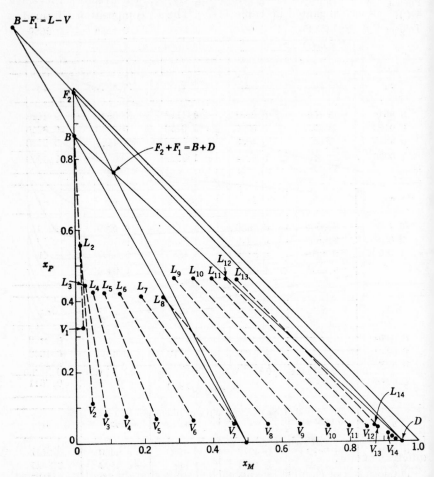

FIG. 11-35. Plot of stage compositions and material-balance lines from Example 11-2. The L's above L_{14} and the V's above V_{16} are omitted.

lengthy when nonideal solutions are involved. Therefore, for the purposes of illustration, the vapor rate at only one point in the column (just below F_1) will be checked. A corrected V_7 will be calculated, and the effect of omitting heats of solutions will be estimated.

Equation (10-8) shows the form of the enthalpy balance in the bottom section of the column. When the balance is drawn between stages 7 and

TABLE 11-7. STAGE-TO-STAGE CALCULATIONS FOR EXAMPLE 11-2

Component	x_1	α_1	$\alpha_1 x_1$	y_1	$0.5555y_1$
Phenol	0.8651	0.076	0.0657	0.3239	0.1799
MCH	0.0013	2.74	0.00356	0.0175	0.00972
Toluene	0.1336	1.00	0.1336	0.6586	0.3659
	1.0000		0.20286	1.0000	

x_2	α_2	$\alpha_2 x_2$	y_2	$0.5555y_2$	x_3
0.5644	0.100	0.0564	0.1118	0.0621	0.4466
0.0103	2.205	0.0227	0.0450	0.0250	0.0256
0.4253	1.00	0.4253	0.8432	0.4684	0.5278
1.0000		0.5044	1.0000		1.0000

α_3	$\alpha_3 x_3$	y_3	$0.5555y_3$	x_4	α_4
0.113	0.0505	0.0800	0.0444	0.4289	0.114
2.08	0.0532	0.0842	0.0468	0.0474	2.05
1.00	0.5278	0.8358	0.4643	0.5237	1.00
	0.6315	1.0000		1.0000	

$\alpha_4 x_4$	y_4	$0.5555y_4$	x_5	α_5	$\alpha_5 x_5$
0.0489	0.0730	0.0406	0.4251	0.111	0.0472
0.0972	0.1451	0.0806	0.0812	2.01	0.1632
0.5237	0.7819	0.4343	0.4937	1.00	0.4937
0.6698	1.0000		1.0000		0.7041

y_5	$0.5555y_5$	x_6	α_6	$\alpha_6 x_6$	y_6
0.0670	0.0372	0.4217	0.109	0.0460	0.0612
0.2318	0.1288	0.1294	1.98	0.2562	0.3411
0.7012	0.3895	0.4489	1.00	0.4489	0.5977
1.0000		1.0000		1.1648	1.0000

$0.5555y_6$	x_7	α_7	$\alpha_7 x_7$	y_7	$0.5555y_8$
0.0340	0.4185	0.107	0.0448	0.0558	0.0310
0.1895	0.1901	1.93	0.3669	0.4568	0.2537
0.3320	0.3914	1.00	0.3914	0.4874	0.2707
	1.0000		0.8031	1.0000	

TABLE 11-7. STAGE-TO-STAGE CALCULATIONS FOR EXAMPLE 11-2 (*Continued*)

x_8	α_8	$\alpha_8 x_8$	y_8
0.4156	0.105	0.0436	0.0513
0.2543	1.87	0.4755	0.5600
0.3301	1.00	0.3301	0.3887
1.0000		0.8492	1.0000

It can be seen from Fig. 11-35 that it is advantageous to switch difference points at this stage. Therefore, the fresh feed is inserted on stage 8.

y_8	$0.6296 y_8$	x_9	α_9	$\alpha_9 x_9$	y_9
0.0513	0.0323	0.4679	0.092	0.0430	0.0522
0.5600	0.3526	0.2867	1.87	0.5361	0.6502
0.3887	0.2447	0.2454	1.00	0.2454	0.2976
1.0000		1.0000		0.8245	1.0000

$0.6296 y_9$	x_{10}	α_{10}	$\alpha_{10} x_{10}$	y_{10}	$0.6296 y_{10}$
0.0329	0.4684	0.087	0.0408	0.0477	0.0300
0.4094	0.3435	1.83	0.6286	0.7330	0.4615
0.1874	0.1881	1.00	0.1881	0.2193	0.1379
	1.0000		0.8575	1.0000	

x_{11}	α_{11}	$\alpha_{11} x_{11}$	y_{11}	$0.6296 y_{11}$	x_{12}
0.4658	0.085	0.0396	0.0451	0.0284	0.4640
0.3956	1.77	0.7002	0.7971	0.5019	0.4360
0.1386	1.00	0.1386	0.1578	0.0993	0.1000
1.0000		0.8784	1.0000		1.0000

α_{12}	$\alpha_{12} x_{12}$	y_{12}	$0.6296 y_{12}$	x_{13}	α_{13}
0.084	0.0390	0.0434	0.0273	0.4630	0.082
1.74	0.7586	0.8452	0.5321	0.4662	1.71
1.00	0.1000	0.1114	0.0701	0.0708	1.00
	0.8976	1.0000		1.0000	

$\alpha_{13} x_{13}$	y_{13}
0.0380	0.0419
0.7972	0.8800
0.0708	0.0781
0.9060	1.0000

One more stage would reduce the toluene to less than the specified amount in D. The phenol concentration is much too high, however. So insert F_2 on stage 13 and continue calculations.

TABLE 11-7. STAGE-TO-STAGE CALCULATIONS FOR EXAMPLE 11-2 (*Continued*)

y_{13}	$1.124y_{13}$	x_{14}	α_{14}	$\alpha_{14}x_{14}$	y_{14}
0.0419	0.0471	0.0469	0.52	0.0244	0.0213
0.8800	0.9891	0.8712	1.19	1.0367	0.9070
0.0781	0.0878	0.0819	1.00	0.0819	0.0717
1.0000		1.0000		1.1430	1.0000

$1.124y_{14}$	x_{15}	α_{15}	$\alpha_{15}x_{15}$	y_{15}	$1.124y_{15}$
0.0239	0.0237	0.64	0.0152	0.0135	0.0152
1.0195	0.9016	1.15	1.0368	0.9202	1.0343
0.0806	0.0747	1.00	0.0747	0.0663	0.0745
	1.0000		1.1267	1.0000	

x_{16}	α_{16}	$\alpha_{16}x_{16}$	y_{16}	$1.124y_{16}$	x_{17}
0.0150	0.67	0.0100	0.0090	0.0101	0.0099
0.9164	1.13	1.0355	0.9294	1.0446	0.9268
0.0686	1.00	0.0686	0.0616	0.0692	0.0633
1.0000		1.1141	1.0000		1.0000

α_{17}	$\alpha_{17}x_{17}$	y_{17}	$1.124y_{17}$	x_{18}	α_{18}
0.71	0.0070	0.0063	0.0071	0.0069	0.73
1.13	1.0473	0.9371	1.0533	0.9354	1.12
1.00	0.0633	0.0566	0.0636	0.0577	1.00
	1.1176	1.0000		1.0000	

$\alpha_{18}x_{18}$	y_{18}	$1.124y_{18}$	x_{19}	α_{19}	$\alpha_{19}x_{19}$
0.0050	0.0045	0.0051	0.0049	0.74	0.0036
1.0476	0.9435	1.0605	0.9426	1.11	1.0465
0.0577	0.0520	0.0584	0.0525	1.00	0.0525
1.1103	1.0000		1.0000		1.1026

y_{19}	$1.124y_{19}$	x_{20}	α_{20}	$\alpha_{20}x_{20}$	y_{20}
0.0033	0.0037	0.0035	0.75	0.0026	0.0024
0.9491	1.0668	0.9490	1.11	1.0534	0.9545
0.0476	0.0535	0.0476	1.00	0.0476	0.0431
1.0000		1.0000		1.1036	1.0000

TABLE 11-7. STAGE-TO-STAGE CALCULATIONS FOR EXAMPLE 11-2 (*Continued*)

$1.124y_{20}$	x_{21}	α_{21}	$\alpha_{21}x_{21}$	y_{21}
0.0027	0.0025	0.76	0.0019	0.0017
1.0729	0.9550	1.11	1.0600	0.9598
0.0484	0.0425	1.00	0.0425	0.0385
	1.0000		1.1044	1.0000

NOTE: Stage 13 is the correct solvent entry point even though the toluene specification is reached before the phenol specification. If stage 12 is used, the correct phenol concentration is reached long before the toluene specification is met.

8, it becomes

$$V_7 = \frac{B(h_8 - h_1) + q_r}{H_7 - h_8} \tag{10-8}$$

Before values of h_1, H_7, and h_8 can be calculated, it is necessary to determine the temperatures of stages 1, 7, and 8 by bubble-point calculations with the following simplified form of Eq. (11-8):

$$p = \sum_{i=1}^{3} y_i p = \sum_{i=1}^{3} \gamma_i x_i p_i^* \tag{11-8}$$

Activity coefficients would be obtained most conveniently from the plot of γ's vs x described in Sec. 11-2. However, such a plot will not be constructed here because Redlich-Kister constants are available. A least-squares fit of the three sets of binary data provides the following binary constants for the MCH (1)–toluene (2)–phenol (3) system.

$$\begin{array}{lll} B_{12} = 0.0946 & C_{12} = -0.00687 & D_{12} = -0.0105 \\ B_{13} = 0.712 & C_{13} = 0.124 & D_{13} = 0.224 \\ B_{23} = 0.403 & C_{23} = 0.0534 & D_{23} = -0.02508 \end{array}$$

These constants do not predict the ternary data with a high degree of accuracy mainly because of inadequacies in the MCH-phenol system data which caused the least-squares procedure to calculate values of B_{13}, C_{13}, and D_{13} which do not adequately represent the binary data and therefore cannot be expected to give a good representation of the ternary data. Nevertheless the ternary predictions obtained with these constants are adequate for the purposes of this example. Use of these constants with the ternary forms of Eq. (2-53) gives the following values of γ in the streams of interest:

Stream	γ_M	γ_T	γ_P
L_1	4.43	1.89	1.01
L_7	1.60	1.17	1.42
L_8	1.55	1.16	1.45

Trial-and-error solution of Eq. (11-8) with those activity coefficients gives temperatures of 326, 257, and 252°F, respectively, for stages 1, 7, and 8. These temperatures are based on the stage liquid compositions tabulated in Table 11-7 and stage pressures of 21.5, 19.4, and 19.0 psia, respectively. These stage pressures were obtained by assuming a 60 per cent stage efficiency and applying the 0.2-psia drop per actual tray assumed in the problem specification.

The correlation of Pierotti et al. (36) will be used to approximate the partial molar heats of solution. The constants for the following equation are given in Table 11-3:

$$\log \gamma_1{}^\infty = K_1 + Bn_p + \frac{C_1}{n_p + 2} + D(n_1 - n_2)^2 \quad (11\text{-}5)$$

where the subscripts 1 and 2 refer to solute and solvent, respectively. For both MCH and toluene in phenol, $n_p = 1.0$, $n_1 = 7$, and $n_2 = 6$. The predicted infinite dilution activity coefficients are as follows:

	\multicolumn{4}{c}{$\log \gamma^\infty$}			

	$\log \gamma^\infty$			
	25°C	50°C	75°C	90°C
MCH in phenol	1.175	1.025	0.928	0.840
Toluene in phenol	0.600	0.508	0.427	0.361

Plotting these data in accordance with Eq. (2-61) provides the following partial molar heats of solution for MCH (1) and toluene (2) at infinite dilution in phenol (3):

$$L_{13}{}^\infty = 4520 \text{ Btu/lb mole}$$
$$L_{23}{}^\infty = 2790 \text{ Btu/lb mole}$$

Constants are not provided by Pierotti et al. for the prediction of infinite dilution activity coefficients of phenol in MCH and toluene or for MCH and toluene in each other. It will therefore be necessary to assume that

$$L_{31}^{\infty} = L_{13}^{\infty} = 4520$$
$$L_{32}^{\infty} = L_{23}^{\infty} = 2790$$
$$L_{12}^{\infty} = L_{21}^{\infty} = 0$$

The last equality assumes that MCH and toluene form ideal solutions and involves a negligible error. The magnitude of the error involved in the first two equalities is not known.

Substitution of the above L^{∞}'s in Eq. (11-22) provides approximate values of partial molar heats of solution at any desired liquid composition. Because of the equalities assumed in the previous paragraph it can be seen that all the $(L_{ir}^{\infty} - L_{ri}^{\infty})$ and $(L_{ji}^{\infty} - L_{ij}^{\infty})$ terms are zero and Eq. (11-22) reduces to

$$L_r = \tfrac{1}{2}(1 - x_r) \sum_i x_i(L_{ir}^{\infty} + L_{ri}^{\infty}) - \tfrac{1}{2} \sum_{\substack{i \neq j \\ i \neq r}} (L_{ji}^{\infty} + L_{ij}^{\infty})x_i x_j$$

For example, the calculation of L_1 in the MCH (1)–toluene (2)–phenol (3) system at the composition of L_8 is as follows:

$$L_1 = \tfrac{1}{2}(1 - x_1)[x_2(L_{21}^{\infty} + L_{12}^{\infty}) + x_3(L_{31}^{\infty} + L_{13}^{\infty})]$$
$$- \tfrac{1}{2}(L_{32}^{\infty} + L_{23}^{\infty})x_2 x_3$$
$$= (1 - 0.2543)[0.3301(0) + 0.4156(4520)] - (2790)(0.3301)(0.4156)$$
$$= 1018 \text{ Btu/lb mole of MCH}$$

The partial molar enthalpy $\bar{h}$ is obtained from Eq. (11-11).

$$\bar{h}_{i,n} = h_{i,n} + L_{i,n}$$

Values of each of the quantities in Eq. (11-11) are shown in Table 11-8, which illustrates the calculation of stream enthalpies. The enthalpies are referred to a reference state of the pure liquids at 77°F. Comparison of the sums of the last two columns in both the stage 1 and 8 calculations shows the error caused by the omission of heats of solution. Ideal solutions were assumed for the stage 7 vapor.

Before these stream enthalpies can be substituted in Eq. (10-8), the reboiler heat load must be calculated by an over-all enthalpy balance. A bubble-point calculation on the overhead product (essentially a binary mixture) gives an outlet condenser temperature of 215°F. Since the overhead product is 95 per cent MCH and since MCH and toluene boil so closely together, the dew point of the material will not differ widely from the bubble point; that is, the top stage temperature will not be much above 215°F, and the sensible heat involved in the condensation of the overhead vapor will be negligible. Neglecting the small heat of solution in MCH-toluene mixtures, the approximate condenser load is

$$q_c = (0.95)(472.4)(13,550) + (0.05)(472.4)(14,650) = 6,427,000$$

where 13,550 and 14,640 are the latent heats of vaporization of MCH and toluene at 215°F, and 472.4 is the top vapor rate per 100 moles of feed. The small amount of phenol in the overhead stream is neglected.

TABLE 11-8. CALCULATION OF STREAM ENTHALPIES IN EXAMPLE 11-2

Component	$h_{i,1}$	$L_{i,1}$	$\bar{h}_{i,1}$	$x_{i,1}$	$x_{i,1}h_{i,1}$	$x_{i,1}\bar{h}_{i,1}$
MCH	12,760	3580	16,340	0.0013	20	20
Toluene	10,690	2090	12,780	0.1336	1,430	1,710
Phenol	13,270	50	13,320	0.8651	11,480	11,520
					12,930	13,250

Component	$h_{i,8}$	$L_{i,8}$	$\bar{h}_{i,8}$	$x_{i,8}$	$x_{i,8}h_{i,8}$	$x_{i,8}\bar{h}_{i,8}$
MCH	8,540	1020	9,560	0.2543	2170	2430
Toluene	7,190	300	7,490	0.3301	2370	2470
Phenol	9,690	1210	10,900	0.4156	4027	4530
					8570	9430

Component	$H_7 = \bar{H}_7$	y_7	y_7H_7
MCH	22,090	0.4568	10,090
Toluene	21,370	0.4874	10,420
Phenol	32,190	0.0558	1,800
			22,310

Both F_2 and F_1 were assumed to be under sufficient pressure to maintain them in the liquid state at the temperature of the stage in which they are introduced. The temperature of F_1 is approximately 252°F, the temperature of stage 8. To bypass the necessity of a bubble-point calculation for the temperature of stage 13 where F_2 enters, the four known temperatures were plotted vs stage number and the stage 13 temperature of approximately 234°F read from the curve. The feed enthalpies can now be calculated as follows:

$$F_2h_{F_2} = 330(8850) = 2,919,000 \text{ Btu/100 moles of } F_1$$
$$F_1h_{F_1} = 0.5(100)(8540) + 0.5(100)(7190) = 786,000$$

The enthalpy of D is

$$Dh_D = (0.95)(52.1)(6770) + (0.05)(52.1)(5530) = 350,000$$

An over-all balance gives q_r:

$$q_r = Dh_D + Bh_B + q_c - F_1 h_{F_1} - F_2 h_{F_2}$$
$$= 350,000 + (377.9)(13,250) + 6,427,000 - 786,000 - 2,919,000$$
$$= 8,079,000 \text{ Btu}/100 \text{ moles of fresh feed}$$

When ideal solutions are assumed, Eq. (10-8) gives

$$V_7 = \frac{(377.9)(8,570 - 12,930) + 8,079,000}{22,310 - 8570} = 468$$

When partial molar heats of solution are included in the calculation of stream enthalpies,

$$V_7 = \frac{(377.9)(9,430 - 13,250) + 8,079,000}{22,310 - 9,430} = 515$$

Omission of heat effects of mixing would cause a 9 per cent error in the calculated V_7 according to the approximate calculations performed above. A constant V of 472.4 was assumed in the stage-to-stage calculations performed previously.

NOMENCLATURE

a = proportionality constant between A and T in Eq. (11-31).

$a_i = \gamma_i x_i$ = activity of component i.

A = constant in symmetrical two-suffix Margules equations. Constant is the same for both components. A_i also used to denote first constant in Antoine equation [Eq. (11-42)].

A_{12}, A_{21} = constants in three-suffix Margules and Van Laar equations.

$A_{12}, B_2, C_1, D, F_2, K_1$ = constants defined in conjunction with Eqs. (11-4) and (11-5).

b, c = proportionality constants between Margules constants and absolute temperature. See Eqs. (11-32) and (11-33).

B = heavy product rate; moles per unit time. B_i also used to denote second virial coefficient for component i [Eq. (11-6)] and second Antoine constant [Eq. (11-42)].

B_{ij}, C_{ij}, D_{ij} = constants in Redlich-Kister equation. See Eq. (2-40).

C = number of components. Also used to denote weight per cent in azeotrope in Fig. 11-12.

d = differential operator.

D = light product rate; moles per unit time.

$\exp$ = exponential term.

$\bar{f}_i$ = fugacity of component i in solution.

f_i° = fugacity of component i in the standard state.

F = feed rate; moles per unit time. F_1 denotes lower feed and F_2 upper feed if two feeds are used.

$\bar{F}_i$ = partial molar free energy of component i in the solution.

$F_i^E = \bar{F}_{i,\text{nonideal}} - \bar{F}_{i,\text{ideal}}$ = excess molar free energy.

h = molar enthalpy of actual liquid mixture.

h^* = molar enthalpy of liquid mixture if it were ideal.

h_i = molar enthalpy of component i as a pure liquid.

$\bar{h}_i$ = partial molar enthalpy of component i in a liquid solution.

H = molar enthalpy of actual vapor mixture.

H^* = molar enthalpy of vapor mixture if it were ideal.

H_i = molar enthalpy of component i as a pure vapor.

$\bar{H}_i$ = partial molar enthalpy of component i in a gas solution.

$K_i = y_i/x_i$ = equilibrium distribution coefficient of component i. Superscript ∞ denotes distribution coefficient at infinite dilution.

L = heavy phase rate; moles per unit time. Primes denote section of column with L' referring to section between feeds and L'' to section below lower feed.

$L_i = \bar{h}_i - h_i$ = partial molar heat of solution of component i.

L_{ij}^{∞} = partial molar heat of solution of i at infinite dilution in j.

n = subscript which denotes a stage in general. Also n is used as exponent on (T/T_0) term in Eqs. (11-35) and (11-36) where it can take on values from zero to 1.0.

n_1, n_2, n_p = carbon numbers defined in conjunction with Eqs. (11-4) and (11-5).

N = total number of equilibrium stages including partial reboiler if any but excluding partial condenser if any.

N_i^u = degrees of freedom or independent variables associated with unit. See Chap. 3.

p = total pressure.

p_i = partial pressure of component i.

p_i^* = vapor pressure of component i.

q = integral heat of solution per pound mole of solution. Same as the q's used in Sec. 2-2.

q_i = integral heat of solution per mole of component i.

q_r, q_c = reboiler and condenser heat loads.

r = subscript denoting reference component for relative volatilities. Also used as a subscript to denote any given component in a multicomponent mixture.

R = gas law constant, 1.987. Also used in Eq. (11-38) to represent a certain grouping of terms.

T = absolute temperature.

V = light phase rate; moles per unit time. Primes denote section of column with V' referring to section between feeds and V'' to section below lower feed.

V_i^L = molar volume of component i in liquid state. See Eq. (11-6).

x = mole fraction in liquid.

y = mole fraction in vapor.

Greek Symbols

$\alpha_{i\text{-}r} = K_i/K_r$ = relative volatility of i referred to r.

$\alpha_{i\text{-}r}^{\infty}$ = relative volatility at infinite dilution.

$\gamma_i = a_i/x_i$ = activity coefficient of component i.

γ_{ij}^{∞} = activity coefficient of i at infinite dilution in j.

∂ = partial differential operator.

$\partial, \Delta, |\Delta|$ = symbols used in conjunction with Fig. 11-12.

440 DESIGN OF EQUILIBRIUM STAGE PROCESSES

REFERENCES

1. Amundson, N. R., and A. J. Pontinen: *Ind. Eng. Chem.*, **50**, 730 (1958).
2. Amundson, N. R., A. J. Pontinen, and J. W. Tierney: *A.I.Ch.E. J.*, **5**, 295 (1959).
3. Atkins, G. T., and C. M. Boyer: *Chem. Eng. Progr.*, **45**, 553 (1949).
4. Benedict, M., and L. C. Rubin: *Trans. A.I.Ch.E.*, **41**, 353 (1945).
5. Buell, C. K., and R. G. Boatwright: *Ind. Eng. Chem.*, **39**, 695–705 (1947).
6. Carlson, H. C., and A. P. Colburn: *Ind. Eng. Chem.*, **34**, 581 (1942).
7. Chambers, J. M.: *Chem. Eng. Progr.*, **47**, 555 (1951).
8. Coates, J.: *Chem. Eng.*, **67**(10), 121 (1960).
9. Colburn, A. P., and E. M. Shoenborn: *Trans. A.I.Ch.E.*, **41**, 421 (1945).
10. Colburn, A. P.: *Can. Chem. Process Inds.*, **34**, 286 (1950).
11. Coulson, E. A., and E. F. G. Herington: *J. Chem. Soc.*, 1947, p. 597.
12. Dodge, B. F.: "Chemical Engineering Thermodynamics," pp. 106 and 388, McGraw-Hill Book Company, Inc., New York, 1944.
13. Drickamer, H. G., G. G. Brown, and R. R. White: *Trans. A.I.Ch.E.*, **41**, 555 (1945).
14. Drickamer, H. G., and H. H. Hummel: *Trans. A.I.Ch.E.*, **41**, 607 (1945).
15. Dunn, C. L., R. W. Millar, G. J. Pierotti, R. N. Shiras, and M. Souders: *Trans. A.I.Ch.E.*, **41**, 631 (1945).
16. Dunn, C. L., and C. E. Liedholm: *Petrol. Refiner*, **31**, 104 (1952).
17. Ewell, R. H., J. M. Harrison, and L. Berg: *Ind. Eng. Chem.*, **36**, 871 (1944).
18. Ewell, R. H., and L. M. Welch: *Ind. Eng. Chem.*, **37**, 1224 (1945).
19. Fenske, M. R., C. S. Carlson, and D. Quiggle: *Ind. Eng. Chem.*, **39**, 1322 (1947).
20. Gerster, J. A., T. S. Mertes, and A. P. Colburn: *Ind. Eng. Chem.*, **39**, 797 (1947).
21. Robinson, C. S., and E. R. Gilliland: "Elements of Fractional Distillation," 4th ed., p. 204, McGraw-Hill Book Company, Inc. New York, 1950.
22. Hachmuth, K. H.: *Chem. Eng. Progr.*, **48**, 617 (1952).
23. Happel, J., P. W. Cornell, DuB. Eastman, M. J. Fowle, C. A. Porter, and A. H. Schutte: *Trans. A.I.Ch.E.*, **42**, 189 (1946).
24. Horsley, L. H.: "Azeotropic Data," Advances in Chemistry Series, American Chemical Society, Washington, D.C.
25. Joffe, J.: *Ind. Eng. Chem.*, **47**, 2533 (1955).
26. Jordon, D., J. A. Gerster, A. P. Colburn, and K. Wohl: *Chem. Eng. Progr.*, **46**, 601 (1950).
27. Karr, A. E., E. G. Scheibel, W. M. Bowes, and D. F. Othmer: *Ind. Eng. Chem.*, **43**, 961 (1951).
28. Licht, W., Jr., and C. G. Denzler: *Chem. Eng. Progr.*, **44**, 627 (1948).
29. Lippincott, S. B., and M. M. Lyman: *Ind. Eng. Chem.*, **38**, 320 (1946).
30. Lyster, W. N., S. L. Sullivan, Jr., D. S. Billingsley, and C. D. Holland: *Petrol. Refiner*, **38**(6), 221 (1959); **38**(7), 151 (1959); **38**(10), 139 (1959).
31. Mertes, T. S., and A. P. Colburn: *Ind. Eng. Chem.*, **39**, 787 (1947).
32. Newman, M., C. B. Hayworth, and R. E. Treybal: *Ind. Eng. Chem.*, **41**, 2039 (1949).
33. Ostwald, W.: "Lehrbuck der allgemeinen Chemie," W. Englemann, Leipzig, 1896–1902.
34. Othmer, D. F., and E. H. Ten Eyck, Jr.: *Ind. Eng. Chem.*, **41**, 2897 (1949).
35. Perry, J. H., "Chemical Engineers' Handbook," 3d ed., pp. 629–655, McGraw-Hill Book Company, Inc., New York, 1950.
36. Pierotti, G. J., C. H. Deal, and E. L. Derr: *Ind. Eng. Chem.*, **51**, 95 (1959).

37. Prigogine, I.: "The Molecular Theory of Solutions," 1st ed., chap. XV, Interscience Publishers, Inc., New York.
38. Ratliff, R. A., and W. B. Strogel: *Petrol. Refiner*, **33**(5), 151 (1954).
39. Richards, A. R., and E. Hargreaves: *Ind. Eng. Chem.*, **36**, 805 (1944).
40. Scheibel, E. G., and D. Friedland: *Ind. Eng. Chem.*, **39**, 1329 (1947).
41. Scheibel, E. G.: *Chem. Eng. Progr.* **44**, 927 (1948).
42. Sesonske, A.: Ph.D. Dissertation, University of Delaware.
43. Severns, W. H., Jr., A. Sesonske, R. H. Perry, and R. L. Pigford: *A.I.Ch.E. J.*, **1**, 401 (1955).
44. Skolnik, H.: *Ind. Eng. Chem.*, **43**, 172 (1951).
45. Smith, R. B., and T. Dresser: *Chem. Eng. Progr.*, **44**, 789 (1948).
46. Steinhauser, H. H., and R. R. White: *Ind. Eng. Chem.*, **41**, 2912 (1949).
47. Swietoslawski: "Ebbulliometry," Chemical Publishing Company, Inc., New York, 1937.
48. White, R. R.: *Trans. A.I.Ch.E.*, **41**, 539 (1945).

CHAPTER 12

MULTICOMPONENT SEPARATIONS
Rigorous Method for Extraction

Multicomponent extraction calculations differ from those for the vapor-liquid processes (distillation, absorption, and stripping) in several respects. First, since two liquid phases are involved, calculation of the distribution coefficient between the two liquid phases involves two liquid activity coefficients instead of one.

$$K_i = \frac{y_i}{x_i} = \frac{\gamma_i^L}{\gamma_i^V} \tag{12-1}$$

where y and V refer to the raffinate phase and x and L to the extract phase. It can be seen that in a multicomponent system, the calculation of the raffinate composition in equilibrium with a known extract composition is iterative in nature. The activity coefficients can be replaced in (12-1) by their correlations as a function of their respective phase compositions by means of Eq. (2-53). Equation (12-1) is then written for each component to give a set of C simultaneous equations which are neither linear nor explicit in the y_i's and which contain logarithm terms because of the nature of Eq. (2-53). Such an equation set is difficult to solve. Methods of solution have been investigated by Boberg (1). The distribution coefficients can, of course, be graphically correlated with composition in ternary and quaternary systems. The degrees of freedom in a three-phase system are C-1 in number. For a ternary system at a given temperature (or pressure), it is necessary to specify only one concentration variable to fix all the phase concentrations and distribution coefficients. Therefore, K's can be plotted vs one concentration variable in a ternary system. In quaternary systems at constant temperature, two concentration variables must be specified and the K's must be plotted vs one concentration variable at constant values of a second one.

A second major difference between liquid-liquid extraction and the vapor-liquid processes is in the nature of the temperature profile. The temperature profile in an extraction tower or train is fixed a priori to take maximum advantage of the solubility behavior of the system. The temperature may be constant throughout the tower, or a temperature gradient may be maintained from one end to the other to induce reflux.

In either case, the stage temperatures are not determined by the stage compositions and cannot be calculated from the compositions by dew or bubble points. Therefore, the originally specified stage temperatures remain constant throughout the iterative calculations which have as their aim the calculation of stage compositions and phase rates corresponding to the specified operating conditions. Solubilities rather than enthalpies control the phase rates, and enthalpy balances cannot be used to calculate the rate profiles. Enthalpy balances must, of course, be used to size the heat-exchange equipment which is used to maintain the desired stage temperatures, but they are not part of the stage-to-stage calculation itself.

The mass transfer between the two phases may be almost unidirectional (similar to absorption and stripping), or the transfer may be almost equal in both directions (similar to distillation). The solubility behavior determines which situation will exist. Complete immiscibility of the solvent and major raffinate component would cause the transfer to be unidirectional in nature. If the solvent components show a high degree of miscibility with the feed components, transfer in both directions will occur.

Extraction is also similar to absorption and stripping in that it is usually not convenient to specify product rates as is often done in distillation. It would be difficult to calculate the solvent/feed (or lean-oil/feed) ratio which corresponds to the specified product rates. Convenient specifications for extraction processes are discussed in Sec. 3-5.

All multistage, multicomponent calculation methods can be considered to consist of two major parts, or loops in computer logic terms. The first loop calculates the stage compositions and phase rates which correspond to a given temperature profile. In the calculation methods described in Chap. 10 for distillation, complete convergence is never desired in this loop. The first set of stage compositions obtained in the first or inside loop are used by the second loop to correct the original temperature profile after which the calculations are returned to the first loop. Convergence is obtained when the temperature profile becomes constant. Since the temperature profile never changes in extraction calculations, it can be seen that only the first of these two major loops will be involved and that it will be necessary to obtain complete convergence in this loop without correction of the temperature profile. That is, it is necessary to calculate exactly the stage compositions and phase rates which correspond to the originally specified temperature profile.

Multicomponent (four or more components) extraction calculations have not been developed to the same degree as for distillation. Powers (5) has shown how four-component systems can be handled graphically. The equilibrium data are plotted as a function of two concentrations as described before. The graphical constructions are performed on two ternary diagrams which are two of the faces in the four-component tetrahedron. The material-balance lines and tie lines drawn on the two ternary diagrams are projections of the actual lines which exist within the tetrahedron. The principles of construction are those discussed in Chap. 7 for three-component systems. Smith and Brinkley (6) proposed an analytical calculation procedure which can be applied to any number of components if correlations are available to provide K values as a function of phase compositions. Essentially, their method reduced to a search for an L_1 composition and accompanying V and L profiles which, when used in a stage-to-stage calculation upward through the column, would reproduce the specified solvent composition. The specified variables to define the problem were those listed in Sec. 3-5 for extraction processes. Stage temperatures rather than heat leaks were specified for each stage to fix the temperature profile. The iteration variables which were assumed to start each iteration were as follows:

1. Extract product (L_1) composition
2. Extract product (L_1) rate
3. Linear extract rate (L) profile

The raffinate rate V profile was calculated from the assumed L profile. Stage-to-stage calculations were made from the bottom up in the usual manner (alternate use of the equilibrium and material-balance relationships). The calculated y_n's and x_n's, however, were never normalized. The reason for not normalizing the calculated phase compositions will be made obvious later by the derivation of Eq. (12-10). The stage-to-stage calculation provided a calculated solvent composition. The errors in the calculated x_S's were then used with Eq. (12-10) to calculate those stage compositions which if used in another stage-to-stage calculation with the same rate and K profiles would reproduce the specified solvent composition without error. Smith and Brinkley then modified the assumed phase-rate profiles slightly by making material balances around each stage to obtain a new set of V and L profiles which were consistent in a material-balance sense with the corrected stage compositions. The stage-to-stage calculation was then repeated with the new L_1 composition and the new rate profiles. The errors in the calculated x_S's were again used with Eq. (12-10) to give new stage compositions and in turn new rate profiles. After a few trials the corrected stage compositions and rate profiles converged to constant values. The final result of this iteration was a set of composition and phase-rate profiles which satisfied all equilibrium and material-balance requirements but which failed to meet the criteria that Σy and Σx must equal 1.0 in each stream throughout the column. (The enthalpy-balance requirements can be assumed to be satisfied, since the load on the heaters or coolers used to maintain the specified stage temperatures can be adjusted independently.)

Smith and Brinkley made several iterations as described in the previous paragraph. The final Σx_1 obtained from each iteration was plotted against the corresponding L_1 rate. The L_1 rate at which Σx_1 became 1.0 was assumed as the final answer. The final criterion of convergence, in other words, was $\Sigma x_1 = 1.0$. Even with $\Sigma x_1 = 1.0$, Smith and Brinkley found that some of the intermediate-phase compositions did not sum to 1.0. This shortcoming was attributed to errors in the determination or the graphical representation of the experimental K values. It was hypothesized that a perfect numerical solution (Σy's and Σx's = 1.0 in all streams) would not be possible unless the equilibrium data were exactly correct. Subsequent work by Friday (4) has shown this rationalization to be incorrect. For any given set of equilibrium data, correct or incorrect, a perfect numerical solution can be obtained. A perfect solution is defined as one where all the equilibrium, material-balance, and enthalpy-balance restrictions are met and all stream compositions sum to 1.0 without normalization. For any given set of equilibrium data there is only one perfect solution. The calculation method of Smith and Brinkley failed to obtain the perfect solution because no procedure was provided for the correction of the individual phase rates to force the Σy's and Σx's to unity in each stage. Friday (4) recalculated the same example problem and obtained a perfect solution. The equations and procedure used will be described in the next two sections. Before that is done, however, the equation used by Smith and Brinkley to correct the stage compositions will be derived. Although this equation is more cumbersome to use than the equations derived in the next section, its presentation is valuable for instructional purposes because it illustrates clearly some of the computational difficulties which arise in multicomponent systems (vapor-liquid or liquid-liquid) where one or more components have large K values.

Error Equation. For the purposes of this derivation, let x_n and y_n refer to the correct concentrations, that is, those concentrations on stage n which correspond to the specified temperature profile and the assumed rate profiles. The assumed extract product concentration for any given component and all the stage liquid concentrations calculated from it will contain an error e_x which is defined by the following equation.

$$\text{Calculated concentration} = x_n + e_{x_n} \qquad (12\text{-}2)$$

The equilibrium relation for a given component on stage 1 is

$$y_1 = K_1 x_1 \qquad (12\text{-}3)$$

However, when the stage-to-stage calculations are started, the correct x_1 is not known and a value equivalent to $x_1 + e_{x_1}$ is assumed. Therefore, the equilibrium relation should be written as

$$y_1 + e_{y_1} = K_1(x_1 + e_{x_1}) \tag{12-4}$$

Subtraction of (12-3) from (12-4) gives

$$e_{y_1} = K_1 e_{x_1}$$

Since the "correct" x_1's are not known at this point, the K_1 in Eqs. (12-3) and (12-4) must be an assumed value or evaluated from the assumed x_1's. In either case, the error e_{y_1} is related to the error e_{x_1} by the K_1 used.

The material balance around stage 1 is

$$L_2 x_2 = V_1 y_1 + L_1 x_1 - V_0 y_0$$

The most complicated process which will be discussed is shown in Fig. 12-1. For such a process, $V_0 y_0$ is related to $L_1 x_1$ by

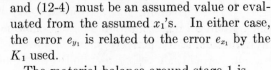

$$V_0 y_0 = \frac{R'(1 - g)}{1 + R'} L_1 x_1 + F' y_{F'} \tag{12-5}$$

where

$$g = \frac{S_E x_{S_E}}{L_1 x_1} \tag{12-6}$$

and represents a recovery factor for a given component in the extract solvent-recovery device. Substitution in the material balance around stage 1 gives

$$x_2 = (S_1 + k) \frac{L_1}{L_2} x_1 - \frac{F' y_{F'}}{L_2} \tag{12-7}$$

where

$$S_1 = \frac{K_1 V_1}{L_1}$$

and

$$k = 1 - \frac{R'(1 - g)}{1 + R'} = \frac{1 + gR'}{1 + R'}$$

Because of the error in the assumed x_1, Eq. (12-7) must be written as

$$x_2 + e_{x_2} = (S_1 + k) \frac{L_1}{L_2} (x_1 + e_{x_1}) - \frac{F' y_{F'}}{L_2} \tag{12-8}$$

Fig. 12-1. Extraction process with two feeds and reflux at both ends.

Subtraction of (12-7) from (12-8) gives

$$e_{x_2} = (S_1 + k) \frac{L_1}{L_2} e_{x_1}$$

Note that the feed terms cancel, since the specified feed concentration $y_{F'}$ contains no error.

Repetition of this procedure with the material balance

$$L_3 x_3 = V_2 y_2 + L_1 x_1 - V_0 y_0$$

gives the following relation between e_{x_3} and e_{x_1}:

$$e_{x_3} = (S_1 S_2 + kS_2 + k) \frac{L_1}{L_3} e_{x_1} \qquad (12\text{-}9)$$

Since all feed terms cancel, it is now possible to write the equation for an N stage column by analogy.

$$e_{x_{N+1}} = (S_1 \cdots S_N + k\phi) \frac{L_1}{L_{N+1}} e_{x_1} \qquad (12\text{-}10)$$

where ϕ is defined as

$$\phi = (S_2 \cdots S_N + S_3 \cdots S_N + \cdots + S_{N-1} S_N + S_N + 1)$$

If raffinate reflux is not used, then $e_{x_{N+1}} = e_{x_s}$ and Eq. (12-10) relates the error in the calculated solvent composition to the error in the assumed x_1. If raffinate reflux is used, then e_{x_s} can be related to e_{x_1} by

$$S e_{x_s} = D e_{y_N} + L_1 e_{x_1} \qquad (12\text{-}11)$$

where

$$e_{y_N} = K_N (S_1 \cdots S_{N-1} + kS_2 \cdots S_{N-1} + \cdots + kS_{N-2} S_{N-1}$$
$$+ kS_{N-1} + k) \frac{L_1}{L_N} e_{x_1} \qquad (12\text{-}12)$$

Once e_{x_1} has been obtained from e_{x_s}, equations similar to (12-10) and (12-12) can be written for each stage and used to correct the calculated stage concentrations. Application of the error equation in this manner provides those stage compositions which correspond to the stripping factors used. A new stage-to-stage calculation beginning with the corrected x_1's would reproduce the specified x_s's exactly if the same stripping factors were used. If the x_1's and all the other stage compositions sum to 1.0, then the stripping factors used are the correct ones. If not, other profiles must be assumed and the calculations repeated.

Equations (12-10) and (12-12) can be applied to any equilibrium-stage processes which do not have side streams. The side-stream concentrations contain the same errors as the interstage stream from which the side stream is withdrawn, and therefore the side-stream terms do not cancel in the derivation as do the feed terms. Equations similar to (12-10) and (12-12) can be derived for side-stream processes. It should be noted that $k = 1.0$ for distillation, absorption, and extraction processes without extract reflux.

The error equations show the proportionality between a component concentration on a given stage to the error in the assumed x_1 for that component. The e_{x_n} for a component with stripping factors less than unity will not differ widely from the e_{x_1} for that component. The stage concentrations obtained for such components in the stage-to-stage calculation are stable; that is, they do not take on unreasonable values. On the other hand, components with large stripping factors are unstable. In extraction where the stage temperature cannot be adjusted to make the y's sum to 1.0, the calculated stage concentrations of the unstable components will, if not normalized, become negative or greater than 1.0 within a few stages from the bottom. If the assumed x_1 is too small, the concentrations will go negative; if the assumed x_1 is too large, the calculated mole fractions will become very large. If the calculated mole fractions are normalized (or if the stage temperature is adjusted as it is in distillation) to make the calculated compositions sum to 1.0, the mushrooming errors in the unstable component concentrations will affect the stable components. Smith and Brinkley therefore found it necessary to calculate each component through the column separately. Stage compositions were never normalized, and therefore the wildness of the components with large K values had no effect on the low K value components (solvents). The low K value components were stable enough to permit the selection of K values from a knowledge of their stage concentrations. For this reason, the graphical correlations for K values should always be plotted vs the solvent component concentrations or other components with low K values.

12-1. Stripping-factor Equations

A stage-to-stage calculation can be made a stage at a time through alternate use of the equilibrium and material-balance relationships. This procedure was followed in the examples in Chaps. 10 and 11. An alternate to this procedure is the combination of the equilibrium and material-balance equations for a given component throughout the column into a single analytical expression which relates the concentration of a component on any given stage to its concentration in the bottom-product stream. Such an expression will be derived for the process pictured in Fig. 12-1. The process pictured in Fig. 12-1 is the most complicated process discussed in this book, and once the equation has been derived for it, the forms for simpler processes such as distillation and absorption can be written immediately by inspection.

The derivation is started by writing a component balance around stage 1.

$$L_2 x_2 = V_1 y_1 + L_1 x_1 - V_0 y_0$$

Substituting for V_0y_0 from Eq. (12-5) and replacing y_1 with K_1x_1 permit the balance to be rewritten as

$$L_2x_2 = K_1V_1x_1 + kL_1x_1 - F'y_{F'}$$

Multiplying the first term on the right by L_1/L_1 and defining

$$S_1 = \frac{K_1 V_1}{L_1}$$

give

$$L_2x_2 = (S_1 + k)L_1x_1 - F'y_{F'} \tag{12-13}$$

Moving up one stage and performing the same operations on the component balance

$$L_3x_3 = V_2y_2 + L_1x_1 - V_0y_0$$

give

$$L_3x_3 = S_2L_2x_2 + kL_1x_1 - F'y_{F'}$$

Substitution for x_2 from (12-13) gives

$$L_3x_3 = (S_1S_2 + kS_2 + k)L_1x_1 - (S_2 + 1)F'y_{F'} \tag{12-14}$$

By analogy it is now possible to write the following equation which relates the component concentration in the overflow from the feed stage to the concentration in the extract phase in the first stage.

$$
\begin{aligned}
L_{M+1}x_{M+1} \\
= (S_1 \cdots S_M + kS_2 \cdots S_M + \cdots + kS_{M-1}S_M + kS_M + k)L_1x_1 \\
- (S_2 \cdots S_M + S_3 \cdots S_M + \cdots + S_{M-1}S_M + S_M + 1)F'y_{F'} \\
\tag{12-15}
\end{aligned}
$$

After the intermediate feed F is crossed, another series of stripping-factor products is formed as a multiplier of the Fy_F term. The equation which relates the concentration in the L_{N+1} stream to x_1 is as follows.

$$
\begin{aligned}
L_{N+1}x_{N+1} \\
= (S_1 \cdots S_N + kS_2 \cdots S_N + \cdots + kS_{N-1}S_N + kS_N + k)L_1x_1 \\
- (S_2 \cdots S_N + S_3 \cdots S_N + \cdots + S_{N-1}S_N + S_N + 1)F'y_{F'} \\
- (S_{M+2} \cdots S_N + S_{M+3} \cdots S_N + \cdots + S_{N-1}S_N + S_N + 1)Fy_F \\
\tag{12-16}
\end{aligned}
$$

For the sake of convenience in writing Eq. (12-16) the following definitions will be used:

$$\phi = (S_2 \cdots S_N + S_3 \cdots S_N + \cdots + S_{N-1}S_N + S_N + 1) \tag{12-17}$$

$$\psi = (S_{M+2} \cdots S_N + S_{M+3} \cdots S_N + \cdots + S_{N-1}S_N + S_N + 1) \tag{12-18}$$

Equation (12-16) can now be written as

$$L_{N+1}x_{N+1} = (S_1 \cdots S_N + k\phi)L_1x_1 - \phi F'y_{F'} - \psi Fy_F \quad (12\text{-}19)$$

If the entering solvent is presaturated with raffinate product (raffinate reflux), $L_{N+1}x_{N+1}$ will not be identical with Sx_S. Equation (12-19) can be extended to Sx_S in the following manner: A component balance around the column gives

$$L_{N+1}x_{N+1} = V_N y_N + L_1 x_1 - Fy_F - V_0 y_0 \quad (12\text{-}20)$$

Equation (12-5) can be used to eliminate $V_0 y_0$. The $V_N y_N$ term can be replaced by means of a balance around the reflux mixer.

$$V_N y_N = \frac{1 + R}{R} L_{N+1}x_{N+1} - \frac{1 + R}{R} Sx_S \quad (12\text{-}21)$$

Substitution of (12-5) and (12-21) in (12-20) gives

$$L_{N+1}x_{N+1} = (1 + R)Sx_S - kRL_1x_1 + RFy_F + RF'y_{F'}$$

Substitution for $L_{N+1}x_{N+1}$ in Eq. (12-19) and solving the resulting expression for L_1x_1 give the following relationship between x_1 and x_S for any given component:

$$L_1x_1 = \frac{(\phi + R)F'y_{F'} + (\psi + R)Fy_F + (1 + R)Sx_S}{S_1 \cdots S_N + k\phi + kR} \quad (12\text{-}22)$$

If the specified R, k, $F'y_{F'}$, Fy_F, and Sx_S values are substituted in Eq. (12-22), and if temperature, rate, and concentration profiles are assumed to fix the stripping factors on each stage, it is possible to solve directly for the x_1 which must be assumed to reproduce the specified solvent concentration in a stage-to-stage calculation up the column. The x_1 obtained will not be the correct one, of course, unless the assumed profiles are correct. Also, the stage compositions will not sum to 1.0 unless the profiles are correct.

The forms of Eq. (12-22) which apply to simpler processes can be written by inspection. If raffinate reflux is not used, $R = 0$,

$$L_{N+1}x_{N+1} = Sx_S$$

and Eq. (12-19) is applicable. If extract reflux is not used, $R' = 0$ and $k = 1.0$. Equation (12-15) is applicable to a process with one feed F'

at the bottom and extract reflux if N is substituted for M in the subscripts. If extract reflux is used with one feed introduced on some intermediate stage, Eq. (12-19) with the $\phi F' y_{F'}$ term omitted is applicable.

A simple extraction column with one feed at the bottom is analogous to a simple absorption column, and since $k = 1$ while $R' = R = F y_F = 0$, $F' y_{F'} = V_0 y_0$, and $L_{N+1} x_{N+1} = S x_S$, Eq. (12-22) reduces to

$$L_1 x_1 = \frac{\phi V_0 y_0 + L_{N+1} x_{N+1}}{S_1 \cdots S_N + \phi} \qquad (12\text{-}23)$$

Equation (12-23) is identical with the equation presented by Edmister (3) for a stripper and is analogous to his equations for absorbers, which are based upon numbering the stages from the top rather than the bottom as done in this book. For a one-feed distillation column with total condenser, $k = 1.0$ while $F' = V_0 = S = 0$ and Eq. (12-22) reduces to

$$L_1 x_1 = \frac{(\psi + R) F x_F}{S_1 \cdots S_N + \phi + R} \qquad (12\text{-}24a)$$

The y_F was changed to x_F, which is more common in distillation nomenclature.

Equation (12-24a) applies to a column with a total condenser because Eq. (12-16) extends only to L_{N+1} and if it is to be applied to a distillation column, it is necessary that $x_{N+1} = y_D$; that is, the condenser must be a total one. The equation for a column with partial condenser is obtained by writing an over-all component balance

$$D y_D = D K_{N+1} x_{N+1} = F x_F + L_1 x_1$$

and substituting for x_{N+1} from Eq. (12-16) to give

$$x_1 = \frac{(\psi + 1/S_{N+1}) F x_F}{L_1 (S_1 \cdots S_N + \phi + 1/S_{N+1})} \qquad (12\text{-}24b)$$

where $S_{N+1} = K_{N+1} D / L_{N+1}$ and equals $1/R$ for a total condenser.

The equations derived above are called stripping-factor equations. They relate a concentration in L_1 to a concentration in L_{N+1} and therefore perform the same function as a stage-to-stage calculation. If the derivation had been made from the top of the column down, the K, L, and V for each stage would have grouped themselves in the ratio L/KV which has been defined as the absorption factor.

12-2. Material-balance Equations

The material-balance equations which relate $L_1 x_1$ to other streams in the column become somewhat complicated if extract reflux is used.

Therefore it is worthwhile to digress briefly for the purpose of deriving several of the more useful balances.

The amount of a given component in the extract from stage 1 is represented by $L_1 x_1$. From the definition of g, the amount of that component in the recovered solvent S_E is

$$S_E x_{S_E} = g L_1 x_1 \tag{12-6}$$

Also,

$$\sum_{i=1}^{C} (g x_1)_i = \frac{S_E}{L_1} \sum_{i=1}^{C} (x_{S_E})_i$$

or

$$S_E = L_1 \sum_{i=1}^{C} (g x_1)_i \tag{12-25}$$

Since

$$S_E = L_1 - (1 + R')B$$

then

$$B = \frac{L_1}{1 + R'} \left[1 - \sum_{i=1}^{C} (g x_1)_i \right] \tag{12-26}$$

The stream V_0 is given by

$$V_0 = R'B + F' \tag{12-27}$$

and

$$V_0 y_0 = \frac{R'(1 - g)}{1 + R'} L_1 x_1 + F' y_{F'} \tag{12-5}$$

A component balance around the solvent-recovery device and the extract reflux divider,

$$L_1 x_1 = S_E x_{S_E} + (1 + R')B x_B$$

can be rearranged to give

$$B x_B = \frac{1 - g}{1 + R'} L_1 x_1 \tag{12-28}$$

A component balance around the column relates L_1 to the raffinate composition as follows:

$$F y_F + S x_S + V_0 y_0 = D y_D + L_1 x_1$$

Substitution for $V_0 y_0$ from (12-5) gives

$$F' y_{F'} + F y_F + S x_S = D y_D + k L_1 x_1 \tag{12-29}$$

The D can be eliminated with an over-all quantity balance.

$$D = F + S + V_0 - L_1 \tag{12-30}$$

Equations (12-26) and (12-27) can be combined to give an expression for

V_0 which when substituted in (12-30) gives

$$D = F' + F + S + jL_1 \tag{12-31}$$

where j represents a grouping of terms as shown by the following equation:

$$j = \frac{R'}{1 + R'} \left[1 - \sum_{i=1}^{C} (gx_1)_i \right] - 1.0 \tag{12-32}$$

Substitution of (12-31) in (12-29) and rearrangement give the desired expression for L_1.

$$L_1 = \frac{F'(y_{F'} - y_D) + F(y_F - y_D) + S(x_S - y_D)}{kx_1 - jy_D} \tag{12-33}$$

If extract reflux is not used, $R' = 0$ and the denominator reduces to $x_1 - y_D$.

12-3. Example Calculation

The calculation method described in this chapter utilizes the equilibrium data in the form of distribution coefficients $K_i = y_i/x_i$, where y_i and x_i refer to the raffinate and extract, respectively. Distribution coefficient data for multicomponent systems are scarce. A considerable number of tie lines have been determined by Brancker et al. (2) for the acetone–chloroform–water–acetic acid system at 25°C, and Smith (7) has calculated additional tie lines. Because of the extensive tie-line data this system will be used to illustrate the calculation method. As mentioned previously, distribution coefficients in a four-component system can be graphically correlated with two-phase concentrations, and the K values for this system are plotted vs the water and acetic acid extract concentrations in Figs. B-6 and B-7. These two solvent components were chosen because they have small K values and are therefore more stable in the stage-to-stage calculations than the two-feed components.

Example 12-1. Calculate the recovery at 25°C of acetone (20 weight per cent) from chloroform (80 weight per cent) with a mixed solvent composed of water (65 weight per cent) and acetic acid (35 weight per cent). Use a column with five equilibrium stages and introduce all the feed on the middle stage. Use a 1:1 solvent/feed ratio and an extract reflux ratio of $R' = 10$. The diagram in Fig. 12-1 applies to this process if N and M are set equal to 5 and 2, respectively, and F' and R are set equal to zero.

Solution. The solution to this problem was approximated in Example 8-5. The specifications for this process are discussed in Sec. 3-5 and are tabulated again below for the sake of convenience.

Specifications		N_i^u
Pressure in each stage		N
Temperature in each stage		N
Temperature and pressure in extract reflux divider		2
Temperature and pressure of outlet streams from solvent recovery device		4
S	C	$+\ 2$
F	C	$+\ 2$
Total number of stages		1
Location of feed stage		1
Reflux ratio		1
Recovery of each component in solvent-recovery device	C	
	$\overline{3C + 2N + 13}$	

A tower pressure of 1 atm will be specified, and all stage temperatures will be 25°C. The same pressure and temperature will be assumed for the divider. The temperature and pressure of the outlet streams from the solvent-recovery device are not important in this solution except for the temperature of the reflux, which must be 25°C. The solvent and feed will enter the column at 25°C and the column pressure. The solvent and feed compositions and rates have been specified in the problem statement as were the number of stages, the feed-stage location, and the extract reflux ratio. Recovery factors g in the solvent-recovery device will be assumed to be 0.01 for the acetone and chloroform and 0.98 for the water and acetic acid. These are outright guesses and in an actual design would have to be checked by separation calculations on the solvent-recovery device. The recovery device cannot be designed until the rate and composition of L_1 are known, and these quantities will not be known until the extraction calculations are made. So to get started it is necessary arbitrarily to assume values of g for each component.

The results of the short-cut calculations in Example 8-5 will be used as the initial assumptions for L_1 and the x_1's even though the short-cut predictions were not accurate owing to the unpredictable nature of the phase-rate profiles in this particular problem. The assumed x_1's along with other needed values are tabulated below.

Component	Assumed x_1	g	gx_1	k
Acetone	0.2255	0.01	0.00226	0.1
Chloroform	0.0074	0.01	0.00007	0.1
Water	0.5562	0.98	0.54508	0.9818
Acetic acid	0.2109	0.98	0.20668	0.9818
	1.0000		0.75409	

The L_1 predicted by the short-cut method was 109.9 lb per 100 lb of F. Equations (12-26) and (12-27) provide the corresponding B and V_0.

$$B = \frac{L_1}{1 + R'} \left[1 - \sum_{i=1}^{c} (gx_1)_i \right] = \frac{109.9}{11} (1 - 0.75409) = 2.457$$

$$V_0 = R'B = 10(2.457) = 24.57$$
$$S_E = L_1 - V_0 - B = 109.9 - 24.6 - 2.5 = 82.8$$
$$D = F + S + V_0 - L_1 = 100 + 100 + 24.6 - 109.9 = 114.7$$

For the process in question, $F'y_{F'} = R = 0$, and

$$k = \frac{1 + 10g}{11}$$
$$\phi = S_2 S_3 S_4 S_5 + S_3 S_4 S_5 + S_4 S_5 + S_5 + 1$$
$$\psi = S_4 S_5 + S_5 + 1$$

Equation (12-22) therefore reduces to

$$L_1 x_1 = \frac{\psi F y_F + S x_S}{S_1 S_2 S_3 S_4 S_5 + k\phi} \qquad (12\text{-}22)$$

This equation provides that $L_1 x_1$ which corresponds to the stripping factors used. However, if a stage-to-stage calculation is started with the x_1's provided by (12-22), the build-up of errors due to small rounding errors in the calculated x_1's will result in unreasonable stage concentrations. This computational difficulty was discussed in detail in the derivation of Eq. (12-10). The necessity for a stage-to-stage calculation can be eliminated by the application of Eq. (12-22) to $L_2 x_2$, $L_3 x_3$, etc., as well as $L_1 x_1$. The necessary forms are as follows:

$$L_2 x_2 = \frac{(S_3 S_4 S_5 + S_4 S_5 + S_5 + 1) V_1 y_1 + (S_4 S_5 + S_5 + 1) F x_F + S x_S}{S_2 S_3 S_4 S_5 + S_3 S_4 S_5 + S_4 S_5 + S_5 + 1}$$

$$L_3 x_3 = \frac{(S_4 S_5 + S_5 + 1) V_2 y_2 + (S_4 S_5 + S_5 + 1) F x_F + S x_S}{S_3 S_4 S_5 + S_4 S_5 + S_5 + 1}$$

$$L_4 x_4 = \frac{(S_5 + 1) V_3 y_3 + S x_S}{S_4 S_5 + S_5 + 1}$$

$$L_5 x_5 = \frac{V_4 y_4 + S x_S}{S_5 + 1}$$

and

$$V_n y_n = S_n L_n x_n$$

The use of the above equations depends upon the prior assumption of V, L, and K profiles. The results of the short-cut calculations in Example 8-5 were used to furnish end values for V, L, and the component

K values. Initial values for the three intermediate stages were obtained by linear interpolation. The initial assumptions are listed below.

Stage	L	V	K			
			Acetone	Chloro-form	Water	Acid
$N+1$	100.0					
5	102.0	114.7	2.9	13.0	0.083	0.32
4	104.0	116.7	2.7	13.12	0.0817	0.307
3	105.9	118.7	2.5	13.25	0.0805	0.295
2	107.9	20.6	2.3	13.37	0.0792	0.282
1	109.9	22.6	2.1	13.5	0.078	0.27
0		24.6				

In order to investigate the effect of the K profiles on the convergence, Friday (4) in his solution of this example made several iterations in which the V and L profiles were held constant and only the K profiles were allowed to vary from trial to trial. In other words he calculated those stage compositions which corresponded to the specified temperature profile and the assumed V and L profiles listed above. This solution satisfied the equilibrium and material-balance restrictions but failed to meet the criteria that $\Sigma y = 1.0$ and $\Sigma x = 1.0$ on each stage. Four trials were made with the initial V and L profiles listed above. The behavior of the stage compositions in these four trials is typified by the normalized stage 1 extract compositions listed below.

Component	Normalized x_1's				
	Assumed	1	2	3	4
Acetone	0.225	0.240	0.205	0.170	0.158
Chloroform	0.007	0.007	0.004	0.002	0.001
Water	0.556	0.542	0.570	0.595	0.607
Acetic acid	0.211	0.211	0.221	0.233	0.234

It can be seen that the stage 1 composition was approaching some constant value. The Σy_n and Σx_n for each trial are given by

$$\sum y_n = \frac{\Sigma V_n y_n}{\text{assumed } V_n} \tag{12-34}$$

and

$$\sum x_n = \frac{\Sigma L_n x_n}{\text{assumed } L_n} \tag{12-35}$$

where $\Sigma V_n y_n$ and $\Sigma L_n x_n$ refer to the calculated values obtained with the stripping-factor equations. The Σy_n's and Σx_n's obtained in the four trials of the first iteration are shown below.

Stage	Σy			
	1	2	3	4
4	1.053	1.003	1.004	1.006
3	1.031	0.947	0.991	0.987
2	0.686	0.552	0.504	0.494
1	0.703	0.583	0.512	0.497

Stage	Σx			
	1	2	3	4
5	1.071	1.037	1.039	1.042
4	1.043	0.973	1.024	1.023
3	0.932	0.941	0.937	0.931
2	0.931	0.940	0.930	0.888
1	1.008	1.006	0.973	0.954

Obviously the criteria that $\Sigma y_n = 1.0$ and $\Sigma x_n = 1.0$ will not be met with the L and V profiles assumed in the first iteration.

Friday used the calculated $\Sigma L_n x_n$'s from the fourth trial of the first iteration as the assumed L profile for the second iteration. The $\Sigma L_n x_n$ on any stage is related to the assumed L_n and the Σx_n by Eq. (12-35). Equations (12-26) and (12-27) were used to obtain the new B and V_0 values. The new V profile was then calculated by material balances around the bottom end of the column.

The use of the calculated $\Sigma L_n x_n$'s from each iteration as the starting assumptions for the next iteration proved to be an effective convergence procedure. Convergence was extremely slow in this example owing to the large extract reflux ratio which magnified small changes in the calculated B rates. Friday made 20 iterations (equivalent to 20 assumed L and V profiles) before the Σy's and Σx's throughout the column fell between 0.998 and 1.002. Each iteration required fewer trials to obtain essentially constant stage compositions at the assumed V and L profiles. The first iteration took 4 trials, the second 3, the third and fourth took 2 each, and subsequent iterations required only 1 trial each for a total of 27 trials. Actually, convergence would have been faster if the multiple trials had not been made in the first few iterations. This is shown in

Fig. 12-2, where the calculated $L_1 = \Sigma L_1 x_1$ is plotted vs the trial number. Note how in iterations 1, 3, and 4 the L_1 from the last trial in the iteration is less than that obtained in the first trial in the iteration.

As mentioned previously, final convergence was extremely slow in this example because of the large extract reflux ratio which magnified the small changes in B from trial to trial and prevented the V_0, V_1, and V_2 rates from settling down. Each rate after the fourth iteration (eleventh

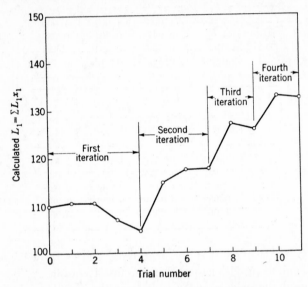

FIG. 12-2. Calculated L_1 rates for the first 11 trials. The temperature and assumed V and L profiles were held constant within each iteration, and only the K profiles varied. The calculated L profile from the last trial in each iteration was used as the initial assumption for the next profile.

trial) approached its final value asymptotically. This is illustrated in Table 12-1, which lists some of the trial values. The only Σy and Σx listed are for stage 1, since they (together with Σy_2) were the last to converge to unity. Note how much more quickly the rates above the feed stage became essentially constant. The problem would have converged much faster if the extract reflux had been replaced with a second feed of fixed rate and composition.

Table 12-2 lists the final unnormalized stage compositions and the final K values. The rates from trial 27 in Table 12-1 and the compositions in Table 12-2 comprise an essentially perfect solution of the example problem. Note the large changes in the K values from stage 1 to stage 2 and the large difference between V_0 and V_1. These changes are a result of introducing a reflux stream which is far removed from the two-liquid-

TABLE 12-1. TYPICAL TRIAL RESULTS FROM EXAMPLE 12-1

	Trial number					
	12	15	18	21	24	27
Σy_1	0.931	0.953	0.980	0.988	0.993	0.998
Σx_1	1.030	1.006	1.004	1.002	1.001	1.000
L_1	136.8	142.4	144.3	145.4	146.4	146.6
L_2	97.7	97.4	97.3	97.2 —————————————→		
L_3	95.1 ——————————————————————————————————→					
L_4	105.4 —————————————————————————————————→					
L_5	107.6 —————————————————————————————————→					
B	4.448	4.994	5.118	5.205	5.300	5.320
V_0	44.48	49.44	51.18	52.05	53.00	53.20
V_1	5.42	4.50	4.17	3.97	3.83	3.81
V_2	2.82	2.14	1.92	1.78	1.70	1.67
V_3	113.1	112.4	112.3	112.2	112.1	112.0
V_4	115.3	114.6	114.4	114.3	114.2	114.2
V_5	107.7	107.1	106.8	106.7	106.6	106.6
S_E	87.8	88.0	88.0	88.1	——————————————→	

TABLE 12-2. FINAL UNNORMALIZED STAGE COMPOSITIONS IN EXAMPLE 12-1

x_1	K_1	y_1	x_2	K_2	y_2
0.325	1.55	0.504	0.069	3.1	0.213
0.066	5.5	0.364	0.024	28.7	0.698
0.436	0.166	0.072	0.648	0.033	0.021
0.173	0.333	0.058	0.259	0.255	0.066
1.000		0.998	1.000		0.998

x_3	K_3	y_3	x_4	K_4	y_4
0.054	3.35	0.180	0.047	3.6	0.169
0.023	32.5	0.731	0.028	25.5	0.722
0.660	0.029	0.019	0.616	0.0355	0.022
0.263	0.26	0.069	0.309	0.285	0.088
1.000		0.999	1.000		1.001

x_5	K_5	y_5
0.038	3.75	0.143
0.030	24.5	0.742
0.607	0.035	0.021
0.325	0.290	0.094
1.000		1.000

phase region.   The addition point formed when V_0 and L_2 are mixed is still inside the region of immiscibility but close to the solubility surface.   Consequently, most of the reflux V_0 goes into the extract phase and the raffinate rate is sharply reduced.   Also the concentration of L_1 differs appreciably from that of L_2 owing to the solution of a large part of V_0.   This causes the big change in K values.   These effects are analogous to the situation created when a solvent rate much larger than the V_{N-1} rate is used without raffinate reflux.   The unsaturated solvent dissolves a large part of V_{N-1} and causes V_N to differ sharply from V_{N-1} in rate and composition.   Short-cut methods which depend upon the use of good average values for K's and phase rates are apt to be quite inadequate in such cases unless the designer has had previous experience with the system and is able to predict the changes beforehand.

NOMENCLATURE

C = number of components.

B = extract product rate; moles, weight, or volume per unit time.

D = raffinate product rate; moles, weight or volume per unit time.

e = calculated concentration minus correct concentration where the "correct" concentration is the one which corresponds to the temperature and rate profiles used.

F = feed rate; moles, weight, or volume per unit time.   If two feeds are used, feed at bottom end is denoted by a prime.

g = $S_E x_{S_E}/L_1 x_1$ = fraction of a given component in L_1 which is removed with S_E.

$$j = [R'/(1 + R')]\left[1 - \sum_{i=1}^{C} (gx_1)_i\right] - 1.0.$$

k = $(1 + gR')/(1 + R')$.

K_i = y_i/x_i = equilibrium-distribution coefficient for component i between the two liquid phases.

L = extract phase rate; moles, weight, or volume per unit time.   Subscripts denote stage number.

m = subscript which denotes a stage below the intermediate feed stage.

M = number of equilibrium stages below the feed stage.   Feed stage is $M + 1$.

n = subscript which denotes a stage above the intermediate feed stage.

N = total number of equilibrium stages in column.

R = external reflux ratio at top or raffinate end of column; amount refluxed/amount of D.

R' = external reflux ratio at bottom or extract end of column; amount refluxed/amount of B.

S = fresh solvent rate; moles, weight, or volume per unit time.

S_n = $K_n V_n/L_n$ = stripping factor for stage n, where n denotes any stage in column.

S_E = solvent-rich material recovered in the solvent-recovery equipment; moles, weight, or volume per unit time.

V = raffinate phase rate; moles, weight, or volume per unit time.   Subscripts denote stage number.

x = concentration in extract phase.   Units consistent with units on rates.   Subscripts denote stage number or stream.

y = concentration in raffinate phase. Units consistent with units on rates. Subscripts denote stage number or stream.

Greek Symbols

γ_i = liquid activity coefficient of component i. Superscripts L and V denote extract and raffinate phases, respectively.

Σ = summation operator.

ϕ = defined by Eq. (12-17).

ψ = defined by Eq. (12-18).

REFERENCES

1. Boberg, T. C.: Ph.D. Thesis, University of Michigan, Ann Arbor, Mich., 1960.
2. Brancker, A. V., T. G. Hunter, and A. W. Nash: *J. Phys. Chem.*, **44**, 683 (1940).
3. Edmister, W. C.: *A.I.Ch.E. J.*, **3**, 165 (1957).
4. Friday, J. R.: Ph.D. Research, Purdue University, Lafayette, Ind., April, 1961.
5. Powers, J. E.: *Chem. Eng. Progr.*, **50**, 291 (1954).
6. Smith, B. D., and W. K. Brinkley: *A.I.Ch.E. J.*, **6**, 451 (1960).
7. Smith, J. C.: *Ind. Eng. Chem.*, **36**, 68 (1944).

CHAPTER 13

MULTICOMPONENT SEPARATIONS
Rigorous Method for Absorption

The rigorous calculation of an absorber column is basically more difficult than for distillation and extraction columns. All multicomponent, multistage separation problems can be considered to consist of two major parts. In computer phraseology, the program to solve such a problem is said to contain two major loops. The first part (or inside loop) is concerned with the calculation of those phase rates and stage compositions which correspond to a given temperature profile. The second part (or outside loop) uses the results of the first part to calculate a new, more nearly correct temperature profile. The calculations are then returned to the inside loop, where the new temperature profile is used to obtain a new set of phase rates and stage compositions. The two loops are used alternately until no change in the temperature profile occurs with further iteration. Convergence is obtained in the inside loop when those phase rates which cause all the stage compositions to sum to unity are found. Convergence in the outside loop occurs when no further change in the temperature profile occurs from trial to trial.

Extraction calculations are basically more simple than absorption calculations because only the inside loop is involved. The temperature profile in an extraction column is specified a priori to take maximum advantage of the solubility relationships and remains unchanged throughout the calculations. Once convergence has been obtained in the inside loop (phase-rate profiles are found which cause all the stage compositions to sum to unity), the solution of an extraction problem is finished.

Distillation calculations involve both loops, but convergence in the inside loop can be made easier by the specification of the end-product rates which fix the ends of the phase-rate profiles. Specification of the end-product rates in absorption (and extraction) is not convenient. The fresh-feed and lean-oil (or solvent) rates are always specified, but it is practically impossible to specify a priori a pair of product rates which can be obtained from the feed rates with a reasonable number of stages, temperature, etc.

The reader of Chap. 10 will note that the distillation methods described there never make more than one trial in the inside loop for any given temperature profile; that is, convergence is never obtained in the inside loop. The stage compositions obtained in this trial are used by the outside loop to calculate new stage temperatures by bubble-point calculations. The calculations are then returned to the inside loop where these new stage temperatures are used along with the stage compositions to obtain new rate profiles by enthalpy balances. The new temperature and rate profiles are then used to calculate new stage compositions. Such a procedure will not always work in absorption calculations. Bubble- and dew-point calculations in absorbers are so dependent upon the concentrations of the very light and very heavy components, respectively, that the calculation of new stage temperatures by bubble or dew points is unreliable. For example, a small error in the methane concentration in the stage liquid in a hydrocarbon absorber can cause an error of several degrees in the bubble point. Consequently, a temperature profile obtained from bubble points based on the calculated liquid compositions of the first trial is apt to be quite erratic. On the other hand, if enough trials are made within the inside loop to converge to that set of phase rates and stage compositions which correspond to the given temperature profile (Σy and $\Sigma x = 1.0$ in all stages), the final stage compositions cannot be used to obtain a new temperature profile by dew- or bubble-point calculations, since these calculations must reproduce those stage temperatures which were used to calculate the stage compositions. The example problem in the next section illustrates this fact and shows how enthalpy balances can be used in an absorber to calculate a new temperature profile.

Besides these inherent computational difficulties encountered in absorption, the nature of the systems and the operating conditions encountered complicate the equilibrium and enthalpy-balance relationships. The wide range in molecular weights in absorber systems together with the elevated operating pressures often used makes nonideal-solution behavior as much a factor in absorber calculations as it is in extraction and nonideal distillation systems. Ideal-solution K values are usually quite inaccurate, particularly for the lightest components. Enthalpy balances based upon the assumption of ideal solutions normally give rate profiles that fail to check with plant data. The source of the equilibrium and enthalpy data may have a more important effect upon the answers obtained than does the calculation method used. Landes and Bell (11) discuss the importance of good data in the design of hydrocarbon absorbers. They suggest the data of Webber (18) for heavy oil systems. Other data have been presented by Kirkbride and Bertetti (9). The Benedict-Webb-Rubin equation (2) or the DePriester charts (5) can be used for light oil systems. The NGAA convergence-pressure charts (7) cover a wider molecular-weight range. Landes and Bell point out that the Benedict-Webb-Rubin equation does not always agree with the NGAA charts. Solomon (16) and Organick and Brown (13) have presented correlations for the prediction of K values in absorber systems which take into account the aromatic content of the absorber oil. Arnold (1) has presented charts for the correction of methane K values at high pressure.

Landes and Bell suggest that more accurate enthalpy balances are obtained through the use of heats of absorption for the phase change and sensible heats for no phase change instead of total phase enthalpies. They used a combination of Buthod's (4) curves and the experimental data of Budenholzer et al. (3) for sensible-heat data for gases. Maxwell (12) is a good source for liquid sensible-heat data. Heats of absorption are more difficult to obtain. Landes and Bell suggest Scheibel (15), Peters (14), and the Kellogg total heat charts (7) as sources and present a table which compares heat-of-absorption data taken from various sources.

Absorber design calculations are usually based upon the Kremser equation (10) as modified by Souders and Brown (17). The major difference in the various design methods published for absorption lies in the manner of selection of effective absorption factors. The methods of Horton and Franklin (8) and Edmister (6) were illustrated in Sec. 8-3. Landes and Bell (11) present a graphical method for the determination of effective absorption factors.

Complete convergence procedures for rigorous stage-to-stage calculations have not been published for absorption as they have been for distillation. The methods of Chap. 10 are not directly applicable because of the previously mentioned difficulty in the correction of the temperature profile. A rigorous solution of an absorber problem is presented in the next section, where convergence is obtained by simple iteration. A real need exists for a more effective convergence procedure.

13-1. Example Calculation

The basic equation to be used in the following example problem is that of Edmister (6).

$$L_1 x_1 = \frac{\phi V_0 y_0 + L_{N+1} x_{N+1}}{S_1 \cdots S_N + \phi} \qquad (12\text{-}23)$$

This equation was derived in Sec. 12-1 as a special form of the more general equation derived for more complicated extraction processes. Equation (12-23) relates the amount of the given component which leaves in the rich oil ($L_1 x_1$) to the amounts which enter in the fresh feed ($V_0 y_0$) and the lean oil ($L_{N+1} x_{N+1}$) in terms of the stripping factors on each stage. For any given set of component stripping factors, Eq. (12-23) provides that value of L_1x_1 with which a stage-to-stage calculation must be started if the concentration of that component in the lean oil (x_{N+1}) is to be reproduced. Theoretically, the set of L_1x_1's obtained from the application of (12-23) to each of the components could be used as a starting point in a stage-to-stage calculation which would produce a complete set of stage compositions which correspond to the assumed stripping factors. Computational difficulties, however, make this procedure impractical. An inspection of Eq. (12-10) shows that e_{x_n} will become very large for a component with large K values (large stripping factors) even though the error (e_{x_1}) in the starting bottom concentration is very small. For a component like methane, it is impractical to carry enough digits to reduce e_{x_1} to a small enough value to make e_{x_n} negligible. This difficulty can be avoided by applying (12-23) to the calculation of the liquid overflow from each stage. For a four-stage absorber the equation must be written four times as follows. The general component subscript i is omitted for convenience.

$$L_1x_1 = \frac{(S_2S_3S_4 + S_3S_4 + S_4 + 1)V_0y_0 + L_{N+1}x_{N+1}}{S_1S_2S_3S_4 + S_2S_3S_4 + S_3S_4 + S_4 + 1} \tag{13-1a}$$

$$L_2x_2 = \frac{(S_3S_4 + S_4 + 1)V_1y_1 + L_{N+1}x_{N+1}}{S_2S_3S_4 + S_3S_4 + S_4 + 1} \tag{13-1b}$$

$$L_3x_3 = \frac{(S_4 + 1)V_2y_2 + L_{N+1}x_{N+1}}{S_3S_4 + S_4 + 1} \tag{13-1c}$$

$$L_4x_4 = \frac{V_3y_3 + L_{N+1}x_{N+1}}{S_4 + 1} \tag{13-1d}$$

The vapor stream in equilibrium with a given liquid is obtained conveniently with

$$V_ny_n = S_nL_nx_n \tag{13-2}$$

For any given temperature profile, there is only one set of vapor and liquid profiles which will permit all the stage compositions to sum to unity. Since these vapor and liquid profiles will not be known, the first few trial applications of (13-1) and (13-2) will produce $\Sigma L_nx_{i,n}$'s and $\Sigma V_ny_{i,n}$'s that differ from the liquid and vapor rates assumed at the start of the trial. Convergence in the following example to the correct set of rate profiles is obtained by simply using the phase rates obtained in a trial as the assumptions for the next trial.

Once the solution for a given temperature profile has converged, enthalpy balances are used to calculate a new temperature profile for the next iteration. Enthalpy balances in an absorber column are, of necessity, of an iterative nature. The temperature of neither leaving stream is known, and it is necessary to assume one temperature to start the enthalpy-balance calculations. If the enthalpy of the residual gas stream V_N is small compared with the rich oil stream L_1, it is most convenient to assume the temperature of the top stage. An error of several degrees in the temperature of V_N will have only a small effect on the calculated temperature of L_1 which is obtained by an over-all enthalpy balance.

$$L_1 h_1 = V_0 H_0 + L_{N+1} h_{N+1} - V_N H_N \qquad (13\text{-}3)$$

The $V_0 H_0$ and $L_{N+1} h_{N+1}$ terms are fixed by the problem specifications. Assumption of t_N fixes $V_N H_N$ and permits the calculation of $L_1 h_1$. The temperature t_1 which corresponds to this calculated enthalpy is found by trial and error. Once t_1 is known, the term $V_1 H_1$ can be calculated and $L_2 h_2$ calculated from an enthalpy balance around stage 1. The temperature of stage 2 is found from the calculated $L_2 h_2$ by trial and error. Enthalpy balances around the individual stages in succession will provide stage temperatures up through the column. The balance around stage $N - 1$ will give the temperature of the top stage. If this temperature is the same as the originally assumed t_N, the calculated stage temperatures are correct and can be used in the next iteration. If the assumed and calculated t_N do not agree, it is necessary to assume a new t_N and repeat the enthalpy balances. The number of repetitions necessary to obtain a check depends upon the relative magnitudes of the terms $V_N H_N$ and $L_1 h_1$.

Once a new temperature profile is determined by the enthalpy balances, the second iteration is made to calculate the corresponding profiles and stage compositions. These rate profiles and compositions are then used in the enthalpy balances to correct the temperature profile for the third iteration. The calculations are repeated until no further change in the temperature occurs. In the example problem below, the temperature profile obtained after the third iteration (third assumed temperature profile) essentially checked the profile obtained in the second iteration, and therefore the fourth iteration was unnecessary. Four trials were necessary in the first iteration to obtain convergence, while only three trials in the second and two trials in the third iteration were required. As the temperature profile approaches the correct one, the number of trials to calculate the corresponding rate profiles and stage compositions decreases.

The procedure described above is tedious. It is probably not necessary to obtain convergence in each trial. Enthalpy balances based upon the results of the first or at least the second trial in each iteration should provide essentially the same temperature profile for the next iteration as obtained from the fourth or third trial. Despite its tedious nature, the example problem below like the extraction example in the previous chapter represents an approach to the multicomponent multistage separation problem which differs somewhat from the methods described in Chap. 10 for distillation.

The solution of the following example problem assumes ideal-solution behavior. Sources of data for a more correct solution are listed in the previous section. The assumption of component stripping factors for the various stages is more difficult when K values are a function of composition as well as temperature and pressure. Concentration profiles must then be assumed to fix the stripping factors. The assumption of concentration profiles can be bypassed by following the procedure illustrated by the extraction example in the previous chapter.

Example 13-1. Rework Example 8-1 by a rigorous stage-to-stage calculation. The absorber in Example 8-1 had four equilibrium stages and operated at 60 psia. The wet gas and lean oil both enter at 90°F.

Solution. The wet gas and lean oil rates and compositions are tabulated below with the temperature and rate profiles obtained in the short-cut solution of Example 8-1.

Component	Specified			Short-cut results			
	V_0y_0	x_{N+1}	$L_{N+1}x_{N+1}$	Stage	t	V	L
C_1	28.5	0	0	V_0	90	100.0	
C_2	15.8	0	0	1	135	86.20	155.2
C_3	24.0	0	0	2	121	74.30	141.4
$n\text{-}C_4$	16.9	0.02	2.21	3	109	64.05	129.5
$n\text{-}C_5$	14.8	0.05	5.52	4	99	55.22	119.2
Oil		0.93	102.67	L_{N+1}	90		110.4
	100.0	1.00	110.4				

The number of design variables subject to control by the designer has been shown in Sec. 3-5 to be $2C + 2N + 5$. Convenient specifications are as follows:

Specifications	$N_i{}^u$	
Pressure in each stage		N
Heat leak in each stage		N
V_0	C	$+ 2$
L_{N+1}	C	$+ 2$
Number of stages		1
	$2C + 2N + 5$	

A negligible pressure drop across each stage and adiabatic operation will be assumed to fix the pressure and heat leak in each stage. The other required specifications have been made.

The solution will be simplified by the assumption of ideal solutions. The K values for the assumed stage temperatures were read from the nomograph in Appendix B (Fig. A-3). The K values and the corresponding stripping factors are as follows:

Component	K values (60 psia)				Stripping factors			
	135°	121°	109°	99°	S_1	S_2	S_3	S_4
C_1	43.0	41.5	40.0	38.5	23.89	21.81	19.78	17.83
C_2	10.4	9.6	8.7	8.05	5.777	5.046	4.303	3.729
C_3	4.00	3.55	3.12	2.81	2.222	1.866	1.543	1.302
n-C_4	1.40	1.20	1.00	0.865	0.7777	0.6307	0.4946	0.4007
n-C_5	0.50	0.42	0.34	0.29	0.2777	0.2208	0.1682	0.1343
Oil	0.036	0.0265	0.0196	0.0155	0.0200	0.01393	0.00969	0.00718

It is convenient to calculate beforehand the stripping-factor functions required in Eq. (13-1).

Component	$S_4 + 1$	$S_3 S_4 + S_4 + 1$	$S_2 S_3 S_4 + S_3 S_4 + S_4 + 1$	$S_1 S_2 S_3 S_4 + S_2 S_3 S_4 + S_3 S_4 + S_4 + 1$
C_1	18.83	371.5	8063	191,800
C_2	4.729	20.78	101.7	569.5
C_3	2.302	4.311	8.060	16.39
n-C_4	1.401	1.599	1.724	1.821
n-C_5	1.134	1.157	1.162	1.163
Oil	1.007	1.007	1.007	1.007

The calculation of a more nearly correct set of rate profiles for the assumed temperature profile is summarized in Table 13-1. The numerator of the appropriate form of Eq. (13-1) is shown for each stage, but the denominator must be obtained from the tabulation preceding this paragraph. The $V_n y_n$'s are obtained from the $L_n y_n$'s by means of Eq. (13-2). The required stripping factors for the trial are tabulated at the beginning of the example.

The x_n's tabulated in Table 13-1 are calculated from the accompanying $L_n x_n$'s and, of course, sum to 1.0. The Σx_n's calculated from

$$\sum x_n = \frac{\Sigma L_n x_n}{L_n}$$

where L_n is the liquid overflow from stage n assumed to start the trial,

TABLE 13-1. STAGE-TO-STAGE CALCULATION IN EXAMPLE 13-1
(First trial in first iteration)

Component	$L_{N+1}x_{N+1}$	V_0y_0	$(S_2S_3S_4 + \cdots + 1)V_0y_0 + L_{N+1}x_{N+1}$	L_1x_1
C_1	0	28.5	229,800	1.20
C_2	0	15.8	1,607	2.82
C_3	0	24.0	193.4	11.80
$n\text{-}C_4$	2.21	16.9	31.35	17.22
$n\text{-}C_5$	5.52	14.8	22.72	19.54
Oil	102.67	0	102.67	101.96
	110.4	100.0		154.54

x_1	V_1y_1	$(S_3S_4 + S_4 + 1)V_1y_1 + L_{N+1}x_{N+1}$	L_2x_2	x_2
0.0078	28.62	10,630	1.32	0.0090
0.0183	16.30	338.7	3.33	0.0227
0.0764	26.22	113.0	14.02	0.0957
0.1114	13.39	23.62	13.70	0.0935
0.1264	5.426	11.80	10.15	0.0693
0.6597	2.039	104.72	104.00	0.7098
1.0000	92.00		146.52	1.0000

V_2y_2	$(S_4 + 1)V_2y_2 + L_{N+1}x_{N+1}$	L_3x_3	x_3	V_3y_3
28.77	541.7	1.46	0.0105	28.84
16.80	79.45	3.82	0.0276	16.45
26.16	60.22	13.97	0.1008	21.56
8.640	14.31	8.95	0.0646	4.426
2.241	8.061	6.97	0.0503	1.172
1.449	104.13	103.41	0.7462	1.002
84.06		138.58	1.0000	73.45

$V_3y_3 + L_{N+1}x_{N+1}$	L_4x_4	x_4	V_4y_4	y_4
28.84	1.53	0.0120	27.32	0.4887
16.45	3.48	0.0272	12.97	0.2320
21.56	9.37	0.0732	12.19	0.2180
6.64	4.74	0.0370	1.898	0.0339
6.69	5.90	0.0461	0.792	0.0142
103.67	102.95	0.8045	0.739	0.0132
	127.96	1.0000	55.91	1.0000

do not sum to 1.0 as shown by the following tabulation:

Stage	Assumed L	Calculated	
		ΣLx	Σx
1	155.2	154.5	0.996
2	141.4	146.5	1.036
3	129.5	138.6	1.070
4	119.2	128.0	1.073

These Σx's (or the comparison between the assumed and calculated L's) show that the assumed rate profiles are not the ones which correspond to the assumed temperature profile. In the absence of a convergence device which would permit the calculation of the correct profiles, the calculated ΣLx's and ΣVy's from the trial shown in Table 13-1 were used as the assumed rate profiles for the second trial in the first iteration. (The group of trials made with a given temperature profile is termed an iteration.) The use of the calculated profiles from one trial as the assumed profiles in the second trial results in an asymptotic approach to convergence; that is, the calculated profiles in any iteration do not oscillate about the final values. This is shown in Table 13-2, which summarizes some of the results of the three iterations made in the solution of this example.

Once the rate profiles in any iteration have become constant, a check must be made on the assumed temperature profile. Since all the Σx's and Σy's are 1.0, dew or bubble points on the calculated stream compositions will reproduce the assumed temperature profile. Therefore enthalpy balances must be used to calculate new stage temperatures. The rates and compositions from the final trial were used in the balances. It is probable that the rates and compositions from an earlier trial would have produced essentially the same new temperature profile. The differences in the calculated rates and compositions between the first and last trials in any iteration are generally small.

The enthalpy-balance calculations are started by assuming that $t_4 = 99°F$. The enthalpies of the two entering streams V_0 and L_{N+1} and the residue gas stream V_4 are shown below. The molar enthalpies were obtained from Figs. B-4 and B-5. Ideal-solution behavior was assumed. The calculations are based on a V_0 of 100 moles.

$$H_4 = \sum_{i=1}^{C} V_4 y_{i,4} (H_i)_{99°F} = 552,300 \text{ Btu/100 moles of feed}$$

$$H_0 = \sum_{i=1}^{C} V_0 y_{i,0} (H_i)_{90°F} = 1,265,500 \text{ Btu/100 moles of feed}$$

$$h_{N+1} = \sum_{i=1}^{C} L_{N+1}x_{i,N+1}(h_i)_{90°F} = 1,624,700 \text{ Btu/100 moles of feed}$$

$$\begin{aligned}
L_1h_1 &= V_0H_0 + L_{N+1}h_{N+1} - V_4H_4 \\
&= 1,265,500 + 1,624,700 - 552,300 \\
&= 2,337,900 \text{ Btu/100 moles of } V_0
\end{aligned}$$

The temperature of stage 1 which corresponds to the $t_4 = 99°F$ assumption is found by a trial-and-error search for the temperature where $\sum_i h_{i,1}L_1x_{i,1} = 2,337,900$. The first-stage temperature was found to be 132.5°F. The enthalpy of V_1 is given by $\sum_i H_{i,1}V_1y_{i,1}$. When the $H_{i,1}$'s are read from the graph at 132.5°F, $V_1H_1 = 1,201,700$. A balance around stage 1 gives L_2h_2.

$$\begin{aligned}
L_2h_2 &= V_1H_1 + L_1h_1 - V_0H_0 \\
&= 1,201,700 + 2,337,900 - 1,265,500 \\
&= 2,274,100 \text{ Btu/100 moles of } V_0
\end{aligned}$$

A trial-and-error search gives $t_2 = 133°F$. At this temperature

$$V_2H_2 = 1,052,300$$

and a balance around stage 2 gives L_3h_3.

$$\begin{aligned}
L_3h_3 &= V_2H_2 + L_2h_2 - V_1H_1 \\
&= 1,052,300 + 2,274,100 - 1,201,700 \\
&= 2,124,700 \text{ Btu/100 moles of } V_0
\end{aligned}$$

The temperature which corresponds to the enthalpy is $t_3 = 124.5°F$.

Repetition of the calculations for the third stage gives $L_4h_4 = 1,942,200$ and a corresponding t_4 of 113°F. Since 99°F was originally assumed, it is necessary to repeat the calculation of the temperature profile. A t_4 of 112°F was therefore assumed, and the over-all enthalpy balance repeated. The 13°F change in t_4 changed t_1 by less than 1°F. Therefore, the balances around each stage were not repeated and a profile of $t_1 = 132$, $t_2 = 133$, $t_3 = 124$, and $t_4 = 112°F$ was assumed for the second iteration. The rate profiles from trial 4 in the first iteration in Table 13-2 were used with this temperature profile to make the trial 1 calculations in the second iteration.

The temperature profiles obtained at the end of the second and third iterations are shown in Table 13-2. The temperatures from the third iteration essentially agreed with those from the second, so a fourth iteration was not made. Note that the number of trials in an iteration decreases as the assumed temperature profiles approach the correct one. This is due to the fact that as the changes in the temperature profile become smaller, the rate profiles originally assumed in the first trial in the iteration more nearly correspond to the correct ones.

TABLE 13-2. SUMMARY OF RESULTS FOR EXAMPLE 13-1

	Assumed	Iteration 1				Iteration 2			Iteration 3	
		Trial 1	2	3	4	1	2	3	1	2
L_1	155.2	154.5	153.3	152.8	152.7	151.2	151.2	151.3	152.3	152.6
L_2	141.4	146.5	146.7	146.4	146.4	141.6	140.3	140.0	140.9	141.3
L_3	129.5	138.6	140.1	140.2	140.1	137.1	136.0	135.8	135.1	135.1
L_4	119.2	128.0	130.5	131.1	131.2	130.5	129.9	129.7	128.8	128.7
V_1	86.2	92.0	93.3	93.61	93.66	90.4	89.1	88.7	88.6	88.8
V_2	74.3	84.1	86.8	87.35	87.40	85.9	84.9	84.5	82.9	82.6
V_3	64.0	73.4	77.1	78.30	78.56	79.3	78.8	78.4	76.6	76.1
V_4	55.2	55.9	57.0	57.57	57.71	59.2	59.2	59.1	58.1	57.8
t_1	135	←――――――――――――――→				132	――――――→		130	―――→ 130
t_2	121	←――――――――――――――→				133	――――――→		125	―――→ 126
t_3	109	←――――――――――――――→				124	――――――→		118	―――→ 118
t_4	99	←――――――――――――――→				112	――――――→		110	―――→ 109

Normalized x_1's

	Assumed	Iteration 1				Iteration 2			Iteration 3	
		Trial 1	2	3	4	1	2	3	1	2
C_1		0.0078	0.0073	0.0071	0.0071	0.0072	0.0074	0.0075	0.0076	0.0076
C_2		0.0183	0.0171	0.0167	0.0166	0.0168	0.0171	0.0173	0.0177	0.0179
C_3		0.0764	0.0720	0.0704	0.0699	0.0679	0.0683	0.0688	0.0713	0.0720
n-C_4		0.1114	0.1112	0.1108	0.1107	0.1073	0.1070	0.1070	0.1085	0.1088
n-C_5		0.1264	0.1275	0.1278	0.1279	0.1278	0.1275	0.1273	0.1269	0.1268
Oil		0.6597	0.6649	0.6672	0.6678	0.6730	0.6727	0.6721	0.6681	0.6669

Normalized y_4's

	Assumed	Iteration 1				Iteration 2			Iteration 3	
C_1		0.4887	0.4798	0.4764	0.4749	0.4625	0.4625	0.4631	0.4706	0.4725
C_2		0.2320	0.2310	0.2302	0.2299	0.2239	0.2228	0.2229	0.2253	0.2260
C_3		0.2180	0.2273	0.2302	0.2312	0.2317	0.2307	0.2300	0.2263	0.2251
n-C_4		0.0339	0.0362	0.0376	0.0382	0.0486	0.0497	0.0495	0.0445	0.0434
n-C_5		0.0142	0.0135	0.0135	0.0136	0.0171	0.0176	0.0177	0.0170	0.0168
Oil		0.0132	0.0122	0.0121	0.0122	0.0162	0.0167	0.0168	0.0163	0.0162

A fourth iteration should result in no change in the temperature profile. The rate profiles and compositions obtained in the fourth iteration would simultaneously satisfy all material- and enthalpy-balance requirements. Also, all Σx_n's and Σy_n's would equal 1.0 without normalization. Therefore, the solution can be accepted as the correct one for the equilibrium and enthalpy data used.

NOMENCLATURE

C = number of components.

e = calculated concentration minus correct concentration where the "correct" concentration is the one which corresponds to the temperature and rate profiles used.

h = molar liquid enthalpy of total liquid, Btu/lb mole.   Subscripts refer to stage' number.

h_i = molar liquid enthalpy of component i, Btu/lb mole.   Second subscript refers to stage number.

H = molar vapor enthalpy of total vapor, Btu/lb mole.   Subscripts refer to stage number.

H_i = molar vapor enthalpy of component i, Btu/lb mole.   Second subscript refers to stage number.

$K_i = y_i/x_i$ = equilibrium distribution coefficient for component i.

L = heavy phase rate; moles per unit time.   Subscript refers to stage number.

N = total number of equilibrium stages.

$N_i{}^u$ = degrees of freedom in process.   See Chap. 3.

$S_{i,n} = (K_iV/L)_n$ = stripping factor for component i on stage n.

t = temperature, °F.   Subscript refers to stage number.

V_n = light phase rate; moles per unit time.   Subscript refers to stage number.

$x_{i,n}$ = mole fraction of component i in heavy phase on stage n.

$y_{i,n}$ = mole fraction of component i in light phase on stage n.

Greek Symbols

Σ = summation operator.

ϕ = defined by Eq. (12-17).

REFERENCES

1. Arnold, J. H.: *Chem. Eng. Progr. Symposium Ser.*, **48**(3), 83 (1952).
2. Benedict, M., G. B. Webb, and L. C. Rubin: *Chem. Eng. Progr.*, **47**, 419, 449, 571, 609 (1951).
3. Budenholzer, R. A., D. F. Botkin, B. H. Sage, and W. N. Lacey: *Ind. Eng. Chem.*, **34**, 878 (1942).
4. Buthod: *Oil Gas J.*, Sept. 29, 1949.
5. DePriester, C. L.: *Chem. Eng. Progr. Symposium Ser.*, **49**(7), 1 (1953).
6. Edmister, W. C.: *A.I.Ch.E. J.*, **3**, 165 (1957).
7. "Engineering Data Book," 7th ed., Natural Gasoline Supply Men's Association, Tulsa, Okla., 1957.
8. Horton, G., and W. B. Franklin: *Ind. Eng. Chem.*, **32**, 1384 (1940).
9. Kirkbride, C. G., and J. W. Bertetti: *Ind. Eng. Chem.*, **35**, 1242 (1943).
10. Kremser, A.: *Natl. Petrol. News*, **22** (May 21, 1930).
11. Landes, S. H., and F. W. Bell: *Petrol. Refiner*, **39**(6), (1960).
12. Maxwell, J. B.: "Data Book on Hydrocarbons," 1st ed., D. Van Nostrand Company, Inc., Princeton, N.J., 1950.
13. Organick, E. I., and G. G. Brown: *Chem. Eng. Progr. Symposium Ser.*, **48**(2), 97 (1952).
14. Peters: *Petrol. Refiner*, **28**, 109 (1949).
15. Scheibel, E. G., and F. J. Jenny: *Petrol. Refiner*, **26** (1947).
16. Solomon, E.: *Chem. Eng. Progr. Symposium Ser.*, **48**(3), 93 (1952).
17. Souders, M., Jr., and G. G. Brown: *Ind. Eng. Chem.*, **24**, 519 (1932).
18. Webber, C. E.: *Am. Inst. Mining Met. Engrs., Tech. Publ.*, 1940, p. 1252.

CHAPTER 14

TRAY HYDRAULICS
Bubble-cap Trays

WILLIAM L. BOLLES
Monsanto Chemical Company

The bubble-cap tray is a device employing bubble caps for effecting an approach to an equilibrium stage between vapor and liquid flowing countercurrent to each other in a bubble-tray column. The bubble caps are *caps* or inverted *cups* located above *risers* through which vapor enters from below the *tray* and is dispersed under the surface of the liquid by means of *slots* in the caps. A photograph of a typical bubble-cap tray is shown in Fig. 14-1, and a photograph of some typical bubble caps, in Fig. 14-2.

The bubble-cap tray is the best known vapor-liquid contacting device. Through the years it has been a standard for the chemical and petroleum industry, and a majority of the existing commercial vapor-liquid contacting devices, such as fractionators and absorbers, contain bubble-cap trays. Because of the widespread acceptance of bubble-cap trays and the wealth of operating experience developed on them through the years, designers have in the past been wary of specifying alternate contacting devices having relatively unknown hydraulic characteristics.

There are other types of vapor-liquid contacting devices, as, for example, perforated trays. Recently there has been a significant shift from bubble-cap to perforated and proprietary trays in the design of new columns, primarily because of their lower cost. However, there are two unique advantages of bubble-cap trays which should be considered:

1. The fixed-seal arrangement of bubble-cap trays enables them to be operated over a wider range of conditions while maintaining relatively constant efficiency.

2. The greater wealth of published performance data on and experience with bubble-cap trays supports their choice when the process risk is likely to be high. However, it should be recognized that these considerations do not always lead to the bubble-cap tray.

FIG. 14-1. Typical bubble-cap tray.   (*Courtesy of F. W. Glitsch and Sons, Dallas, Tex.*)

FIG. 14-2. Some typical bubble-caps.   (*Courtesy of F. W. Glitsch and Sons, Dallas, Tex.*)

The primary purpose of this chapter is to present a quantitative treatment of vapor and liquid hydraulics on bubble-cap trays and to show how these relationships can be used in the commercial design of this type of equipment. The tray dynamics technology presented is also of importance in the prediction of performance (rating) of existing bubble-cap trays. Chapter 15 will then present a similar treatment for perforated trays. Finally, Chap. 16 will cover tray efficiencies, which are applicable to both bubble-cap and perforated trays.

TRAY DYNAMICS

14-1. General Concepts

The contacting action of a properly designed bubble-cap tray results in a highly turbulent liquid-vapor mass, or "froth" of high interfacial

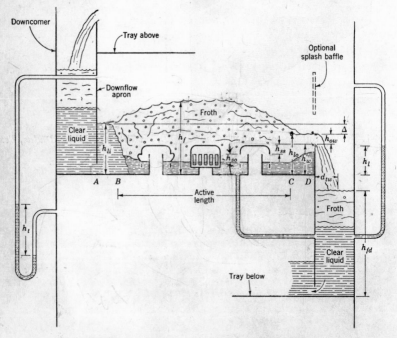

Fig. 14-3. Schematic diagram—bubble-cap-tray dynamics.

area. Net liquid movement is in crossflow to the ascending vapor stream. The froth is usually liquid-continuous and may be characterized by transient "surging" and "backmixing," but it serves the primary function of providing high interfacial area and turbulent mass transport essential to high tray efficiencies.

For clarity of presentation and to enable rational tray dynamic calculations, the complex fluid mechanics of the bubble-cap tray must be considered from a simplified steady-state point of view. The schematic diagram in Fig. 14-3 represents such a simplification.

Liquid descends from the tray above through a downcomer and onto the tray at point A. Between points A and B, essentially clear liquid is in crossflow, since there is some clearance between the downflow apron and the first row of bubble caps. *The distance between A and B will vary according to design conditions but is usually quite short. *Thus, aeration may occur immediately adjacent to the downcomer apron. The height of clear liquid h_{li} at point A, then, is the equivalent of liquid in the aerated mass.

The active portion of the tray lies roughly between points B and C. This is the section occupied by the caps and is termed *active area*. "Bubbling area" or "allocated cap area" are also used to denote this area.

In the active area tray action is characterized by the aerated mass, or froth. The observable height of this mass is designated as h_f. The density of the mass varies with height, being greatest near the tray floor; however, it is convenient to utilize a height-average density of ρ_f. There is also an effective hydrostatic head of the aerated mass h_l which can be measured by a floor manometer as shown in Fig. 14-3. If the weight of the vapor portion of the mass is neglected, h_l may be thought of as a "settled" height of clear liquid, having a density of ρ_l.

The active area and aerated mass of the tray are of primary importance in mass-transfer calculations. A *relative froth* density ϕ (dimensionless) can be defined as follows:

$$\phi = \frac{h_l}{h_f} = \frac{\rho_f}{\rho_l} \tag{14-1}$$

The holdup of vapor in the aerated mass is

$$\frac{\text{Volume vapor}}{\text{Volume mixture}} = \frac{h_f - h_l}{h_f} = 1 - \phi \tag{14-2}$$

and the holdup liquid is

$$\frac{\text{Volume liquid}}{\text{Volume mixture}} = \frac{h_l}{h_f} = \phi \tag{14-3}$$

The *residence time of vapor* in the aerated mass is, then,

$$t_v = \frac{h_f A_a (1 - \phi)}{12Q}$$

$$= \frac{h_f (1 - \phi)}{12 U_a} = \left(\frac{h_l}{12 U_a}\right)\left(\frac{1 - \phi}{\phi}\right) \tag{14-4}$$

Similarly, the *residence time of liquid* in the aerated mass is

$$t_l = \frac{h_f A_a \phi}{12q} = \frac{h_l}{12u_a} \qquad (14\text{-}5)$$

Returning to Fig. 14-3, it should be apparent that a liquid gradient Δ will build up across the tray in order to overcome the flow resistance imposed by the bubble caps and the aerated mass. This liquid gradient is measured in terms of the difference in the equivalent height of clear liquid between the inlet, h_{li}, and the outlet, h_{lo}, sides of the tray.

After the aerated mass has crossed the tray, collapse begins when the mass leaves the bubble-cap zone at point C. A calming section between point C and the outlet weir (point D) provides time for collapse, but some foam may be carried into the downcomer. In this calming section the liquid height is h_{lo}. The distance between C and D will vary according to the foam stability of the system and whether an exit splash baffle (for foam backup) is used. In practice, most liquid-vapor systems allow close spacing between points C and D without danger of excess foam being carried into the exit downcomer. In flowing over the weir, the liquid is "thrown," and the *throw over the weir* d_{tw} is the distance beyond the weir where the waterfall strikes the material in the downcomer.

Secondary froth formation occurs in the downcomer as a result of froth carry-over, liquid splashing, and generally turbulent conditions. The vapor must disengage from the froth in the downcomer to return upward through the free portion of the downcomer. Otherwise, vapor may be carried down with the liquid to the tray below. The result of this is both a lowering of efficiency and an increase in pressure drop. The height of froth (aerated mass) in the downcomer is denoted by h_{fd}.

The vapor from the tray below, meanwhile, enters the risers of the bubble caps, reverses its direction, flows downward through the annular spaces while depressing the liquid level, and then passes out through the slots. The *slot opening* h_{so} is the distance which the liquid is depressed below the top of the slots. The vapor issues from the slots in the form of bubbles (low flow) or in continuous channels (high flow). The distance between the top of the slot and the *equivalent* liquid level is termed the *dynamic slot seal*. The *vapor pressure drop* of the tray h_t is measured between the vapor space above and below the tray (Fig. 14-3). The rising vapor stream may carry liquid droplets to the tray above, which is called liquid *entrainment*.

Because of the liquid gradient across the tray, the hydrostatic liquid head on the downflow side of the tray is less than that at the upstream side. Therefore, more vapor flows through the downstream bubble caps than through those upstream. This causes variations in vapor distribution across the tray.

As noted above, this description of tray dynamics is somewhat idealized. Tray action is rarely at a steady state and usually is hidden behind metal walls. However, many observations of operating trays have been made, and the simplifications given here are reasonable approximations to the observed phenomena. Furthermore, these simplifications stand the test of permitting tray design variables to be correlated in such a way as to enable confident prediction of performance.

Further study of Fig. 14-3 reveals the reason for the dependability of the bubble-cap tray. The overflow weir can be set at such a height to produce a positive seal of the slots so as to force the vapor to contact the liquid. Furthermore, the maintenance of this liquid seal is not dependent on the flow of vapor; therefore, good vapor-liquid contacting is achieved at low as well as high vapor rates. In addition, the contacting of vapor and liquid is not greatly influenced by the flow of liquid, and both low and high liquid rates can be accommodated with good efficiencies.

14-2. Foaming

The foaming properties of a system are important in tray dynamics (32). *Foamability* refers to the degree of expansion of liquid when aerated; the foam† so produced is a function primarily of liquid physical properties, although method and degree of aeration also exert their influence. *Foam stability* refers to the rate of collapse of foam after aeration is stopped. Foamability and foam stability are not necessarily related; for example, a system may produce a large amount of foam which collapses rapidly.

A certain degree of foaming is desirable in order that high interfacial area for contacting is achieved. Excessive foaming, however, results in a build-up on the tray which may cause excessive entrainment and flooding. Also, foam-producing impurities can concentrate at the interface and act as mass-transfer barriers. Ideally, the designer would control foamability (by use of surfactants) to obtain an optimum balance between interfacial area and mass-transfer rate.

Foam stability must be considered in downcomer design and desirably should be low. Stable foams passing to the downcomer require high residence times (large downcomer volumes) for the necessary collapse. It is likely that most of the reported problems of foaming in commercial equipment could be traced to excessive stability, i.e., insufficient downcomer capacity. (Many of the problems are associated with absorbers where liquid capacity is a general limitation.)

Prediction of foamability and foam stability is much more difficult

† In this book "foam" refers to aeration due primarily to physical property (surface phenomena) effects whereas "froth" is aeration due primarily to vapor-agitation effects. The terms are to a degree interchangeable in tray dynamics.

than it would appear at first glance. Simple laboratory tests do not always correlate well with plant observations and at best serve only to indicate relative foaming properties. There is a great deal of interest in developing a satisfactory and standardized "foam index." At the present time, however, only general observations are possible.

It is convenient to consider "foaming" and "nonfoaming" systems in regard to tray dynamics. Into the former category fall systems with surfactants, such as dilute aqueous solutions of alcohols and ketones. Into the latter category fall the usual hydrocarbon systems such as cyclo-hexane–n-heptane, isobutane–n-butane, and benzene-toluene. The air-water system so commonly used in tray dynamic experiments foams even less than the "nonfoamers" as categorized above. Mixtures such as absorption oils can have varying foaming tendencies, depending on the nature of minor impurities.

Prediction of a foaming system is difficult. The usual practice is to compare laboratory aeration test results obtained on systems known to be "foamers" and "nonfoamers." The test liquid is placed in a glass column equipped with a sintered glass distributor. Nitrogen (or other suitable gas) is passed through the liquid, and foam height (or density) recorded as a function of gas rate. Suitable standard "foamers" are

800 ppm isopropanol in water
200 ppm methylisobutyl ketone in water
1.0 volume per cent n-butanol in water

It is difficult to reproduce foaming tests on the same system but in different equipment; hence, no data are included here. There are times also when a system exhibiting definite foaming tendencies in a small laboratory column may foam very little on the plant scale. Much work remains before a suitable foam-test method will be available. Observations of foaming tendencies in small laboratory plate columns (e.g., Oldershaw columns) are of value in that the equilibrium vapor and liquid are in contact. As in the case of the simple dynamic test mentioned above, large wall effects in the small column may indicate foaming not later found in the large column. Zuiderweg and Harmens (40), working with various laboratory distillations, have observed that when there is an *increase* in surface tension from top to bottom of the column, there are *increased* foaming tendencies. Such observations have yet to be correlated with plant performance.

14-3. Flooding

Flooding on a bubble-cap tray may be brought on by either excessive *entrainment* or *liquid backup* in the downcomer. The true point of flooding is difficult to determine experimentally, and "maximum capacity" is usually synonymous with an incipient flooding condition brought on by either of the two phenomena noted above. Regardless of cause, the onset of flooding is detected by (1) a sharp increase in pressure drop and (2) a sharp decline in efficiency.

Observations of flooding due to entrainment show that the aerated mass reaches the tray above. This may happen when the downcomers are only half full of liquid; however, the effect of "choking" the tray with foam is to increase pressure drop sharply, which in turn causes downcomer backup to increase to the point of flooding the downcomers.

Alternately, the downcomer may fill with liquid because of inadequate handling capacity before entrainment becomes excessive. At such a time the aerated mass level may not be near the tray above. However, the rising level of liquid first in the downcomer and then on the tray soon "chokes" the tray, and flooding results.

Fair and Matthews (13) developed a generalized correlation of flooding in bubble-cap trays by means of "liquid-vapor flow" and "vapor capacity" parameters. The liquid-vapor flow parameter is defined as

$$F_{lv} = \frac{w}{W} \sqrt{\frac{\rho_v}{\rho_l}} \tag{14-6}$$

This parameter accounts for liquid flow effects on the tray and in reality is a ratio of liquid/vapor kinetic-energy effects. The capacity parameter is defined as

$$C_{sb} = U_n \sqrt{\frac{\rho_v}{\rho_l - \rho_v}} \tag{14-7}$$

The latter parameter was originally developed by Souders and Brown (35) and derived from a consideration of droplet suspension in a gas stream. The vapor velocity term U_n is based on the "net area" of the tray available for liquid disengagement; normally this area is the difference between total area and area blocked off by one downcomer. Unusual baffling effects on the tray can reduce this area.

The available data on flooding were then correlated by means of these two parameters. The resulting generalized correlation is presented in Fig. 14-4. The data used came from Jones and Pyle (21), Mayfield et al. (26), Gerster et al. (15), Rhys and Minich (29), and Clay et al. (7). Additional data were included in a later publication (12). The values of the capacity parameter C_{sb} in Fig. 14-4 are higher in general than those originally recommended by Souders and Brown. The latter values are not based on a flooding condition but rather on a conservative condition where entrainment is negligible.

The curves in the correlation are general and are therefore subject to

some limitations, which are

1. System is low to nonfoaming.
2. Weir height is less than 15 per cent of tray spacing.
3. Bubbling area occupies most of area between weirs.
4. Liquid surface tension is 20 dynes/cm.

With the above qualifications, the flooding correlation presented in Fig. 14-4 may be expected to be accurate within about ± 15 per cent.

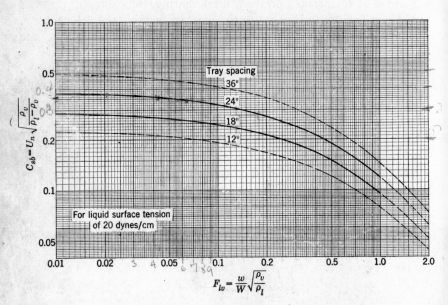

FIG. 14-4. Generalized correlation of flooding on bubble-cap trays. (*Fair and Matthews, Petrol. Refiner, April, 1958, p. 153. Copyright by Gulf Publishing Co., Houston, Texas, 1958.*)

Fair (12) has developed the effect of liquid surface tension on flood capacity for sieve trays. The final correlation, Eq. (15-3a), applies also to bubble-cap trays and is as follows:

$$\frac{C_{sb}}{(C_{sb})_{\sigma=20}} = \left(\frac{\sigma}{20}\right)^{0.2} \tag{14-8}$$

14-4. Entrainment

Entrainment is defined as the liquid carried with the vapor from one tray to the tray above. Entrainment is detrimental for two reasons. First, it lowers plate efficiency, since it carries liquid from a tray of lower volatility to a tray containing liquid of higher volatility, thereby diluting the effect of fractionation. Second, entrainment may also be detrimental since it can carry nonvolatile constituents upward with the result that nonvolatile impurities in the feed may be carried up the column to contaminate the distillate.

The deleterious effect of entrainment can be related to lowered tray efficiency by means of the Colburn equation (8):

$$\frac{E_a}{E_{mv}} = \frac{1}{1 + eE_{mv}/L} \tag{14-9}$$

A method for correlating entrainment in bubble-cap columns was developed by Fair and Matthews (13). Under entrainment conditions liquid is carried from tray to tray by the vapor, exchanging with liquid above each tray. The gross liquid downflow then includes the normal "dry" value L plus the entrainment e. A fractional entrainment ψ, based on gross downflow, is defined as follows:

$$\psi = \frac{e}{L + e} \tag{14-10}$$

which can be substituted in Eq. (14-9):

$$\frac{E_a}{E_{mv}} = \frac{1}{1 + E_{mv}[\psi/(1 - \psi)]} \tag{14-11}$$

When there is little or no entrainment, $\psi \sim 0$ and $E_a \sim E_{mv}$. When the entrainment-flood point is approached, $\psi \to 1$ and $E_a \to 0$.

Following the above approach, Fair and Matthews took the available data on entrainment (1, 4, 5, 19, 21, 27, 34, 36) and developed a generalized correlation of entrainment, which is presented in Fig. 14-5.

The use of the flow parameter F_{lv} in Fig. 14-5 should be noted. Since the quantity of entrainment is a general function of vapor rate, an increase in the liquid/vapor ratio results in a decrease in the fraction of liquid entrained. However, this is not a simple proportionality, since effects of liquid rate on contacting action must be considered. These effects include vapor "jetting" at shallow seals, splashing, and frothing action.

According to the original reference (13), this correlation represents experimental data with an accuracy of about ± 20 per cent.

14-5. Slot Opening

The method for calculating slot opening was originally developed by Rogers and Thiele (31), and simplified by Winn (38) and Bolles (3). All these methods are based on applying the orifice equation

$$U_s = K_s \sqrt{2g \left(\frac{\rho_l - \rho_v}{\rho_v} \right) \left(\frac{h}{12} \right)} \tag{14-12}$$

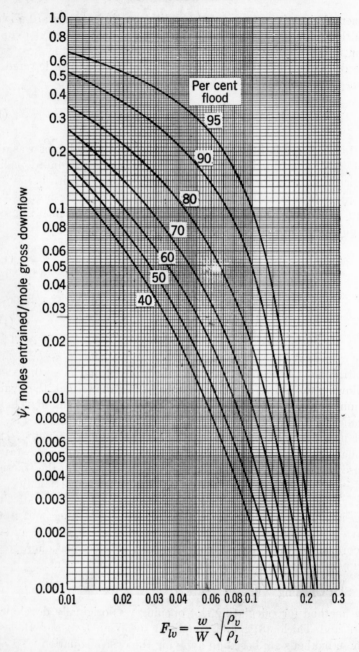

Fig. 14-5. Generalized correlation of entrainment on bubble-cap trays. (*Fair and Matthews, Petrol. Refiner, April,* 1958, *p.* 153. (*Copyright by Gulf Publishing Co., Houston, Texas,* 1958.)

to a differential element of slot area, followed by integration over the open slot area until the vapor load has been satisfied.

The generalized equation recommended for slot opening in trapezoidal slots is as follows:

$$Q = 2.36 \frac{A_s}{h_{sh}} \sqrt{\frac{\rho_l - \rho_v}{\rho_v}} \left[\frac{2}{3} \left(\frac{R_s}{1 + R_s} \right) h_{so}^{3/2} + \frac{4}{15} \left(\frac{1 - R_s}{1 + R_s} \right) \frac{h_{so}^{5/2}}{h_{sh}} \right] \quad (14\text{-}13)$$

where

$$R_s = \frac{d_{st}}{d_{sb}} \quad (14\text{-}14)$$

The orifice coefficient K_s employed is 0.51, which was determined experimentally by Rogers and Thiele. Trial and error is necessary, of course, to solve Eq. (14-13) for slot opening.

Maximum slot capacity is obtained by the substitution of h_{sh} for h_{so}, which results in

$$Q_{max} = 2.36 A_s \sqrt{h_{sh} \left(\frac{\rho_l - \rho_v}{\rho_v} \right)} \left[\frac{2}{3} \left(\frac{R_s}{1 + R_s} \right) + \frac{4}{15} \left(\frac{1 - R_s}{1 + R_s} \right) \right] \quad (14\text{-}15)$$

Rectangular slots are a special case of trapezoidal slots in which $R_s = 1$, for which Eq. (14-13) reduces to

$$Q = 0.79 \frac{A_s}{h_{sh}} \sqrt{\frac{\rho_l - \rho_v}{\rho_v}} h_{so}^{3/2} \quad (14\text{-}16)$$

This equation can be solved explicitly for slot opening:

$$h_{so} = 1.17 \left(\frac{\rho_v}{\rho_l - \rho_v} \right)^{1/3} \left(\frac{Q h_{sh}}{A_s} \right)^{2/3} \quad (14\text{-}17)$$

Maximum capacity of rectangular slots is, then,

$$Q_{max} = 0.79 A_s \sqrt{h_{sh} \left(\frac{\rho_l - \rho_v}{\rho_v} \right)} \quad (14\text{-}18)$$

Triangular slots are a special case of trapezoidal slots in which $R_s = 0$, for which Eq. (14-13) reduces to

$$Q = 0.63 \frac{A_s}{(h_{sh})^2} \sqrt{\frac{\rho_l - \rho_v}{\rho_v}} h_{so}^{5/2} \quad (14\text{-}19)$$

This equation can also be solved explicitly for slot opening:

$$h_{so} = 1.20 \left(\frac{\rho_v}{\rho_l - \rho_v} \right)^{1/5} (h_{sh})^{4/5} \left(\frac{Q}{A_s} \right)^{2/5} \quad (14\text{-}20)$$

Maximum capacity of triangular slots is, then,

$$Q_{max} = 0.63 A_s \sqrt{h_{sh} \left(\frac{\rho_l - \rho_v}{\rho_v} \right)} \tag{14-21}$$

Because of the complexity of the above equations, and especially

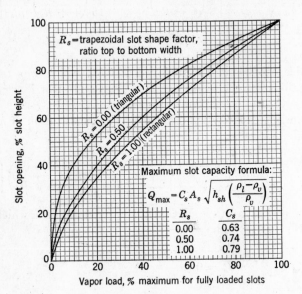

Fig. 14-6. Generalized correlation for slot opening.

because of the trial and error required to solve for slot opening in trape-zoidal slots, a chart was prepared from Eq. (14-13) as a generalized cor-relation for slot opening. This chart, presented in Fig. 14-6, applies to rectangular slots, triangular slots, and any shape of trapezoidal slots.

14-6. Liquid Crest over Weir

The height of liquid over straight weirs is estimated by the Francis weir formula:

$$h_{ow} = 0.48 \left(\frac{q'}{l_w} \right)^{2/3} \tag{14-22}$$

When the weir is located so as to discharge into a segmental downcomer, the tower wall has a constricting effect on the flow of liquid over the weir, and a correction factor F_w should be introduced:

$$h_{ow} = 0.48 F_w \left(\frac{q'}{l_w} \right)^{2/3} \tag{14-23}$$

Bolles (2) derived an equation for the weir constriction correction factor as follows:

$$\frac{q'}{(L_w)^{2.5}} = 485 \left[\frac{\sqrt{1 - (R_w)^2/(F_w)^3} - \sqrt{1 - (R_w)^2}}{(R_w)(F_w)} \right]^{3/2} \quad (14\text{-}24)$$

This equation cannot be solved directly for F_w, and so a convenient working chart is presented in Fig. 14-7.

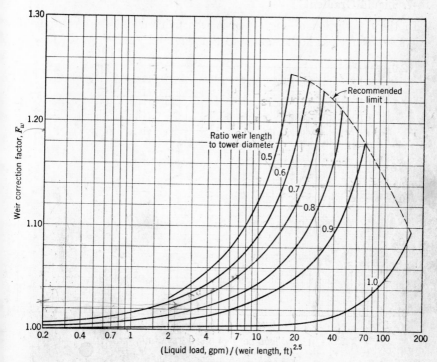

FIG. 14-7. Weir correction factor for segmental downcomers. [*Bolles, Petrol. Refiner,* **25,** 613 (1946). *Copyright by Gulf Publishing Co., Houston, Texas,* 1946.]

The height of liquid over notched weirs, with notches running full, is given (with trial and error) by Davies (11):

$$q' = 1.20 \frac{l_w}{h_n} [h_{ow}^{5/2} - (h_{ow} - h_n)^{5/2}] \quad (14\text{-}25)$$

However, if the notches are running less than full, an alternate formula must be employed:

$$h_{ow} = 0.96 \left(\frac{q' h_n}{l_w} \right)^{2/5} \quad (14\text{-}26)$$

Where circular weirs are encountered and $h_{ow} < 0.2d_w$,

$$h_{ow} = 0.18 \left(\frac{q'}{d_w}\right)^{2/3} \tag{14-27}$$

When $0.2d_w < h_{ow} < 1.5d_w$,

$$h_{ow} = 0.13 \left(\frac{q'}{d_w^2}\right)^2 \tag{14-28}$$

14-7. Liquid Gradient

In the description of Fig. 14-3, it was pointed out that a gradient exists between clear liquid inflow to the tray and clear liquid outflow to

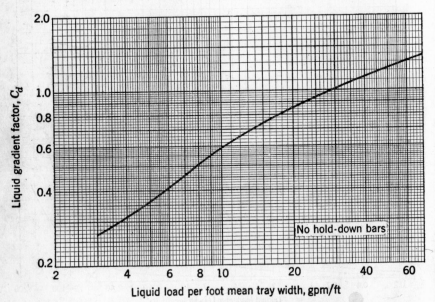

FIG. 14-8. Liquid gradient factor.

the downcomer. This is designated *liquid gradient* and is caused by frictional effects of the ascending vapor, tray floor and walls, baffles, and bubble caps.

Work on liquid gradient has been published by Davies (10), Klein (23), and Kemp and Pyle (22). Davies has developed a method of general utility based on a fundamental analysis of fluid hydraulics supplemented with experimental data, and his method is summarized here. The liquid gradient (uncorrected for vapor load) is given by the following equation:

$$\sqrt{\Delta'} \left\{ \Delta' \left[3 \left(\frac{N_r}{2} - 1\right) + \frac{2}{1 + \frac{1}{4}R_{cc}^2} \right] + 3N_r[h_{lo} + h_{sc}(R_{rc} - 1)] \right\}$$
$$= \frac{N_r \sqrt{N_r} \sqrt{1 + \frac{1}{4}R_{cc}} \, q'}{2.4C_d l_c} \tag{14-29}$$

For bubble caps placed on equilateral triangular pitch, $R_{cc} = 1$ and the above equation reduces to

$$\sqrt{\Delta'}\left\{\Delta'\left(\frac{3N_r}{2} - 1.4\right) + 3N_r[h_{lo} + h_{sc}(R_{rc} - 1)]\right\} = \frac{N_r\sqrt{N_r}\,q'}{2.15C_d l_c} \quad (14\text{-}30)$$

The liquid gradient factor C_d is given in Fig. 14-8. The vapor load correction factor C_v is given in Fig. 14-9 and applied as follows:

$$\Delta = C_v\,\Delta' \quad (14\text{-}31)$$

It should be apparent that these equations are rather cumbersome for frequent and repetitive use and that trial and error is necessary for solution for liquid gradient. Bolles (3) has developed a convenient solution based on algebraic manipulations and two simplifying assumptions. One is that equilateral-triangular cap pitch is used, and the other is that the ratio of riser to cap diameter is 0.7. The former assumption covers the majority of cap layouts, and the latter is reasonable for most cap designs (including the optimum).

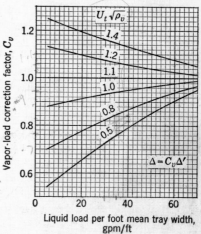

FIG. 14-9. Vapor-load correction factor for liquid gradient.

The final liquid gradient correlations are presented in Figs. 14-10 to 14-13. These charts cover the entire practical range of tray design variables for equilateral-triangular cap pitch. The liquid gradient obtained from these charts, however, must be corrected for vapor load by means of Fig. 14-9 and Eq. (14-31).

The original Davies method for estimating liquid gradient agrees with experimental measurements to within about ±25 per cent, and the Bolles charts agree with the Davies equations within 2 per cent.

14-8. Vapor Pressure Drop

The pressure drop for vapor flow from one tray to the next is designated h_t and has the value that would be read from a manometer tapped to the vapor space of adjacent trays (see Fig. 14-3). h_t is conveniently segregated into three flow resistances:

$$h_t = h_{cd} + h_{so} + h_{al} \quad (14\text{-}32a)$$

where h_{cd} is the drop through the dry caps, h_{so} is the drop through the

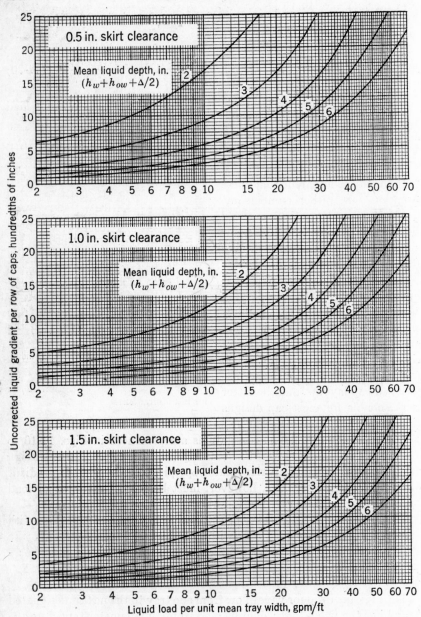

Fig. 14-10. Liquid gradient chart—cap spacing 25 per cent cap diameter. (Equilateral triangular cap pitch only.)

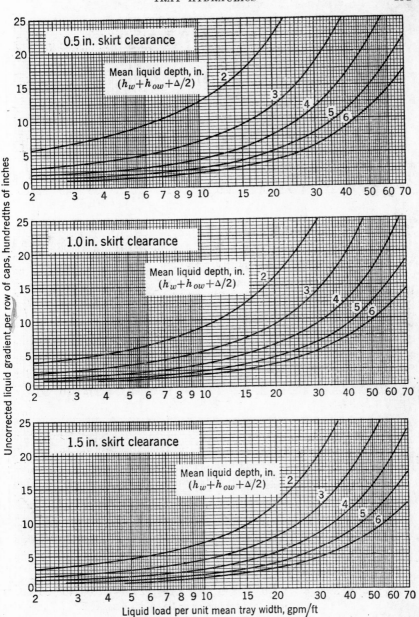

FIG. 14-11. Liquid gradient chart—cap spacing 31.25 per cent cap diameter. (Equilateral triangular cap pitch only.)

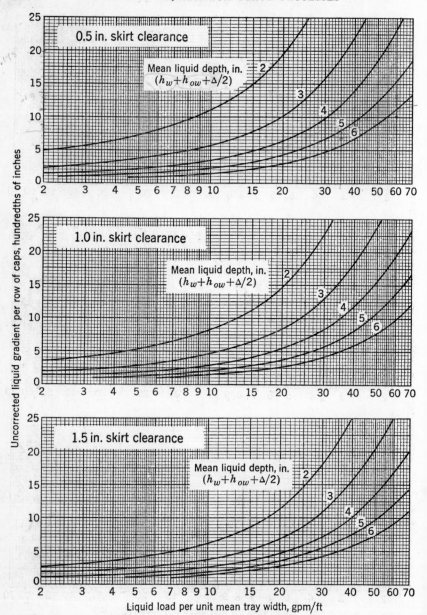

Fig. 14-12. Liquid gradient chart—cap spacing 37.5 per cent cap diameter. (Equilateral triangular cap pitch only.)

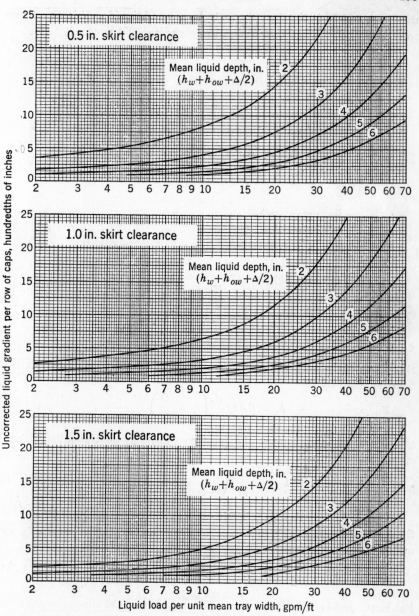

FIG. 14-13. Liquid gradient chart—cap spacing 50 per cent cap diameter. (Equilateral triangular cap pitch only.)

wet slots, and h_{al} is the drop through the aerated liquid. It is sometimes convenient to combine the dry cap drop and the slot drop:

$$h_c = h_{cd} + h_{so} \tag{14-32b}$$

The classical method for computation of dry-cap drop is that of

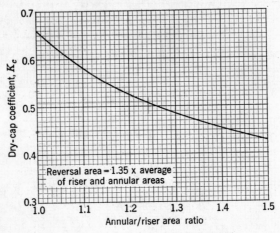

FIG. 14-14. Dry-cap head-loss coefficient.

Dauphine (9). According to this method, the head loss through the risers is given by

$$h_r = 0.111 \frac{d_r}{\rho_l}\left(\sqrt{\rho_v}\,\frac{Q}{A_r}\right)^{2.09} \tag{14-33}$$

and the head loss through the passage consisting of the reversal and annulus is given by

$$h_{ra} = \frac{0.68}{\rho_l}\left(\frac{2a_r^2}{a_{er}a_c}\sqrt{\rho_v}\,\frac{Q}{A_r}\right)^{1.71} \tag{14-34}$$

Bolles (3) found that practically the same results can be obtained for the sum of h_r and h_{ra} by means of treating the dry cap as an orifice by means of the equation

$$h_{cd} = K_c \frac{\rho_v}{\rho_l}\left(\frac{Q}{A_r}\right)^2 \tag{14-35}$$

For round bubble caps with a reversal area about 1.35 times the average of riser and annular areas (which is optimum), K_c is given by Fig. 14-14. This simplified method gives dry-cap drops within ± 10 per cent of the Dauphine equations.

Equations for slot opening have already been presented. For slots not fully loaded, the head loss accompanying vapor flow through the slots would be equal to the slot opening if clear liquid existed around the slots. Actually, the liquid around the slots is aerated and there is some difference. However, it is more convenient to assume that the slot drop is equal to the slot opening and lump the difference into the aerated liquid drop term.

When the slots become loaded, the slot head loss becomes equal to the slot height. For vapor flow greater than that for loading, slot drop is increased only by the height of the shroud ring if the cap design includes clearance under the skirt. *If the caps are flush*, the increased pressure drop can be estimated from the equation

$$\frac{h_{so,\text{overloaded}}}{h_{sh}} = \left(\frac{Q_{\text{overloaded}}}{Q_{\max}}\right)^2 \tag{14-36}$$

The head loss through the aerated liquid is calculated by means of the equation

$$h_{al} = \beta h_{ds} \tag{14-37}$$

where β is called the *aeration factor* and h_{ds} is the *dynamic slot seal*:

$$h_{ds} = h_{ss} + h_{ow} + \frac{\Delta}{2} \tag{14-38}$$

β takes into account energy loss due to bubble formation, frictional resistance of flow through the aerated mass, the difference between slot drop and slot opening, and static head effects. In the past, many designs have implicitly assigned values of unity to β; i.e., wet tray pressure drop has been assumed equal to the dynamic slot seal. Such an assumption is unnecessarily conservative.

The aeration factor can be determined from experimental pressure-drop data by means of combining Eqs. (14-32) and (14-37):

$$\beta = \frac{h_t - h_{cd} - h_{so}}{h_{ds}} \tag{14-39}$$

By means of the above equation, aeration factors were computed from the extensive bubble-cap-tray data of Kemp and Pyle (22). It was then found that an approximate correlation exists between β and a vapor flow parameter:

$$F_{va} = U_a \sqrt{\rho_v} \tag{14-40}$$

The resulting correlation is presented in Fig. 14-15.

When operation is close to the flood point, pressure drop increases markedly owing to entrainment. Fair, in Chap. 15, considers the effect of entrainment on pressure drop in sieve trays. The same method, summarized by Eq. (15-12), can be used for bubble-cap trays.

The method given here for calculating vapor pressure drops has been compared with the experimental data presented by Kemp and Pyle (22), Atteridg et al. (1), and Jones and Pyle (21). The accuracy was generally within ±15 per cent, with those deviations greater than 15 per cent almost always on the conservative side.

14-9. Downcomer Dynamics

There are four tray dynamic factors pertaining to downcomers: (1) head loss for outflow from downcomer, (2) total liquid backup in the

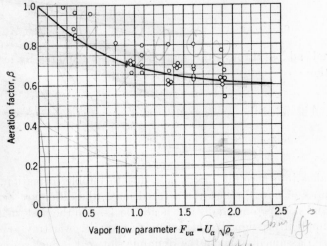

FIG. 14-15. Aeration factor for bubble-cap trays.

downcomer, (3) residence time of liquid in downcomer, and (4) liquid throw over weir.

The liquid leaving the downcomer is restricted somewhat by being forced to flow under the downflow apron (see Fig. 14-3). The liquid head loss for flow under the downflow apron is calculated from the orifice equation with a coefficient of 0.6:

$$q = 0.6A_{da} \sqrt{\frac{2gh_{da}}{12}} \tag{14-41}$$

or

$$h_{da} = 0.03 \left(\frac{q'}{100A_{da}}\right)^2 \tag{14-42}$$

If a tray inlet weir is used after the downcomer, the head loss is about 20 per cent greater than calculated from Eq. (14-42), with A_{da} being the minimum area of flow, whether under the downflow apron or between the apron and the weir. Analysis of the liquid hydraulics in Fig. 14-3 will show that the height of clear liquid backup in the downcomer can be obtained by

$$h_{dc} = h_t + h_w + h_{ow} + \Delta + h_{da} \tag{14-43}$$

For this equation, h_t is obtained from Eq. (14-32), h_w is specified, h_{ow} is obtained from Eqs. (14-23), etc., Δ is obtained from Figs. 14-10 through 14-13, and h_{da} is obtained from Eq. (14-42).

It should be noted that h_{dc} in Eq. (14-43) is given in terms of clear liquid. The actual height of *aerated liquid* in the downcomer is

$$h_{fd} = \frac{h_{dc}}{\phi} \tag{14-44}$$

It is common to assume a downcomer froth density ϕ of 0.5 minimum, as noted below.

A certain residence time of liquid in the downcomer is necessary in order to allow collapse of foam. The foam may be created by general turbulence in the downcomer, or it may actually be carried across the tray outlet weir and into the downcomer. If the residence time is too short, as noted earlier, vapor may be carried down to the next tray and danger of choking arises.

The liquid in the downcomer is aerated to some extent and has characteristics analogous to the aerated mass on the tray. Thus, the relative foam density in the downcomer is

$$\phi = \frac{\rho_f}{\rho_l} = \frac{h_{dc}}{h_{fd}} \tag{14-45}$$

and the volume of aerated mass is

$$V_{dc} = \frac{A_d h_{fd}}{12} \tag{14-46}$$

where A_d, the downcomer area at the top, is adjusted if the downcomer is sloped. The true residence time of aerated mass in the downcomer is

$$t_{dc} = \frac{A_d h_{fd}/12}{q/\phi} = \frac{A_d h_{dc}}{12q} \tag{14-47}$$

Experimental values of the relative foam density in downcomers are also lacking. Obviously, the specific foam density ρ_f varies from a value of approximately ρ_l at the bottom of the downcomer to a much lower value at the foam-vapor interface. A conservative average value of the relative density ϕ, widely used in the design of both foaming and non-foaming systems, is $\phi = 0.5$. For low-foaming systems, a higher value of ϕ is reasonable.

It has been assumed by some that liquid thrown too far over the weir of a segmental downcomer impinges against the tower wall, thereby

interfering with the free upflow of vapor disengaging in the downcomer.
Bolles (3) has derived an equation for the trajectory of clear "liquid
thrown" over a straight weir. This equation is

$$d_{tw} = 0.8 \sqrt{h_{ow}h_{ff}} \qquad (14\text{-}48)$$

h_{ff} is measured from the top of the weir down to the level of aerated mass
in the downcomer:

$$h_{ff} = h_{ts} + h_w - h_{fd} \qquad (14\text{-}49)$$

"Liquid throw" is no longer considered to be a limiting factor in tray
design, especially if "antijump" baffles are provided on center downcomers. This will be discussed in more detail in the later section on
tray design.

TRAY DESIGN

The literature on bubble-cap tray design is voluminous. Among the
more significant works are those of Cicalese et al. (6), Davies (11), Perry
(28), Robinson and Gilliland (30), and Bolles (3). Also significant are
several recent reviews (16, 14, 18). In order that a perspective of the
over-all problem can be obtained, the reader is directed to Table 14-1,

TABLE 14-1. PROCESS DESIGN SPECIFICATION CHECK LIST
A. *Over-all column design*
1. Column diameter
2. Number of trays
3. Tray spacing
4. Feed and drawoff locations
5. Operating temperatures and pressures
6. Materials of construction
B. *Tray design*
1. Tray type
2. Active area
3. Bubble-cap size, design, and number
4. Cap layout (pitch and spacing)
5. Cap skirt clearance and static slot seal
6. Cap stepping (if desired)
7. Vapor riser dimensions
8. Tray baffles and calming zones
9. Downflow area
10. Downcomer type and clearance
11. Tray inlet weir
12. Tray outlet splash baffle
13. Outlet weir type
14. Outlet weir height and length
15. Tray drain holes
16. Tray and weir levelness
17. Materials of construction

which lists most of the factors which must be determined before a bubble-
cap tray column can be released for mechanical design.

Among these are nine principal independent variables:

1. Tower diameter
2. Tray spacing
3. Tray type
4. Downflow area
5. Bubble-cap size
6. Cap spacing
7. Cap skirt clearance
8. Outlet weir height
9. Cap stepping

The object of good tray design is to determine the optimum combination of these nine variables. This optimum is that combination which:

1. Ensures required performance under design conditions
2. Provides required flexibility to handle anticipated varying loads
3. Achieves the above at minimum cost

In addition, it is necessary to consider materials of construction, tray construction, and a number of miscellaneous factors.

14-10. Materials of Construction

Bubble-cap trays have been constructed of cast iron, sheet carbon steel, and sheet metal of various alloys. Cast-iron caps and trays used to be common. However, they are very heavy and require heavy foundations and tower structures. Furthermore, cast-iron caps are thick and are consequently wasteful of column cross-sectional area.

Sheet-metal trays and caps overcome these disadvantages, as well as being cheaper. Carbon-steel trays are considerably cheaper than cast iron. Furthermore, pressed caps and trays of 11 to 13 per cent chrome in the gauge metal thicknesses are competitive with cast-iron trays. As a result, cast-iron trays are seldom specified any more.

Nearly all the common metals and alloys have been used in the construction of sheet-metal trays and caps. Low-carbon steel is the usual standard for noncorrosive, ordinary-temperature service. The use of other metals and alloys is dictated by the conditions of corrosion and temperature expected. In order to indicate the effect of materials of construction on cost, the following relative tray costs are illustrative of some of the common metals and alloys:

Material	Relative tray cost
Carbon steel	1
11–13% chrome type 410	2
18-8 type 304	$2\frac{1}{2}$
18-8 moly type 316	$3\frac{1}{2}$
Monel	4

14-11. General Tray Types

General tray types are classified by the mode of liquid flow on the tray. The more common and recommended tray types are illustrated in Fig. 14-16.

The crossflow tray, wherein the liquid flows directly across the tray, is the most common tray type. This represents the simplest tray construction and is consequently the most economical to fabricate. In

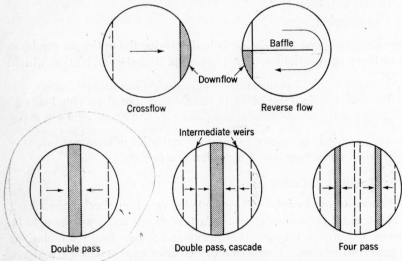

FIG. 14-16. Recommended general tray types.

addition, the long liquid path across the tray contributes to high plate efficiency.

In the reverse-flow tray, the downcomers are all located on one side of the column and the liquid is forced to flow around a center baffle, reversing its direction at the far side of the tray. The reverse-flow tray provides more cap area at the expense of downcomer area. It should be evident that the liquid path is quite long, and hence this tray tends toward high liquid gradients. The reverse flow tray is advantageous only for very low liquid/vapor ratios.

In the double-pass tray, the liquid flow is split into two portions, each flowing across half the tray. Not only is the liquid load per unit tray width cut in half, but the length of each liquid path is also cut in half. As a result, it should be evident that the double-pass tray provides considerably more liquid capacity and less liquid gradient than the single crossflow tray. Consequently, the double-pass tray is particularly advantageous for high liquid/vapor ratios or large tower diameters.

However, it should be realized that the double-pass design costs slightly more than a crossflow tray of the same diameter. Also, the shorter liquid path of double-pass trays reduces tray efficiency as compared with crossflow trays.

If even more liquid capacity is required, intermediate weirs may be provided to give the double-pass, cascade tray, where the tray floor is "stepped" at two elevations. This cuts the liquid path for each bubble-cap section in half again, as compared with the basic double-pass tray.

For very large tower diameters it may be necessary to employ four-pass or even six-pass trays. An odd number of passes is not recommended because of problems in liquid distribution.

In addition to the tray types shown in Fig. 14-16, numerous other types have been employed (11, 30, 37). However, they are recommended only in special cases.

The most suitable type of tray depends on the liquid/vapor ratio and the tower diameter. The various types of trays can be listed in order of a general increase of either or both of these variables as follows:

1. Reverse flow
2. Crossflow
3. Double pass
4. Double pass, cascade
5. Four pass

A guide for tentative selection of tray type is given in Table 14-2. The ranges given in Table 14-2 are approximate and are intended for tentative selection only. The final selection of tray type should be based upon the results of complete tray dynamic calculations for the specific application.

TABLE 14-2. GUIDE FOR TENTATIVE SELECTION OF TRAY TYPE

Estimated tower diam, ft	Range of liquid capacity, gpm			
	Reverse flow	Crossflow	Double pass	Cascade double pass
3	0–30	30–200		
4	0–40	40–300		
6	0–50	50–400	400– 700	
8	0–50	50–500	500– 800	
10	0–50	50–500	500– 900	900–1400
12	0–50	50–500	500–1000	1000–1600
15	0–50	50–500	500–1100	1100–1800
20	0–50	50–500	500–1100	1100–2000

14-12. Tray Construction

Bubble-cap trays are of either removable or nonremovable construction. The most common type of removable tray is designed to be passed in sections through manholes. The bubble caps are removable and are bolted in place. The tray parts are also bolted together and designed to be removable through manholes which are located in the column shell.

This type of tray is easily installed, maintained, and repaired and is recommended for most applications. However, it is limited to columns $2\frac{1}{2}$ ft in diameter and larger, as this is about the minimum space in which a man can work inside a column.

The so-called "cartridge" tray was designed to fill the need of removable bubble-cap trays for smaller column diameters (3). These trays are installed in columns which are flanged at the top. The trays are built in cartridges which are designed to be slipped in and stacked on one another in the column.

Trays can also be designed to be removed by dismantling shell flanges. These trays are usually built in one piece, and the column flanged every tray or in sections containing several trays. This type of construction is expensive in both original and maintenance cost and is not recommended except for columns under $2\frac{1}{2}$ ft.

In the case of nonremovable construction, the trays are welded into the column. In the larger installations, a manway can be provided in the column and passageways built into the trays so a man can work his way down through the column for inspection and minor repairs. This represents the cheapest construction, but tray inspection, cleaning, or repair is difficult and costly.

14-13. Tray Spacing

Tray spacings from 6 to 54 in. have been used in commercial columns. Tray spacing has an important effect on flooding and entrainment. Under flood-limiting conditions there is a reciprocal relationship between tray spacing and tower diameter. For example, with a design at the flood point, any reduction in tray spacing must be compensated for by an increase in tower diameter in order to avoid flooding.

It should be apparent that there is a theoretical optimum tray spacing based on tray dynamic considerations which gives minimum column cost. Bolles (2) illustrated how this optimum can be determined and showed in an example that the economic optimum occurred at a tray spacing of about 24 in.

Actually, it is usually mechanical factors which decide tray spacing. For example, it is necessary to provide sufficient space between trays so as to facilitate inspection and repairs. In columns 5 ft and larger in diameter, it is recommended that the tray spacing be a minimum of 24 in. in order to permit workmen to crawl between trays. In columns 4 ft and below in diameter, it is not necessary to be able to crawl between trays and a tray spacing of 18 in. can be permitted.

Sometimes consideration of the number of trays in the column may affect the selection of tray spacing. In the case of fractionations requiring a very large number of trays, smaller tray spacings may be considered in order to avoid splitting the fractionator into two towers.

With fractionators requiring a very small number of trays, on the other hand, consideration may be given to tray spacings greater than 24 in.

Another special case involves the design of columns to be installed inside existing buildings. Here available headroom may fix allowable tray spacing.

Still another situation is encountered in the case of cryogenic columns, as, for example, in oxygen plants. Here the heat leak through insulation is of such paramount importance that tray spacings as low as 6 in. are common.

14-14. Downcomers

Circular downflow pipes used to be a common form of downcomer. This type of downcomer provides very low downflow area and poor vapor disengaging space. Circular pipe downcomers usually constitute the first bottleneck to column capacity and consequently are seldom used.

The segmental downcomer is the type generally recommended. This type provides maximum utilization of column area for downflow and results in greatest downflow capacity. In order further to ensure maximum downflow capacity it is recommended that the tray support rings not extend through the downcomers.

The downflow apron constitutes the internal wall of the segmental downcomer. Sometimes this baffle is inclined so that the top is farther from the tower wall than the bottom so as to provide more inlet area for vapor disengagement in the case of foaming systems.

A few notes on practical ranges of downcomer dimensions may be helpful. In the design of downcomers for crossflow trays the optimum weir length is usually 60 to 75 per cent of tower diameter. In segmental downcomers for double-pass trays, the optimum weir length is usually 50 to 60 per cent of tower diameter. The optimum width of center downcomers for double-pass trays is usually 8 to 12 in.

The overflow weir constitutes the inlet to the downcomer. The simplest and most satisfactory design consists of a straight weir formed as an extension of the downflow apron.

The height of liquid over the weir should be at least $\frac{1}{4}$ in. in order to provide adequate liquid distribution. In the case of lower liquid loads, notched weirs are recommended in order to improve distribution. However, the depth of the notches should not exceed ½ in. to provide greater capacity in case high liquid loads are encountered.

Some designers specify overflow weirs with a vertical adjustment of 1 to 2 in. This provides for adjustment of slot seal to permit extended operation at high plate efficiency with low loads and to provide greater flexibility for new services. It also provides insurance against errors in judgment in the original weir setting and facilitates initial leveling of the weir.

The height of the overflow weir is determined on the basis of providing adequate dynamic seals. This is discussed later in connection with bubble-cap design.

The "liquid throw" over weirs, as shown in Fig. 14-3, should receive special attention in the case of center and off-center downcomers of multi-pass trays, where two liquid streams from opposing weirs are "thrown" toward each other. If the liquid throw, as calculated by Eq. (14-48), is more than one-half the width of center downcomers, "antijump" baffles should be provided. An antijump baffle consists of a vertical sheet of metal located in the center of and down the length of the downcomer. The baffle extends vertically for some distance both above and below the overflow weirs, thereby preventing liquid overflowing one weir from "jumping" over the opposing weir.

Liquid throw over weirs in side downcomers is not considered to be a limiting factor.

There must be sufficient residence time in the downcomer to allow foam collapse and vapor disengagement. It is common practice to base this residence time on total downcomer volume. Davies (11) reports a study of flooding towers in which the maximum residence time of any of the towers flooding was 4 sec. In this study the residence time was based on the total downcomer volume contained between trays, and the calculated height of clear liquid averaged about two-thirds of the tray spacing. This corresponds to a maximum *true* residence time of two-thirds of 4 sec, or less than 3 sec for any of Davies' flooding towers.

The minimum allowable residence time depends on the foamability of the system. A design minimum of 3 sec true residence time, t_{dc}, as calculated by Eq. (14-47) is recommended except in the case of systems of high foamability, where the minimum allowable should be extended to about 5 sec.

It is desirable to minimize the head loss of liquid leaving the downcomer. This makes it desirable to leave adequate clearance under the downflow apron [see Eq. (14-42)]. However, some downflow apron seal is necessary to ensure that vapor cannot pass up through the downcomer.

The amount of apron seal required on larger columns is greater in order to allow for departures from true levelness of trays. The following seals (height of weir minus clearance under apron) are recommended:

Weir to apron distance, ft	Apron seal, in.
Below 6	0.5
6–12	1.0
Above 12	1.5

As was pointed out in the section on flooding, if the height of froth (aerated mass) in the downcomer exceeds the height of the downcomer or rises above the top of the overflow weir, flooding will be induced. Hence the tray and downcomer must be designed so this does not occur [see Eq. (14-44)].

14-15. Bubble Caps

Types of Bubble Caps. A great variety of types of bubble caps have been employed, some of which are illustrated in Fig. 14-2. However, the round, bell-shaped cap is most widely used. This type of cap, which is illustrated in Fig. 14-17, with trapezoidal slots, shroud ring, and removable mounting, is recommended for most applications.

Fig. 14-17. Recommended type of bubble cap. (*Courtesy of F. W. Glitsch and Sons, Dallas, Tex.*)

Sizes of Bubble Caps. Small bubble caps provide relatively more slot area. This is demonstrated in the following tabulation, where the ratio of slot to allocated cap areas is shown for the three standard bubble-cap sizes at the four cap spacings, expressed as ratio of clearance to cap diameter:

Nominal cap size, in.	Ratio slot to allocated cap area†			
	0.25	0.3125	0.375	0.50
3	0.39	0.35	0.32	0.27
4	0.36	0.33	0.30	0.25
6	0.29	0.26	0.24	0.20

† See Fig. 14-18.

Small bubble caps also provide for more flexible arrangement and result in less end wastage and liquid-distribution areas. This is especially true in the case of small columns. In addition, smaller caps are more easily covered by liquid, which promotes "froth blanketing" with resulting improved mass transfer and higher efficiencies.

There are two advantages of large bubble caps. One is that they give lower liquid gradients. The other is that they are cheaper for the same amount of tray area. However, when total tray costs are developed on the basis of equal slot area and proportionate cap spacing, there is very little difference in total tray cost. This is demonstrated by the following figures, developed for standard caps spaced at 25 per cent of cap diameter:

Cap size, in.	Relative tray cost per unit tray area	Ratio slot to allocated cap area	Relative tray cost per unit slot area
6	1.0	0.29	3.5
4	1.2	0.36	3.4
3	1.35	0.39	3.5

As a rough guide for the selection of cap diameter, it is recommended that the cap size be chosen on the basis of column size as follows:

Tower diam, ft	Cap diam, in.
2.5–5	3
4 and up	4

However, it should be realized that experimentation with tray layouts may indicate the advisability of deviating from the above guide.

Caps up to 6 in. have been used, but they are not generally recommended because of their lower efficiency.

Bubble-cap Design. *Wall thicknesses* which are commonly used in the fabrication of bubble caps are:

Gauge†	In.
10	0.134
12	0.109
14	0.078
16	0.062

† U.S. Standard.

For alloy construction, 16-gauge metal is recommended for both caps and risers. This thickness provides sufficient structural strength and rigidity while at the same time keeping the cost to a minimum.

In the case of carbon-steel caps, heavier thicknesses are recommended because of the relatively cheap material of construction. Also, some corrosion may be encountered, even where not originally expected.

Trays fitted with 12-gauge carbon-steel caps cost only about 7 to 9 per cent more than those with 16-gauge caps and will provide an additional corrosion allowance of about 0.05 in. Therefore, 12-gauge thickness is generally recommended for carbon-steel caps and risers.

Designers used to devote a great deal of attention to *slot design*. However, recent evidence has shown that slot design is not an important factor in tray performance as long as the tray is well loaded and clear liquid submergence is adequate. For example, bubble caps with no slots at all have given a remarkably creditable account of themselves!

The most commonly used slot shapes are triangular, rectangular, trapezoidal, or minor variations thereof. Bolles (3) has indicated that rectangular and trapezoidal slots are preferred because of better loading characteristics.

Slot widths ranging from ⅛ to ½ in. or more have been employed. Narrow slots theoretically favor plate efficiency because the vapor is broken up into smaller bubbles, thereby promoting greater interfacial contact area. Wide slots, on the other hand, are somewhat cheaper to fabricate. A reasonable compromise in mean slot width is about ¼ in.

Slot heights commonly used include the range from ½ to 2 in. Low slot heights limit slot capacity. On the other hand, there is a limit to maximum useful slot height because at the greater height other tray dynamic factors become controlling. The recommended slot height is in the range of 1 to 1½ in., depending on the cap size, according to the following guide:

Cap size, in.	Slot height, in.
3	1.00
4	1.25
6	1.50

Slots punched in caps so as to leave the bottom of the slots open, that is, with no shroud ring, are used by some designers in the belief that cap pressure drop will be lower when the slot capacity is exceeded. However, the incremental pressure drop per tray resulting from a shroud ring amounts in liquid head to only the height of the shroud ring itself, which is negligible. Caps with open bottom slots, on the other hand, have weak teeth and are subject to deformation in normal handling. Therefore a shroud ring is recommended on all caps, the height of the ring being about ¼ in.

The diameter of the *risers* determines the distribution of cap area between riser area and annular area. Bolles (3) has shown the optimum annular/riser area ratio at the point of minimum pressure drop to be in the range 1.1 to 1.4. Riser height is determined by skirt clearance and riser-slot seal.

The *reversal area* is that area between the top of the riser and the under-surface of the roof of the cap. It is recommended that this area be greater than either the riser or the annular area in order to avoid a flow restriction. This can be proved by analysis of the Dauphine equations.

The *riser-slot seal* is the vertical distance between the top of the riser and the top of the slots. Some seal is necessary in order that the liquid will not splash down the risers during normal operation. One-half inch is regarded as a reasonable seal for caps in the 3- to 6-in. range.

The use of hold-down bars used to be a common method of *mounting* bubble caps. However, Good et al. (17) and Davies (10) have clearly shown that hold-down bars are very undesirable because of their contribution to high liquid gradients.

It is recommended that the caps be mounted by means of bolts. It is also recommended that the fastening be designed so as to permit removal of both caps and risers for maintenance. Cap designs are available so that the installation or removal of caps can be carried out entirely from above the tray, which is a great convenience.

The number of different bubble-cap designs which have been employed is almost endless. In order to simplify tray design and increase interchangeability, Bolles (3) proposed a set of *standard bubble caps*. These are presented in Table 14-3. There are basically six different bubble caps consisting of three sizes for carbon steel and three sizes for alloy construction. This range of cap sizes will fit a wide range of column diameters. Variable riser heights are provided in order to permit variation in skirt height for optimum tray dynamics. The other variables in cap design were chosen in accordance with the preceding discussion.

Skirt Clearance. The lower limit of cap skirt height is obtained when the shroud ring fits flush with the tray floor. However, it has been found advisable to employ a skirt height of at least ½ in. in order to provide for vapor overload, to avoid excessive liquid gradient, and to provide a space for the settling of sediments.

There is also an upper practical limit to skirt height when the liquid can pass under the caps without adequate contact with the vapor. Good performance has been obtained with skirt heights up to 1½ in.

The recommended range of skirt height is from ½ to 1½ in. The selection of skirt height within this range involves determining the optimum balance of tray dynamic factors.

Slot Seal. Some slot seal is necessary to force the vapor to bubble through and contact the liquid. Larger slot seals promote higher plate efficiencies, but at the expense of pressure drop. High slot seals are practical for operations at high pressure, but in the case of vacuum operation the seal must be kept low in order to keep pressure drop within reasonable bounds.

Some designers base their tray design on a selection of *static seal*. This

TABLE 14-3. STANDARD BUBBLE-CAP DESIGNS

Material	Carbon steel			Alloy steel		
Nominal size, in.	3	4	6†	3	4	6†
Cap:						
U.S. Standard gauge	12	12	12	16	16	16
OD, in.	3.093	4.093	6.093	2.999	3.999	5.999
ID, in.	2.875	3.875	5.875	2.875	3.875	5.875
Height over-all, in.	2.500	3.000	3.750	2.500	3.000	3.750
Number of slots	20	26	39	20	26	39
Type of slots‡	Trpzl.	Trpzl.	Trpzl.	Trpzl.	Trpzl.	Trpzl.
Slot width, in.:						
Bottom	0.333	0.333	0.333	0.333	0.333	0.333
Top	0.167	0.167	0.167	0.167	0.167	0.167
Slot height, in.	1.000	1.250	1.500	1.000	1.250	1.500
Height shroud ring, in.	0.250	0.250	2.250	0.250	0.250	0.250
Riser:						
U.S. Standard gauge	12	12	12	16	16	16
OD, in.	2.093	2.718	4.093	1.999	2.624	3.999
ID, in.	1.875	2.500	3.875	1.875	2.500	3.875
Standard heights, in.:						
0.5-in. skirt height	2.250	2.500	2.750	2.250	2.500	2.750
1.0-in. skirt height	2.750	3.000	3.250	2.750	3.000	3.250
1.5-in. skirt height	3.250	3.500	3.750	3.250	3.500	3.750
Riser-slot seal, in.	0.500	0.500	0.500	0.500	0.500	0.500
Cap areas, in.²:						
Riser§	2.65	4.80	11.68	2.65	4.80	11.68
Reversal	3.99	7.30	17.40	4.18	7.55	17.80
Annular	3.05	5.99	13.95	3.35	6.38	14.55
Slot	5.00	8.12	14.64	5.00	8.12	14.64
Cap	7.50	13.15	29.0	7.07	12.60	28.3
Area ratios:						
Reversal/riser	1.50	1.52	1.49	1.58	1.57	1.52
Annular/riser	1.15	1.25	1.20	1.26	1.33	1.25
Slot/riser	1.89	1.69	1.25	1.89	1.69	1.25
Slot/cap	0.67	0.62	0.50	0.71	0.65	0.52

† Not generally recommended because of lower efficiency.

‡ The trapezoidal shape is frequently modified by rounding at top and bottom to facilitate die stamping.

§ Allowing for $\frac{3}{8}$-in. hold-down bolt.

can be misleading, however, because the actual slot seal is dependent on the liquid height over the weir and liquid gradient as well as the static seal. Therefore, the seal should be selected on the basis of the *dynamic slot seal*, as defined by Eq. (14-38).

The selection of dynamic slot seal might be based on an analysis of its effect on pressure drop and plate efficiency. The latter will be covered in a later chapter. As a rough guide the following seals have been found reasonable:

Pressure of operation	Dynamic slot seal, in.
Vacuum (50–200 mm Hg abs)	0.5–1.5
Atmospheric	1.0–2.5
50–100 psig	1.5–3.0
200–500 psig	2.0–4.0

The heights of overflow weirs are determined on the basis of providing adequate dynamic seal.

Slot Capacity. Equations for estimating the liquid depression, or slot opening, for bubble caps were developed previously. These equations are useful primarily for their contribution to the calculation of total tray pressure drop. They serve another useful purpose, however. Overloaded slots can indicate a "jetting" flow regime on the tray, depending upon clear liquid level, and thus can indicate a high-entrainment condition. Slots are considered overloaded when the calculated slot opening exceeds the actual slot height. However, many commercial fractionators have demonstrated good efficiencies even though the slots were calculated as overloaded, provided that all the other tray dynamic factors were within recommended limits.

Optimum contacting action exists when the clear liquid height at the overflow weir h_{lo} is greater than the height of the bubble caps. This condition allows sufficient slot submergence for the formation of a "froth blanket," which prevents excessive entrainment. Such a regime approximates that of a sieve tray. Overloaded slots do not appear deleterious when a high liquid level is maintained.

Davies (11) has called attention to the fact that low vapor loads lead to undesirable flow pulsation. This occurs when the slot opening is somewhat less than 0.5 in., which should be avoided.

14-16. Tray Layout

Analysis of Tray Areas. The area on a bubble-cap tray can be classified into four categories as shown in Fig. 14-18. The allocated cap area is that area allocated to the bubble caps, sometimes referred to as the active tray area. This is defined as the agglomeration of hexagonal areas surrounding the bubble caps, the width of each hexagon equal to the cap pitch.

The liquid-distribution area includes those areas between the allocated cap area and the overflow weir on one side and the downflow apron on the other. One area is for inlet distribution, and the other for outlet disengaging. In the case of cascade trays, this area also includes the inactive space on either side of intermediate weirs.

The downflow area includes that area on both sides of the tray. In the case of double-pass trays, this area includes, of course, the center as well as side downcomers.

The remaining space around the sides of the tray is the end wastage area, which consists of that area between the allocated cap area and the tower wall.

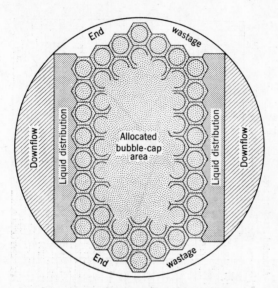

FIG. 14-18. Classification of tray area.

The distribution of area according to these four classifications varies with the tower diameter, type of tray, and other factors. However, on the basis of experience certain practical ranges have been determined, and these are summarized in Table 14-4.

TABLE 14-4. TYPICAL DISTRIBUTION OF AREAS AS PER CENT OF TOWER AREA
(Allocated cap area is determined by difference)

Tower diam, ft	Downflow area		Liquid-distribution area			End wastage
	Crossflow	Double pass	Crossflow	Double pass	Cascade double	
3	10–20		10–25			10–30
4	10–20		8–20			7–22
6	10–20	20–30	5–12	15–20		5–18
8	10–20	18–27	4–10	12–16		4–15
10	10–20	16–24	3– 8	9–13	20–30	3–12
12	10–20	14–21	3– 6	8–11	15–25	3–10
15	10–20	12–18	2– 5	6– 9	12–20	2– 8
20		10–15		5– 7	9–15	2– 6

For well-designed trays, the allocated cap area is usually between 60 and 70 per cent of tower area.

Cap Pitch and Spacing.   Triangular cap pitch is recommended because this gives the maximum number of bubble caps per unit tray area for any given cap spacing.

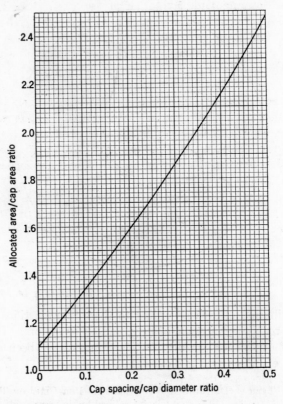

FIG. 14-19. Effect of cap spacing on area requirements.

The orientation should be so that the rows are normal to liquid flow. With the rows parallel to the liquid flow the liquid gradient will be less, but there is too great a possibility that the liquid near the tray floor can flow across the tray without contacting vapor.

The spacing of the bubble caps has a pronounced effect on the required column size.   This should be clearly evident by inspection of Fig. 14-19. The cap area here is defined as the cross-sectional area contained within the outside diameter of the cap.

If the caps are too close together, there will not be sufficient space for vapor discharge from the slots.   On the other hand, if the caps are too far apart, it is possible for the liquid to flow between the caps without satisfactory contact with the vapor. The practical range of cap spacing is between 1 and 3 in., measured between outside diameters of adjacent caps. The selection of cap spacing within this range involves a determination of the optimum balance of tray dynamic factors.

Inlet and Intermediate Weirs. Inlet weirs are specified by some designers to provide a positive seal against vapor flow up the downcomer and to distribute liquid more uniformly across the tray. A positive seal can almost always be accomplished by proper choice of clearance below the downflow apron, and there is no liquid-distribution problem on bubble-cap trays. Inlet weirs represent a waste of space and cost and are not recommended. Intermediate weirs are recommended to reduce the vapor-distribution ratio (to be discussed) when other measures are insufficient. However, it is important that the intermediate weir be designed to function as a true weir and not submerged. Therefore, it is necessary that the weir height be greater than the calculated liquid height immediately downstream.

Cap Stepping. Bubble-cap "stepping" means locating caps in "steps" of different elevations. The purpose of cap stepping is to match cap elevations more or less to the liquid gradient so as to improve vapor distribution (25).

The bubble caps can be stepped at two or more cap elevations by choosing caps with different riser heights from Table 14-3. The stepped sections are not necessarily divided into equal numbers of rows. In the case of double-pass trays, dividing the stepped rows equally would result in greater liquid gradient across the section next to the side weirs because of the curvature of the tower wall. Instead, the stepped sections should be divided so as to equalize the liquid gradient across each of the stepped sections.

In the stepping of caps on cascade trays, it is important that allowance be made for the fact that the liquid height over intermediate weirs is different from that over end weirs. The setting of the weirs and the stepping of caps on cascade trays should be made on the basis of equalizing dynamic submergences.

Redistribution Baffles. With uniform cap spacing, it is not possible to locate all caps at the end of rows close to the column wall. Thus, liquid will tend to short-circuit those rows with excessive end clearances. Therefore, it is important to close these gaps by the installation of redistribution baffles, which are recommended at the end of every row where the end space to tower wall exceeds the normal cap spacing by 1 in. The clearance between these baffles and the cap should be equal to the cap spacing. The height of the redistribution baffle should be approximately twice the height of clear liquid in order to allow for froth height.

Reverse-flow Baffles.   Reverse-flow baffles are required in the center of reverse-flow trays.   However, it must be realized that the baffle will not function properly unless it is high enough to prevent short-circuiting of liquid and foam.

It is recommended that the height of the reverse-flow baffle be at least twice the highest calculated height of clear liquid on the tray to allow for froth.

Minimum Clearances.   It is desirable that the clearance between bubble caps and tower wall be as small as possible in order to permit maximum utilization of tray area.   However, there is a limit to this clearance because of the interference of tray support rings.   It is usually possible to design for a minimum clearance of about $1\frac{1}{2}$ in. to column wall.

Similarly, it is desirable that the clearance between the last row of caps and weirs be as small as possible.   However, if this clearance is too close, the vapor bubbling from the slots will blow liquid over the weir and produce an uncontrolled effect on slot submergence.   In order to avoid this, it is recommended that the minimum clearance between caps and weirs be 3 in.

Similarly, it is recommended that the minimum clearance between caps and downflow apron also be 3 in.

However, it should be noted that if these recommended clearances are not met, the results will not be catastrophic.   Therefore, smaller clearances may be considered in order to avoid having to increase the column diameter.

14-17. Vapor Capacity

Approach to Flooding.   The flood-point correlation presented in Fig. 14-4 is, of course, not absolutely accurate.   Since flooding results in a completely inoperable column, some safety factor is necessary.   In addition, as flooding is approached, tray efficiency falls rapidly, and more can be lost than gained.

Generally, it is recommended that commercial columns for new services be sized so as not to exceed 80 to 85 per cent of the flooding velocity so predicted.   If flood-point data are available from another commercial column in the same general service, the new column may be designed for 85 to 90 per cent of flooding.

In addition, if there is danger of encountering foaming conditions in regular operation, the design flooding velocity should be based on about 75 per cent of the velocity obtained from Fig. 14-4.

Entrainment.   One method of selecting the optimum approach to flooding is to balance the cost of column size against the cost of extra trays to compensate for the detrimental effect of entrainment on efficiency.   Studies of this type have been reported by Bolles (3) and Zenz (39).   However, it is advisable that the liquid entrainment ratio ψ not exceed about 0.15 lb entrainment per pound of gross liquid downflow, or plate efficiency will suffer unduly.

Another factor which should be borne in mind in connection with entrainment is the limitation of nonvolatile impurities in the distillate. This is usually not a controlling factor in setting the maximum limit of entrainment, but it can become so in special cases, particularly when there are a very few plates in the rectification section.

If there is any doubt concerning contamination of product due to entrainment, the degree of contamination should be calculated by the method of Sherwood and Jenny (33).

Still another factor to consider in connection with entrainment is the possible carry-over of liquid entrainment to the overhead condenser. This is undesirable because it results in loss of reflux control. However, this entrainment can be mostly eliminated by providing a disengaging space above the top plate of about twice the tray spacing. In special cases consideration might be given to installation of a mist separator.

14-18. Vapor Distribution

The distribution of vapor among individual caps and rows of caps is very important, since it has been shown that poor vapor distribution is one of the principal causes of poor plate efficiencies with bubble-cap trays (6). Detailed studies of the effect of liquid gradient on vapor distribution are reported by Kemp and Pyle (22) and Bolles (3).

The latter study was based on analysis of the performance of standard 4-in. carbon-steel bubble caps loaded with benzene vapor at atmospheric pressure. The resulting performance calculated by the equations previously presented is shown in Fig. 14-20. Here plotted as a function of vapor load per cap are slot opening, slot load, and cap pressure drop. It will be noted that the slots of the cap are fully loaded at a vapor flow of 0.77 ft³/sec and that above this point vapor is blowing under the shroud ring.

Consider now the vapor distribution obtained on a tray with several rows of caps in the limiting case of zero liquid gradient. If these caps were loaded to 100 per cent of the slot capacity, the cap performance would be described by the dotted lines drawn on Fig. 14-20. Under these conditions, the pressure drop through all caps would be 2.3 in. and the vapor distribution would be perfectly uniform among all rows of caps.

Actually there will always be some liquid gradient; otherwise liquid would not flow across the tray. The liquid gradient and its effect on cap performance are illustrated in Fig. 14-21 for a tray with bubble caps of constant skirt height. Because of the varying liquid depth, the vapor flows most easily through the outlet row of caps and least through the inlet row.

It should be evident that the total tray pressure drops through all rows

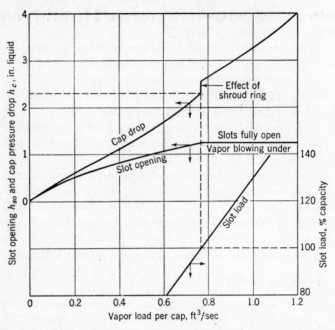

FIG. 14-20. Effect of vapor load on bubble-cap performance.

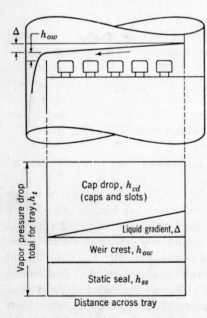

FIG. 14-21. Effect of liquid gradient on tray dynamics.

of bubble caps are equal. On the other hand, the vapor-pressure drop through any one row of caps is described by the equation

$$h_t = h_{cd} + h_{ss} + h_{ow} + \Delta \quad (14\text{-}50)$$

where Δ is the liquid gradient from the weir to that particular row.

h_t, h_{ss}, and h_{ow} are constants for the tray. Therefore, h_{cd} must vary with Δ, row by row, as shown in Fig. 14-21. Since h_{cd} varies by the row, the vapor load per cap also varies from row to row according to the cap performance curve, Fig. 14-20.

Further study of Fig. 14-21 will reveal that the effect of liquid gradient should be analyzed, not in terms of absolute gradient, but rather as the ratio of liquid gradient to mean cap pressure drop.

This factor is called the *vapor-distribution ratio* (3), and is defined by the equation

$$R_{vd} = \frac{\Delta}{h_c} \qquad (14\text{-}51)$$

The effect of liquid gradient on vapor distribution has been studied over a wide range of vapor-distribution ratios (3). For example, consider the vapor distribution obtained on a tray with a vapor-distribution ratio R_{vd} of 0.50. For this purpose the same conditions used in Fig.

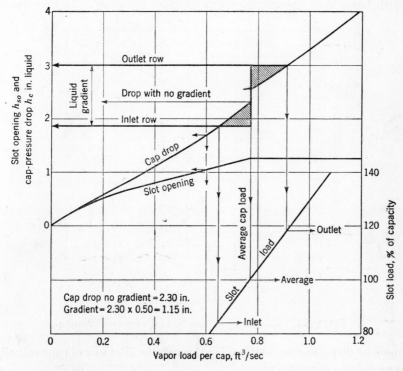

FIG. 14-22. Effect of liquid gradient on vapor distribution.

14-20 will be employed. The caps are loaded to an average cap slot load of 100 per cent, where the average cap pressure drop is 2.3 in. The liquid gradient is 1.15 in. as shown in Fig. 14-22.

A mathematical analysis of this diagram will reveal that the cap loads will line out when the shaded triangular areas between the inlet and outlet rows are equal. Under these conditions it will be noted that the vapor load through the inlet row of caps is 84 per cent whereas the load on the outlet row is 119 per cent of slot capacity.

The above procedure was repeated for vapor-distribution ratios of 0.1, 0.2, 1.0, and 2.3, the results of which are shown in Fig. 14-23. As the vapor-distribution ratio increases, the degree of maldistribution also increases. When the vapor-distribution ratio reaches 1.0, the inlet row is 62 per cent loaded whereas the outlet row is loaded to 141 per cent. When the vapor-distribution ratio reaches 2.3, the load on the outlet row reaches 188 per cent while the inlet row passes no vapor at all.

If the vapor-distribution ratio is increased beyond 2.3, liquid will begin to pour down the risers of inlet row of caps. This is called *dumping*, which has a very detrimental effect on plate efficiency. In the case of

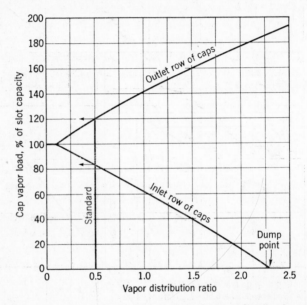

Fig. 14-23. Vapor-distribution ratio and vapor distribution.

crossflow trays, any liquid dumping through the inlet row of caps actually bypasses two complete trays.

With high values of vapor-distribution ratio there is considerable flow of vapor *across* the tray because the rows of caps with the heaviest vapor loads alternate from tray to tray. Furthermore, this crossflow of vapor is countercurrent to the liquid flow. With a significant amount of maldistribution, particularly if the tray spacings are small, the crossflow of vapor can reach such an extent that friction between countercurrent streams of vapor and liquid adds to the liquid gradient, resulting in even greater vapor maldistribution, etc. These phenomena have been treated in detail by Kemp and Pyle (22) and photographed by Good, Hutchinson, and Rousseau (17).

Investigation of bubble-cap tray columns exhibiting poor performance because of high liquid gradients has always revealed vapor-distribution ratios in excess of 0.5. Therefore, this is recommended as the normal design limit.

A few comments should be made in regard to the application of vapor-distribution ratio to special cases. One special case is where the caps are *stepped*. Here it is advisable to calculate the vapor-distribution ratio for all combinations of pairs of rows bordering the stepped sections in order to find the worst vapor-distribution ratio on the tray. Also, it is necessary that the *effective* gradient be employed, which is the liquid gradient between the rows in question minus the amount of stepping.

Another special case arises in connection with analysis of vapor-distribution ratio for a cascade tray. Here, again, it is advisable to calculate the vapor-distribution ratios for all extreme row combinations. However, in this case the *effective* gradient is most conveniently obtained as the difference in dynamic submergences for the rows in question, which includes the static seal, height over weir, and gradient to the particular row.

Further details on calculating vapor-distribution ratio in these special cases is given by Bolles (3).

Some designers try to improve vapor distribution by blanking off some of the caps on the tray or intentionally designing caps with high pressure drop. These measures do improve vapor distribution, but they have the disadvantage of increasing pressure drop and reducing tray capacity.

It is quite possible to obtain satisfactory vapor distribution and still retain a tray with low pressure drop and high capacity by employing other techniques. One is to decrease liquid gradient by (1) employing wider cap spacings or (2) providing higher skirt clearances. Another is to employ intermediate weirs (cascade trays). Still another is to step the caps at several levels.

14-19. Miscellaneous Design Factors

Tray Deflection. Every tray will deflect under its own weight and the weight of the liquid on the tray. Excessive tray deflection is undesirable because it increases the submergence of the caps in the center of the tray, thereby contributing to vapor maldistribution. On the other hand, the specification of design deflection should not be too stringent or it will add to the cost of the tray.

An analysis of vapor-distribution ratio will show that a deflection of $\frac{1}{8}$ in. will produce a negligible effect on vapor distribution. In the case of large towers, a design deflection of $\frac{1}{4}$ in. might be permitted.

Tray Levelness. Any departures from tray levelness will cause vapor maldistribution. The recommended tolerance for tray levelness is a maximum of $\frac{1}{4}$ in. measured from the lowest to highest points. This amount will produce only a negligible effect on vapor distribution.

The overflow weir must also be level or the liquid will not be distributed uniformly over the tray. The recommended tolerance for weir levelness is $\frac{1}{8}$ in. maximum from low to high points.

Tray Leakage. The joints of removable trays will leak under hydrostatic test unless they are packed. Actually, it is possible to have many holes and crevices on the tray and still not show leakage under normal operation. However, if the differential vapor pressure across the tray is unusually high, leakage during operation is definitely possible.

Liquid leakage is detrimental to plate efficiency because it permits a portion of the liquid to leave the tray without properly contacting vapor. The effect on efficiency depends on the amount of leakage, the place of leakage, and the normal liquid load. In the case of vacuum distillation, where the liquid load per unit tray area is low, leakage can have a serious effect on fractionating efficiency. Even where leakage itself is not critical, many designers include a leakage specification in order to ensure the general quality of fitting and workmanship.

A leakage specification in common use calls for a water test in which the level must not fall more than 1 in. from the top of the weir in 20 min with the drain holes plugged (11). This specification is tight enough to satisfy practically all services but is still attainable if reasonable care is exercised in the design, fabrication, and installation of the tray. However, if leakage is not critical, a more liberal specification may be adopted.

In order to meet a tight leakage specification it is necessary to pack all joints. Woven-asbestos packing material is commonly used because of its inert chemical properties. However, care must be exercised in the design, selection, and installation of the packing to ensure that bits or parts of the packing will not come loose to plug other parts of the equipment and piping.

Recent trends are toward elimination of gasketing in sectional trays except in cases of vacuum service or where liquid flows are relatively small. Where gasketing is not used, adequate metal thickness and bolting should be provided to achieve reasonably tight metal-to-metal contact.

Tray Drainage. Tray drain holes are necessary to permit the draining of liquid holdup from a column after shutdown. If the size of these holes is too small, they will plug with sediment and polymer. However, if they are too large, excessive dumping of liquid during normal operation may occur. It is recommended that the drain holes be between $\frac{3}{8}$ and $\frac{5}{8}$ in. in diameter.

Drainage of bubble-cap trays has been studied by several investigators (3, 20, 24). The number and area of drain holes in a tray should be determined on a basis of the desired draining time. Consider, first, the draining of a single tray. In this case, the velocity of flow through the drain holes is given by the orifice equation:

$$u = 0.6 \sqrt{\frac{2gh}{12}} \tag{14-52}$$

Initially, the head is equal to the height of the weir, which may be substituted in the equation to provide the initial drain rate. As time goes on, however, the head reduces and consequently the draining rate also diminishes. An application of calculus will show that the average drain rate is equal to half the initial rate. Therefore, the drain time for a single tray can be calculated as twice the liquid volume on the tray divided by the initial drain rate.

In the case of a fractionating column, trays are arranged above one another, and every tray must drain through all other trays below it. Unfortunately, it is not possible to develop an explicit solution of the differential equations expressing the drain rate for N_t trays arranged in a vertical column. However, it has been found empirically that the drain time for N_t trays is approximately $(\sqrt{2}/2)N_t$ times the drain time for a single tray. Combining these relations then permits the derivation of an equation for the total drain time of a bubble-tray fractionating column as follows:

$$t_d' = \frac{N_t \sqrt{h_w}}{3a_w} \tag{14-53}$$

Lockhart (24) has refined this relationship to include the effect of liquid viscosity and to provide a better fit to experimental data. This is recommended in cases where it is important to predict the exact drain time.

Most designers, however, simply provide approximately 4 in.² of hole area for every 100 ft² of tray area. This will result, for example, in the draining of a 50-tray column with a 4-in. weir in about 8 hr, provided the liquid kinematic viscosity is not greater than 50 centistokes.

Usually the majority of the drain holes are placed at the overflow weir, so that any leakage during regular operation will not impair performance. However, at least some of the holes should be provided at potential low spots in order to permit complete drainage of the column.

Feed Entries. The main feed and reflux feed entries to bubble trays require some consideration. Normally, the feed is introduced either in a downcomer or on the upstream side of the tray. Many special devices have been designed for this purpose, including spray nozzles, perforated "sprinklers," and weir boxes. However, it is felt that a simple entry nozzle will generally be equally effective.

However, one precaution is worthy of mention here.   A feed consisting in part or whole of vapor should not be introduced in a downcomer or it may cause the downcomer to flood.

Trays to Check.   As a general rule, it is recommended that at least two trays be subjected to a complete tray dynamic check: the top tray and the bottom tray of the column.   This is recommended because the liquid and vapor loads are usually sufficiently different on these trays to make individual checking advisable.

Sometimes it is also advisable to check the trays immediately above and below the feed, or side drawoff, if there is sufficient variation in loads at these points.

In the case of double-pass trays, it is advisable to check both side and center downcomer trays.

If more than one tray design is used in a given tower, it is also advisable to check the tray immediately above and below the design change.

Loads to Check.   Obviously the tray should be checked at the maximum anticipated vapor and liquid loads.   After a satisfactory tray arrangement is developed, it is also advisable to check the tray performance under conditions of minimum anticipated loads.

Design for Load Variations.   Bubble-cap trays provide exceptionally wide flexibility, and a single tray design is capable of handling wide variations in loads.

However, some cases are encountered in which liquid and vapor loads in different sections of the tower may be so different as to make a single tray design for use throughout the column inadvisable.

For these cases there are several possible solutions.   One is to blank off some rows of caps on the trays in the column where the lower vapor loads are anticipated.   Another means is to design entirely different trays for different sections of the column, but all based on the same tower diameter.   Still another expedient is to vary the column diameter, employing entirely different tray designs for the different sections.

Summary of Tray Design Standards.   A convenient summary of recommended tray design standards is given in Table 14-5.

14-20. Achieving Optimum Design

At the start of the discussion on tray design, it was pointed out that the object of good tray design is to determine the optimum design, which means selection of the optimum combination of design elements.   In this connection a list of nine specific tray design variables which must be optimized was given.

The technique of squeezing maximum capacity into minimum tray area is based on the most efficient utilization of each of the four area classes shown in Fig. 14-18.   Theoretically, this is achieved when all clearances are pushed to the limit and all tray dynamic variables are

TABLE 14-5. SUMMARY OF RECOMMENDED TRAY DESIGN STANDARDS

Materials of construction:

Type	Sheet metal
Material	Determined by corrosion conditions

Tray type:

General use	Crossflow
Very low L/V ratio	Reverse-flow
High L/V or large columns	Double-pass
Very high L/V	Double-pass, cascade
Very large columns	3- or 4-pass

Downcomers and weirs:

Downcomer type	Segmental
Downflow apron	Vertical
Weirs for normal loads	Straight
Weirs for low loads	Notched

Bubble caps:

Nominal size for:	
2.5- to 5-ft columns	3 in.
Columns 4 ft. and up	4 in.
Design	See Table 14-4
Pitch	Equil. triangular, rows normal to flow
Spacing	1–3 in.
Skirt clearance	0.5–1.5 in.
Mounting	Removable design
Clearances:	
Cap to tower wall	1.5 in. min†
Cap to weir	3 in. min†
Cap to downflow apron	3 in. min†

Tray dynamics:

Approach to flood point	80–85%
Mean slot opening:	
Maximum	100%† slot height, no overload
Minimum	0.5 in.
Mean dynamic slot submergence:	
Vacuum operation	0.5–1.5 in.
Atmospheric	1.0–2.5 in.
50–100 psig	1.5–3.0 in.
200–500 psig	2.0–4.0 in.
Vapor-distribution ratio (Δ/h_{cd})	0.5 max
Height aerated mass in downcomers	100%‡ downflow height, max
Downflow residence time (true):	
All but high foamability	3 sec, min
High foamability	5 sec, min
Entrainment ratio ψ	0.15 max
Pressure drop	As limited by process

Tray spacing:

For towers 2.5–4 ft	18 in.§
For towers 5–20 ft	24 in.§

† Preferred, but not critical if all the other standards are met.
‡ Assuming safety factor is included with ϕ in Eq. (14-44).
§ Generally recommended, but modified if special conditions warrant.

TABLE 14-5. SUMMARY OF RECOMMENDED TRAY DESIGN STANDARDS (*Continued*)

Miscellaneous design factors:

Inlet weirs	Not recommended
Intermediate weirs	Minimum height > height liquid downstream
Reverse-flow baffles	Minimum height twice clear liquid
Redistribution baffles:	
Location	All rows where end space is 1-in. > cap spacing
Clearance to caps	Same as cap spacing
Height	Twice height clear liquid
Downflow apron seal:	
Weir to baffle < 6 ft	0.5 in.
Weir to baffle 6–12 ft	1.0 in.
Weir to baffle > 12 ft	1.5 in.
Tray design deflection under load	⅛ in.
Drain holes:	
Size	⅜–⅝ in.
Area	4 in.²/100 ft² tray area
Leakage	Max fall 1.0 in. from top of weir in 20 min with drain holes plugged
Construction tolerances:	
Tray levelness	¼-in. max
Weir levelness	⅛-in. max

bottlenecked simultaneously. Under these conditions each of the four tray area classifications are squeezed to the minimum.

For example, the minimum number of bubble caps are employed so as to push the slot opening to 100 per cent of slot height. The caps are spaced as close to the tower wall as possible and as close together as can be done without exceeding the upper limit of vapor-distribution ratio.

Smaller caps are used if necessary for more efficient cap arrangement. In order to permit closer cap spacings, skirt clearance is increased to reduce liquid gradient and the caps are stepped to compensate for the remaining liquid gradient. The downcomers are reduced in size until one or more limits are reached involving liquid height in downcomers, downcomer residence time, or liquid throw over weirs. The tray type is also changed, if advantageous.

In this manner the tower diameter may be successively reduced until the flood point is reached, until entrainment becomes intolerable, or until a combination of other tray dynamic factors shows that the absolute minimum tower size has been achieved.

In judging of optimum tray design, it should be remembered that the smallest column diameter does not necessarily correspond to the minimum over-all cost. For example, small bubble caps may add to the tray cost. Also, complicated tray types are more costly. Furthermore, it should be remembered that the length of the liquid path affects the tray efficiency. Finally, tray spacing affects column costs.

It should be apparent, then, that the real criterion for optimum tray design is the achievement of minimum installed tray cost per theoretical plate.

A detailed example of how these principles are applied in practice to achieve an optimum commercial design is given by Bolles (3).

14-21. Operating Flexibility

It may be interesting to note the qualitative effect of liquid and vapor loads on bubble-cap tray performance as limited by tray dynamics. This is illustrated by Fig. 14-24, where the limit of each tray dynamic factor is sketched in for a typical bubble-cap tray.

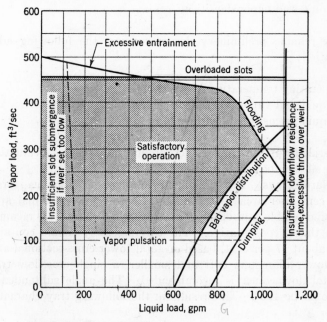

Fig. 14-24. Typical bubble-cap-tray performance chart.

An area of satisfactory operation is shown surrounded by excessive entrainment, overloaded slots, flooding, insufficient downflow residence time, excessive throw over weir, bad vapor distribution, dumping, and vapor pulsation.

With our present knowledge of bubble-tray dynamics, it is not possible to predict the exact boundaries of the area of satisfactory operation.

However, tests of commercial towers have shown that most well-designed bubble-cap tray columns will maintain an essentially flat efficiency curve under vapor loads ranging from 15 to 85 per cent of the flood point.

14-22. Preliminary Tray Design

Frequently it is desirable to estimate, within a minimum of time and effort, the size of a bubble-cap tray column. Such estimates are frequently required in connection with preliminary plant studies and process economic evaluations. Also, it is expedient to conduct a preliminary tray design as the first step of detailed tray design. A preliminary process design of a bubble-cap tray column can be made with a minimum of data. The following information must be available:

1. Liquid and vapor mass-flow rates
2. Liquid and vapor densities

A more reliable preliminary design requires the following additional information:

1. Operating temperature and pressure
2. Liquid and vapor molecular weights
3. Liquid viscosity and surface tension

In general, even the roughest of preliminary process evaluations will provide all the above information.

Standard Trays. For preliminary design, a bubble-cap tray with 10 per cent riser area (based on total tower cross-sectional area) and 4-in. round bubble caps of the design given in Table 14-4 is recommended because this is conservative. The tray assembly should contain a slot area of about 17 per cent, based on total tower cross-sectional area.

For tower diameters 6 ft and smaller, a single-crossflow tray with segmental downcomer is recommended. The angle subtended by the weir should be 100°. This results in the following tray characteristics:

Downflow area	12% of tower area
Maximum active area	76% of tower area
Weir length	77% of tower diam
Max liquid flow path	64% of tower diam

For tower diameters greater than 6 ft, a double-flow tray is recommended for preliminary design. The angle subtended by the side weirs should be 77°. The center downcomer should have approximately the same area as the combined side downcomers. The following tray characteristics are typical:

Downflow area	12% of tower area
Maximum active area	76% of tower area
Side weir length	62% of tower diam
Center weir length	97% of tower diam
Max liquid flow path	30% of tower diam

Tray spacing may be varied, but 18 or 24 in. will fit most cases in practice. Mean dynamic slot submergence h_{ds} should be based on operating pressure as follows:

	In.
Vacuum operation	1.0
Atmospheric	1.5
50–100 psig	2.0
200–500 psig	3.0

The preferred tray and cap material is carbon steel if process conditions permit.

Flooding. The required tower diameter for the specified liquid and vapor flows is obtained from Fig. 14-4 and corrected for surface tension by Eq. (14-8), if desired. It should be noted that vapor velocity in the capacity parameter is based on tower cross section minus one downcomer area (i.e., 88 per cent of tower area for preliminary designs) and that the chart applies to nonfoaming systems. For foaming systems, flooding velocity obtained from the chart should be multiplied by 0.75.

It is recommended that preliminary designs be based on 75 to 80 per cent of flooding, which allows some leeway for unknowns.

Pressure Drop. Pressure drop per tray can be estimated from the following equation:

$$h_t = 0.53 \frac{\rho_v}{\rho_l} \left(\frac{Q}{A_r}\right)^2 + 0.8h_{ds} + 1.0 \qquad (14\text{-}54)$$

EXAMPLE PROBLEM

14-23. Statement of Problem

The problem is to design a bubble-cap tray for a benzene-toluene fractionator to produce 40,000 lb/hr of benzene product while operating at atmospheric pressure and a reflux ratio (reflux to net overhead) of 5:1.

In the example problem only the top tray will be analyzed. In an actual design other critical trays should also be analyzed (see text). The approach and method, however, are the same.

The available data on the top tray are as follows:

Material	Essentially benzene
Molecular weight	78.1
Operating pressure, psia	14.7
Operating temperature, °F	176
Liquid density, lb/ft³	43.3
Vapor density, lb/ft³	0.168
Liquid surface tension, dynes/cm	21
Max liquid load	200,000 lb/hr or 577 gpm
Max vapor load	240,000 lb/hr or 397 cfs

14-24. Preliminary Tray Design

First, the tower diameter is estimated on the basis of a reasonable approach to flooding. The liquid-vapor flow parameter is calculated from Eq. (14-6):

$$F_{lv} = \frac{200,000}{240,000} \sqrt{\frac{0.168}{43.3}} = (0.833)(0.0622) = 0.052$$

Assuming a tray spacing of 24 in. (to be confirmed later), Fig. 14-4 gives a capacity parameter C_{sb} of 0.35. Checking the effect of surface tension by Eq. (14-8) gives no measurable change:

$$C_{sb} = 0.35(^{21}\!/_{20})^{0.2} = (0.35)(1.01) = 0.35$$

Solving Eq. (14-7) for flooding vapor velocity gives

$$U_n = \frac{0.35}{\sqrt{0.168/(43.3 - 0.168)}} = \frac{0.35}{0.0625} = 5.6 \text{ fps}$$

The system is nonfoaming (hydrocarbons), and following the recommendation of estimating column size on the basis of 75 per cent of flooding, the required net tray area is

$$A_n = \left(\frac{1}{0.75}\right)\left(\frac{397 \text{ cfs}}{5.6 \text{ fps}}\right) = 94.5 \text{ ft}^2$$

Noting that the recommended standard tray has a downflow area 12 per cent of tower area gives the required tower of

$$A_t = \frac{94.5}{0.88} = 107.5 \text{ ft}^2$$

and a tower diameter of

$$D_t = \sqrt{\frac{(4)(107.5)}{3.14}} = 11.7 \text{ ft}$$

This is rounded off to a 12-ft diameter.

Consulting Table 14-5 shows that the assumption of 24-in. tray spacing was reasonable.

Consulting Table 14-2 indicates that a double-pass tray is probably best, and the text recommends a tray of the following approximate characteristics:

Riser area	10% of tower area
Downflow area	12% of tower area
Maximum active area	76% of tower area
Side weir length	62% of tower diam
Center weir length	97% of tower diam
Max liquid flow path	30% of tower diam

The text recommends a dynamic slot submergence of 1.5 in. for atmospheric-pressure towers, and so application of Eq. (14-54) gives an estimated tray pressure drop of

$$h_t = 0.53 \frac{0.168}{43.3} \left(\frac{397}{11.3}\right)^2 + (0.8)(1.5) + 1.0 = 4.73 \text{ in. liquid}$$

14-25. Tray Layout

The tray layout is based on the results of the preliminary design as follows:

Tower diameter	12 ft
Tray type	Double-pass
Tray spacing	24 in.

Bubble caps of 4-in. diameter are selected as recommended by Table 14-5. Since the service is noncorrosive, the standard carbon-steel cap is selected from Table 14-3.

The remainder of the tray layout is developed on the basis of trial and error. First, a tray layout is assumed. Then all the tray dynamic factors are calculated for this layout. Next, the dynamic results are studied in relation to the layout and decisions are made regarding improvements in the layout. Then another tray layout is developed, tray dynamics calculated, etc. This procedure is repeated until a satisfactory design is reached.

For the example calculations to follow, only one tray layout will be studied. In order to make the example more interesting, it will be assumed that several successive layouts have already been studied. Let us also assume that these results have indicated the advisability that the next layout should be based on the following:

Length of side weir	50% tower diam
Width of center weir	9 in.
Cap pitch	Equilateral triangular
Cap spacing	25% cap diameter
No. caps per tray	560
Cap skirt clearance	1.0 in.
Static slot seal	0.5 in.

The following computations are now made to locate the downcomers:

Tower diameter	12 ft or 144 in.
Tower area	113.1 ft²
Side downcomers:	
Weir length (144 × 0.50)	72 in.
Downcomer width (from standard table of circular segments)	6.8% D_t
(144 × 0.068)	9.8 in.
Downcomer area (144 × 0.50)	2.8% A_t
Per downcomer (113.1 × 0.028)	3.2 ft²
Per tray	6.4 ft²
Center downcomers:	
Length	12.0 ft
Width	9 in. or 0.75 ft
Area (12 × 0.75)	9 ft²

The resulting tray layout is shown in Fig. 14-25.

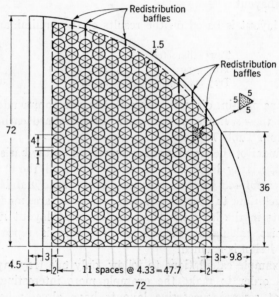

FIG. 14-25. Tray layout in example problem.

The weir setting is found as follows:

	In.
Skirt clearance	1.00
Shroud ring	0.25
Slot height	1.25
Static seal	0.50
Total — Height of weir	3.00

The downflow apron seal is set at 1.0 in., and the area under the baffle determined as follows:

Clearance under baffle (3.00 − 1.00)	2.00 in.
Area under baffle (both passes):	
Side downcomer (2.00 × 72) ($\frac{2}{144}$)	2.0 ft²
Center downcomer (2.00 × 144) ($\frac{2}{144}$)	4.0 ft²

Based on 560 bubble caps in the 12-ft column, the following critical tray areas are obtained:

	Per cap, in.²	Per tray, ft²		Avg % of tower area
		Side DC	Center DC	
Riser area	4.80	18.7	18.7	16.5
Slot area	8.12	31.5	31.5	27.8
Downflow area		6.4	9.0	6.8
Active area		97.7	97.7	86.5
Area under apron		2.0	4.0	2.7
Tower area		113.1	113.1	100.0
Net area		104.1	106.7	93.3

14-26. Tray Dynamics Calculations

Tray dynamics calculations on the top tray then proceed as follows, assuming a side-downcomer tray.

Foaming. Since the materials involved are hydrocarbons, the system is classified as a nonfoamer.

Flooding. The net tray area in the case of a side-downcomer tray is the total tower area less the area of the center downcomer, or

$$A_n = 113.1 - 9.0 = 104.1$$

and the vapor velocity based on net area is

$$U_n = \frac{397}{104.1} = 3.81 \text{ fps}$$

The flooding vapor velocity was previously estimated at 5.6 fps, so the approach to flooding is

$$\frac{3.81}{5.6} \times 100 = 68\%$$

Entrainment. The liquid-vapor flow parameter F_{lv} (already calculated) is 0.052. With this and the value of 68 per cent of flooding,

Fig. 14-5 gives the liquid entrainment ratio ψ as 0.04 mole/mole gross downflow.

Slot Opening. The chosen bubble caps have trapezoidal slots with a shape factor

$$R_s = \frac{d_{st}}{d_{sb}} = \frac{0.167}{0.333} = 0.50$$

From Fig. 14-6, the maximum slot capacity is

$$Q_{max} = (0.74)(31.5)\sqrt{(1.25)\frac{43.3 - 0.168}{0.168}} = 417 \text{ cfs}$$

The vapor load is then $(397/417)(100) = 95$ per cent of slot capacity. Referring again to Fig. 14-6, the mean slot opening is 97 per cent of slot height, or

$$h_{so} = (1.25)(0.97) = 1.21 \text{ in.}$$

Liquid Crest over Weir. The liquid load for each side weir is $577/2 = 288$ gpm. Application of Eq. 14-22 gives

$$h_{ow} = 0.48(^{288}\!/_{2})^{2/3} = 1.21 \text{ in.}$$

This is then corrected for tower wall constriction by means of Fig. 14-7:

$$\frac{\text{Liquid load, gpm}}{(\text{Weir length, ft})^{2.5}} = \frac{288}{(6)^{2.5}} = 3.3$$
$$F_w = 1.04$$
$$h_{ow} \text{ (corrected)} = (1.21)(1.04) = 1.26 \text{ in.}$$

Liquid Gradient. Liquid gradient is estimated on the basis of Figs. 14-9 and 14-10. The arithmetic average flow width is 9.5 ft, and the liquid load per unit width is $288/9.5 = 30$ gpm/ft. The mean liquid depth is based on the estimated liquid gradient:

$$h_l = 3.0 + 1.26 + \frac{1.3}{2} = 5.0$$

Referring to Fig. 14-10, the chart for 1.0-in. skirt clearance gives a liquid gradient (uncorrected) of 0.090 in. per row, which results in 1.08 in. for the 12 rows. Figure 14-9 then gives

$$U_t\sqrt{\rho_v} = \frac{397}{113.1}\sqrt{0.168} = 1.44$$
$$\Delta = C_v\Delta' = (1.18)(1.08) = 1.28 \text{ in.}$$

Thus the estimated mean liquid depth was correct and a recalculation is not necessary.

Vapor-pressure Drop. The dry cap head loss is calculated by Eq. (14-35) in conjunction with Fig. 14-14:

$$h_{cd} = 0.50 \frac{0.168}{43.3} \left(\frac{397}{18.7} \right)^2 = 0.88 \text{ in. liquid}$$

The mean dynamic slot seal is calculated with Eq. (14-38):

$$h_{ds} = 0.50 + 1.26 + \frac{1.28}{2} = 2.40$$

The vapor-flow parameter from Eq. (14-40) is

$$F_{va} = \frac{397}{97.7} \sqrt{0.168} = 1.67$$

and the aeration factor β from Fig. 14-15 is 0.63. According to Eq. (14-37), the head loss through aerated liquid is

$$h_{al} = (0.63)(2.40) = 1.51$$

Finally, the total tray pressure drop is calculated by Eq. (14-32):

$$h_t = 0.88 + 1.21 + 1.51 = 3.60 \text{ in. liquid}†$$

Vapor-distribution Ratio. The dry-cap drop is calculated with Eq. (14-32b):

$$h_c = 0.88 + 1.21 = 2.09 \text{ in. liquid}$$

The vapor-distribution ratio is then determined by means of Eq. (14-51):

$$R_{vd} = \frac{1.28}{2.09} = 0.61$$

Downcomer Dynamics. The head loss due to liquid flow under the downflow apron is given by Eq. (14-42):

$$h_{da} = 0.03 \left(\frac{288}{100 \times 1.0} \right)^2 = 0.25 \text{ in. liquid}$$

The height of clear liquid backup in the downcomer is obtained by Eq. (14-43):

$$h_{dc} = 3.60 + 3.00 + 1.26 + 1.28 + 0.25 = 9.39 \text{ in.}$$

and the height of aerated liquid by Eq. (14-44):

$$h_{fd} = \frac{9.39}{0.50} = 18.8 \text{ in.}$$

† The calculated pressure drop of the final tray is less than that of the preliminary tray because of the higher riser area.

The residence time of aerated mass in the downcomer is estimated by Eq. (14-47):

$$t_{dc} = \frac{(3.2)(9.39)}{(12)(0.64)} = 3.9 \text{ sec}$$

The "liquid throw" over the weir is not calculated, as it is not critical in side-downcomer trays.

14-27. Evaluation of Results

The tray dynamics are summarized in Table 14-6, along with a comparison against the recommended standards. It will be seen here that all tray dynamic factors are comfortably within the recommended range with the exception of the vapor-distribution ratio, which is excessive and should be corrected. This can easily be done on the final tray layout by stepping the inlet seven rows of caps so they are 0.5 in. higher than the outlet five rows, which will bring the vapor-distribution ratio down to about 0.4.

TABLE 14-6. SUMMARY OF TRAY DYNAMICS–EXAMPLE PROBLEM
(For 12'-0" tray layout)

	Standard	Actual
% of flooding	80 max	68
Entrainment ratio ψ	0.15 max	0.04
Slot opening, % slot height	100 max	97
Mean dynamic slot seal, in.	1.0–2.5	2.4
Vapor-distribution ratio R_{vd}	0.50 max	0.61†
Height froth in downcomer, % downflow height	100 max	70
Downcomer residence time, sec	3 min	3.9

† Factors outside recommended range.

The revised tray layout, with stepped caps, in a 12-ft-diameter tower would be entirely satisfactory. However, since the approach to flooding, entrainment ratio, and other dynamic factors are so conservative, it appears possible to reduce the tower size. Straight scale-up of the approach to flooding from 68 to 80 per cent indicates a tower diameter of 11 ft. In order to avoid overloading the slots, it would be desirable to have 530 four-in. bubble caps on the tray. This cannot quite be done in an 11-ft tower.

A subsequent tray layout demonstrated that 470 four-in. bubble caps can be placed in 11 rows on a double-pass tray in an 11 ft, 0 in. tower while maintaining the same cap spacing and adequate downflow areas. The 11 ft, 0 in. tray shows a calculated loading 80 per cent of flooding and a satisfactory balance of all the other dynamic factors except slot capacity. However, as was discussed in the text, a small slot overload is quite acceptable provided the other dynamic factors are within the recommended limits. Therefore, it is concluded that 11 ft, 0 in. is the optimum tower diameter for the bubble-cap tray.

NOMENCLATURE

a = area, in.^{2}

a_c = inside cross-sectional area of cap, in.^{2}

a_{cr} = annular area (between cap and riser) per cap, in.^{2}

a_r = riser area per cap, in.^{2}

a_w = weep hole area, in.^{2}/100 ft^2 tray area.

A = area, ft^2.

A_a = active, or "bubbling" area of tray (generally $A_t - 2A_d$), ft^2.

A_d = downcomer area, cross-sectional area for total liquid downflow, ft^2.

A_{da} = minimum area under downflow apron, ft^2.

A_n = net cross-sectional area for vapor flow above the tray (generally $A_t - A_d$), ft^2.

A_r = total riser area per tray, ft^2.

A_s = total slot area per tray, ft^2.

A_t = total tower cross-sectional area, ft^2.

C_d = liquid gradient factor, dimensionless.

C_s = coefficient in maximum slot capacity formula, dimensionless.

C_{sb} = vapor capacity parameter, as defined by Souders and Brown in Eq. (14-7), fps.

C_v = vapor load correction factor for liquid gradient, fractional.

d = distance or diameter, in.

d_c = inside diameter of cap, in.

d_r = inside diameter of risers, in.

d_s = slot width, in.

d_{sb} = slot width at bottom, in.

d_{st} = slot width at top, in.

d_{tw} = liquid throw over weir, in.

d_w = diameter of circular weir, in.

D = distance or diameter, ft.

D_f = total flow width across tray normal to flow, ft.

D_t = tower diameter (ID), ft.

e = liquid entrainment, lb moles/hr.

E_a = local wet (with entrainment) efficiency, fractional.

E_{mv} = local dry (Murphree vapor) plate efficiency, fractional.

F_{lv} = liquid-vapor flow parameter, defined by Eq. (14-6), dimensionless.

F_{va} = vapor flow parameter based on active area, defined by Eq. (14-40), dimensionless.

F_w = weir constriction correction factor, fractional.

g = acceleration of gravity, 32.2 ft/sec^2.

h = height or head, in.

h_{al} = head loss due to aerated liquid, in. liquid.

h_c = cap head loss (through riser, reversal, annulus, and slots), in. liquid.

h_{cd} = dry cap head loss, excluding slots, in. liquid.

h_{da} = head loss due to liquid flow under downflow apron, in. liquid.

h_{dc} = height of clear liquid in downcomer, in.

h_{ds} = dynamic slot seal, in.

h_f = height of froth (aerated mass) on tray, in.

h_{fd} = height of froth (aerated mass) in downcomer, in.

h_{ff} = height of liquid free fall in downcomer measured from weir, in.

h_l = equivalent height of clear liquid on tray, in.

h_{li} = height of clear liquid at inlet side of tray, in.

h_{lo} = height of clear liquid at overflow weir, in.

h_n = height of weir V notches, in.

h_{ow} = height of liquid crest over weir, measured from top of weir (straight or circular weirs) or from bottom of notches (V-notch weirs), in.

h_r = head loss through risers, in. liquid.

h_{ra} = head loss through reversal and annulus, in. liquid.

h_{sc} = cap skirt clearance, in.

h_{sh} = slot height, in.

h_{so} = height of slot opening, in., or slot head loss, in. liquid.

h_{ss} = static slot seal, in.

h_t = total vapor head loss (pressure drop) per tray, in liquid.

h_{ts} = tray spacing, in.

h_w = height of weir above tray floor, in.

H = height, ft.

H_{ts} = tray spacing, ft.

K_c = dry-cap head-loss coefficient, dimensionless.

K_s = slot orifice coefficient, dimensionless.

l = length, in.

l_c = total free width between caps normal to flow, in.

l_w = weir length, in.

L = liquid flow rate, lb moles/hr.

L_w = weir length, ft.

N_r = number of rows of caps normal to flow.

N_t = number of trays in column.

q = liquid flow rate, cfs.

q' = liquid flow rate, gpm.

Q = vapor flow rate, cfs.

R_{cc} = ratio of distance between caps on parallel liquid pass to that on oblique liquid pass.

R_{sd} = ratio of distance between caps to cap diameter.

R_{rc} = ratio of distance between risers to distance between caps.

R_s = trapezoidal slot shape ratio, d_{st}/d_{sb}.

R_{vd} = vapor-distribution ratio = Δ/h_c, dimensionless.

R_w = ratio of weir length to tower diameter, D_w/D_t.

t = time, sec.

t_{dc} = residence time of aerated mass in downcomer, sec.

t_l = residence time of liquid in aerated mass on tray, sec.

t_v = residence time of vapor in aerated mass on tray, sec.

t' = time, hr.

t'_d = time to drain tower, hr.

u = liquid velocity, fps.

u_a = liquid velocity based on active area A_a, fps.

U = vapor velocity, fps.

U_a = vapor velocity based on active area A_a, fps.

U_n = vapor velocity based on net area A_n, fps.

U_s = vapor velocity through slots, fps.
U_t = superficial vapor velocity based on total area A_t, fps.
V = vapor flow rate, lb moles/hr.
V_{dc} = volume of aerated mass in downcomer, ft^3.
w = liquid flow rate, lb/hr.
W = vapor flow rate, lb/hr.

Greek symbols

β = aeration factor, dimensionless.
Δ = liquid gradient for tray or tray section, in.
Δ' = liquid gradient uncorrected for vapor load, in.
ρ_f = average foam density, lb/ft^3.
ρ_l = density of clear liquid lb/ft^3.
ρ_v = vapor density, lb/ft^3.
σ = liquid surface tension, dynes/cm.
σ_w = surface tension of water, dynes/cm.
ϕ = relative froth density, ratio of foam density to clear liquid density.
ψ = liquid entrainment ratio, lb/lb (or mole/mole) gross liquid downflow.

REFERENCES

1. Atteridg, P. T., E. J. Lemieux, W. C. Schreiner, and R. A. Sundback: *A.I.Ch.E. J.*, **2**, 3 (1956).
2. Bolles, W. L.: *Petrol. Refiner*, **25**, 613 (1946).
3. Bolles, W. L.: *Petrol. Processing*, **11**(2), 64; **11**(3), 82; **11**(4), 72; **11**(5), 109 (1956).
4. Brook, W. E., D. E. Honnold, W. C. Cunningham, and R. L. Huntington: *Petrol. Engr.*, **27**(8), C-32 (1955).
5. Chillas, R. B., and H. M. Weir: *Trans. A.I.Ch.E.*, **22**, 79 (1929).
6. Cicalese, J. J., J. A. Davies, P. J. Harrington, G. S. Houghland, A. J. L. Hutchinson, and T. J. Walsh: *Petrol. Refiner*, **26**, 431 and 495 (1947).
7. Clay, H. A., T. Hutson, and L. D. Kleiss: *Chem. Eng. Progr.*, **50**, 517 (1954).
8. Colburn, Allan P.: *Ind. Eng. Chem.*, **28**, 526 (1936).
9. Dauphine, T. C.: Sc.D. Thesis, Massachusetts Institute of Technology, Cambridge, Mass., 1939.
10. Davies, James A.: *Ind. Eng. Chem.*, **39**, 774 (1947).
11. Davies, James A.: *Petrol. Refiner*, August, 1950, p. 93, and September, 1950, p. 121.
12. Fair, J. R.: *Petro/Chem Engr.*, **33**(10), 45 (1961).
13. Fair, J. R., and R. L. Matthews: *Petrol. Refiner*, April, 1958, p. 153.
14. Fryback, M. G., and J. A. Hufnagle: *Ind. Eng. Chem.*, **52**(8), 654 (1960).
15. Gerster, J. A., N. N. Hochgraf, A. G. Laverty, L. E. Scriven, and F. W. Wallis: Paper presented at Boston A.I.Ch.E. Meeting, December, 1956.
16. Gerster, J. A.: *Ind. Eng. Chem.*, **52**(8), 645 (1960).
17. Good, A. J., M. H. Hutchinson, and W. C. Rousseau: *Ind. Eng. Chem.*, **34**, 1445 (1942).
18. Haines, Jr., H. W.: *Ind. Eng. Chem.*, **52**(8), 662 (1960).
19. Holbrook, George E., and Edwin M. Baker: *Trans. A.I.Ch.E.*, **30**, 520 (1934).
20. Huitt, J. L., W. C. Ziegenhain, F. C. Fowler, and R. L. Huntington: *Petrol. Refiner*, **28**(11), 143 (1949).
21. Jones, J. B., and C. Pyle: *Chem. Eng. Progr.*, **51**, 424 (1955).

22. Kemp, Harold S., and Cyrus Pyle: *Chem. Eng. Progr.*, **45**(7), 435 (1949).
23. Klein, J. H.: Sc.D. Thesis, Massachusetts Institute of Technology, Cambridge, Mass., 1950.
24. Lockhart, F. J.: *Petrol. Refiner*, **35**(11), 165 (1956).
25. May, J. A., and Joseph C. Frank: *Chem. Eng. Progr.* **51**, 189 (1955).
26. Mayfield, F. D., W. L. Church, A. C. Green, D. C. Lee, and R. W. Rasmussen: *Ind. Eng. Chem.*, **44**, 2238 (1952).
27. Peavy, C. C., and E. M. Baker: *Ind. Eng. Chem.* **29**, 1056 (1937).
28. Perry, John H.: "Chemical Engineers' Handbook," 3d ed., pp. 597–602, McGraw-Hill Book Company, Inc., New York, 1950.
29. Rhys, C. O., and H. L. Minich: Paper presented at Los Angeles A.I.Ch.E. Meeting, March, 1949.
30. Robinson, Clark Shove, and Edwin Richard Gilliland: "Elements of Fractional Distillation," 4th ed., 403–470, McGraw-Hill Book Company, Inc., New York, 1950.
31. Rogers, M. C., and E. W. Thiele: *Ind. Eng. Chem.*, **26**, 524 (1934).
32. Schutt, H. C.: *Petrol. Refiner*, **24**, 249 (1945).
33. Sherwood, T. K., and F. J. Jenny: *Ind. Eng. Chem.*, **27**, 265 (1935).
34. Simkin, D. J., C. P. Strand, and R. B. Olney: *Chem. Eng. Progr.*, **50**, 565 (1954).
35. Souders, Mott, and George Granger Brown: *Ind. Eng. Chem.*, **26**, 98 (1934).
36. Strang, L. C.: *Trans. Inst. Chem. Engrs.* (*London*), **12**, 169 (1934).
37. White, Robert R.: *Petrol. Processing*, **2**, 147 and 228 (1947).
38. Winn, F. W., and G. J. Keller: *Petrol. Refiner*, July, 1955, p. 111.
39. Zenz, F. A.: *Petrol. Refiner*, **36**(3), 179 (1957).
40. Zuiderweg, F. J., and A. Harmens: *Chem. Eng. Sci.*, **9**, 89 (1958).

CHAPTER 15

TRAY HYDRAULICS
Perforated Trays

JAMES R. FAIR

Monsanto Chemical Company

In Chap. 14 the hydraulics of bubble-cap trays were treated in some detail. It was pointed out that many existing commercial vapor-liquid contacting devices, such as fractionators and absorbers, contain bubble-cap trays. Such popularity of bubble-cap trays was attributed to the wealth of operating experience developed during their use through the years; designers chose not to experiment with alternate contacting devices having relatively unknown hydraulic characteristics.

In recent years the pressure of tightening plant profitabilities has obliged designers to consider simpler and cheaper contacting devices. In so doing, the designers have converged on the simplest possible device employing liquid crossflow—the perforated tray. The resulting experience gained with perforated trays now enables design with confident prediction of performance. Accordingly, most new designs today specify some type of perforated tray instead of the traditional bubble-cap tray.

Like bubble-cap trays, perforated trays are available in many shapes and forms. The most common perforated tray is termed a *sieve tray*. This tray is comprised of a flat metal sheet perforated with round holes and suitably supported in the column. It is connected with one or more downcomers for handling liquid discharge and may contain weirs and baffles for directing vapor and liquid flows. Typical sieve trays are shown in Fig. 15-1.

In this chapter primary emphasis is given to sieve trays, i.e., simple perforated trays with liquid crossflow. Hydraulic relationships are developed, and their application to commercial design is shown. The reader will recognize that the treatment here is parallel to that for bubble-cap trays and that, in some cases, it contains duplicate material. Such duplication is held to a minimum and is condoned only for the sake of clarity and completeness of the individual chapters.

The author wishes to thank Mr. J. S. Moczek for his helpful suggestions.

FIG. 15-1a. Typical sieve tray (48-in. diameter).   (*Courtesy Fractionation Research, Inc.*)

FIG. 15-1b. Typical sieve tray (30-in.-diameter cartridge tray).   (*Courtesy Fritz W. Glitsch and Sons, Inc.*)

Mention here will also be made of proprietary perforated trays. In some of these devices, liquid and vapor compete for the same openings; such devices are related to packed columns in which liquid and vapor are in counterflow rather than in crossflow. There are other proprietary devices which have downcomers and crossflow of liquid; however, their vapor openings are not simple and may contain variable-opening orifices, or "valves." The several proprietary devices are mentioned in Secs. 15-18 and 15-19 but in general are outside the scope of the present work.

TRAY DYNAMICS

15-1. General Concepts

The liquid-vapor contacting action of a sieve tray is similar to that of a bubble-cap tray having several inches of running slot submergence, i.e., having sufficient liquid holdup to minimize discontinuities of the caps themselves. In normal operation, vapor flow through the perforations expands the liquid into a turbulent, surging mass called "froth." In moving across the tray floor, this mass provides high interfacial area and turbulence for efficient vapor-liquid mass transfer.

The complex fluid mechanics of the sieve tray must be considered from a simplified steady-state point of view. The schematic diagram in Fig. 15-2 represents such a simplification. The reader will recognize similarities between this diagram and that for bubble-cap trays (Fig. 14-3).

Liquid descends from the tray above through a downcomer and onto the tray at point A. A separate weir (not shown) may be used to ensure a liquid seal at A for all flow conditions. Between points A and B, essentially clear liquid is in crossflow, since this portion of the tray is not perforated. The distance between A and B will vary according to design conditions but often is quite short. Thus, aeration may occur immediately adjacent to the downcomer apron (or to the inlet weir, if used). The height of clear liquid h_{li} then is the equivalent of liquid in the aerated mass.

The active portion of the tray lies roughly between points B and C. (Aeration occurs outside the actual perforated zone because of turbulence.) In the active area tray action is characterized by the aerated mass, or froth. The visually observable height of this mass is designated as h_f. The density of the mass varies with height, being greatest near the tray floor; however, it is convenient to utilize a height-average density of ρ_f. There is also an effective hydrostatic head of the aerated mass h_l which can be measured by a floor manometer as shown in Fig. 15-2. If the weight of the vapor portion of the mass is neglected, h_l may be thought of as a "settled" height of clear liquid having a density of ρ_l.

The volume of aerated mass on the tray is of primary importance in mass-transfer calculations. Application of its dimensions and density for obtaining liquid and vapor residence time has been shown in Eqs. (14-1) to (14-5). 477

The aerated mass moves across the tray, and collapse begins when the perforations end at point C. A calming section between point C and the outlet weir (point D) may be provided for partial froth collapse but generally is not needed. Thus, aeration often occurs immediately

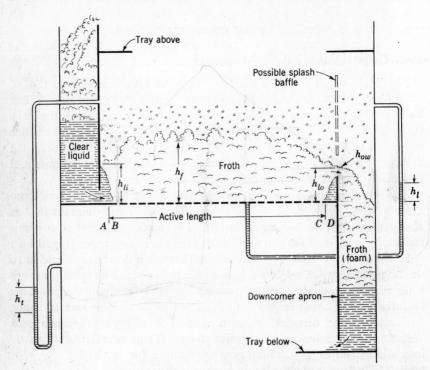

FIG. 15-2. Tray dynamics schematic diagram.

adjacent to the outlet weir, and the overflow material is in reality aerated mass. There is, however, an equivalent height of clear liquid h_{lo} between points C and D, calculated as the sum of the outlet weir height plus the crest of equivalent liquid flowing over the weir.

Secondary froth formation normally occurs in the downcomer as a result of liquid splashing and general turbulent conditions. Downcomer design must account for this condition and allow space for froth collapse; otherwise, froth density may be too low for adequate liquid outflow. The result of this is both a lowering of efficiency and an increase in pressure drop.

Vapor ascending from the tray below passes through the perforations and contacts liquid in the aerated mass. In this vapor flow lie two important differences between sieve and bubble-cap trays:

1. For the sieve tray, vapor emerges from a large number of small openings primarily in a vertical direction.

2. For the sieve tray, there is no built-in liquid seal and only vapor flow can prevent liquid passage through the holes.

As noted above, this description of tray dynamics is idealized. Many supporting observations and studies have been made in open water-air simulators and in commercial towers equipped with inspection windows. As a result, the simplifications given here are known to be reasonable approximations to the observed phenomena. And, as stated for bubble-cap trays, the simplifications permit design and physical-property variables to be correlated in such a way as to enable confident prediction of performance.

15-2. Foaming

When a gas or vapor is passed through a liquid, the expanded material is called a "froth" as noted above. If the expansion is related more to liquid physical properties than to method and degree of aeration, the material is called a "foam." Obviously, the terms are to a degree interchangeable. Of importance here is that foamability, as a physical property, be recognized and taken into account in sieve tray design.

In Chap. 14, simple laboratory tests were suggested for those cases where foamability is likely to be a problem in bubble-cap tray design. This problem may be manifested in expansion tendency (vapor flow limitation) or excessive stability (liquid downflow limitation). For sieve tray design, recourse should be made to the same laboratory tests (or, of course, to applicable experience gained in handling similar liquids).

15-3. Flooding

The initial step in sieve tray design is the determination of tower diameter. This diameter is first related to vapor capacity at its upper limiting condition—the point of incipient flooding. Separate flooding calculations must be made for points in the column having significant variations in liquid and vapor volumetric flow rates in order that the limiting tower size can be found. The result of this flooding analysis is a tentative diameter–tray-spacing combination, subject to adjustment in subsequent design calculations.

Flooding on a sieve tray may be brought on either by excessive *entrainment* or by *liquid backup* in the downcomer. The true point of flooding is difficult to determine experimentally, and "maximum capacity" is usually synonymous with an incipient flooding condition brought on by either of these phenomena. Regardless of cause, the onset of flooding is detected by (1) a sharp increase in pressure drop and (2) a sharp decline in efficiency.

Observations of flooding by entrainment show that liquid may be thrown or "jetted" to the tray above or it may be carried with froth expanding to the full tray spacing. At high volumetric vapor/liquid ratios, liquid may be blown completely clear of the tray ("blowing") and carried over in part to the tray above. When any of these phenomena occur, the downcomers may be only partly filled with liquid; however, the effect of "choking" the tray with entrainment is to increase pressure drop sharply, which in turn causes backup in the downcomers to increase to the point that they too are flooded.

Alternately, the downcomers may fill with liquid because of inadequate handling capacity before entrainment becomes excessive. However, the increasing liquid holdup on the tray soon causes "choking" with aerated mass, and excessive entrainment begins.

Thus, either vapor- or liquid-capacity limitations can lead to incipient flooding. The former is more usual and should be checked initially. But both will be limiting at true flooding.

As for bubble-cap trays, a flow parameter is used for sieve tray design:

$$F_{lv} = \frac{LM_l}{VM_v} \sqrt{\frac{\rho_v}{\rho_1}} = \frac{w}{W} \sqrt{\frac{\rho_v}{\rho_l}} \qquad (15\text{-}1)$$

This parameter accounts for liquid flow effects on the tray and in reality is a ratio of liquid/vapor kinetic-energy effects. For a capacity parameter, defined *for the flood point*, the Souders-Brown (17) coefficient is adapted:

$$C_{sb} = U_n \sqrt{\frac{\rho_v}{\rho_l - \rho_v}} \qquad (15\text{-}2)$$

The vapor velocity term U_n is based on the "net area" of the tray available for liquid disengagement; normally this area is the difference between total area and area blocked off by the downcomer(s). Unusual baffling effects on the tray can reduce this area; for example, if a splash baffle is used at the outlet weir, only the active area is then available for liquid disengaging and the velocity term in Eq. (15-2) should then be U_a instead of U_n.

The author recently developed a generalized correlation for sieve tray flooding (5). The parameters F_{lv} and C_{sb} are shown to be related to hole area, tray spacing, and system surface tension. A design chart based on the work is given in Fig. 15-3. Supporting experimental data are given in the original reference. Accuracy of predicting the flood point is about 10 per cent.

Figure 15-3 is based on small ($<\frac{1}{4}$-in.) holes and hole/active-area ratios of at least 0.10. Recent data (3) indicate that larger hole sizes may give higher entrainment; hence extrapolation of Fig. 15-3 to larger

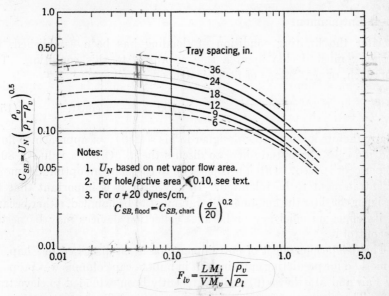

FIG. 15-3. Flooding capacity, sieve trays (surface tension = 20 dynes/cm).

hole sizes may result in optimistic designs. For hole/active-area ratios less than 0.10 it is suggested that flooding limits be reduced as follows:

Hole/active-area ratio	C_{sb}/C_{sb} from Fig. 15-3
0.10	1.00
0.08	0.90
0.06	0.80

Further restrictions on Fig. 15-3 are:

1. Weir height is less than 15 per cent of tray spacing. Higher weirs seriously impair vapor capacity as read from the chart.

2. System is low to nonfoaming. If foaming is present, actual capacity may be 75 per cent or less than that indicated from the chart.

Thus, the *flooding vapor velocity* is calculated as follows:

$$U_{n,\text{flooding}} = C_{sb} \left(\frac{\rho_l - \rho_v}{\rho_v} \right)^{1/2} \qquad (15\text{-}3a)$$

with C_{sb} obtained from Fig. 15-3 and corrected for surface tension as follows:

$$\frac{C_{sb}}{(C_{sb})_{\sigma=20}} = \left(\frac{\sigma}{20}\right)^{0.2} \tag{15-3b}$$

and subject to the other restrictions noted above.

15-4. Entrainment

After the limiting condition of flooding has been established, the designer must decide upon a safe approach to this condition. This approach, or percentage of flood, is defined as follows:

$$\% \text{ flood} = \frac{U_{n,\text{design}}}{U_{n,\text{flooding}}} \times 100 \qquad \frac{L}{V} = \text{const.} \tag{15-4}$$

Many factors influence the designer in his selection of approach to flood, and it is not appropriate here to discuss them. Often a value of 80 to 85 per cent is used. Often it is possible to calculate an optimum amount of entrainment to be tolerated. In any case, it is important that the designer consider the amount of entrainment to be handled, either because of its effect on efficiency or because collection devices may be needed to control it.

The deleterious effect of entrainment has been discussed in Chap. 14. A method for predicting entrainment in bubble-cap columns was proposed by Fair and Matthews (6) and has recently been extended to sieve trays (5). The concept of fractional entrainment ψ, defined as

$$\psi = \frac{e}{L + e} \tag{15-5}$$

is retained. When substituted in the Colburn equation [Eq. (14-9)], the effect of ψ on dry efficiency is as follows:

$$\frac{E_a}{E_{mv}} = \frac{1}{1 + E_{mv}[\psi/(1 - \psi)]} \tag{15-6}$$

When there is little or no entrainment, $\psi \sim 0$ and $E_a \sim E_{mv}$. When the entrainment flood point is approached, $\psi \to 1.0$ and $E_a \to 0$.

A design chart for estimating entrainment is given in Fig. 15-4. Based on the work of several investigators (5), this chart predicts experimental data with an accuracy of about ± 20 per cent.

It should be clear that some designs may very well call for a significant amount of entrainment. Zenz (22) has calculated that values of ψ in the 0.10 to 0.20 range represent optimum conditions. As Fig. 15-4 shows, sieve trays operating at high liquid/vapor ratios may be operated quite close to the flood point before significant entrainment occurs. Accord-

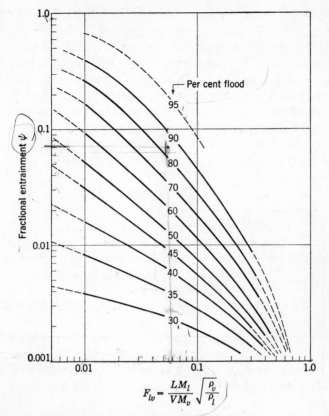

$$F_{lv} = \frac{LM_l}{VM_v}\sqrt{\frac{\rho_v}{\rho_l}}$$

Fig. 15-4. Fractional entrainment, sieve trays.

ingly, optimum values lose their significance under such conditions, since the flood point itself cannot be predicted with high accuracy.

15-5. Weeping

Just as entrainment represents an upper limit to tray operation, excessive flow of liquid through the perforations represents a lower limit. If a tray is to have a broad operating range between these limits, i.e., high turndown,† the designer must pay special attention to weeping.

Liquid passage through the tray may occur to some extent at all vapor rates, but as the rate is reduced, the passage becomes pronounced at the "weep point." This point is difficult to ascertain. Visual observation is inconclusive, and recourse is often made to a change of slope of the pres-

---

† "Turndown" is a term used by designers to denote ratio of maximum allowable (i.e., flooding) to minimum allowable operating throughput. The term is rather loosely used.

sure-drop–vapor-velocity curve.   Weeping may be fairly uniform across
the tray, or it may be localized near the point of liquid entry to the tray.
Transient weeping is also encountered owing to inertia of liquid splashing
in the aerated zone.   It is important to note that even though some tray
bypassing results from weeping, some mass transfer occurs in the vapor
zone.   The influence of weeping on tray efficiency depends on the fraction
of total liquid downflow that weeps; thus, for cases of low liquid flow a
small amount of weeping can be relatively serious.

As vapor rate is reduced below the weep point, serious liquid drainage
begins at the "dump point."   Dumping is easily observed visually and is
characterized by a definite drop in tray efficiency.   Below the dump point

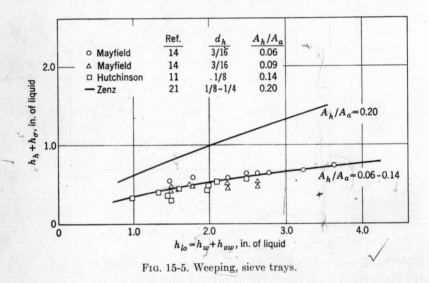

FIG. 15-5. Weeping, sieve trays.

operation may be unstable and efficiency so low that effective separation
is difficult, if not impossible.

Correlations presented in this chapter are based on measurement or
observation of the weep point.   Frequently the experimental data are
not precise and may not distinguish between weeping and dumping.   If a
tray is rated for minimum throughput coinciding with the correlation
weep point, lower throughput with adequate mass transfer can be maintained only if total liquid crossflow is relatively high.

At steady state, liquid will not drain through the perforations as long
as surface-tension effects and vapor-pressure drop through the perforations are present to prevent it.   It may be postulated that

$$h_h + h_\sigma \geq h_l \qquad (15\text{-}7)$$

is a sufficient condition to prevent drainage. Here, h_h is the pressure drop due to vapor flow through the perforations and h_σ is the head of liquid necessary to overcome surface tension of liquid over the perforations. Thus, at zero vapor velocity

$$h_\sigma \geqq h_l$$

for no drainage. This represents a familiar observation: For small hole sizes a certain head of liquid is required before any drainage can take place.

Weep points have been observed visually in the tray simulator studies of Hutchinson et al. (11), Mayfield et al. (14), and Zenz (21). Their data are plotted in Fig. 15-5 and cover both the air-water and air-menthanol systems. Data are lacking to permit further generalization.

Figure 15-5 is based on correlation according to Eq. (15-7), with equivalent liquid head taken as that calculated at the weir. As will be shown later, aeration effects actually give lower values of liquid head throughout the tray, but such a refinement is not needed for a crude correlation such as that in Fig. 15-5.

Lacking further weeping data, the designer can use Fig. 15-5 to obtain a rough estimate of the lower operating limits of sieve trays. The surface tension head is calculated from the simplified dimensional equation

$$h_\sigma = \frac{0.040\sigma}{\rho_l d_h} \qquad (15\text{-}8)$$

15-6. Liquid Crest over Weir

In Sec. 15-1 it was noted that liquid or froth entering the downcomer has an equivalent clear liquid crest h_{ow} over the outlet weir. The calculation of this weir crest is handled the same as for bubble-cap trays, and use should be made of Eqs. (14-22) to (14-28) as appropriate.

15-7. Vapor-pressure Drop

The total pressure drop for vapor flow from one tray to the next has the value that would be read from a manometer connected to the vapor spaces of the adjacent trays (see Fig. 15-2). It is convenient to consider this pressure drop as consisting in two resistances in series:

1. Drop due to vapor flow through the perforations
2. Drop due to the hydrostatic head of aerated mass on the tray

Such an approach oversimplifies the physical situation, since interaction effects are present, but is convenient and surprisingly reliable for design purposes.

Dry-tray Drop.   The calculation of pressure drop through the perforations, "dry-tray drop," is based on the simple orifice equation

$$U_h = C_{vo}\left(\frac{2g\rho_l h_h}{12\rho_v}\right)^{1/2} \tag{15-9}$$

which may be rearranged for direct calculation of dry-tray drop:

$$h_h = \frac{6U_h^2\rho_v}{gC_{vo}^2\rho_l} = 0.186\frac{\rho_v}{\rho_l}\left(\frac{U_h}{C_{vo}}\right)^2 \tag{15-10}$$

A great deal of experimental work has been carried out to obtain values of the dry orifice coefficient C_{vo}. Its value has been found to be a function of velocity of approach, hole-diameter/tray-thickness ratio, Reynolds number through the hole, condition of hole "lip," and other less important variables.   Simple correlation is made difficult not only because of the many variables but also because the exponent in Eq. (15-9) is not really constant at $\frac{1}{2}$.

Leibson, Kelley, and Bullington (13) have made a careful study of literature values of the orifice coefficient and have developed a correlation suitable for design purposes.   Their correlation is presented in Fig. 15-6 and should be used in conjunction with Eq. (15-10) for calculating dry-tray drop.

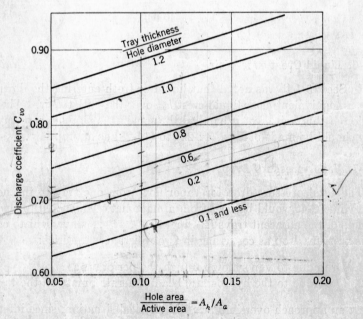

FIG. 15-6. Discharge coefficients for vapor flow, sieve trays.   [*I. Liebson, R. E. Kelley, and L. A. Bullington, Petrol. Refiner,* **36**(2), 127 (*February,* 1957); **36**(3), 288 (1957).]

When entrainment is significant ($\psi > 0.10$), the two-phase flow effect increases pressure drop through the perforations. Correction is made by analogy to two-phase pipe flow where the vapor phase is continuous:

$$\frac{h_{h,\text{wet}}}{h_{h,\text{dry}}} = 15X + 1.0 \qquad (15\text{-}11)$$

where the parameter X is defined as

$$X = \frac{\psi}{1 - \psi} F_{lv} \qquad (15\text{-}12)$$

and will be recognized as approximating the correlating parameter often used in two-phase pipe flow calculations.

Aerated Liquid Drop. Pressure losses above the perforations include those for forming bubbles ("surface tension loss" h_σ), those for flow through the aerated mass, and those for static effects in the vapor space. In practice, static vapor effects are neglected, except when vapor density is large in relation to liquid density. The other losses are grouped together and correlated by means of an *aeration factor* β:

$$h_l = \beta(h_w + h_{ow}) \qquad (15\text{-}13)$$

where $h_w + h_{ow}$ is the operating liquid seal at the tray outlet weir in terms of clear liquid. The term h_l represents a hydrostatic head which can be measured by manometers attached to the tray floor. This term is also important in mass-transfer calculations, since it represents a liquid holdup, or residence, on the tray.

Values of the aeration factor are usually obtained from measurements of total pressure drop:

$$\beta = \frac{h_t - h_h}{h_w + h_{ow}} \qquad (15\text{-}14)$$

Design correlations of β are, accordingly, subject to vagaries of dry-tray pressure-drop correlations as well as unknown interactions of dry and aerated effects.

A more direct and satisfactory approach to evaluating β is based on direct measurement of relative froth density ϕ. This density is obtained from visual observation of froth height combined with measurement of h_l by tray manometers:

$$\phi = \frac{h_l}{h_f} \qquad (15\text{-}15)$$

Hutchinson et al. (11) developed a useful theoretical relationship between ϕ and β:

$$\beta = \frac{\phi + 1}{2} \qquad (15\text{-}16)$$

and applied it successfully to measurements of sieve tray pressure drop and froth height.

A careful study of relative froth density in an air-water simulator was made by Foss and Gerster (7). Their data are plotted in Fig. 15-7 as a function of the kinetic-energy parameter F_{va}, defined as

$$F_{va} = U_a \rho_v^{1/2} \qquad (15\text{-}17)$$

Also included in Fig. 15-7 is an aeration factor curve calculated according to Eq. 15-16. Studies by the A.I.Ch.E. Research Committee (1, 2) have shown that the parameter F_{va} is satisfactory for generalizing ϕ data to

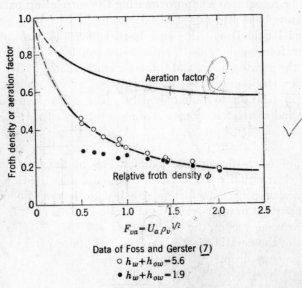

Data of Foss and Gerster (7)
○ $h_w + h_{ow} = 5.6$
● $h_w + h_{ow} = 1.9$

FIG. 15-7. Aeration factor, sieve trays.

systems other than air-water. Accordingly, the upper curve in Fig. 15-7 is suitable for design pressure-drop calculations involving the aeration factor β.

Total Pressure Drop. Total pressure drop across the tray is calculated from the expression

$$h_t = h_h + \beta(h_w + h_{ow}) \qquad (15\text{-}18)$$

where h_h is obtained from Eq. (15-10) plus Fig. 15-6 and β is obtained from Fig. 15-7.

The method given here for calculating vapor-pressure drop has proved to be reliable for commercial design of sieve tray columns. In comparison with literature data, the method checks within ±15 per cent the measurements of Jones and Pyle (11a), Hutchinson et al. (11), and

Mayfield et al. (14), with calculated values tending to be high at high liquid submergence.

15-8. Liquid Gradient

The head required to produce aerated mass crossflow is termed liquid gradient. In Fig. 15-2, this head is $h_{li} - h_{lo}$. If the head is great, vapor maldistribution and liquid weeping can occur, as noted in Chap. 14. In general, liquid gradient problems are much less severe in sieve tray design than in bubble-cap tray design, and accordingly larger tower diameters are possible without the expense of double flow paths. However, for long flow paths and high liquid rates, liquid gradient on sieve trays should be checked. The general criterion for *absence* of gradient difficulties (i.e., tray stability) is taken from bubble-cap tray design practice (Chap. 14):

$$(h_{li} - h_{lo}) < 0.5h_h \qquad (15\text{-}19)$$

where $h_{li} - h_{lo} = \Delta$ = hydraulic gradient.

The only published experimental data on sieve tray gradient are those of Hucks and Thomson (9) and Hutchinson et al. (11). These data have been analyzed by Hughmark and O'Connell (10) in terms of a friction factor for flow of an aerated mass in an open channel. The friction factor is defined as

$$f = \frac{12gR_h\Delta}{U_f{}^2 L_f} \qquad (15\text{-}20)$$

and this factor is correlated against a Reynolds modulus as in pipe flow

$$Re_h = \frac{R_h U_f \rho_l}{\mu_l} \qquad (15\text{-}21)$$

In Eqs. (15-20) and (15-21), R_h is the hydraulic radius of the aerated mass, defined as follows:

$$R_h = \frac{\text{cross section}}{\text{wetted perimeter}} \quad \text{ft}$$
$$= \frac{h_f D_f}{2h_f + 12D_f} \qquad (15\text{-}22)$$

where D_f is the arithmetic average between tower diameter D_t and weir length L_w and h_f is estimated from Eqs. (15-15) and (15-16) and Fig. 15-7. The velocity of the aerated mass is the same as for the clear liquid:

$$U_f = \frac{12q}{h_f \phi D_f} = \frac{12q}{h_l D_f} \qquad (15\text{-}23)$$

The relationship between f and Re_h is shown in Fig. 15-8, which is based on the paper of Hughmark and O'Connell (10). The effect of weir height on friction factor appears to be related to degree of mixing on the tray, but this requires further confirmation. Figure 15-8 is recommended for design purposes. Hydraulic gradient is calculated directly from Eq. (15-20).

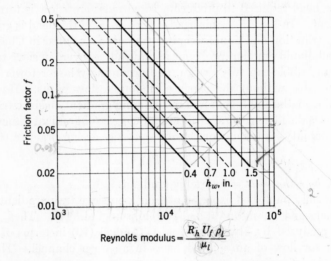

FIG. 15-8. Friction factor for froth crossflow, sieve trays.

15-9. Downcomer Dynamics

The problem of liquid capacity is the same for sieve trays as for bubble-cap trays. The tentative tray design evolved from flooding, weeping, and pressure-drop considerations must be checked for liquid-handling capacity. If this capacity is inadequate, operating difficulties with liquid gradient, downcomer backup, or flooding may be encountered. The reader will recognize that both sieve and bubble-cap trays employ the same downcomer design; hence the only real differences in liquid handling occur in the aerated zone.

As for bubble-cap trays, downcomer design involves the following considerations: (1) weir crest for liquid entering downcomer, (2) residence time of liquid in downcomer, (3) head loss of liquid leaving downcomer and flowing under downcomer apron, (4) total liquid backup as a criterion of approach to flooding. Items 1, 2, and 3 are covered specifically in Chap. 14, pages 486–488 and 496–498. Item 4, liquid backup, requires only minor adaptation to the case of sieve trays.

Downcomer backup is obtained from a simple pressure balance (see Fig. 15-2) which results in the equation

$$h_{dc} = h_t + h_w + h_{ow} + \Delta + h_{da} \tag{15-24}$$

where h_t is obtained from Eq. (15-18), h_w is specified, h_{ow} is obtained from Eqs. (14-22) to (14-28), Δ is obtained from Eq. 15-20, and h_{da} is obtained from Eq. (14-42).

It should be noted that h_{dc} in Eq. (15-24) is given in terms of clear liquid. Actually, the downcomer contains an aerated mass of height greater than h_{dc}. If this mass reaches the level of the tray above, flooding is approached. Usual practice is to assume a downcomer froth density [Eq. (14-44)] of 0.5 minimum. Accordingly, design should call for

$$h_{ts} > 2h_{dc}$$

which has been found conservative in all but those cases involving systems of high foamability. Further downcomer dynamic considerations are given in Sec. 14-9.

15-10. Liquid Mixing

As noted earlier, the aerated mass does not always move across the tray in plug flow. The back-mixing which occurs may be considerable, especially at high vapor rates and liquid holdups (high weir and/or low liquid rates). This mixing is of concern in that it influences concentration driving forces for mass transfer. The various models for predicting liquid mixing and its effect on efficiency are discussed in Chap. 16.

TRAY DESIGN

15-11. General Considerations

The foregoing treatment on tray dynamics has, for convenience, included comments and recommendations for the process design of sieve trays. In the present sections supplementary material is given to assist the process designer with "hardware" aspects and to enable him to summarize his calculations in the form of a finished design job.

Several items must be specified by the process designer. These items are summarized in Table 15-1. Furthermore, the specifications must permit "satisfactory operation" in a zone bounded by liquid-vapor flow restrictions, as shown qualitatively in Fig. 15-9. The over-all optimum design of the sieve tray column embraces a great many considerations, and the important ones emerge only after the designer has a specific problem at hand.

As noted earlier, sieve trays and bubble-cap trays have many features in common. The reader should refer to Chap. 14 for details on selection of tray spacing, selection of tray type, and downcomer design. Material given here is selected for special emphasis or unique application regarding sieve trays.

TABLE 15-1. PROCESS DESIGN SPECIFICATION CHECK LIST FOR A
SIEVE TRAY COLUMN

Over-all column design:
1. Diameter
2. Number of trays
3. Tray spacing
4. Feed and drawoff locations
5. Operating temperatures and pressures
6. Materials of construction

Tray design:
1. Liquid flow arrangement
2. Active area
3. Free (hole) area
4. Hole size, pitch, and pattern
5. Hole blanking
6. Tray baffles and calming zones
7. Downcomer area
8. Downcomer type and clearance
9. Tray inlet arrangement
10. Outlet weir type
11. Outlet weir dimensions
12. Tray thickness
13. Materials of construction
14. Tray and weir levelness

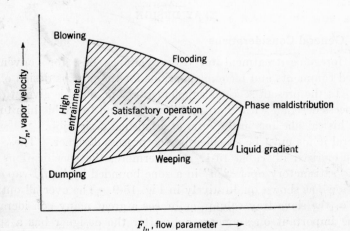

FIG. 15-9. Effect of liquid and vapor load on sieve tray performance.

15-12. Area Terms

A number of different area designations are used in the design of sieve trays. Although these areas are defined where needed in the text and nomenclature, it is convenient to summarize them at this point.

Tower area A_t is the total, or superficial, internal cross-sectional area of the tower.

Downcomer area A_d is the cross-sectional area at the top of the downcomer(s). For a segmental downcomer, it is the area of the segment formed by the overflow weir and the tower wall.

Net area A_n is the area for vapor flow above the tray. In the common single crossflow tray, $A_n = A_t - A_d$. If a splash baffle is located above the overflow weir (Fig. 15-2), $A_n = A_t - 2A_d$.

Active area A_a is the general area where aeration occurs. It is not limited entirely to the perforated zone, since turbulence effects carry aeration several inches past the perforations. For most design cases, A_a is taken as the total area between inlet and outlet weirs, and when straight segmental downcomers are used with single crossflow, $A_a = A_t - 2A_d$.

Hole area A_h is the total area open to vapor flow. In a tray layout, A_h is equal to the total opening in the perforated sheet metal used, minus blockage due to tray support members and special blanking strips.

Open area is a ratio used only in identifying perforated sheet metal material. For a given section of material, open area refers to the ratio of hole area to total area and can be calculated from the following equations:

Equilateral triangular pitch:

$$\text{Open area} = 0.905 \left(\frac{\text{hole diameter}}{\text{pitch}}\right)^2$$

Square pitch:

$$\text{Open area} = 0.785 \left(\frac{\text{hole diameter}}{\text{pitch}}\right)^2$$

15-13. Materials of Construction

Perforated trays may be fabricated from a wide range of materials, depending upon both the nature of the process fluids and the availability of fabrication techniques. Most metals and commercial alloys can be perforated conveniently in the multiple punch presses employed by tray fabricators. In noncorrosive service, the use of stainless-steel perforated sheeting is recommended as a safeguard against hole enlargement by erosion and against rusting before initial startup or during shutdowns. Type 410 stainless steel is widely used and does not increase tray cost significantly because of the small metal thickness needed. Downcomers, weirs, support beams, etc., may be specified of carbon steel if suitable for the materials handled.

Gasketing material, where employed, will also depend upon process conditions, with woven asbestos being generally satisfactory. Recent trends are toward elimination of gasketing in sectional trays except in cases where liquid flows are small. Where gasketing is not used, adequate bolting should be provided to achieve reasonably tight metal-to-metal contact.

15-14. Tray Layout

Hole diameters of $\frac{1}{8}$ to $\frac{1}{2}$ in. are commonly employed in sieve tray service, with larger hole sizes offering possible advantages in severe fouling applications. Small holes ($\frac{1}{8}$ to $\frac{1}{4}$ in. diameter) with the direction of punching oriented in the direction of vapor flow are often preferred. In usual practice, tray thickness of 12 or 14 U.S. Standard gauge is employed, except for carbon steel, where 10-gauge thickness has been used.

In the selection of small hole sizes, consideration should be given to minimum hole size limitations of punch dies, which depend on the thickness and type of material perforated. For small hole sizes and in the usual range of open area applying to sieve trays, a safe general rule is that the hole size should not be less than the sheet thickness in the case of carbon steel or not less than 1.25 times the sheet thickness when of stainless steel. Nickel, monel, and admiralty metal have hole-diameter limitations more nearly approaching those of carbon steel, while the requirements for Inconel are similar to those applying to stainless steel. With dies of advanced design, the limitations applying to hard materials are somewhat less stringent. For example, $\frac{1}{8}$-in. perforations can be provided in 14 U.S. Standard gauge (0.078 in.) stainless steel, and $\frac{3}{16}$-in. holes in 10 U.S. Standard gauge (0.141 in.) stainless steel. Thus, in cases where it is desirable to exceed the limitations expressed by the general rule given above, the matter should be referred to the tray fabricator.

Stock perforated metal is listed in terms of open area. Equations given above can be used to estimate this area and for determining hole pitch and blanking required for specified hole areas. A preliminary hole pitch is calculated by considering as perforated area the difference between the active area and the area covered by tray supports, tray rings, etc. If the calculated pitch does not correspond to a commercial standard, the next smaller pitch is selected and the (net) perforated area is calculated to determine the amount of blanking required. Some tray fabricators prefer that the designer specify hole diameter and total hole area, with a tolerance of 1 per cent, preferably 5 per cent. This provides leeway for using die arrangements not corresponding to stock perforated metal.

In the event that the hole area requirements vary considerably within the tower, it may be desirable to employ more than one pitch and/or hole diameter in order to avoid excessive tray blanking. The blanking should not exceed about 25 per cent of the tray area, except when it is intentionally provided for temporary operation at reduced rates.

Blanking strips should be distributed uniformly over the active area, except when used for calming zones at the weirs. In order to avoid "dead spots" in the active area, it is suggested that the width of the blanking strips not exceed about 7 per cent of the tray diameter for small towers or about 5 per cent of the diameter for large towers.

Blanking strips should be fastened so that a fairly tight seal is provided, removal is easy, and the perforated sheet not be distorted or buckled. Fastening with machine screws or similar attachments is perhaps the simplest.

Perforated sheet metal, available from several manufacturers, comes in standard widths of 36 and 48 in. and in standard lengths of 96, 120, and 144 in. For special requirements, other widths and lengths can be provided. Tray support members can be spaced at 12, 16, 18, or 24 in. for minimum waste of metal. In some designs these members are formed from blank margins of perforated sheets and thus eliminate the need for separate supports. It should be kept in mind that the spacing and type of tray support members employed with sectional trays will be influenced by accessibility through manways.

15-15. Summary of Effects of Tray Design Variables

The process designer has several tray design variables that he must specify (Table 15-1). Not all these variables are independent, but each will be discussed separately for its general effect on performance.

Liquid-flow Arrangement. A long liquid-flow path contributes to high tray efficiency. However, the long path also contributes to hydraulic gradient and possible tray instability. Shortening the path by splitting the liquid into two parts (double-crossflow tray) adds to the cost of tray construction.

Active Area. A large active area contributes to low vapor velocity and mass-transfer coefficients. Excessive vapor rates through a small active cross section cause entrainment, flooding, and low interfacial areas.

Hole Area. A high hole area contributes to weeping. A low hole area contributes to stable tray operation but increases pressure drop, and, below a given value, increases entrainment.

Hole Size, Pitch, and Pattern. Small hole size contributes to good gas dispersion, low pressure drop, low weeping, and low entrainment. Large holes are less easily fouled.

Hole pitch and pattern are associated with active area and hole areas as discussed above.

Hole Blanking. Blanking a portion of the perforations (i.e., covering with sheet metal) enables reduction of open area of perforated sheets to the desired value and provides flexibility for future increases in vapor throughput. However, if the blanked areas are localized and constitute a significant portion of the perforated area, excessive entrainment can result. Blanked area should normally not exceed about 25 per cent of the active area.

Tray Baffles and Calming Zones. Baffles for directing liquid movement on the tray are not required unless a reverse-flow tray is used. Calming zones at liquid entry and exit are not usually required.

Downcomer Area. A high downcomer area contributes to low liquid velocities and good froth collapse. A low downcomer area permits greater utilization of the tower area for vapor throughput.

Downcomer Type and Clearance. Sloped segmental downcomers are more expensive than straight segmental downcomers and consume tray area that might be used for perforations if the downcomer base has ample flow area. However, for foaming systems the increased inlet area of the sloped downcomer is helpful for vapor disengagement from liquid.

In small towers circular downcomers are cheaper than segmental downcomers but may represent wasted tray area for contacting.

A high clearance under the downflow baffle can allow vapor passage up the downcomer unless a seal is provided. A low clearance contributes toward undesirable backup of liquid in the downcomer.

Tray Inlet Weir. A tray inlet weir provides a positive seal against vapor flow up the downcomer and effects some distribution of liquid flowing to the tray. The disadvantages of the inlet weir are the additional expense, the possibility of plugging due to deposits of solids from the liquid, and the possibility of weeping due to impact of liquid on the tray.

Tray Outlet Splash Baffle. A splash baffle blocks froth movement into the downcomer and increases liquid holdup on the tray. On the other hand, it increases pressure drop and entrainment and thus limits vapor-handling capacity.

Outlet Weir Type. A straight weir is cheap and readily adapted to a segmental downcomer. At very low weir crests, however, the notched weir is more likely to maintain uniform liquid overflow.

Outlet Weir Height and Length. Pressure drop, liquid holdup, and weeping tendency increase with weir height, but much less than in direct proportion. Higher weirs contribute to back-mixing of liquid and may be desirable for special holdup requirements (e.g., chemical reaction on tray).

15-16. Operating Flexibility

In general, sieve trays can be designed for a wide range of operating rates. Hole area can be chosen such that weeping occurs only at very low vapor rates. Entrainment characteristics are such that operation can be maintained very close to the flood point. Thus, operating ranges from 30 to 40 to 90 to 95 per cent of flooding are often possible without sacrifice in tray efficiency.

For vacuum columns where pressure drop is critical, narrower operating ranges must be expected. Low hole area to prevent weeping creates excessive dry-tray pressure drop, and the required compromise may increase minimum operating rate to 50 per cent of flooding or higher.

There frequently is some misinterpretation regarding the need for operating flexibility in tray columns. For distillation operations, loading can be maintained at low feed rates by effectively varying reflux and boilup ratios. For absorption operations the changed liquid/vapor ratio may be a help because of its effect on the absorption factor, although there is at the same time an effect on flood point. The reasoning for absorbers also applies to strippers where steam or other extraneous gas is used; for reboiled strippers, however, sensitivity to load changes may be quite great.

15-17. Design Procedure and Example Calculation

The example used for bubble-cap tray design will now be recalculated on the basis of using sieve trays. The reader may recall that the benzene-toluene system is involved, and he should refer to Chap. 14, page 527, for problem details. The numbered steps below constitute a procedure which is general for all new sieve tray designs. If the problem involves rerating an existing tower, adjustments in the procedure should be obvious.

1. Calculate flow parameter. From Eq. (15-1)

$$F_{lv} = \frac{200,000}{240,000} \left(\frac{0.168}{43.3}\right)^{\frac{1}{2}} = 0.052$$

2. Choose tray and spacing. For initial design choose $\frac{3}{16}$-in. holes, hole area/tower area = 0.10, 14 U.S. Standard gauge tray material (0.078 in.), 2-in. weir height, and 24-in. tray spacing. Although tower diameter will probably be large, attempt to use a *single crossflow* tray for minimum cost. Also use *segmental downcomers, straight weirs,* and weir length equaling 77 per cent of tower diameter. For this case the downflow segment is 12 per cent of tower area.

This column will operate under *nonfouling* and *noncorrosive* conditions; thus, it may be desirable in later calculations to reduce the hole size to

$\frac{1}{8}$ in.   In addition, the system handled is nonfoaming and no difficulties with excessive entrainment are anticipated.

3. Calculate tower diameter.   From Fig. 15-3, C_{sb} for flooding = 0.36. Since surface tension $\sim$ 20 dynes/cm, no correction is needed.   System also meets other requirements of Fig. 15-3.

Use 85 *per cent of flooding* for design.   Then, by Eqs. (15-3) and (15-4),

$$U_n = \frac{(0.36)(0.85)}{[0.168/(43.3 - 0.17)]^{1/2}} = 4.90 \text{ fps}$$

For Q = 397 cfs and A_d = 0.12A_t,

$$A_t = \frac{397}{(0.88)(4.90)} = 92.1 \text{ ft}^2$$

for which a tower diameter of 10.8 ft. is calculated.   The usual practice is to round off to the next $\frac{1}{2}$-ft size.   Hence, *tower diameter* = 11.0 *ft.*

4. Tabulate areas:

$$A_t = 121(0.785) = 95.0 \text{ ft}^2$$
$$A_d = 0.12(95.0) = 11.4 \text{ ft}^2$$
$$A_n = 0.88(95.0) = 83.6 \text{ ft}^2$$
$$A_a = 0.76(95.0) = 72.2 \text{ ft}^2$$
$$A_h = 0.10(95.0) = 9.5 \text{ ft}^2$$

5. Adjust flow conditions:

$$U_n = \frac{397}{83.6} = 4.75 \text{ ft/sec}$$

Approach to flood: $\frac{4.75}{4.90}(85) = 82\%$

6. Calculate entrainment.   From Fig. 15-4, for F_{lv} = 0.052 and 82 per cent of flooding, ψ = 0.064.   Total entrainment by Eq. (15-5) is

$$e = \frac{0.064}{1 - 0.064}\frac{200,000}{78.1} = 175 \text{ moles/hr}(13,700 \text{ lb/hr})$$

7. Calculate pressure drop.   For a hole/active-area ratio of

$$\frac{9.5}{72.2} = 0.132$$

and for a tray-thickness/hole-diameter ratio of 0.078/0.188 = 0.41, an orifice coefficient of 0.75 is read from Fig. 15-6.   A hole velocity of 397/9.5 = 41.8 fps is then calculated.   Finally, the dry-tray drop is obtained from Eq. (15-10):

$$h_h = 0.186\frac{0.168}{43.3}\left(\frac{41.8}{0.75}\right)^2 = 2.24 \text{ in. liquid}$$

For a liquid flow rate of 577 gpm, and for a weir length of

$$0.77(132) = 102 \text{ in.}$$

weir crest is calculated from the Francis equation for straight weirs [Eq. (14-22)]:

$$h_{ow} = 0.48 \left(\frac{q'}{l_w}\right)^{\frac{2}{3}} = 0.48 \left(\frac{577}{102}\right)^{\frac{2}{3}} = 1.53 \text{ in. liquid}$$

[The correction factor for tower wall constriction (Fig. 14-7) is not significant.]

The F-factor through the active area is

$$F_{va} = \frac{397}{72.2}(0.168)^{\frac{1}{2}} = 2.25$$

From Fig. 15-7, the aeration factor $\beta = 0.58$. Hence, wet-tray drop is

$$h_l = 0.58(2.0 + 1.53) = 2.05 \text{ in. liquid}$$

and total tray pressure drop [Eq. (15-18)] is

$$h_t = 2.24 + 2.05 = 4.29 \text{ in. liquid}$$

8. Calculate weep point. Pressure drop for bubble formation is obtained from Eq. (15-8):

$$h_\sigma = \frac{0.040(21)}{43.3(0.188)} = 0.10 \text{ in. liquid}$$

Comparing $h_h + h_\sigma = 2.24 + 0.10 = 2.34$ with

$$h_w + h_{ow} = 2.0 + 1.53 = 3.53$$

in Fig. 15-5, operation is found to be well above the weep point.

By trial and error, it is found that the tray can be operated down to 50 per cent of design rate [turndown = 1.0/0.50(0.82) = 2.5] before weeping begins. Since pressure drop is probably not too critical in this particular design case, a smaller hole area could be specified. For example, if the hole area were 7.2 ft² ($A_h/A_a = 0.10$), the dry-tray drop would be 3.90 in. liquid and operation could then go down to 35 per cent of design rate [turndown = 1.0/0.35(0.82) = 3.5].

9. Check liquid-handling capacity. Downcomer velocity, calculated on the basis of clear liquid and $A_d = 11.4$ ft², is

$$\frac{577}{(450)(11.4)} = 0.11 \text{ fps}$$

which for a downcomer half full gives an indicated residence time of 1.0/0.11 = 9 sec. Although this is well above the minimum value of 3 sec (Chap. 14), the downcomer level must be checked.

Liquid gradient calculations require an approximate value of the froth height. This can be obtained from Eqs. (15-15) and (15-16):

$$h_f = \frac{h_l}{2\beta - 1} = \frac{2.05}{2(0.58) - 1} = 12.8 \text{ in.}$$

Equations (15-20) through (15-23) are then used to compute Δh_c:

$$U_f = \frac{(12)(1.28)(12)}{2.05(117)} = 0.77 \text{ fps}$$

where the width D_f is the arithmetic average of weir length (102 in.) and tower diameter (132 in.).

$$R_h = \frac{(12.8)(117)}{[2(12.8) + 117]12} = 0.87 \text{ ft}$$

$$Re_h = \frac{(0.87)(0.77)(43.3)}{(0.32)(6.72)(10^{-4})} = 1.35 \times 10^5$$

A liquid viscosity of 0.32 centipoise is used in Eq. (15-23).

From Fig. 15-8,

$$f \sim 0.02$$

Since the weir-to-wall distance of a 12 per cent downflow segment is 18 per cent of the tower diameter, the net distance between weirs is $0.64(11.0) = 7.0$ ft. Assuming this value of $L_f = 7.0$, Δ is calculated from Eq. 15-20:

$$\Delta = \frac{0.02(0.77)^2(7.0)(12)}{(32.2)(0.87)} = 0.036 \text{ in.}$$

This low value of Δ shows that liquid gradient will not be a problem.

For downcomer backup estimation, the pressure drop for flow under the downcomer apron is required. Using 1.5-in. clearance, the area under the downflow apron is

$$A_{da} = \frac{1.5 \times 102}{144} = 1.06 \text{ ft}^2$$

The downcomer apron pressure drop is [Eq. (14-42)]

$$h_{da} = 0.03 \left(\frac{q'}{100 A_{da}} \right)^2 = 0.03 \left(\frac{577}{106} \right)^2 = 0.89 \text{ in.}$$

Finally, downcomer backup is calculated [Eq. (15-24)]:

$$\begin{aligned} h_{dc} &= h_l + h_w + h_{ow} + \Delta + h_{da} \\ &= 4.3 + 2.0 + 1.5 + 0.04 + 0.9 \\ &= 8.7 \text{ in.} \end{aligned}$$

This value of h_{dc} is safely below the recommended maximum of 12 in.

(one-half tray spacing). It may be concluded that the tray design contains ample liquid-handling capacity.

10. Summarize process design of tray:

a. Single-pass tray with chord-type overflow weir. Tray material is type 410 stainless steel. Tray supports, downcomer baffle, and other auxiliaries are to be of carbon steel. No inlet weir is required. Downcomers are to be segmental, nonsloped, nonrecessed. No blanking is required.

b. Pertinent dimensions are:

Tower diameter, ft	11.0
Tray spacing, in.	24
Active area, ft²	72.2
Hole area, ft²	9.5
Downflow area, ft²	11.4
Hole area/tower area	0.10
Hole area/active area	0.13
Hole size, in.	$\frac{3}{16}$
Weir length, in.	102
Weir height, in.	2
Downcomer clearance, in.	1.5
Tray thickness, U.S. Standard gauge	14

c. The hole pattern will probably be $^{15}\!\!/_{32}$ in. triangular pitch, although this may require adjustment to match the mechanical design of the tray. The tray levelness tolerance should be set at $\pm\frac{1}{8}$ in.

PROPRIETARY TRAYS

15-18. Crossflow Trays

The important proprietary perforated trays embodying froth crossflow are generally known as *valve trays*. The trays are similar to sieve trays with one important exception: The perforations are covered with liftable lids or "valves" which rise and fall with variations in vapor flow. The lids thus act as check valves to limit liquid weeping or dumping at low vapor rates. Accordingly, the chief advantage of valve trays is that high efficiency can be maintained over a wide range of operating throughputs. Design turndown ratios (page 547) can be as high as 10.

There are three major suppliers (proprietors) of valve trays (8, 12, 15), each having a variety of designs available. The perforations in the tray are large—typically 1.5 in. in diameter for circular perforations (8, 12) and 6 in. long for rectangular-slot perforations (15). The lids are larger than the perforations and either may move within spiders clinched to the tray deck or may be guided and stopped by integral legs which fit into the perforations. The lids are permitted a vertical movement of $\frac{1}{4}$ to

½ in. and may have varying weights depending on system properties and desired operating flexibility.

Tray dynamics and design follow the same principles applying to sieve trays; however, test data covering performance are not generally available. Accordingly, detailed design and fabrication of valve trays are combined, and the tray cost includes a certain amount of engineering labor. It is usually necessary for the process designer to submit duty specifications to the proprietor; suggestions on this procedure have been presented by Thrift (19) and Winn (20).

In summary, valve trays provide more operating flexibility than sieve trays because of their variable-orifice characteristics. Since they are more complex mechanically, their fabrication is somewhat more expensive than sieve trays. Savings in engineering labor may be possible with valve trays, since the proprietor contributes to this function. On the other hand, he retains knowledge of design methods, and this may handicap the user in correlating plant performance data. Other features, such as entrainment, flooding, resistance to fouling, etc., are not likely to be greatly different between valve trays and sieve trays.

15-19. Counterflow Trays

In counterflow perforated trays, liquid and vapor compete for the same openings. The tray occupies the full tower cross section and has no weirs, baffles, or other attachments. Because of their simplicity, the trays are cheapest of the various perforated devices on a unit cross-section basis.

There are two major suppliers (proprietors) of counterflow trays, one (16) offering a flat tray perforated with rectangular slots and the other (18) offering a corrugated tray with small (⅛ to ¼ in.) round perforations. There are obvious variations of these basic types, such as flat trays with large (½ to 1 in.) circular perforations and trays in which the openings represent spacings between parallel metal bars. In general the total hole area is 15 to 30 per cent of the total tower cross section.

The hydraulic characteristics of counterflow trays are somewhat similar to those of conventional packed columns. Liquid holdup and interfacial area are strong functions of vapor rate; hence, the operating range for high efficiency tends to be narrow. In operation, liquid dumps momentarily through one or more sections of the tray and the locations of liquid passage move about the tray in a random fashion. The corrugated trays have better control of liquid holdup and discharge and hence have a broader operating range. A unique advantage of counterflow trays is their self-cleaning ability, which has proved advantageous in fouling services.

As for crossflow trays, proprietors usually must be consulted for tray design.   Use of the trays involves a fee which offsets the lower fabrication cost.   At the same time, a certain amount of engineering is provided by the proprietor.

In summary, counterflow trays have a narrow operating range (turn-down ratio of 2 or less) and are sensitive to load changes.   They occupy the full tower cross section and thus provide slight vapor-capacity advantages.   In fouling services they operate more cleanly than the crossflow devices.   Design methods are retained by the proprietor, and this may be a handicap to the user in analyzing plant performance of the trays.   Cost, efficiency (at load), pressure drop, and entrainment are not greatly different from crossflow perforated trays.

NOMENCLATURE

A_a = active, or "bubbling," area of tray (generally $A_t - 2A_d$), ft².
A_d = downcomer area, cross-sectional area for total liquid downflow, ft².
A_{da} = minimum area under downflow apron, ft².
A_h = net perforated area of tray, ft².
A_n = net cross-sectional area for vapor flow above the tray (generally $A_t - A_d$), ft².
A_t = total tower cross-sectional area, ft².
C_{sb} = vapor capacity parameter, as defined in Eq. (15-2), fps.
C_{vo} = vapor discharge coefficient for dry tray.
d_h = diameter of perforation, in.
D_f = total flow width across tray normal to flow, ft.
D_t = tower diameter (ID), ft.
e = liquid entrainment, lb moles/hr.
$\bar{e}$ = entrainment ratio, lb liquid/lb dry vapor.
E_a = local wet (with entrainment) efficiency, fractional.
E_{mv} = local dry (Murphree) vapor plate efficiency, fractional.
f = friction factor for froth crossflow, Eq. (15-20).
F_{lv} = liquid-vapor flow parameter, defined by Eq. (15-1), dimensionless.
F_{va} = vapor flow parameter based on active area, defined by Eq. (15-17).
F_w = weir constriction correction factor, fractional.
g = acceleration of gravity, 32.2 ft/sec².
h_{da} = head loss due to liquid flow under downflow apron, in. liquid.
h_{dc} = height of clear liquid in downcomer, in.
h_f = height of froth (aerated mass) on tray, in.
h_{fd} = height of froth (aerated mass) in downcomer, in.
h_h = head loss due to vapor flow through perforations, in. liquid.
h_l = equivalent height of clear liquid on tray, in.
h_{li} = height of clear liquid at inlet side of tray, in.
h_{lo} = height of clear liquid at overflow weir, in.
h_{ow} = height of liquid crest over weir, measured from top of weir (straight or circular weirs) or from bottom of notches (V-notch weirs), in.
h_t = total vapor head loss (pressure drop) per tray, in. liquid.
h_{ts} = tray spacing, in.
h_w = height of weir above tray floor, in.
h_σ = head loss due to bubble formation, in. liquid.
l_w = weir length, in.

L = liquid flow rate, lb moles/hr.
L_f = length of flow path, ft.
L_w = weir length, ft.
M_l = molecular weight of liquid.
M_v = molecular weight of vapor.
q = liquid flow rate, cfs.
q' = liquid flow rate, gpm.
Q = vapor flow rate, cfs.
R_h = hydraulic radius for froth crossflow, ft.
u_a = liquid velocity based on active area A_a, fps.
U_a = vapor velocity based on active area A_a, fps.
U_f = velocity of froth crossflow, fps.
U_h = vapor velocity through perforations, fps.
U_n = vapor velocity based on net area A_n, fps.
U_t = superficial vapor velocity based on total area A_t, fps.
V = vapor flow rate, lb moles/hr.
w = liquid flow rate, lb/hr.
W = vapor flow rate, lb/hr.
X = parameter for two-phase flow, dimensionless.

Greek Symbols

β = aeration factor, dimensionless.
Δ = liquid gradient for tray or tray section, in.
μ_l = viscosity of liquid, lb/ft-sec.
ρ_f = average froth density, lb/ft³.
ρ_l = density of clear liquid, lb/ft³.
ρ_v = vapor density, lb/ft³.
σ = liquid surface tension, dynes/cm.
σ_w = surface tension of water, dynes/cm.
ϕ = relative froth density, ratio of froth density to clear liquid density.
ψ = liquid entrainment ratio, lb/lb (or mole/mole) gross liquid downflow.

REFERENCES

1. American Institute of Chemical Engineers Research Committee: "Tray Efficiencies in Distillation Columns," Final Report, University of Delaware, Dec. 1, 1958.
2. American Institute of Chemical Engineers Research Committee: Final Report, University of Michigan, Ann Arbor, Mich., Mar. 1, 1960.
3. Bain, J. L., and M. Van Winkle: *A.I.Ch.E. Journal*, **7**, 363(1961).
4. Colburn, A. P.: *Ind. Eng. Chem.*, **28**, 526 (1936).
5. Fair, J. R.: *Petro./Chem. Eng.*, **33**(10), 45(1961).
6. Fair, J. R., and R. L. Matthews: *Petrol. Refiner*, **37**(4), 153 (1958).
7. Foss, A. S., and J. A. Gerster: *Chem. Eng. Progr.*, **52**, 28 (1956).
8. Glitsch, F. W., and Sons, Inc., Dallas, Tex. Also B. J. Robin, *Brit. Chem. Eng.*, **4**, 351 (1959).
9. Hucks, R. T., and W. P. Thomson: M.S. Thesis, Massachusetts Institute of Technology, Cambridge, Mass., 1951.
10. Hughmark, G. A., and H. E. O'Connell: *Chem. Eng. Progr.*, **53**, 127 (1957).
11. Hutchinson, M. H., A. G. Buron, and B. P. Miller: Paper presented at Los Angeles A.I.Ch.E. Meeting, May, 1949.

11a. Jones, J. B., and C. Pyle: *Chem. Eng. Progr.*, **51**, 424 (1955).
12. Koch Engineering Co., Wichita, Kans.   Also G. C. Thrift, *Oil Gas J.*, **52**(51), 165 (1954).
13. Leibson, I., R. E. Kelley, and L. A. Bullington: *Petrol. Refiner*, **36**(2), 127 (February, 1957); **36**(3), 288 (1957).
14. Mayfield, F. D., W. L. Church, A. C. Green, D. C. Lee, and R. W. Rasmussen: *Ind. Eng. Chem.*, **44**, 2238 (1952).
15. Nutter Engineering Co., Amarillo, Tex.   Also I. E. Nutter, *Oil Gas J.*, **52**(51), 165 (1954).
16. Shell Development Co. (Emeryville, Calif.) Staff: *Chem. Eng. Progr.*, **50**, 57 (1954).   Also J. A. Samaniego, *Oil Gas J.*, **52**(51), 161 (1954).
17. Souders, M., and G. G. Brown: *Ind. Eng. Chem.*, **26**, 98 (1934).
18. Stone and Webster Eng. Corp., Boston, Mass.   Also M. H. Hutchinson and R. F. Baddour, *Chem. Eng. Progr.*, **52**, 503 (1956).
19. Thrift, G. C.: *Petrol. Refiner*, **39**(8), 93 (1960).
20. Winn, F. W.: *Petrol. Refiner*, **39**(10), 145 (1960).
21. Zenz, F. A.: *Petrol. Refiner*, **33**(2), 99 (1954).
22. Zenz, F. A.: *Petrol. Refiner*, **36**(3), 179 (1957).

CHAPTER 16

TRAY EFFICIENCY
Vapor-Liquid Systems

The first 13 chapters of this book have dealt with the calculation of the number of hypothetical equilibrium stages required to effect a desired separation between components. Chapters 14 and 15 have described the design of the actual contacting devices which are used to accomplish the separations in plant equipment. The last major part of the over-all design to be discussed is the methods whereby the required number of hypothetical equilibrium stages N is related to the required number of actual stages N_t.

An equilibrium stage by definition produces an exit vapor stream which is in thermodynamic equilibrium with the exit liquid stream. It is impossible for an actual plant sieve or bubble-cap tray installation of any size to produce two such equilibrium streams. However, small trays on which the liquid is almost completely mixed can approach the performance of the hypothetical equilibrium stage. The liquid composition across such a tray would be constant and equal to the exit composition. The vapor entering the tray would be of uniform composition, since it comes from the completely mixed tray below. If enough residence or contact time for the vapor is provided, and if the resistance to mass and heat transfer is small enough, the exit vapor can approach a temperature and composition which is in equilibrium with the liquid even though all the vapor does not contact all the liquid. The approach to equilibrium on such a tray has been expressed by Murphree (14) as the ratio of the actual change in a vapor concentration through the tray to the change which would have occurred if the vapor had actually reached a state of equilibrium with the liquid.

$$E_{MV} = \frac{y_n - y_{n-1}}{y_n^* - y_{n-1}} \qquad (16\text{-}1)$$

where the subscripts n and $n - 1$ refer to the outlet and inlet vapor streams, respectively. In this chapter it will be necessary to distinguish between the vapor concentration y that actually exists and the vapor concentration y^* which would exist if the exit vapor were in equilibrium

570

with the actual exit liquid concentration x. The Murphree *tray* efficiency can also be expressed in liquid terms:

$$E_{ML} = \frac{x_{n+1} - x_n}{x_{n+1} - x_n^*}$$ (16-2)

where x_n^* is the liquid concentration which would exist if the exit liquid were in equilibrium with the actual exit vapor y.

The Murphree tray efficiency can represent physical reality only in the special case of completely mixed tray liquids where there is only one value for y_n and for x_n. When the liquid on the tray is not completely mixed, Eqs. (16-1) and (16-2) cannot apply except at a point in the liquid pool. *Point* Murphree efficiencies can be defined as

$$E_{OG} = \frac{y - y_{n-1}}{y^* - y_{n-1}}$$ (16-3)

and

$$E_{OL} = \frac{x_{n+1} - x}{x_{n+1} - x^*}$$ (16-4)

where the lack of a subscript denotes the actual vapor or liquid concentration at a given point in the pool. Again, the y^* is the vapor concentration in equilibrium with x while x^* is the liquid concentration in equilibrium with y. The y does not equal y_n, since the exit vapor is not uniform because the liquid concentration changes across the tray ($x \neq x_n$).

The fundamental approach to prediction of tray efficiencies involves the prediction of point efficiencies from the mass-transfer characteristics of the fluid systems and the physical dimensions of the tray. The point efficiencies are then related to the tray efficiencies by some assumed mathematical model. The tray efficiencies can be related in turn to the over-all column efficiency, which is defined as

$$E_o = \frac{N}{N_t}$$ (16-5)

The bulk of this chapter will be devoted to the development of this fundamental approach. Before this is done, however, simpler empirical ways of estimating E_o will be discussed.

16-1. Experimental Efficiencies

The best way to determine the efficiency of a given tray with a given fluid system is to measure it experimentally. This is seldom done on a laboratory scale before the separation equipment is built. The methods available for the prediction of the efficiency are usually good enough to make such experimental programs unnecessary. Instead, the experimental determinations of tray efficiencies usually are made in plant test runs on operating equipment. Product streams are analyzed to determine the separation actually accomplished between two key components in N_t stages. The number of equilibrium stages N required to accomplish the same separation is then calculated, and the over-all column efficiency obtained with Eq. (16-5). The E_o obtained will depend upon which two components were used to define the separation. Also, the numerical value of E_o will depend upon the equilibrium data used to calculate N. Reported over-all efficiencies should always be accompanied by the method used to obtain K values.

Organizations which utilize separation equipment usually accumulate such experimental data so that the design engineer can quickly select an accurate efficiency by simply comparing his system and tray design to a similar situation on which experience has been obtained. If an "experience file" is not available, the engineer may be able to find a similar case in the rather extensive published literature on tray efficiencies. Perry (18) provides compilations of such data. A great deal of care must be exercised in the selection of an efficiency in this manner because of the large number of variables which can affect it. The new tray must be similar to the tray on which the efficiency was measured. Particular attention should be paid to the comparison of tray diameters, tray spacing, number of passes, and height of liquid over the tops of the slots (or above the perforations in sieve trays). Also, the fluid systems should be similar and the pressure of operation the same. The stage temperature is a function of column pressure, and viscosity and surface tension are strongly affected by temperature. Viscosity plays a very important role and probably is the most important single factor in the difference between the relatively high efficiencies obtained in ordinary distillation and the low efficiencies obtained in vacuum distillation, absorption, extraction, and some extractive distillation processes. In distillation systems where the boiling-point range is not large, all the components are near their normal boiling points and have viscosities in the range of 0.2 to 0.3 centipoise (18). Increased pressure means higher temperatures, lower viscosities, and usually increased efficiencies. In absorption, extraction, and some extractive distillation processes the bulk of the liquid-phase components are far removed from their boiling points, the viscosities are high, and the tray efficiencies are much lower. The relative volatilities are also important. A high volatility is equivalent to low solubility in the liquid phase and a high liquid-phase resistance to mass transfer. Vapor velocity through the cap slots seems to have little effect as long as the tray is in stable operation and entrainment is not excessive. The tendency of the new system to foam should be considered because of the effect on entrainment. If all these factors are considered, there should be little danger involved in the estimation of the efficiency from performance in a closely similar situation. However, the efficiency should always be checked with at least one and preferably all of the correlations presented below.

EMPIRICAL CORRELATIONS

Two correlations which predict tray efficiencies as a function of liquid viscosity and component relative volatility have found wide use. Neither involves the basic mass-transfer approach, and they are therefore simple to use. However, the correlations are empirical in that they represent adequately only those systems upon which they are based, and therefore their use for other systems may be doubtful. Also they utilize only one or two of the many variables involved and therefore oversimplify a complex problem. On the other hand they are easy to use and if used intelligently will generally provide a good, quick estimate of the efficiency.

16-2. Drickamer-Bradford Correlation

Drickamer and Bradford (6) utilized plant test data on 54 refinery fractionating columns to develop the simple correlation between E_o and viscosity shown in Fig. 16-1. The authors realized that E_o is not a func-

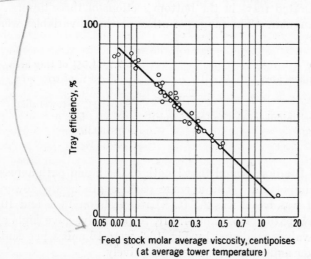

Fig. 16-1. Drickamer-Bradford correlation for over-all column efficiency E_o. The points represent plant tests on 54 refinery columns. [*H. G. Drickamer and J. R. Bradford, Trans. A.I.Ch.E.*, **39**, 319 (1943).]

tion of only one property but found that a simple correlation between E_o and μ_L was satisfactory for most petroleum systems on the usual run of trays. The use of viscosity was justified by noting its appearance in correlations for molecular diffusivity and mass-transfer functions. Also, viscosity varies dependably with temperature and molecular weight. Since the change in liquid viscosity with temperature and molecular weight is much larger than for the vapor viscosity, μ_L was chosen as the single correlating variable. The feed stream was chosen as the most representative stream in the column. The feed-stream composition is used to calculate μ_L from

$$\mu_L = x_1\mu_1 + x_2\mu_2 + \cdot\cdot\cdot + x_C\mu_C \qquad (16\text{-}6)$$

The μ_L's are evaluated at the temperature obtained by an arithmetic average of the top and bottom stage temperatures. The viscosity plots in Figs. 16-2 and 16-3 were prepared by Drickamer and Bradford from literature data.

For absorbers the rich oil composition is used instead of the feed. The average tower temperature is obtained as for distillation.

Drickamer and Bradford attributed most of the scatter of points in Fig. 16-1 to irregularities in the plant test data. These irregularities masked the effect of variations in tray design. The trays tested were both single and double pass. All were bubble-cap trays with the exception of 11 perforated trays in the bottom sections of three towers. All caps were rectangular. The ranges of other pertinent variables were as follows:

Riser area	7.2–11.5% of tray area
Liquid downflow area	4.5–9.9% of tray area
Weir height	1.5–2.75 in.
Weir length	65–80% of tower diam
Submergence (middle of slot to top of weir)	⅞–2⅛ in.
Plate spacing	18–30 in.
Tower diam	4–7.5 ft

Drickamer and Bradford found that length of the liquid path across the tray was also important. The average path length in the towers on which Fig. 16-1 was based was 2.5 ft. Later tests on 13.5- and 10.5-ft towers with flow paths of 5.5 and 4.0 ft, respectively, gave higher efficiencies. The values predicted by Fig. 16-1 were 18 and 8 per cent too low for the larger and smaller towers, respectively.

Literature data on 30 commercial towers operating on nonhydrocarbon systems were found to agree quite well with Fig. 16-1. Despite the agreement, Drickamer and Bradford cautioned against the extension of the correlation to nonhydrocarbon systems or fractionating equipment widely different from the usual refinery columns. One of the major reasons for restricting the correlation to relatively narrow boiling hydrocarbon mixtures is the effect of relative volatility or solubility on the tray efficiency for a given component. High volatility is more or less synony-

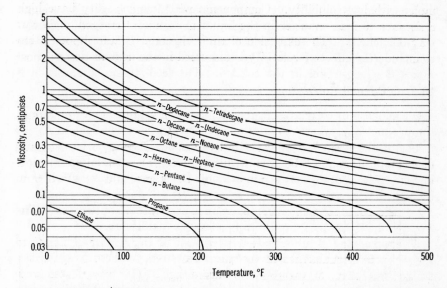

FIG. 16-2. Normal hydrocarbon viscosities as presented by Drickamer and Bradford. Olefins are approximately 15 per cent more viscous than the corresponding paraffin. Branched-chain hydrocarbons are approximately 10 per cent less viscous than the corresponding normal compound. [*H. G. Drickamer and J. R. Bradford, Trans. A.I.Ch.E.*, **39**, 319 (1943).]

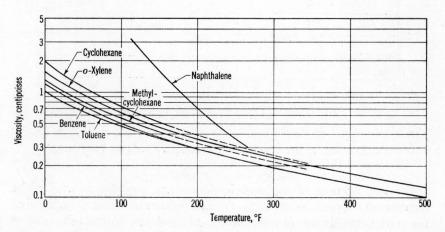

FIG. 16-3. Viscosities of cyclic hydrocarbons as presented by Drickamer and Bradford. [*H. G. Drickamer and J. R. Bradford, Trans. A.I.Ch.E.*, **39**, 319 (1943).]

mous with low solubility. Components with low solubility have high resistance to mass transfer in the liquid phase and correspondingly low tray efficiencies. The extension of the Drickamer-Bradford correlation to include the effect of volatility is discussed in the next section.

16-3. O'Connell Correlation

O'Connell (16) correlated E_o for fractionating columns as a function of the product $\alpha\mu_L$, where α is the relative volatility of the light key with respect to the heavy key and μ_L is the average liquid viscosity based on the feed composition and calculated with Eq. 16-6. The correlation is shown in Fig. 16-4. Both α and μ_L are evaluated at a temperature which is an

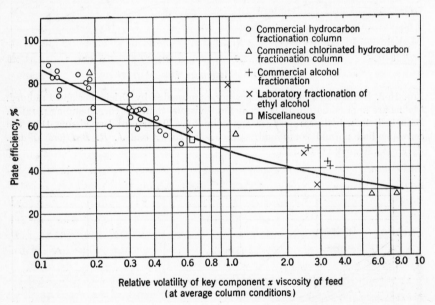

Relative volatility of key component x viscosity of feed
(at average column conditions)

Fig. 16-4. O'Connell plot for over-all column efficiency E_o as a function of relative volatility and viscosity. [*H. E. O'Connell, Trans. A.I.Ch.E.*, **42**, 741 (1946.)]

arithmetic average of top and bottom stage temperatures. By the inclusion of the relative volatility, O'Connell was able to extend somewhat the Drickamer-Bradford correlation to systems with high relative volatility components (low solubilities). Also, the presence of the volatility in the correlation implies that the tray efficiency of the various components in a multicomponent mixture is not the same.

O'Connell also presented a separate correlation for absorbers in which the relative volatility was replaced by a solubility function.

A.I.CH.E. CORRELATION

The four main factors which affect tray efficiency have been listed (3) as follows:

1. Rate of mass transfer in the vapor phase
2. Rate of mass transfer in the liquid phase
3. Degree of liquid mixing on the tray
4. Amount of liquid entrainment between trays

A precise prediction of E_o requires that each of these factors be properly evaluated and taken into account. The A.I.Ch.E. Tray Efficiency Research Program (3) has developed separate correlations for each of these factors and shown how they can be used to predict the efficiency. The major steps in the prediction method are (a) prediction of point Murphree efficiencies from the mass-transfer characteristics of the vapor and liquid, (b) assumption of some liquid mixing model on the tray in order to relate the point efficiency to the Murphree tray efficiency, (c) correction of the Murphree tray efficiency for entrainment, and (d) calculation of over-all column efficiency E_o from the corrected Murphree tray efficiency. The first step utilizes the traditional mass-transfer relationships. Many assumptions and definitions are involved as the correlation structure is developed from the diffusion equations of Maxwell and Stefan. To provide a convenient review for the reader, the development of the mass-transfer equations upon which the A.I.Ch.E. correlation is based will be summarized briefly. For a more complete treatment the reader is referred to Sherwood and Pigford (20) or other similar texts. Once the mass-transfer equations have been reviewed, the derivation of the mathematical relationships whereby the last three steps are accomplished will be presented.

It will be the purpose of the remainder of this chapter to present more completely the background material utilized by the A.I.Ch.E. "Bubble-tray Design Manual" (3). No practical help is offered to the designer, and therefore this material will be of interest only to students in the field. As soon as the mathematical development of the Manual correlations has been completed, the reader will be referred to the Manual and the various final reports from the universities (8, 19, 30) which participated in the program for discussion of the research results and methods for application of the correlations. Before becoming involved in the background derivations it is well to pause at this point and briefly discuss the over-all advantages and limitations of A.I.Ch.E. method as summarized by Prof. J. A. Gerster in his review of this chapter. Gerster emphasized the fact that a designer would be foolish to use the A.I.Ch.E. prediction method for systems (such as light hydrocarbon systems) where a wealth of experimental efficiency data exists. The major usefulness of the method is for systems where no previous experience exists. Also, the prediction method can be used in translating experimental efficiencies from a small column to a large one or for predicting the effect of changes in flow rates or temperatures on the efficiencies in an existing column. Gerster warned of the dangers involved in applying any predicted efficiency to columns which are operated close to the flooding point. In such cases, small changes in throughput can cause large changes in entrainment and thus in efficiency. Gerster also pointed out that the A.I.Ch.E. mixing correction is largely untested and may not adequately represent large-diameter columns where the efficiency increase due to non-mixing can be very large.

16-4. Steady-state Diffusion Equations

The equations for steady-state diffusion in mixtures of gases are based on the kinetic theory of gases and therefore can be considered basic or fundamental. These equations were first derived by Maxwell (13) and later reworked by Stefan (21), who also applied the diffusion equations to liquid mixtures. Lewis and Chang (12) were among the first to apply Stefan's work to engineering problems.

The basic differential equation for unidirectional diffusion of component 1 through a binary gaseous mixture of 1 and 2 is based upon the following hypotheses which have been validated by experimental evidence. Resistance to motion of 1 by steady-state diffusion can be said to be proportional to the following:

1. Length of the path which the molecules of component 1 must traverse
2. Relative velocity of 1 with respect to 2
3. Number of molecules of 1 and 2 in the path of diffusion

The number of molecules of 1 and 2 in the path is proportional to the molar densities $(\rho v)_1$ and $(\rho v)_2$. Therefore, the resistance to diffusion of 1 over a differential length dz can be expressed mathematically as

$$b_{12}(\rho v)_1(\rho v)_2(u_1 - u_2)\, dz$$

where b_{12} is the proportionality factor which is usually termed the coefficient of resistance. If all external forces acting upon the molecules of 1 are neglected, the resistance to the steady-state diffusion of 1 must be equivalent to the driving force for diffusion. The potential gradient which acts as the driving force can be expressed in terms of partial pressure for an ideal gas.

$$dp_1 = b_{12}(\rho v)_1(\rho v)_2(u_1 - u_2)\, dz \qquad (16\text{-}7a)$$

A similar equation can be written for component 2.

$$dp_2 = b_{21}(\rho v)_2(\rho v)_1(u_2 - u_1)\, dz \qquad (16\text{-}7b)$$

For a binary mixture, $dp_1 = -dp_2$, and it can be shown by algebraic addition of Eqs. (16-7a) and (16-7b) that $b_{12} = b_{21}$.

Equation (16-7) has a firm theoretical basis in the kinetic theory of gases. A satisfactory kinetic theory of liquids has not been developed, and the basic differential equation for diffusion in liquids is written as an analogy to the gas equation. In concentration terms, the liquid equation is

$$dc_1 = b'_{12}c_1c_2(u_1 - u_2)\, dz \qquad (16\text{-}8)$$

Returning to the gas equation, (16-7a) can be rewritten as

$$\frac{dp_1}{dz} = b_{12}[(\rho v)_1(\rho v)_2 u_1 - (\rho v)_1(\rho v)_2 u_2]$$

Since $\qquad (\rho v u)_i = \dfrac{N_i}{S} = \dfrac{\text{moles of } i \text{ diffusing}}{\text{hr-ft}^2}$

and since, for an ideal gas,

$$(\rho v)_i = \frac{1}{V_i} = \frac{p_i}{RT} = \frac{\text{moles}}{\text{ft}^3}$$

the equation can be written as

$$\frac{dp_1}{dz} = \frac{b_{12}}{RT}\left(p_2\frac{N_1}{S} - p_1\frac{N_2}{S}\right) \qquad (16\text{-}9)$$

Further development requires that some relation between N_1/S and N_2/S be postulated. The two cases of general interest are (a) equimolar countercurrent diffusion of 1 and 2 and (b) diffusion of 1 through a stagnant layer of 2.

Equimolar Countercurrent Diffusion. The case where both components are diffusing in equal amounts but in opposite directions will be considered first. This model approximates the steady-state situation at any given point on a tray in a binary distillation column. Some of the heavier component in the vapor will diffuse through the vapor to the vapor-liquid interface, condense, and then diffuse into the bulk liquid stream. At the same time, some of the lighter component in the liquid will diffuse in the opposite direction to the interface, vaporize by utilizing the heat of condensation of the condensed heavier component, and then diffuse into the bulk vapor stream. Since vapor is continually entering the tray from below while liquid continually flows down from above, the concentrations at the given point on the tray are maintained at constant values and the diffusion is essentially a steady-state phenomenon at the point. Also, the molar heats of vaporization of the two components are

580 DESIGN OF EQUILIBRIUM STAGE PROCESSES

usually similar. Consequently, the following equations which assume equimolar countercurrent diffusion of components 1 and 2 are a fairly reliable mathematical model for a binary distillation at any point on a tray.

For equimolar countercurrent diffusion, $N_2/S = -N_1/S$. Since $p_2 = p - p_1$, Eq. (16-9) reduces to

$$dz = \frac{RT}{b_{12}p(N_1/S)} dp_1 \qquad (16\text{-}10)$$

Integrating across the vapor film from the bulk vapor stream (where z will be set equal to zero and $p_1 = p_{1,G}$) to the vapor-liquid interface (where $z = z_G$ and $p_1 = p_{1,i}$) gives

$$\frac{N_1}{S} = \frac{RT}{b_{12}p z_G} (p_i - p_G)_1 \qquad (16\text{-}11)$$

In the derivation of (16-10) from the kinetic theory of gases [see page 3 in Sherwood and Pigford (20)], the term $\lambda u/2$ appears in the coefficient of dp_1. The λ is the mean distance a molecule of gas travels in the z direction before colliding with another molecule, and u is the mean speed of the molecules in the z direction. The $\lambda u/2$ is replaced with the symbol D_G, which is termed the *diffusity* or the *diffusion coefficient* for the pair of gases. In terms of Eq. (16-11)

$$D_G = \frac{R^2 T^2}{b_{12}p}$$

so

$$\frac{N_1}{S} = \frac{D_G}{RT z_G} (p_i - p_G)_1 \qquad (16\text{-}12a)$$

Because of the lack of a better procedure, the equation for the equimolar countercurrent diffusion across the liquid film is written by analogy.

$$\frac{N_1}{S} = \frac{D_L}{z_L} (c_L - c_i)_1 \qquad (16\text{-}12b)$$

Diffusion of One Component Only. Equimolar countercurrent diffusion describes distillation quite well where transfer in both directions does occur. In absorption, however, the bulk of the transfer is in one direction, from the vapor to the liquid. A diffusion model which assumes movement of one component through a second stagnant one comes close to a ternary absorption system where two gases are contacted with a nonvolatile liquid absorbent and only one gas is absorbed. The expression which describes the transfer of the diffusing component from the bulk vapor to the vapor-liquid interface is obtained from Eq. (16-9) when $N_2/S = 0$.

$$\frac{dp_1}{dz} = \frac{b_{12}}{RT} \left[\frac{N_1}{S} (p - p_1) \right]$$

Integrating across the vapor film from the bulk vapor phase (where $p_1 = p_{1,G}$ and $z = 0$) to the vapor-liquid interface (where $p_1 = p_{1,i}$ and $z = z_G$) gives

$$\frac{N_1}{S} = \frac{RT}{b_{12}z_G} \ln \frac{p_{2,G}}{p_{2,i}}$$

A logarithmic mean value for the partial pressure of component 2 (which is stagnant and not diffusing) is defined as

$$p_{2,m} = \frac{p_{2,i} - p_{2,G}}{\ln (p_{2,i}/p_{2,G})}$$

Substitution for the logarithmic term in the preceding equation and remembering that $D_G = R^2T^2/b_{12}p$ and $p_2 = p - p_1$ give

$$\frac{N_1}{S} = \frac{D_G p}{RT z_G p_{2,m}} (p_i - p_G)_1 \tag{16-13a}$$

Comparison of (16-13a) with (16-12a) shows that for dilute concentrations of component 1 ($p_{2,m} \cong p$) the two equations become identical. Therefore, absorption systems where the solute is present in small concentrations can be represented by the equimolar countercurrent equation (16-12a).

The equation for one-component diffusion in a liquid can be written by analogy to (16-13a)

$$\frac{N_1}{S} = \frac{D_L \rho_L}{z_L c_{2,m}} (c_L - c_i)_1 \tag{16-13b}$$

In the subsequent development, the ratios $p/p_{2,m}$ and $\rho_L/c_{2,m}$ will be assumed to be unity and Eqs. (16-12a) and (16-12b) will be used to represent the mass-transfer rate of a given component. This assumption means that the mass-transfer correlations developed later will apply only to (a) binary distillation systems in which equimolar transfer is approached and (b) ternary absorption systems in which the oil is nonvolatile, one gas is insoluble in the oil, and the single gas being absorbed is present in very dilute concentrations.

Diffusion Coefficients. The A.I.Ch.E. Manual recommends the Wilke and Lee (29) modification of the Hirschfelder, Bird, and Spotz equation for the calculation of binary gaseous coefficients for molecular diffusion.

$$D_G = \frac{BT^{3/2}[(M_1 + M_2)/M_1 M_2]^{1/2}}{p r_{12}^2(I_D)} \tag{16-14a}$$

where T is the absolute temperature in degrees Kelvin,

$$B = 9.53 \left[4.35 - \left(\frac{M_1 + M_2}{M_1 M_2}\right)^{1/2} \right] \times 10^{-4}$$

and $\qquad r_{12} = \dfrac{(r_0)_1 + (r_0)_2}{2}$ = collision diameter, A

The (I_D) represents the collision integral for diffusion and is a function of the group (Tk/ϵ_{12}), where ϵ/k is the force constant parameter and

$$\frac{\epsilon_{12}}{k} = \sqrt{\frac{\epsilon_1}{k}\frac{\epsilon_2}{k}}$$

Tables of force constants, collision diameters, and collision integrals taken from Wilke and Lee are included in the A.I.Ch.E. Manual. For components not listed, the following equations are recommended by the Manual for the estimation of the force constant ϵ/k and the collision diameter:

$$\frac{\epsilon}{k} = 0.77 T_c$$

$$\frac{\epsilon}{k} = 1.15 T_b$$

$$r_0 = 1.18 V_0^{1/3}$$

where T_c and T_b are the critical and normal boiling temperatures in degrees Kelvin and V_0 is the molar volume in cubic centimeters per gram mole of the liquid at the normal boiling point. A table of molar volumes is included in the Manual.

The Manual discusses the limitations of Eq. (16-14a) and compares its accuracy with the equations of Gilliland (9) and of Arnold (2).

Diffusion coefficients in multicomponent systems are functions of composition and are different for every component. The equation of Wilke (26) is used to calculate the average gaseous diffusivity for component i.

$$D_{Gi} = \frac{1 - y_i}{\displaystyle\sum_{\substack{j=1 \\ j \neq i}}^{c} y_j/D_{ij}} \qquad (16\text{-}14b)$$

where D_{ij} is the diffusion coefficient in the ij binary.

The Wilke (27) method for the estimation of binary liquid diffusion coefficients is also recommended by the Manual.†

$$\frac{(D_L\mu_L/T)_1}{d \ln a_1/d \ln x_1} = \left[\left(\frac{D_L\mu_L}{T}\right)_{2_0} - \left(\frac{D_L\mu_L}{T}\right)_{1_0}\right] x_1 + \left(\frac{D_L\mu_L}{T}\right)_{1_0} \qquad (16\text{-}15a)$$

† Recent work by D. K. Anderson and A. L. Babb [*J. Phys. Chem.* **65**, 1281 (1961); **63**, 2059 (1959); **62**, 404 (1958)] with ideal and nonideal binary systems has illustrated the inadequacy of existing correlations and offers hope for improved prediction methods.

where $(D_L\mu_L/T)_1$ = Stokes-Einstein group for diffusion of component 1
in 2 at concentration x_1

$(D_L\mu_L/T)_{1_0}$ = Stokes-Einstein group for diffusion of component 1
in dilute solution of 1 in 2

$(D_L\mu_L/T)_{2_0}$ = Stokes-Einstein group for diffusion of component 2
in dilute solution of 2 in 1

x_1 = mole fraction of component 1 in mixture

a_1 = activity of component 1 in mixture

The liquid diffusion coefficient at infinite dilution is obtained from the
following equation of Wilke (28):

$$\frac{D_L\mu_L}{T} = \frac{2.87 \times 10^{-7}(XM)^{0.5}}{V_0^{0.6}} \qquad (16\text{-}15b)$$

where X is an association parameter for the solvent and equals 1.0 for a
nonassociated liquid. Tables of values of X and V_0, the molar volume
of the solute, are given in the Manual.

16-5. Mass Transfer between Phases

The equations developed in the previous section describe the transfer
of mass by steady-state molecular diffusion within one stagnant phase.
In the mass-transfer processes of interest to this text, mass is always transferred between phases. The path which a molecule follows in its journey
from one phase to the other can be divided into three parts. Consider
the direction of net transfer for component 1 to be from the liquid to the
vapor phase. The three steps (or resistances) in the path are as follows:

1. Transfer from the bulk liquid phase to the vapor-liquid interface
2. Transfer through the interface
3. Transfer from the interface to the bulk vapor phase

The bulk vapor and liquid phases are usually in motion relative to each
other. The bulk movement within any phase will be more or less turbulent depending upon the velocity of motion. Consequently, the major
part of the transfer within the bulk phase will be by eddies which keep the
bulk phase well mixed. Molecular diffusion is negligible in this region.

The mechanism of mass transfer in the region next to the interface is
not so well defined. The interface between two moving phases cannot be
considered to be stagnant. Therefore, there is no region where eddy
diffusion (or mixing) is completely inoperative and all the mass transfer
is accomplished by molecular diffusion. Undoubtedly eddy diffusion
plays an important and possibly major role all the way to the interface,
while the contribution of molecular diffusion is negligible in the bulk
phase and of a secondary, but not negligible, nature in the more laminar

region adjacent to the interface. This means that the equations of the previous section will not adequately describe the transfer in steps 1 and 3 above. Since our ability to describe mathematically the complex interplay between eddy and molecular diffusivities is limited, we are forced to abandon the theoretical approach to mass transfer between phases and attack the problem in a more empirical manner. The approach of greatest value for design is the "two-film" concept of Whitman.

Two-film Concept. Whitman's (25) "two-film" (or more correctly "two-resistance") model assumes that the resistance to transfer of a molecule from one bulk phase to the other is the sum of two resistances, one in each phase. The model hypothesizes the existence of stagnant "films" adjacent to the interface. Resistance to transfer within the bulk phase is assumed negligible because of eddy mixing. All the resistance to transfer from the bulk phase to the interface (or vice versa) is considered to be in the "film" where transfer occurs solely by molecular diffusion. If this model were true, a knowledge of the film thickness z_G or z_L and the interface partial pressure $p_{1,i}$ or concentration $c_{1,i}$ would permit the calculation of the transfer rate N_1/S to or from the interface by means of Eqs. (16-12). The interface itself is considered to have no resistance; that is, the two phases are considered to be in equilibrium with each other at the interface with $p_{1,i}$ and $c_{1,i}$ representing equilibrium values in the vapor and liquid at the interface.

A better understanding of boundary-layer behavior has caused the two-film model to fall into disrepute as a theoretical model for transfer between phases. As a correlation and design tool it is still widely used for lack of a better procedure. Even as a design tool the concept must be modified considerably to be of practical use. Even if the films existed, their thicknesses would never be known. Also, the interface conditions $p_{1,i}$ and $c_{1,i}$ are practically impossible to determine experimentally. Therefore Eqs. (16-12a) and (16-12b) cannot be applied directly.

To obtain practical correlation and design equations, the concept of two resistances is retained but the fictitious film idea is dropped. The form of Eqs. (16-12) is also retained, but now all the resistance within the phase from the bulk stream up to and including whatever resistance may or may not exist at the interface is included in an empirical resistivity factor (mass-transfer coefficient) to give for component 1

$$\frac{N_1}{S} = k_G(p_i - p_G)_1 \tag{16-16a}$$

and
$$\frac{N_1}{S} = k_L(c_L - c_i)_1 \tag{16-16b}$$

These equations are analogous to Eqs. (16-12a) and (16-12b). Comparison of the equations shows that

$$k_G = \frac{D_G}{RTz_G}$$ (16-17a)

and
$$k_L = \frac{D_L}{z_L}$$ (16-17b)

If Eqs. (16-16a) and (16-16b) are used to represent single-component diffusion through a second stagnant component, comparison with (16-13a) and (16-13b) shows that

$$k_G = \frac{D_G p}{RT z_G p_{2,m}}$$ (16-18a)

and
$$k_L = \frac{D_L \rho_L}{z_L c_{2,m}}$$ (16-18b)

As described at the end of the previous section, the ratios $p/p_{2,m}$ and $\rho_L/c_{2,m}$ will be taken as unity for the remainder of the development, so the mass-transfer coefficients defined by (16-17) and (16-18) can be considered to be the same.

Definition of the mass-transfer coefficient eliminates the distance term z, but the potential terms in (16-16a) and (16-16b) are still expressed in terms of interface concentrations. This difficulty is bypassed by the definition of over-all mass-transfer coefficients as follows:

$$\frac{N_1}{S} = K_{OG}(p^* - p_G)_1$$ (16-19a)

$$\frac{N_1}{S} = K_{OL}(c_L - c^*)_1$$ (16-19b)

The p^* is that partial pressure of the diffusing component which would exist in the bulk vapor phase if the vapor were in equilibrium with the bulk liquid phase. Likewise, c^* is that concentration in the liquid which would exist if the bulk liquid phase were in equilibrium with the bulk vapor phase. Note that Eqs. (16-19a) and (16-19b) retain the general form of Eq. (16-12) but express the potential for mass transfer in a radically different manner. The driving force is no longer expressed as the gradient in concentration or partial pressure between the two ends of the diffusion path but instead as a displacement from equilibrium. The potential terms in (16-19) do revert to a gradient in the special case where one phase (the liquid phase in the first equation) offers no resistance, but otherwise the only remaining similarity between (16-19) and the equations based on the kinetic theory of gases is the general form

$$\text{Rate} = (\text{conductance})(\text{potential}) = \frac{\text{potential}}{\text{resistance}}$$

Equations in a Column. Now that usable (although empirical) correlation forms have been developed for the rate of mass transfer, the next step is to apply them to an actual mass-transfer device such as a bubble tray. One difficulty arises immediately. The area for mass transfer in a contacting device cannot be measured and expressed as a planar cross section of area S. To get around this difficulty, it is customary to define an area term a by the equation

$$dS = aA \, dZ$$

where S = total interfacial area, ft^2

a = interfacial area per unit volume of vapor and liquid holdup, ft^2/ft^3

A = cross-sectional area of contact volume, ft^2

Z = height of contact volume, ft

Writing Eqs. (16-16) and (16-19) in differential form, substituting for S, and replacing the concentrations and partial pressures with mole fraction terms give the following four equivalent expressions for the amount of a given component transferred at steady-state from one phase to the other:

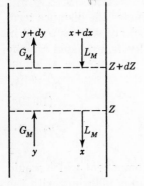

FIG. 16-5. Countercurrent contact between liquid and vapor in a differential element.

$$dN_1 = k_G a(y_i - y)pA \, dZ \qquad (16\text{-}20a)$$
$$= k_L a(x - x_i)(\rho_L)A \, dZ \qquad (16\text{-}20b)$$
$$= K_{OG}a(y^* - y)pA \, dZ \qquad (16\text{-}20c)$$
$$= K_{OL}a(x - x^*)(\rho_L)A \, dZ \qquad (16\text{-}20d)$$

Equation (16-20) expresses the fact that at steady state or whenever the holdup of the transferring component in the regions near the interface is negligible with respect to the total amount transferred, the material leaving one phase must be equal in amount to the material entering the other phase.

Since the interfacial area a cannot be measured, it is necessary to combine it with the mass-transfer coefficients to give new coefficients of $k_G a$, $k_L A$, $K_{OG}a$, and $K_{OL}a$. It is sometimes convenient to express the interfacial area in terms of a unit volume of vapor holdup a' or in terms of a unit volume of liquid holdup $\bar{a}$.

Transfer Unit. It has been found to be convenient in the correlation of mass-transfer data to group several of the variables in Eq. (16-20) together to define a single new quantity. This is accomplished as follows: Consider a differential element of contact volume in a countercurrent vapor-liquid contacting device. A sketch of such an element is shown in Fig. 16-5. If the phase rates are expressed in pound moles per hour per square foot of cross-sectional area, and if the phase rates are assumed to be constant, the following material balance applies for any given component in the differential element:

$$dN = G_M A \, dy = L_M A \, dx$$

Combination of this material balance with each of the four rate equations in (16-20), followed by an integration from $Z = 0$ to $Z = Z$ (G_M, L_M, p, ρ_L, and the mass-transfer coefficients are assumed to be constant), gives

$$\frac{k_G a p}{G_M} Z = \int \frac{dy}{y_i - y} \equiv N_G \tag{16-21a}$$

$$\frac{k_L a \rho_L}{L_M} Z = \int \frac{dx}{x - x_i} \equiv N_L \tag{16-21b}$$

$$\frac{K_{OG} a p}{G_M} Z = \int \frac{dy}{y^* - y} \equiv N_{OG} \tag{16-21c}$$

$$\frac{K_{OL} a \rho_L}{L_M} Z = \int \frac{dx}{x - x^*} \equiv N_{OL} \tag{16-21d}$$

Equation (16-21) defines "the number of transfer units." This concept was introduced by Chilton and Colburn (4) for use in mass-transfer operations where the concept of the equilibrium stage is obviously inconvenient. A transfer unit is accomplished when the change in a stream concentration equals the mean driving force over the interval in which the change in concentration occurs. The integral terms in Eq. (16-21) therefore represent the number of transfer units, and since Z represents the height of the contact volume, the height of a transfer unit must be represented by the reciprocal of the coefficient of Z in each case. For further discussion of the transfer unit concept in design, the reader is referred to Sherwood and Pigford (20). The transfer unit was introduced here because it has been found in the correlation of tray efficiency data that a simpler correlation is obtained if the tray variables are related to the number of transfer units rather than to the mass-transfer coefficients or the stage efficiency directly.

Equations similar to (16-21) can be derived for the case of diffusion of one component only. For these the reader is referred to Sherwood and Pigford. As mentioned previously, for dilute solute concentrations in ternary absorption systems, the equations for one-component diffusion reduce to those derived above for equimolar transfer.

Additivity of Resistances. The "two-resistance" concept in mass transfer postulates two resistances in series, one in the vapor phase and one in the liquid phase. Each of the two resistances is a function of the physical properties within its own phase and largely independent of the physical properties of the other phase. Therefore, the confidence with which mass-transfer phenomena can be predicted is greatly increased if the two resistances are dealt with separately in the correlation. After each resistance (reciprocal of the phase mass-transfer coefficient $k_G a$ or $k_L a$) has been evaluated separately, the two resistances must be added to give the total resistance to transfer from one bulk phase to the other. The total resistance will be the reciprocal of the over-all coefficient $K_{OG} a$ or $K_{OL} a$. The manner in which the individual resistances must be added is developed as follows: First it is assumed that the binary equilibrium curve at the concentrations in question can be adequately represented by a straight line. Then

$$y^* = mx + b \qquad (16\text{-}22)$$

where x is the actual bulk phase liquid concentration for the given component. Also, if the interface is assumed to offer no resistance to mass transfer, y_i is in equilibrium with x_i and

$$y_i = mx_i + b \qquad (16\text{-}23)$$

where the subscript i refers to the interface concentrations of the component in question. The m represents the slope of the equilibrium curve at x and x_i, and b is the intercept which the assumed straight line makes with the y axis. In general, b is not zero and therefore m does not equal the equilibrium-distribution coefficient $K = y^*/x$.

Equations (16-22) and (16-23) permit the following identities to be written:

$$(y^* - y) \equiv (y^* - y_i) + (y_i - y) \equiv m(x - x_i) + (y_i - y)$$

Substitution from Eqs. (16-20a), (16-20b), and (16-20c) gives the following relationship between the over-all resistance $1/K_{OG}$ and the individual phase resistances:

$$\frac{1}{K_{OG} a} = \frac{mp/\rho_L}{k_L a} + \frac{1}{k_G a} \qquad (16\text{-}24)$$

Or starting with

$$(x - x^*) \equiv (x - x_i) + (x_i - x^*) \equiv (x - x_i) + \frac{1}{m}(y_i - y)$$

the following expression for the over-all resistance in liquid terms is obtained:

$$\frac{1}{K_{OL} a} = \frac{1}{K_L a} + \frac{\rho_L}{mp k_G a} \qquad (16\text{-}25)$$

Combination of (16-24) and (16-25) shows the conversion factors necessary to convert the over-all resistance from liquid to vapor terms or vice versa.

$$\frac{1}{K_{OG} a} = \frac{mp/\rho_L}{K_{OL} a} \qquad (16\text{-}26)$$

Since the resistances are related to the number of transfer units by Eq. (16-21),

$$\frac{1}{N_{OG}} = \frac{1}{N_G} + \frac{mG_M}{L_M N_L} = \frac{mG_M}{L_M N_{OL}} \tag{16-27}$$

16-6. Relation between E_{OG} and N_{OG}

As will be shown later, the A.I.Ch.E. research program resulted in correlations for the number of transfer units N_G and N_L in terms of the tray dimensions, flow rates, and phase properties. The predicted N_G and N_L can be combined by means of Eq. (16-27) to give N_{OG} or N_{OL}. It is the purpose of this section to develop a relationship between N_{OG} and the Murphree point efficiency E_{OG}. Section 16-8 will accomplish the same thing for N_{OL} and E_{OL}.

It is necessary to postulate some sort of tray model in order to relate N_{OG} to E_{OG}. The model used by the A.I.Ch.E. Manual is shown in Fig. 16-6. Consider any point in the liquid pool between the downcomer and the overflow weir. Three assumptions are made in this model concerning the flow behavior of the vapor and liquid.

1. The liquid is assumed to be completely mixed in the vertical direction; that is, the liquid from the tray deck to the top of the froth is, at any vertical line on the tray, of constant composition x. No stipulation need be made now concerning liquid mixing in the horizontal direction.

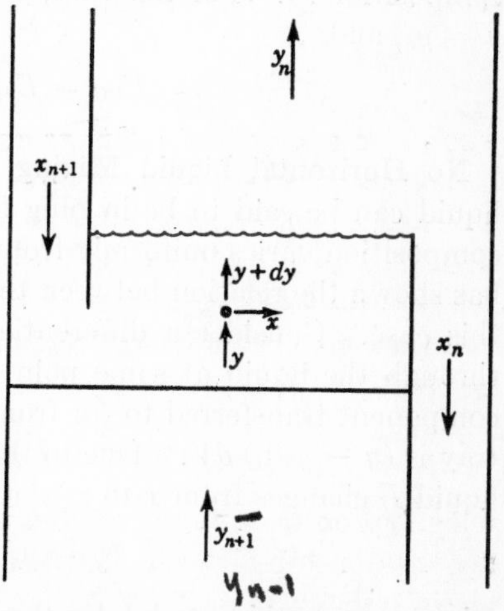

2. The vapor entering through the tray deck is assumed to be of constant composition y_{n-1} across the column cross section. This assumes complete mixing of the vapor between trays $n-1$ and n.

FIG. 16-6. Tray model in vapor terms. The model assumes plug flow of vapor upward through the liquid and no change in liquid composition in the vertical direction.

3. The vapor passing upward through the liquid along any vertical line is in plug flow; that is, there is no vertical (or horizontal) mixing of the vapor while in contact with the liquid.

Subject to these assumptions, Eq. (16-21c) can be integrated along any vertical line on the tray from the tray deck ($Z = 0$, $y = y_{n-1}$, $x = x$) to the top of the froth ($Z = Z$, $y = y$, $x = x$). Since x is constant, y^* is a constant.

$$\frac{K_{OG} a p Z}{G_M} = N_{OG} = -\ln \frac{y^* - y}{y^* - y_{n-1}}$$

The ln term is equivalent to $1 - E_{OG}$, so the equation can be rewritten as

$$1 - E_{OG} = \exp(-N_{OG}) = \exp\left(-\frac{K_{OG}apZ}{G_M}\right) \qquad (16\text{-}28)$$

16-7. Relation between E_{MV} and E_{OG}

The tray model described in the previous section can be used to relate the point efficiency E_{OG} to the tray efficiency E_{MV}. The relationship obtained depends upon the assumption made concerning horizontal mixing of the liquid on the tray. The three cases of interest are discussed below.

Liquid Completely Mixed. If the liquid on the tray is completely mixed, then the entire tray liquid composition is identical with the outlet composition x_n. For this case, y^* and y are constant all across the tray, $y = y_n$, and

$$E_{OG} = E_{MV} = \frac{y_n - y_{n-1}}{y_n^* - y_{n-1}} \qquad (16\text{-}1)$$

No Horizontal Liquid Mixing. If no horizontal mixing occurs, the liquid can be said to be in plug flow across the tray; that is, the liquid composition varies uniformly from x_{n+1} to x_n across the tray. Lewis (11) has shown the relation between the point and over-all tray efficiencies for this case. Consider a differential amount of vapor dV passing upward through the liquid at some point on the tray. The amount of a given component transferred to (or from) this vapor in its passage through the tray is $(y - y_{n-1})\,dV$. Due to this transfer the composition of the entire liquid L changes from x to $x + dx$. The material balance is then

$$(y - y_{n-1})\,dV = L\,dx$$

Division of both V and L by the bubbling area on the tray gives

$$(y - y_{n-1})\,dG_M = L_M\,dx$$

A new variable $dW = dG_M/G_M$ is defined and used to replace dG_M. The result of this step will be to include in the final result the important parameter $\lambda = mG_M/L_M$. The equation can now be rewritten as

$$y - y_{n-1} = \frac{L_M}{G_M}\frac{dx}{dW}$$

Assume that the equilibrium relationship over the range of composition covered by the tray can be satisfactorily approximated by the straight line $y^* = mx + b$. This will generally be a good assumption, since each tray will have its own unique values of m and b and will be concerned with only a small segment of the equilibrium curve. Since $dy^* = m\,dx$, then

$$y - y_{n-1} = \frac{L_M}{mG_M}\frac{dy^*}{dW}$$

the local efficiency in vapor terms,

$$E_{OG} = \frac{y - y_{n-1}}{y^* - y_{n-1}} \tag{16-3}$$

can be solved for y^* and differentiated to give

$$dy^* = \frac{dy}{E_{OG}}$$

at any point on the tray. Substitution for dy^* in the material-balance equation and rearrangement give

$$\lambda E_{OG}\, dW = \frac{dy}{y - y_{n-1}}$$

In order to integrate this equation across the tray it is necessary that E_{OG} be assumed constant for all points on the tray. Integrating from the overflow weir (where W will be taken as zero and $y = y_{W=0}$) across the tray to some point where $W = W$ and $y = y$ gives the following equation:

$$y = y_{n-1} + (y_{W=0} - y_{n-1})\, \exp\,(\lambda E_{OG}W)$$

The average composition y_n of the total vapor from stage n is given by

$$y_n = \int_0^{1.0} y\, dW$$

Substitution for y from the previous equation and subsequent integration (with $y_{W=0}$ and y_{n-1} as constants) give

$$\frac{y_n - y_{n-1}}{y_{W=0} - y_{n-1}} = \frac{1}{\lambda E_{OG}}\, [\exp\,(\lambda E_{OG}) - 1]$$

The E_{OG} in this equation is the point efficiency at the overflow weir where

$$E_{OG} = \frac{y_{W=0} - y_{n-1}}{y_n^* - y_{n-1}}$$

since $y_{W=0}^* = y_n^*$. Substitution for E_{OG} provides, after cancellation of terms, the desired relation between E_{MV} and E_{OG}.

$$E_{MV} = \frac{1}{\lambda}\, [\exp\,(\lambda E_{OG}) - 1] \tag{16-29}$$

The assumptions involved in Eq. (16-29) are summarized below.

1. Plug flow of liquid across the tray (no back mixing)
2. No vertical concentration gradient in the liquid
3. Linear equilibrium relationship ($y^* = mx + b$) over the concentration range on the tray
4. Constant value of E_{OG} across the tray

5. Constant vapor and liquid rates

6. Uniform composition across the column in the vapor entering the tray (y_{n-1} = constant)

It can be seen that reasonable values of λ and E_{OG} can cause E_{MV} to be greater than 1.0 in this model. Lewis (11) tabulates E_{MV} as a function of E_{OG} and λ. For example, with $E_{OG} = 1.0$ and $\lambda = 0.5$, $E_{MV} = 1.3$. Plug flow of liquid across large trays can give tray efficiencies much larger than 100 per cent if the tray hydraulics can be controlled to maintain high point efficiencies throughout the tray.

Partial Liquid Mixing. Several relationships between E_{MV} and E_{OG} which take into consideration the degree of liquid mixing on the tray have been proposed (7, 10, 15, 17, 22). The various approaches to the problem have been summarized by Gerster et al. (8). The correlation used by the

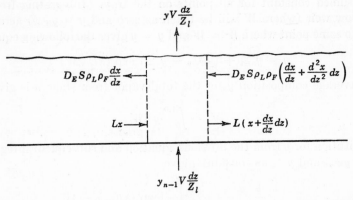

FIG. 16-7. Liquid-mixing model assumed in the derivation of Eq. (16-38).

A.I.Ch.E. Manual is based upon the eddy diffusion concept as developed by Anderson (1) and Wehner and Wilhelm (23). The following derivation of this correlation is taken from Gerster et al.

Consider the vertical differential slice of aerated liquid pictured in Fig. 16-7. Liquid is flowing from left to right across the tray. The total distance from the inlet downcomer to the outlet weir is Z_l ft. The differential slice is located at a distance of z ft from the inlet downcomer. The liquid rate L and the vapor rate V are assumed to be constant. The composition of the liquid entering the differential slice is represented by x. At $z + dz$, the liquid composition is $x + dx$ or, in terms of the concentration gradient, $x + (dx/dz) \, dz$. The amount of vapor passing upward through the differential volume is V multiplied by the fraction of the total tray area covered by the differential slice, that is, $V \, dz/Z_l$.

If no back mixing of the liquid occurred (plug flow), the material transferred between phases in the differential slice would be given by

$$L \left(x + \frac{dx}{dz} dz \right) - Lx$$

To describe the amount of the transferring component which is carried into and out of the differential element by back mixing (or eddy diffusion) an eddy coefficient of mixing is defined by

$$\frac{N}{S} = D_E \frac{dc}{dz}$$

The concentration term (pound moles per cubic foot) can be replaced with $\rho_F \rho_L \, dx$ to give

$$N = D_E S \rho_F \rho_L \frac{dx}{dz}$$

where ρ_F is the froth density in cubic feet of liquid per cubic foot of froth. The driving force in this transfer equation is the concentration gradient: that is, the transfer by mixing of a component is assumed to be proportional to the concentration gradient of that component. At $z = z$, the concentration gradient is dx/dz. At $z = z + dz$, the gradient is

$$\frac{dx}{dz} + \frac{d^2x}{dz^2} dz$$

where d^2x/dz^2 is the rate of change of the gradient with respect to z.

A material balance around the differential slice reduces to

$$D_E S \rho_L \rho_F \frac{d^2x}{dz^2} dz - L \frac{dx}{dz} dz + (y_{n-1} - y) V \frac{dz}{Z_l} = 0$$

Dividing through by $S \rho_F \rho_L \, dz$ gives

$$D_E \frac{d^2x}{dz^2} - \frac{L}{S \rho_F \rho_L} \frac{dx}{dz} + \frac{y_{n-1} - y}{S \rho_F \rho_L} \frac{V}{Z_l} = 0$$

The ratio $L/S \rho_F \rho_L$ has the units feet per hour and is defined as the froth velocity U_f. Making this substitution and replacing the ratio z/Z_l with w reduce the equation to

$$\frac{D_E}{U_f Z_l} \frac{d^2x}{dw^2} - \frac{dx}{dw} + (y_{n-1} - y) \frac{V}{L} = 0 \qquad (16\text{-}30)$$

The term

$$\frac{Z_l}{D_E/U_f}$$

is the ratio of the tray length Z_l to the mixing length D_E/U_f and is defined as the Peclet number

$$N_{Pe} = \frac{U_f Z_l}{D_E}$$

Since the average residence time for the liquid on the tray is $t_l = Z_l/U_f$ the Peclet number can be written in the alternate form

$$N_{\text{Pe}} = \frac{Z_l^2}{D_E t_l} \tag{16-31}$$

The definition of the point efficiency

$$E_{OG} = \frac{y - y_{n-1}}{y^* - y_{n-1}} = \frac{y - y_{n-1}}{(mx + b) - y_{n-1}}$$

is used to replace the $y_{n-1} - y$ term in (16-30). The point efficiency ratio is first modified by defining $x_e^* = (y_{n-1} - b)/m$ to give

$$E_{OG} = \frac{y - y_{n-1}}{m(x - x_e^*)} \tag{16-32}$$

where x_e^* is the composition of the liquid in equilibrium with the entering vapor. If both V and L in Eq. (16-30) are divided by the bubbling area to give G_M and L_M, the differential equation can now be written as

$$\frac{1}{N_{\text{Pe}}} \frac{d^2x}{dw^2} - \frac{dx}{dw} - \lambda E_{OG}(x - x_e^*) = 0 \tag{16-33}$$

The boundary conditions to be used in the solution (23) are $dx/dw = 0$ and $x = x_n$ at $w = 1.0$ (tray outlet). The solution as written by Gerster et al. is then

$$\frac{x - x_e^*}{x_n - x_e^*} = \frac{\exp\left[(\eta + N_{\text{Pe}})(w - 1)\right]}{1 + (\eta + N_{\text{Pe}})/n} + \frac{\exp\left[\eta(1 - w)\right]}{1 + \eta/(\eta + N_{\text{Pe}})} \tag{16-34}$$

where

$$\eta = \frac{N_{\text{Pe}}}{2}\left[\sqrt{1 + \frac{4\lambda E_{OG}}{N_{\text{Pe}}}} - 1\right] \tag{16-35}$$

The last step in the derivation is to relate the ratio $(x - x_e^*)/(x_n - x_e^*)$ to the ratio E_{MV}/E_{OG}. For this purpose, return to Eq. (16-32). The term $y - y_{n-1}$ is the difference between the outlet y and inlet y_{n-1} concentrations at some point on the tray. The inlet concentration is given the stage subscript $n - 1$ because all the vapor from stage $n - 1$ is assumed to be completely mixed and of the average composition y_{n-1} before entering stage n. Likewise the average vapor composition from stage n will be denoted by y_n. The difference between the average compositions $y_n - y_{n-1}$ is given by

$$y_n - y_{n-1} = \int_0^{1.0} (y - y_{n-1})\, dw$$

$$= mE_{OG} \int_0^{1.0} (x - x_e^*)\, dw \tag{16-36}$$

where $w = z/Z_l$ = fraction of the total distance from the tray inlet to

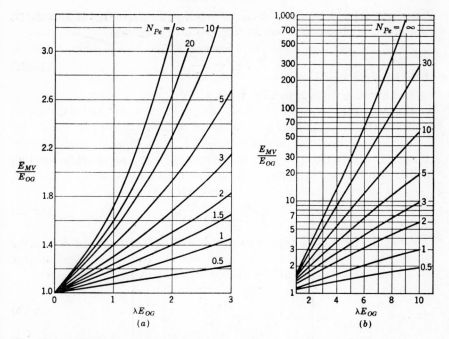

FIG. 16-8. Solution curves for Eq. (16-35). (*From "Bubble-tray Design Manual," American Institute of Chemical Engineers, New York.*)

the tray outlet. Again making use of the assumption that the equilibrium relationship can be expressed as

$$y^* = mx_n + b$$

and
$$y_{n-1} = mx_e^* + b$$

the over-all tray (Murphree) efficiency can be written as

$$E_{MV} = \frac{y_n - y_{n-1}}{(mx_n + b) - y_{n-1}} = \frac{y_n - y_{n-1}}{m(x_n - x_e^*)}$$

Substituting for $y_n - y_{n-1}$ from Eq. (16-36) gives

$$\frac{E_{MV}}{E_{OG}} = \int_0^{1.0} \frac{x - x_e^*}{x_n - x_e^*}\, dw \qquad (16\text{-}37)$$

where $(x_n - x_e^*)$ is a constant, since y_{n-1} is assumed to be a constant with respect to w. Substitution of Eq. (16-37) into Eq. (16-34) provides the equation recommended by the A.I.Ch.E. Manual for partially mixed trays.

$$\frac{E_{MV}}{E_{OG}} = \frac{1 - \exp\left[-(\eta + N_{Pe})\right]}{(\eta + N_{Pe})[1 + (\eta + N_{Pe})/\eta]} + \frac{\exp(\eta) - 1}{\eta[1 + \eta/(\eta + N_{Pe})]} \qquad (16\text{-}38)$$

where η is defined by Eq. (16-35) and N_{Pe} by Eq. (16-31). Numerical solutions of (16-38) have been worked out on a computer for various values of N_{Pe} and λE_{OG} to provide the plot shown in Fig. 16-8.

16-8. Relation between E_{OL} and N_{OL}

The tray model suggested by West, Gilbert, and Shimizu (24) is used by the A.I.Ch.E. Manual to relate the point efficiency and number of transfer units in liquid terms. The model is illustrated in Fig. 16-9. As in the case of the vapor model discussed in Sec. 16-6, the assumptions inherent in the liquid model strain reality somewhat but nevertheless the models provide useful results. The liquid model is sometimes approached when all the resistance to transfer is in the liquid phase.

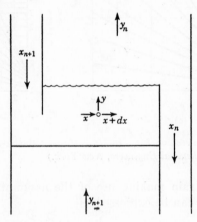

FIG. 16-9. Tray model in liquid terms. The model assumes constant vapor composition along any horizontal plane in the froth.

The liquid composition in the liquid model is not restricted by any assumptions; that is, liquid composition will change in both the vertical and horizontal directions at any point on the tray. The vapor composition, on the other hand, is restricted by the assumption that it will be constant along any horizontal plane through the froth. This improbable situation could obtain only if the vapor bubbles rose more rapidly near the liquid inlet (where the driving force for mass transfer is high) than near the liquid outlet (where the concentration differences are smaller). The assumption made in the vapor model of a constant y_{n-1} across the column cross section is retained in the liquid model.

It should be noted that the vapor and liquid models are not compatible; that is, they cannot exist simultaneously. This fact must be kept in mind if the E_{OG} from the vapor model is to be used in the same equation with an E_{OL} from the liquid model.

The mathematical relationship between E_{OL} and N_{OL} based on the liquid model can be found by integrating Eq. (16-21d) from some point on the tray (where $Z = 0$ and $x = x$) across the tray along a horizontal plane (where y and x^* are constant) to the inlet downcomer (where $Z = Z_l$ and $x = x_{n+1}$).

$$\frac{K_{OL}a\rho_L Z_l}{L_M} = N_{OL} = \ln \frac{x_{n+1} - x^*}{x - x^*}$$

Inverting the ln term and noting that

$$1 - E_{OL} = \frac{x - x^*}{x_{n+1} - x^*}$$

give

$$1 - E_{OL} = \exp(-N_{OL}) = \exp\left(-\frac{K_{OL}a\rho_L Z_l}{L_M}\right) \qquad (16\text{-}39)$$

which is similar in form to Eq. (16-28).

At first glance, the direction of integration above might appear to be wrong. However, in the development of Eq. (16-28), the vapor was assumed to go from y_{n-1} at $Z = 0$ to y at $Z = Z$. If countercurrent contact between the vapor and liquid is to be assumed, the liquid must go from x at $Z = 0$ to x_{n+1} at $Z = Z_l$.

16-9. Relation between E_{ML} and N_{OL}

It is not necessary to relate the liquid efficiencies E_{ML} and E_{OL} because in the A.I.Ch.E. prediction method the liquid efficiencies are always converted immediately to vapor efficiencies by means of the equations to be developed in the next section. The vapor point and tray efficiencies E_{MV} and E_{OG} are related then by means of the equations developed in Sec. 16-7. Therefore, instead of relating E_{ML} to E_{OL}, the equations whereby E_{ML} can be calculated directly from N_{OL} will now be derived. The relation between E_{ML} and N_{OL} depends upon the degree of mixing in the tray liquid. The three cases described in Sec. 16-7 will now be discussed.

Liquid Completely Mixed. When the liquid is completely mixed, $x = x_n$ and

$$E_{OL} = \frac{x_{n+1} - x_n}{x_{n+1} - x^*}$$

while

$$E_{ML} = \frac{x_{n+1} - x_n}{x_{n+1} - x_n^*}$$

The x^* represents the liquid composition which would be in equilibrium with the vapor composition y at some point within the pool whereas x_n^* would be in equilibrium with the total vapor composition y_n which is equivalent to y only at the froth surface in the liquid model. Therefore E_{OL} and E_{ML} are not in general equal in this case.

The relationship between E_{ML} and N_{OL} can be developed from a material balance around the entire tray.

$$L(x_{n+1} - x_n) = K_{OL}a(x_n - x_n^*)\rho_L A Z$$

Since $L/A_a = L_M$, the equation can be rearranged to

$$\frac{L_M}{K_{OL}a\rho_L Z} = \frac{1}{N_{OL}} = \frac{x_n - x_n^*}{x_{n+1} - x_n}$$

Addition of 1.0 to each of the last two terms and combination with the definition of E_{ML} in Eq. (16-2) give

$$E_{ML} = \frac{N_{OL}}{1 + N_{OL}} \tag{16-40}$$

Substitution for N_{OL} from (16-39) would give a relationship between E_{ML} and the point efficiency E_{OL} at the tray outlet. Such a relationship is not needed, since, as mentioned above, values of E_{ML} or E_{OL} obtained from N_{OL} values will always be converted to E_{MV} or E_{OG} before further use.

No Horizontal Liquid Mixing. In this case, plug flow of the liquid is assumed. Integration across the tray from the overflow weir ($Z = 0$ and $x = x_n$) along the horizontal plane at the top of the froth ($y = y_n$ and $x^* = x_n^*$) to the inlet downcomer ($Z = Z_l$ and $x = x_{n+1}$) gives

$$\frac{K_{OL}a\rho_L Z_l}{L_M} = N_{OL} = \ln \frac{x_{n+1} - x_n^*}{x_n - x_n^*}$$

Inverting the ln term and noting that

$$1 - E_{ML} = \frac{x_n - x_n^*}{x_{n+1} - x_n^*}$$

give

$$1 - E_{ML} = \exp(-N_{OL}) = \exp\left(-\frac{K_{OL}a\rho_L Z_l}{L_M}\right) \tag{16-41}$$

This equation has the same form as (16-39), but the similarity ends there. Even if the integration path in the derivation of (16-39) had extended all the way across the tray to give E_{OL} at the tray outlet, the N_{OL} in (16-39) would not equal the N_{OL} in (16-41). The composition limits x_n and x_{n+1} are the same in each case, but the driving force for transfer varies with the horizontal plane chosen for the integration path. For any plane in the pool, $y = y$ and $x^* = (y - b)/m$. At the top of the froth, $y = y_n$ and $x^* = x_n^* = (y_n - b)/m$. Therefore, the driving force for mass transfer is $x - x^*$ in the former case and $x - x_n^*$ in the latter and $x^* \neq x_n^*$

for any plane other than the top one. Equations (16-39) and (16-41) become identical, and $E_{ML} = E_{OL}$ only if no vertical gradient in either the vapor or liquid is assumed in addition to the assumed plug flow of the liquid.

Partial Liquid Mixing. An equation analogous to Eq. (16-38) but in liquid terms is not presented. As mentioned previously, the equations in the next section are used to convert the liquid efficiency to the corresponding vapor efficiency and then Eq. (16-38) is used to take into account partial mixing on the tray.

16-10. Relation between Vapor and Liquid Efficiencies

Direct conversion of E_{OL} to E_{OG} (or E_{ML} to E_{MV}) is permissible, since the liquid and vapor efficiencies expressed in over-all terms are simply two different ways to express the same thing. (This statement is not true for E_L and E_G.) However, care must be taken to ensure that the conversion between the two is not based upon the use of the vapor and liquid models described in the previous sections. The two models are not compatible; that is, they cannot exist simultaneously on a tray and therefore should not be used to relate the over-all vapor and liquid efficiencies. A material balance can be used to develop the relationship without reference to any model. The desired final equations are

$$E_{OG} = \frac{E_{OL}}{E_{OL} + \lambda(1 - E_{OL})} \tag{16-42}$$

and

$$E_{MV} = \frac{E_{ML}}{E_{ML} + \lambda(1 - E_{ML})} \tag{16-43}$$

Inspection of these equations shows that when $\lambda = 1.0$, $E_{OL} = E_{OG}$. When $\lambda > 1.0$, $E_{OL} > E_{OG}$, and for $\lambda < 1.0$, $E_{OL} < E_{OG}$. The same relationships exist between E_{MV} and E_{ML}.

The derivations of Eqs. (16-42) and (16-43) are similar and begin with a material balance.

$$L(x_{n+1} - x_n) = V(y_n - y_{n-1}) \tag{16-44}$$

Equation (16-44) assumes constant phase rates. Since it is a balance around the entire stage, it will involve E_{MV} and E_{ML} and therefore lead to Eq. (16-43). If the n subscripts were dropped, Eq. (16-42) would be obtained.

It will be assumed that the equilibrium curve can be represented by a straight line over the range of composition covered by the tray so that

$$y_n^* = mx_n + b$$

and

$$y_n = mx_n^* + b$$

Since

$$\frac{y_n^* - y_n}{y_n^* - y_n} = 1.0 = \frac{y_n^* - y_n}{mx_n - mx_n^*}$$

the slope of the straight equilibrium line is

$$m = \frac{y_n^* - y_n}{x_n - x_n^*}$$

Multiplying both sides of (16-44) by m/A_a and rearranging give

$$\lambda = \frac{x_{n+1} - x_n}{y_n - y_{n-1}} \frac{y_n^* - y_n}{x_n - x_n^*} \qquad (16\text{-}45)$$

Since

$$\frac{1}{E_{MV}} - 1 = \frac{y_n^* - y_n}{y_n - y_{n-1}}$$

and

$$\frac{1}{E_{ML}} - 1 = \frac{x_n - x_n^*}{x_{n+1} - x_n}$$

Eq. (16-45) can be rewritten in the form of (16-43).

Usually, resistance to transfer will be present in both phases. The two resistances will be represented by N_G and N_L for which correlations are presented later. The individual resistances are generally added to give N_{OG} which has been related to E_{MV}. The E_{MV} can then be used in the entrainment correlation presented in the next section to calculate E_a, the corrected tray efficiency. Sometimes, however, all or almost all the resistance may be centered in one phase. In such cases the procedure and the equations are somewhat modified.

No Liquid-phase Resistance. The degree of liquid mixing has no effect on the efficiency when all the resistance lies in the vapor phase. The lack of dependence of E_{OG} on liquid mixing can be demonstrated by noting that when λ is very small (large vapor-phase resistance),

$$\exp{(\lambda E_{OG})} \cong 1.0 + \lambda E_{OG}$$

and Eq. (16-29), which assumes no liquid mixing, reduces to $E_{MV} = E_{OG}$, which also is true in the vapor model when the liquid is completely mixed. Therefore, Eq. (16-38) is not needed to relate E_{OG} to E_{MV} and the E_{OG} can be used directly in the entrainment correlation presented in the next section.

No Gas-phase Resistance. When all or most of the resistance is centered in the liquid phase, it is usually more convenient to deviate from the usual procedure and calculate N_{OL} and E_{OL} rather than N_{OG} and E_{OG}. To correct for the degree of liquid mixing and for entrainment, however, it is necessary to convert E_{OL} to E_{OG} by Eq. (16-42) before proceeding further. Inspection of Eq. (16-27) shows that when λ is large (large liquid resistance), N_{OG} and therefore E_{OG} will be small.

The degree of mixing has the largest effect when all the resistance lies on the liquid side. When the gas-phase resistance is absent, $N_{OL} = N_L$, $E_{OL} = E_L$, and Eq. (16-40) can be rewritten as

$$N_L = \frac{E_L}{1 - E_L} \qquad (16\text{-}46)$$

Inspection of (16-42) shows that when λ is very large (all liquid-phase resistance), the first term in the denominator will be negligible in comparison with the second and

$$E_{OG} = \frac{E_L}{\lambda(1 - E_L)}$$

or, after combination with (16-46),

$$\lambda E_{OG} = N_L \qquad (16\text{-}47)$$

Also Eq. (16-43) reduces to

$$\lambda E_{MV} = \frac{E_{ML}}{1 - E_{ML}} \qquad (16\text{-}48)$$

Division of (16-48) by (16-47) relates the Murphree tray efficiency in liquid terms to the ratio E_{MV}/E_{OG}.

$$\frac{E_{MV}}{E_{OG}} = \frac{E_{ML}}{N_L(1 - E_{ML})} \qquad (16\text{-}49)$$

The effect of liquid mixing can now be taken into account by means of Fig. 16-8 with N_L substituted for λE_{OG} and values of

$$\frac{E_{ML}}{N_L(1 - E_{ML})}$$

read directly from the ordinate. Values of N_L can be calculated from the empirical research correlations to be presented later.

16-11. Correction for Entrainment

The A.I.Ch.E. Manual recommends the equation of Colburn (5) for the correction of the dry Murphree tray efficiency E_{MV} for entrainment.

$$E_a = \frac{E_{MV}}{1 + e'E_{MV}/L_M} \qquad (16\text{-}50)$$

The derivation of this equation is shown below.

The material balance for a given component in the top section of a column which is operating with no entrainment (dry vapor) is

$$Vy_n = Lx_{n+1} + (V - L)x_D \qquad (16\text{-}51)$$

When entrainment occurs, the material balance must be modified to

$$Vy_n + e_M Vx_n = (L + e_M V)x_{n+1} + (V - L)x_D \qquad (16\text{-}52)$$

where e_M is the entrainment expressed as moles of entrained liquid per mole of dry vapor V. The product $e_M V$ represents the moles of liquid of composition x_n which are carried from stage n to stage $n + 1$ by the vapor. Note that the use of x_n infers completely mixed liquid on stage n.

For a constant overhead product rate D, the liquid carried up by the vapor must be balanced by an equal number of extra moles in L_{n+1}.

Division of Eq. (16-52) by V gives

$$y_n + e_M x_n = \left(\frac{L}{V} + e_M\right) x_{n+1} + \left(1 - \frac{L}{V}\right) x_D \tag{16-53}$$

At this point Colburn defined an "apparent" vapor concentration as

$$Y_n = \frac{L}{V} x_{n+1} + \left(1 - \frac{L}{V}\right) x_D \tag{16-54}$$

When e_M is zero, $Y_n = y_n$ and (16-54) reduces to (16-51). Comparison of (16-52) and (16-54) when e_M is not zero shows that

$$Y_n = y_n - e_M(x_{n+1} - x_n) \tag{16-55}$$

The "apparent" tray efficiency as defined by Colburn in terms of Y is

$$E_a = \frac{Y_n - Y_{n-1}}{y_n^* - Y_{n-1}} \tag{16-56}$$

which is the expression for E_{MV} with Y substituted for y. Substitution for Y_n and Y_{n-1} with Eq. (16-55) provides

$$E_a = \frac{y_n - y_{n-1} - e_M c}{(y_n^* - y_{n-1}) + e_M(x_{n+1} - x_n) - e_M c} \tag{16-57}$$

where $c = (x_{n+1} - x_n) - (x_n - x_{n-1})$. Colburn now wrote Eq. (16-53) for stages n and $n - 1$ and subtracted the second from the first to give

$$x_{n+1} - x_n = \frac{y_n - y_{n-1}}{L/V} - \frac{e_M c}{L/V}$$

which can be used to replace the $(x_{n+1} - x_n)$ term in (16-57). Also the numerator and denominator of (16-57) were divided by $(y_n^* - y_{n-1})$ to get E_{MV} into the equation. These operations provide

$$E_a = \frac{E_{MV} - e_M c/(y_n^* - y_{n-1})}{1 + e_M V E_{MV}/L - e_M V c (e_M + 1)/L(y_n^* - y_{n-1})}$$

If the slopes of the operating and equilibrium lines are not widely different, the successive increments $(x_{n+1} - x_n)$ and $(x_n - x_{n-1})$ will be approximately equal in magnitude and c will be small. Colburn also notes that the presence of c in both the numerator and denominator tends to cancel the effect of that quantity on E_a. Therefore it is usually satisfactory to use the following simplified form:

$$E_a = \frac{E_{MV}}{1 + e_M V E_{MV}/L} \tag{16-58}$$

Division of both V and L by the bubbling area A_a permits the equation to be rewritten in the form shown in Eq. (16-50).

16-12. Over-all Column Efficiency

The most precise way to calculate the required number of trays would be to estimate values of E_a on each tray and utilize these to modify the equilibrium relationships in the stage-to-stage calculations. This can be done on a McCabe-Thiele diagram through the use of a pseudo-equilibrium curve as shown in Fig. 5-22. The position of the pseudo-equilibrium curve is determined by the value of the efficiency applied in each stage. The number of actual stages thus obtained with some

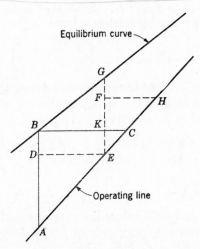

Fig. 16-10. Diagram for derivation of Eq. (16-61). [*W. K. Lewis, Ind. Eng. Chem.,* **28,** 399 (1936).]

constant value of E_a will not be the same as the number obtained if this E_a is used in

$$N_t = \frac{N}{E_a}$$

Obviously, the tray efficiency E_a does not in general equal the column efficiency E_o where E_o is defined by

$$E_o = \frac{N}{N_t} \qquad (16\text{-}5)$$

In fact, the E_a for any given component will have slightly different values on each tray whereas there is only one E_o for that component in a given column.

Lewis (11) has derived an analytical relationship between E_o and E_a. The derivation assumes the equilibrium and operating lines to be straight but not necessarily parallel. Figure 16-10 shows a portion of a McCabe-Thiele diagram which conforms to these assumptions. *ABC* represents an equilibrium stage $(E_a = E_{MV} = 1.0)$, while steps ADE and EFH represent two actual stages both of which have an over-all tray efficiency of $E_a = E_a$. By definition, $E_a = AD/AB$ and also EF/EG, since the same value of E_a is assumed for each of the two stages. From the diagram it can be seen that

$$EK = AB - AD = \left(\frac{1}{E_a} - 1\right) AD$$

The slope of the operating line is $L/V = AD/DE$, while the slope of the equilibrium curve is $m = KG/BK$. From these ratios, $\lambda = KG/AD$ and the following equality can be written.

$$EG = EK + KG = \left(\frac{1}{E_a} - 1 + \lambda\right) AD$$

The actual progress (in vapor terms) made is AD on the first tray and EF on the second. The ratio of the second to the first will be denoted by r, which can be related to E_a as follows.

$$r = \frac{EF}{AD} = \frac{EG(E_a)}{AD} = E_a \left(\frac{1}{E_a} - 1 + \lambda\right)$$

or
$$E_a = \frac{r - 1}{\lambda - 1} \tag{16-59}$$

Since the operating and equilibrium lines are assumed to be straight, the ratio r will be constant for any pair of actual trays considered. Let $AB = a_1$, $AD = d_1$, $EF = d_2$, etc. Then

$$d_2 = d_1 r = a_1 E_a r$$
$$d_3 = d_2 r = a_1 E_a r^2$$

$$\cdot$$
$$\cdot$$
$$\cdot$$

$$d_n = d_{n-1} r = a_1 E_a r^{n-1}$$

The total increase in y on the diagram for trays 1 to n inclusive is related to E_a by

$$d_1 + d_2 + \cdots + d_n = a_1 E_a (1 + r + r^2 + \cdots + r^{n-1}) = a_1 E_a \frac{r^n - 1}{r - 1}$$

Substitution for $(r - 1)$ from Eq. (16-59) gives

$$d_1 + d_2 + \cdots + d_n = \frac{a_1(r^n - 1)}{\lambda - 1} \tag{16-60}$$

Let N represent the total number of equilibrium stages required to accomplish the specified increase in y starting from A on the diagram.

The number of actual trays required to accomplish the same change in y is N_t. From Eq. (16-59) it is noted that $r = \lambda$ when $E_a = 1.0$. Therefore, when $E_a = 1.0$, the change in y over N equilibrium stages is given by

$$\frac{a_1(\lambda^N - 1)}{\lambda - 1}$$

Equating this quantity to the right-hand side of (16-60) gives

$$r^{N_t} = \lambda^N$$

Therefore

$$\frac{\ln r}{\ln \lambda} = \frac{N}{N_t}$$

Substituting for r from (16-59),

$$E_o = \frac{N}{N_t} = \frac{\ln [1 + E_a(\lambda - 1)]}{\ln \lambda} \tag{16-61}$$

This equation must be considered as only a gross approximation for the whole column. The equilibrium curve is usually not straight throughout the column, and more than one operating line will be involved in a complete column. The N_t calculated will be more accurate if Eq. (16-61) is applied to each individual section of the column, since λ will be different in each section.

16-13. Research Results

All the necessary relationships between the various transfer units (N_{OG}, N_G, N_{OL}, and N_L) and the various tray efficiencies have been derived in the previous sections. Also, methods for taking into account the degree of liquid mixing and the amount of liquid entrainment have been presented. Before these equations and methods can be used, however, the designer must be able to predict (a) the rate of mass transfer in the vapor phase, (b) the rate of mass transfer in the liquid phase, (c) the degree of liquid mixing, and (d) the amount of entrainment on a particular tray operating with a particular fluid system under given column conditions. The A.I.Ch.E. Manual presents correlations for the individual prediction of these four quantities. These correlations are shown below but discussed very briefly. For a complete discussion of their sources, their limitations, and their specific uses the reader is referred to the Manual itself.

The basic advance which the A.I.Ch.E. correlation scheme offers over simpler prediction methods lies in the ability to consider separately each of the four major factors which affect the efficiency. For example, operating and tray design variables affect the liquid- and vapor-transfer rates in different ways. Therefore it is advantageous to estimate the vapor and liquid mass-transfer resistances separately and then combine them later. This procedure is analogous to that followed in many prediction schemes for vapor-liquid equilibrium. The nonideality factors of the vapor and liquid phases are considered separately to permit more precise predictions.

Rate of Mass Transfer in the Gas Phase. The relationship between the point efficiency E_{OG} and the number of transfer units N_{OG} was derived in Sec. 16-6 in terms of the vapor model in Fig. 16-6. The final result is shown as Eq. (16-28). If the integration had utilized Eq. (16-21a) instead of (16-21c) the result would have been

$$1 - E_G = \exp\left(-N_G\right) = \exp\left(-\frac{k_G a p Z}{G_M}\right) \qquad (16\text{-}62a)$$

which relates the point efficiency to the number of transfer units when all the mass-transfer resistance is in the vapor phase ($y_i = y^*$). Equation (16-62a) can be written in the alternate form

$$-2.3 \log\left(1 - E_G\right) = N_G = \frac{k_G a' R T \beta}{3600 U_a} = \frac{k_G' a' t_v}{3600} \qquad (16\text{-}62b)$$

where $t_v = \beta/U_a$ = average contact time for the vapor in its passage through the tray liquid. The assumptions involved in Eq. (16-62) have been listed in Sec. 16-6.

The main variables affecting the point efficiencies are (a) the physical characteristics of the tray, (b) the vapor and liquid flow rates, and (c) the mass-transfer characteristics of the fluid phases. Of all the physical dimensions of the tray only outlet weir height was found to be important enough to appear in the final correlation for E_G. (The effect of liquid path length on E_{MV} is taken into effect in the correlation to be presented later for the degree of liquid mixing which occurs on the tray.) Other tray design variables such as slot velocity, cap spacing, shape of caps, etc., were found to be relatively unimportant as far as E_G is concerned.

The mass-transfer characteristics of the vapor phase were found to be satisfactorily represented by the Schmidt number

$$N_{\text{Sc}} = \frac{\mu_G}{\rho_v D_G}$$

The liquid and vapor flow rates were characterized, respectively, by

$$\frac{q'}{l} = \frac{\text{gal of hot liquid}}{(\text{min})(\text{ft})}$$

and

$$F_{va} = U_a \sqrt{\rho_v} = \text{fps} \sqrt{\text{lb/ft}^3}$$

The effect of column pressure (outside its effect on F_{va} and N_{Sc}) was negligible.

The logarithmic relation in Eq. (16-62b) makes it simpler to relate N_G to the important variables rather than E_G. The empirical correlation for N_G suggested by the Manual is

$$N_G = \frac{0.776 + 0.116 h_w - 0.290 F_{va} + 0.0217 q'/l}{(N_{Sc})^{0.5}} \qquad (16\text{-}63)$$

Methods for the estimation of the vapor viscosity and diffusivity in N_{Sc} when experimental values are not available are presented in the Manual.

Equation (16-63) predicts only the vapor-phase mass-transfer resistance $1/N_G$. A correlation for the prediction of $1/N_L$ is given below. For the reasoning behind Eq. (16-63) and the limitations of the correlation, the reader is referred to the Manual.

Rate of Mass Transfer in the Liquid Phase. The relationship between E_{OL} and N_{OL} was derived in Sec. 16-8 and is shown as Eq. (16-39). If the derivation had been started with Eq. (16-21b) instead of (16-21d), the following equation would have been obtained:

$$1 - E_L = \exp(-N_L) = \exp\left(-\frac{k_L a \rho_L Z}{L_M}\right) \qquad (16\text{-}64a)$$

Equation (16-64a) relates the point efficiency to the number of transfer units when resistance to mass transfer exists only in the liquid phase $(x_i = x^*)$. The equation can be written in the alternate form

$$-2.3 \log(1 - E_L) = N_L = \frac{k_L \bar{a} \rho_L \epsilon}{L_M}$$
$$= \frac{k_L \bar{a} t_l}{3600} \qquad (16\text{-}64b)$$

where ϵ is the liquid holdup on the tray in cubic feet per square foot of bubbling area and t_l is the liquid residence time on the tray in seconds. In the case of plug flow of liquid and no vertical concentration gradient in either the liquid or vapor, $E_{ML} = E_{OL}$ (see Sec. 16-9) and E_{ML} can be substituted for E_L in Eq. (16-64b). For the other extreme case of completely mixed liquid,

$$E_{ML} = \frac{N_{OL}}{1 + N_{OL}} \qquad (16\text{-}40)$$

and since $E_{ML} = E_L$ and $N_{OL} = N_L$ for no gas-phase resistance, Eq. (16-64b) can be rewritten

$$\frac{E_L}{1 - E_L} = N_L = \frac{k_L \bar{a} \rho_L \epsilon}{L_M} = \frac{k_L \bar{a} t_l}{3600} \qquad (16\text{-}64c)$$

Equations (16-64b) and (16-64c) relate N_L and E_{ML} for two extreme cases of liquid mixing. The actual situation on the tray will usually be somewhere between these two extremes. For the in-between cases, E_{ML} can be related to N_L by means of Eq. (16-49) as described at the end of Sec. 16-10 for the case of no vapor-phase resistance.

Irrespective of the degree of liquid mixing, $1/N_L$ represents the resistance to mass transfer at a given point in the pool. The A.I.Ch.E. Manual suggests the following empirical correlations for the prediction of N_L for bubble-cap trays:

$$N_L = [(1.065 \times 10^4)D_L]^{0.5}(0.26F + 0.15)t_l \qquad (16\text{-}65a)$$

where D_L is the liquid diffusivity in square feet per hour. The t_l is the liquid contact time in seconds and was correlated by

$$t_l = \frac{37.4h_lZ_l}{q'/l} \qquad (16\text{-}66)$$

where h_l is the liquid holdup on the tray in cubic inches per square inch of bubbling area and Z_l is the length of liquid travel from the inlet downcomer to the outlet weir in feet. The h_l can be calculated from

$$h_l = 1.65 + 0.19h_w - 0.65F_{va} + 0.020\frac{q'}{l} \qquad (16\text{-}67a)$$

All the quantities except D_L in the preceding three equations can be evaluated from the design conditions.

Correlations for N_L and h_l for sieve trays are not given in the A.I.Ch.E. Manual. Based on the data in Fig. III-4, page 46, in the Delaware report, the following equation has been recommended by Gerster† for N_L:

$$N_L = 100D_L{}^{0.5}(0.49F_{va} + 0.17)t_l \qquad (16\text{-}65b)$$

For the liquid holdup on sieve trays,

$$h_l = 0.24 + 0.725h_w - 0.29h_wF_{va} + 0.010\frac{q'}{l} \qquad (16\text{-}67b)$$

can be used as an alternate to Eq. (16-67a).

In the University of Delaware report, Gerster et al. (8) suggest the following form of Eq. (16-65) for all trays (bubble cap and sieve):

$$N_L = [(4.13 \times 10^4)(D_L)]^{0.5}(0.26F + 0.15)t_L$$

Combination of N_G and N_L. Generally resistance to mass transfer will exist in both phases. The N_G and N_L predicted by Eqs. (16-63) and

† Equations (16-65b) and (16-67b) were recommended by J. A. Gerster in a private communication and do not appear in the Delaware report. Equation (16-67b) is based upon the air-water data of Foss and Gerster, *Chem. Eng. Progr.* **52**, 28-J (1956).

(16-65), respectively, are then combined to give the total resistance as shown at the end of Sec. 16-5.

$$\frac{1}{N_{OG}} = \frac{1}{N_G} + \frac{\lambda}{N_L} = \frac{\lambda}{N_{OL}} \qquad (16\text{-}27)$$

The N_{OG} is related to E_{OG} by

$$-2.3 \log (1 - E_{OG}) = N_{OG} \qquad (16\text{-}28)$$

If desired, E_{OL} can be obtained from E_{OG} by

$$E_{OG} = \frac{E_{OL}}{E_{OL} + \lambda(1 - E_{OL})} \qquad (16\text{-}42)$$

For further discussion of the assumptions and limitations of these equations beyond that represented in the previous sections, the reader is referred to the A.I.Ch.E. Manual.

Degree of Liquid Mixing. The relations between E_{MV} and E_{OG} for three different liquid mixing conditions were developed in Sec. 16-7. Generally, the liquid on the tray will be partially mixed and E_{MV} can be related to E_{OG} by Eq. (16-38) or, more easily, by Fig. 16-8. The parameter which characterizes the degree of mixing is the dimensionless Peclet number

$$N_{\text{Pe}} = \frac{Z_l^2}{D_E t_l} \qquad (16\text{-}31)$$

The residence time t_l is given by Eq. (16-66). The correlation developed for D_E by the A.I.Ch.E. research program is as follows. For 3-in.-round bubble caps on 4.5-in. triangular spacing and for sieve trays,

$$(D_E)^{0.5} = 0.0124 + 0.0171U_a + 0.0025\frac{q'}{l} + 0.015h_w \qquad (16\text{-}68)$$

For 6.5-in.-round bubble caps on 8.75-in. triangular spacing, $(D_E)^{0.5}$ was 1.154 times the value predicted by Eq. (16-68).

When λ is so large that the vapor-phase resistance is essentially absent, it may be more convenient to work in terms of the liquid quantities N_L and E_{ML}. The equations to be used in such a case have been developed at the end of Sec. 16-10.

Prediction of Entrainment. The correlations presented above for the prediction of the Murphree tray efficiencies E_{MV} and E_{ML} assume no entrainment. Colburn's equation to correct E_{MV} for entrainment was derived in Sec. 16-11. The Manual suggests that the entrainment be estimated from Fig. 16-11, where $e\sigma$ is plotted vs U_a/S'. The S' is given by

$$S' = \text{tray spacing (in.)} - h_f \qquad (16\text{-}69)$$

where h_f is the visual froth height in inches and can be estimated by

$$h_f = 2.53F_{va}^2 + 1.89h_w - 1.6 \qquad (16\text{-}70)$$

Equation (16-70) is based on limited data and assumes ample downcomer capacity with no foaming. Also the liquid rate should be below 50 gpm/ft, and the liquid-to-vapor density ratio should be above 60 if the equation is used.

Satisfactory entrainment measurements cannot be made in the small-diameter columns used in the A.I.Ch.E. research program. Therefore, the correlation in Fig. 16-11 is based mainly upon a reworking of literature data. The correlation is untested and was recommended by the Manual

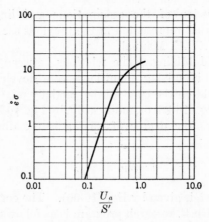

FIG. 16-11. Entrainment correlation. (*From "Bubble-tray Design Manual," American Institute of Chemical Engineers, New York.*)

with reservations. It was presented here only for the sake of completeness. For actual design work, it is recommended that the correlations for flooding and entrainment in Chaps. 14 and 15 be used. Figures 15-3 and 14-4 predict flooding points for sieve and bubble-cap trays, respectively. In Sec. 15-4, Fair suggests that a "per cent flood" of 80 to 85 per cent be used with Figs. 15-4 and 14-5 to predict the fractional entrainment in sieve and bubble-cap trays, respectively. The fractional entrainment can then be converted to the e' in Eq. (16-50) by means of Eq. (15-5) and the bubbling area A_a.

Application of Correlations. The specific details of the application of the correlations and relationships presented in this chapter are adequately described in Sec. IV of the A.I.Ch.E. Manual. Although the efficiency prediction method does not restrict the tray layout, the results may indicate that certain design changes can be made to improve efficiency without adversely affecting operability or throughput capacity. Methods for improving efficiency are discussed in Sec. V in the Manual. The extension of the binary prediction method to multicomponent mixtures is discussed in Sec. VI, and sample calculations are presented in Sec. IX.

NOMENCLATURE

A = cross-sectional area of contact volume, ft^2.

A_a = bubbling area, sq ft.

a = interfacial area per unit volume of gas and liquid holdup, $1/ft$. Also denotes activity in Eq. (16-15a) and a certain line segment in Fig. 16-10 in the derivation of Eq. (16-61).

a' = interfacial area per unit volume of gas holdup, $1/ft$.

$\bar{a}$ = interfacial area per unit volume of liquid holdup, $1/ft$.

b = constant.

b_{12} = coefficient of resistance for diffusion in a binary vapor mixture.

b'_{12} = coefficient of resistance for diffusion in a binary liquid mixture.

C = number of components.

c = concentration, lb moles/ft^3. Subscripts 1 and 2 denote components 1 and 2. Subscripts L and i denote concentrations in the bulk liquid phase and at the interface, respectively. Subscript m denotes mean concentration in film. Superscript * denotes equilibrium value.

c = certain grouping of terms in Eq. (16-57).

D = overhead product rate, lb moles/hr.

D_E = eddy diffusivity, ft^2/sec.

D_G = gas diffusivity, ft^2/hr.

D_L = liquid diffusivity, ft^2/hr.

d = differential operator. Also used to denote certain line segments in Fig. 16-10 in the derivation of Eq. (16-61).

E_a = Murphree tray efficiency corrected for entrainment.

E_G = point efficiency for all vapor-phase resistance.

E_L = point efficiency for all liquid-phase resistance.

E_{ML} = Murphree tray efficiency in liquid terms.

E_{MV} = Murphree tray efficiency in vapor terms.

E_o = over-all column efficiency.

E_{OG} = over-all point efficiency in vapor terms.

E_{OL} = over-all point efficiency in liquid terms.

$\dot{e}$ = entrainment, lb/lb of vapor.

e' = entrainment, moles/(hr)(ft^2).

e_M = entrainment, moles/mole of dry vapor V.

exp = exponent. Denotes exponent of natural logarithm base e.

F_{va} = F factor = $U_a \sqrt{\rho_v}$

G_M = gas flow rate per bubbling area, lb moles/(hr)(ft^2).

h_f = height of froth (aerated mass) on tray, in.

h_l = equivalent height of clear liquid on tray, in.

h_w = outlet weir height, in.

K_{OG} = over-all gas mass-transfer coefficient, lb moles/(hr)(ft^2)(atm).

K_{OL} = over-all liquid mass-transfer coefficient, ft/hr.

K_G = mass-transfer gas-film coefficient, lb moles/(hr)(ft^2)(atm).

k'_G = $k_G RT$ = ft/hr.

k_L = mass-transfer liquid-film coefficient, ft/hr.

L = liquid rate, lb moles/hr.

L_M = liquid flow rate per bubbling area, lb moles/(hr)(ft²).

l = average liquid flow width, ft.

M = molecular weight. Subscripts l and v refer to liquid and vapor, respectively.

m = slope of equilibrium curve = dy/dx.

N = total number of equilibrium stages.

N_t = total number of actual stages.

N_i = rate of diffusion of component i, lb moles/hr.

N_{Pe} = $Z_l^2/D_E t_l$ = Peclet number.

N_G = number of vapor-film transfer units.

N_L = number of liquid-film transfer units.

N_{OG} = number of over-all vapor transfer units.

N_{OL} = number of over-all liquid transfer units.

N_{Sc} = Schmidt number = $\mu_G/\rho_v D_G$ (gas Schmidt number).

p = total pressure, atm. Subscripts 1 and 2 denote partial pressures of components 1 and 2. Subscripts G and i denote partial pressures in bulk vapor phase and at the interface, respectively. Subscript m denotes mean partial pressure in film. Superscript * denotes equilibrium value.

q' = actual clear-liquid flow rate, gpm.

R = gas constant = 0.730 (atm)(ft³)/(lb mole)(°R).

r = ratio defined by Eq. (16-59).

r_0 = collision diameter, angstroms.

S = interfacial area, ft².

S' = distance between the top of froth and the tray above.

T = absolute temperature.

t_l = residence time of liquid in aerated mass on tray, sec.

t_v = residence time of vapor in aerated mass on tray, sec.

U_a = superficial vapor velocity based on bubbling area, fps.

U_f = froth velocity across tray, fph.

u = velocity.

V = vapor rate, lb moles/hr.

V_0 = molar volume at normal boiling point, cm³/g mole.

W = dG_M/G_M. Used in the derivation of Eq. (16-29) to represent a fraction of the vapor rate G_M. Also used in Chaps. 14 and 15 to denote vapor flow rate in lb/hr.

w = z/Z_l. Used in the derivation of Eq. (16-38). Also used in Chaps. 14 and 15 to denote liquid flow rate in lb/hr.

x = mole fraction in liquid phase at a given point. Subscript i denotes value at vapor-liquid interface.

x_n = mole fraction in total mixed liquid from stage n. Subscript $n + 1$ has same meaning for stage $n + 1$.

x^* = mole fraction in a liquid which would be in equilibrium with some given vapor. Subscript e denotes equilibrium with y_{n-1}.

Y = apparent vapor concentration. Defined by Eq. (16-54).

y = mole fraction in vapor phase at a given point. Subscript i denotes value at vapor-liquid interface.

y_n = mole fraction in total mixed vapor from stage n. Subscript $n - 1$ has same meaning for stage $n - 1$.

y^* = mole fraction in a vapor which would be in equilibrium with some given liquid.

Z = height of contact volume, ft.
Z_l = length of liquid travel across tray, ft.
z = distance.
z_G = gas film thickness, ft.
z_L = liquid film thickness, ft.

Greek Symbols

α = relative volatility.
β = gas holdup on tray, ft³/ft² of bubbling area. Also used in Chaps. 14 and 15 as aeration factor as defined by Eqs. (14-37) and (15-13).
ϵ = liquid holdup on tray, ft³/ft² of bubbling area.
ϵ/k = force constant parameter, °K.
η = grouping of terms defined in Eq. (16-35).
$\lambda = mG_M/L_M$.
μ_G = viscosity of vapor phase, lb/(hr)(ft).
μ_L = viscosity of liquid phase, centipoises.
ρ_F = froth density, ft³ of liquid/ft³ of froth.
ρ_v = mass density of vapor phase, lb/ft³.
ρ_V = molar density of vapor phase, lb moles/ft³.
ρ_L = molar density of liquid phase, lb moles/ft³.
σ = liquid surface tension, dynes/cm.

REFERENCES

1. Anderson, J. E.: Sc.D. Thesis, Massachusetts Institute of Technology, Cambridge, Mass., 1954.
2. Arnold, J. H.: *Ind. Eng. Chem.*, **22**, 1091 (1930).
3. "Bubble-tray Design Manual," American Institute of Chemical Engineers, New York.
4. Chilton, T. H., and A. P. Colburn: *Ind. Eng. Chem.*, **27**, 255 (1935).
5. Colburn, A. P.: *Ind. Eng. Chem.*, **28**, 526 (1936).
6. Drickamer, H. G., and J. R. Bradford: *Trans. A.I.Ch.E.*, **39**, 319 (1943).
7. Gatreaux, M. F., and H. E. O'Connell: *Chem. Eng. Progr.*, **51**, 232 (1955).
8. Gerster, J. A., A. B. Hill, N. N. Hochgraf, and D. G. Robinson: "Tray Efficiencies in Distillation Columns," Final Report from University of Delaware, American Institute of Chemical Engineers, New York, 1958.
9. Gilliland, E. R.: *Ind. Eng. Chem.*, **26**, 681 (1934).
10. Kirschbaum, E.: "Distillation and Rectification," p. 276, trans. by M. Wulfing-hoff, Brooklyn Chemical Publishing Co., 1948.
11. Lewis, W. K.: *Ind. Eng. Chem.*, **28**, 399 (1936).
12. Lewis, W. K., and K. C. Chang: *Trans. A.I.Ch.E.*, **21**, 135 (1928).
13. Maxwell, J. C.: "Scientific Papers," vol. 2, Cambridge University Press, New York, 1890.
14. Murphree, E. V.: *Ind. Eng. Chem.*, **17**, 747 (1925).
15. Nord, M.: *Trans. A.I.Ch.E.*, **42**, 863 (1946).
16. O'Connell, H. E.: *Trans. A.I.Ch.E.*, **42**, 741 (1946).
17. Oliver, E. D., and C. C. Watson: *A.I.Ch.E. J.*, **2**, 18 (1956).
18. Perry, J. H., "Chemical Engineers' Handbook," 3d ed., pp. 610–617, McGraw-Hill Book Company, Inc., New York, 1950.
19. Schoenborn, E. M., C. A. Plank, and C. E. Winslow: "Tray Efficiencies in Dis-

tillation Columns," Final Report from North Carolina State College, American Institute of Chemical Engineers, New York, 1959.
20. Sherwood, T. K., and R. L. Pigford: "Absorption and Extraction," 2d ed., McGraw-Hill Book Company, Inc., New York, 1952.
21. Stefan: *Sitzber. Akad. Wiss. Wien*, **63**(2), 63, 323 (1872).
22. Warzel, L. A., Ph.D. Thesis, University of Michigan, Ann Arbor, Mich., 1955.
23. Wehner, J. F., and R. H. Wilhelm: *Chem. Eng. Sci.*, **6**, 89 (1956).
24. West, F. B., W. D. Gilbert, and T. Shimizu: *Ind. Eng. Chem.*, **44**, 2470 (1952).
25. Whitman, W. G.: *Chem. Met. Eng.*, **29**(4) (July 23, 1923).
26. Wilke, C. R.: *Chem. Eng. Progr.*, **45**, 218 (1949).
27. Wilke, C. R.: *Chem. Eng. Progr.*, **46**, 95 (1950).
28. Wilke, C. R., and Pin Chang: *A.I.Ch.E. J.*, **1**, 264 (1955).
29. Wilke, C. R., and C. Y. Lee: *Ind. Eng. Chem.*, **47**, 1253 (1955).
30. Williams, B., and K. Gordon: "Tray Efficiency in Distillation Columns," Final Report from the University of Michigan, American Institute of Chemical Engineers, New York, 1961.

APPENDIX A

A-1. Least-squares Determination of Constants

Repeated measurements of a given quantity will differ from one another because of errors in the experimental method and technique. This variation causes the experimental data to "scatter" around some true value. According to the principle of least squares, the most probable value of the quantity which can be obtained from several experimental measurements of equal absolute precision is that value which causes the sum of the squares of the deviations (differences between the most probable value and the individual experimental points) to be a minimum. The application of this principle to the evaluation of constants in a correlating equation is illustrated below after the manner of Hougen and Watson (1). For the development of the principle of least squares from probability theory, the reader is referred to Mickley, Sherwood, and Reed (2) or other texts which deal with probability.

The correlating equation should first be rearranged (if necessary) to make it linear with respect to the constants which are to be determined. Let $a, b, c, \ldots$ represent the unknown constants which will be used to predict y from the experimentally determined quantities $x_b, x_c, \ldots$ by means of the general linear equation

$$y = a + bx_b + cx_c + \cdots \qquad (A\text{-}1)$$

At any given set of $x_b, x_c, \ldots$ the deviation d of the y predicted by (A-1) from the experimentally determined value y_e is given by

$$d = a + bx_b + cx_c + \cdots - y_e \qquad (A\text{-}2)$$

The sum of the squares of the deviations from n experimental points is

$$\Sigma d^2 = d_1{}^2 + d_2{}^2 + \cdots + d_n{}^2 \qquad (A\text{-}3)$$

where the subscripts refer to the individual experimental points. The principle of least squares states that the best fit of the experimental data is obtained when Σd^2 is a minimum. Since Σd^2 is a function of the unknowns $a, b, c, \ldots$, the derivatives $\partial(\Sigma d^2)/\partial a$, $\partial(\Sigma d^2)/\partial b$, $\partial(\Sigma d^2)/\partial c$, $\ldots$ must be zero when Σd^2 is a minimum. Therefore, from Eq. (A-3),

$$\frac{\partial(\Sigma d^2)}{\partial a} = 2\left(d_1 \frac{\partial d_1}{\partial a} + d_2 \frac{\partial d_2}{\partial a} + \cdots + d_n \frac{\partial d_n}{\partial a}\right) = 0 \qquad (A\text{-}4)$$

Similar equations must be written for each of the other unknown constants. The partial derivatives $\partial d/\partial a$, $\partial d/\partial b$, $\partial d/\partial c$, . . . are evaluated from Eq. (A-2). Since $\partial d/\partial a = 1.0$, substitution in (A-4) gives

$$\frac{\partial(\Sigma d^2)}{\partial a} = 0 = 2(a + bx_b + cx_c + \cdots - y_e)_1(1.0) + \cdots$$
$$+ 2(a + bx_b + cx_c + \cdots - y_e)_n(1.0)$$

Letting i refer to any experimental point, this equation can be written more compactly as

$$\frac{\partial(\Sigma d^2)}{\partial a} = 0 = \sum_{i=1}^{n} (a + bx_b + cx_c + \cdots - y_e)_i \qquad (A\text{-}5a)$$

Similarly, since $\partial d/\partial b = x_b$,

$$\frac{\partial(\Sigma d^2)}{\partial b} = 0 = \sum_{i=1}^{n} (ax_b + bx_b{}^2 + cx_cx_b + \cdots - y_ex_b)_i \qquad (A\text{-}5b)$$

Since $\partial d/\partial c = x_c$,

$$\frac{\partial(\Sigma d^2)}{\partial c} = 0 = \sum_{i=1}^{n} (ax_c + bx_bx_c + cx_c{}^2 + \cdots - y_ex_c)_i \qquad (A\text{-}5c)$$

Similar equations must be written for any other unknown constants. Equations (A-5) can be rewritten with like terms grouped together.

$$na + b\Sigma x_b + c\Sigma x_c + \cdots - \Sigma y_e = 0 \qquad (A\text{-}6a)$$
$$a\Sigma x_b + b\Sigma x_b{}^2 + c\Sigma x_cx_b + \cdots - \Sigma y_ex_b = 0 \qquad (A\text{-}6b)$$
$$a\Sigma x_c + b\Sigma x_bx_c + c\Sigma x_c{}^2 + \cdots - \Sigma y_ex_c = 0 \qquad (A\text{-}6c)$$

where the summations are made from $i = 1$ to $i = n$. The resulting set of simultaneous equations is solved to provide values of a, b, c, ...

Phase-equilibrium data are seldom of the same absolute precision through the range of concentration. Obviously incorrect points should be discarded (given a weight of zero). Otherwise the correlation curve may be so distorted as to be completely unrealistic.

Example A-1. The three-constant Redlich-Kister binary equation can be written as

$$\log \frac{\gamma_1}{\gamma_2} = B_{12}(1 - 2x_1) + C_{12}(6x_1x_2 - 1) + D_{12}(1 - 2x_1)(1 - 8x_1x_2)$$

Derive the equations which when solved will provide values of B_{12}, C_{12}, and D_{12} which will cause the Redlich-Kister equation to predict the most probable value of $\log(\gamma_1/\gamma_2)$ for any given x_1.

Solution. Denote the number of experimental data points (γ_1 and γ_2 vs x) by n. The deviation for any point i will be defined as

$$d_i = \left[B_{12}(1 - 2x_1) + C_{12}(6x_1x_2 - 1) + D_{12}(1 - 2x_1)(1 - 8x_1x_2) - \log \frac{\gamma_1}{\gamma_2} \right]_i$$

where the log (γ_1/γ_2) is the experimental value. Since

$$\frac{\partial d}{\partial B_{12}} = 1 - 2x_1$$

$$\frac{\partial d}{\partial C_{12}} = 6x_1x_2 - 1$$

$$\frac{\partial d}{\partial D_{12}} = (1 - 2x_1)(1 - 8x_1x_2)$$

substitution in Eq. (A-4) provides

$$\sum_{i=1}^{n} \left[B_{12}(1 - 2x_1)^2 + C_{12}(1 - 2x_1)(6x_1x_2 - 1) \right. $$
$$\left. + D_{12}(1 - 2x_1)^2(1 - 8x_1x_2) - (1 - 2x_1) \log \frac{\gamma_1}{\gamma_2} \right]_i = 0$$

$$\sum_{i=1}^{n} \left[B_{12}(1 - 2x_1)(6x_1x_2 - 1) + C_{12}(6x_1x_2 - 1)^2 \right. $$
$$\left. + D_{12}(1 - 2x_1)(1 - 8x_1x_2)(6x_1x_2 - 1) - (6x_1x_2 - 1) \log \frac{\gamma_1}{\gamma_2} \right]_i = 0$$

$$\sum_{i=1}^{n} \left[B_{12}(1 - 2x_1)^2(1 - 8x_1x_2) + C_{12}(6x_1x_2 - 1)(1 - 2x_1)(1 - 8x_1x_2) \right. $$
$$\left. + D_{12}(1 - 2x_1)^2(1 - 8x_1x_2)^2 - (1 - 2x_1)(1 - 8x_1x_2) \log \frac{\gamma_1}{\gamma_2} \right]_i = 0$$

Rearrangement to get the unknown constants outside the summation signs gives the following equations which would be solved simultaneously to give B_{12}, C_{12}, and D_{12}:

$$B_{12}\Sigma(1 - 2x_1)^2 + C_{12}\Sigma(1 - 2x_1)(6x_1x_2 - 1)$$
$$+ D_{12}\Sigma(1 - 2x_1)^2(1 - 8x_1x_2) = \Sigma(1 - 2x_1) \log \frac{\gamma_1}{\gamma_2}$$

$$B_{12}\Sigma(1 - 2x_1)(6x_1x_2 - 1) + C_{12}\Sigma(6x_1x_2 - 1)^2$$
$$+ D_{12}\Sigma(1 - 2x_1)(6x_1x_2 - 1)(1 - 8x_1x_2) = \Sigma(6x_1x_2 - 1) \log \frac{\gamma_1}{\gamma_2}$$

$$B_{12}\Sigma(1 - 2x_1)^2(1 - 8x_1x_2) + C_{12}\Sigma(1 - 2x_1)(6x_1x_2 - 1)(1 - 8x_1x_2)$$
$$+ D_{12}\Sigma(1 - 2x_1)^2(1 - 8x_1x_2)^2 = \Sigma(1 - 2x_1)(1 - 8x_1x_2) \log \frac{\gamma_1}{\gamma_2}$$

A-2. Derivation of General Short-cut Equation (8-3)

The nomenclature for this material is at the end of Chap. 8.

All the equations in the derivation are written for any given component. A component subscript is omitted for the sake of simplicity.

Write a component balance around stage $n + 1$ to obtain

$$L_{n+2}x_{n+2} + V_n y_n = L_{n+1}x_{n+1} + V_{n+1}y_{n+1}$$

Since $y_n = K_n x_n$ and $y_{n+1} = K_{n+1}x_{n+1}$,

$$L_{n+2}x_{n+2} + V_n K_n x_n = L_{n+1}x_{n+1} + V_{n+1}K_{n+1}x_{n+1}$$

Rearranging gives

$$x_{n+2} - \left(\frac{K_{n+1}V_{n+1}}{L_{n+2}} + \frac{L_{n+1}}{L_{n+2}}\right)x_{n+1} + \frac{K_n V_n}{L_{n+2}}x_n = 0$$

The rates and K's within the column section under consideration must be assumed constant if a simple mathematical solution of the difference equation is desired. Making these assumptions and writing the equation in operator form give

$$\left[E^2 - \left(\frac{KV}{L} + 1\right)E + \frac{KV}{L}\right]x_n = 0$$

or

$$\left(E - \frac{KV}{L}\right)(E - 1)x_n = 0$$

The properties of the E operator are discussed by Wylie (3).

Let S_1 and S_2 represent the two roots

$$S_1 = \frac{KV}{L} \qquad \text{and} \qquad S_2 = 1.0$$

The solution is then

$$x_n = c_1(S_1)^n + c_2$$

Since n refers to any stage above the intermediate feed stage (raffinate end in extraction), the subscript on the stripping factor S_1 will be changed to n to denote the upper section stripping factor for the component for which the equation is written. Then for the upper (or raffinate) section,

$$x_n = c_1(S_n)^n + c_2 \tag{A-7a}$$

For the lower (or extract) section by analogy

$$x_m = c_3(S_m)^m + c_4 \tag{A-8a}$$

These equations relate the liquid concentration of the given component to the stage number which appears as the exponent on the stripping factor.

The constants in Eq. (A-7a) can be eliminated as follows: For the given component

$$V_N y_N = D y_N + R D y_N = (1 + R) D y_N$$

By definition,

$$D y_N = (1 - f) A$$

where A is the total amount of the given component entering the column. So

$$y_N = \frac{(1 + R)(1 - f)A}{V_N}$$

and

$$x_N = \frac{(1 + R)(1 - f)A}{K_N V_N}$$

Substitution for x_N in Eq. (A-7a) gives

$$\frac{(1 + R)(1 - f)A}{KV} = c_1 (S_n)^N + c_2 \qquad (A-7b)$$

An expression for x_{N-1} can be obtained from a balance around stage N.

$$y_{N-1} = \frac{L_N x_N}{V_{N-1}} + \frac{V_N y_N}{V_{N-1}} - \frac{L_{N+1} x_{N+1}}{V_{N-1}}$$

Since $L_{N+1} x_{N+1} = S x_s + R D y_N$

and $q_s A = S x_s$

and $D = \dfrac{V_N}{1 + R}$

then $L_{N+1} x_{N+1} = q_s A + \dfrac{R}{1 + R} V_N y_N$

Substituting for x_N, y_N, and $L_{N+1} x_{N+1}$ in the balance around stage N and dropping the subscripts on the rate terms give

$$y_{N-1} = \frac{(1 - f)A}{V^2} \left[V(1 + R) + \frac{L}{K}(1 + R) - RV - \frac{V q_s}{1 - f} \right]$$

Substitution for x_{N-1} in Eq. (A-7a) gives

$$x_{N-1} = \frac{(1 - f)A}{KV^2} \left[V(1 + R) + \frac{L}{K}(1 + R) - RV - \frac{V q_s}{(1 - f)} \right]$$
$$= c_1 (S_n)^{N-1} + c_2 \qquad (A-7c)$$

Subtracting (A-7b) from (A-7c) and solving for c_1 give

$$c_1 = \frac{(1 - f)A[(L/K)(1 + R) - RV - V q_s/(1 - f)]}{KV^2 (S_n^{N-1} - S_n^N)}$$

The constant c_2 can be expressed in terms of c_1 by rearranging (A-7b).

$$c_2 = \frac{(1 + R)(1 - f)A}{KV} - c_1 S_n^N$$

Substituting for c_1 and c_2 in Eq. (A-7a) gives the following equation for the upper (or raffinate) section of the column:

$$x_n = \frac{(1-f)A[(L/K)(1+R) - RV - Vq_s/(1-f)]}{KV^2(S_n{}^N - S_n{}^{N-1})}(S_n{}^N - S_n{}^n)$$
$$+ \frac{(1+R)(1-f)A}{KV} \quad (A-9)$$

A similar procedure can be used to eliminate c_3 and c_4 from Eq. (A-8a). For the given component,

$$L_1 x_1 = R'B x_B + S_E x_{S_E} + B x_B$$

By definition,

$$fA = S_E x_{S_E} + B x_B$$

so

$$L_1 x_1 = R'B x_B + fA$$

Defining

$$g = \frac{S_E x_{S_E}}{L_1 x_1}$$

it can be shown by a material balance that

$$B x_B = \frac{1-g}{1+R'} L_1 x_1$$

and from this

$$x_1 = \frac{(1+R')fA}{(1+gR')L_1}$$

Substituting for x_1 in Eq. (A-8a) and denoting the assumed constant extract rate in the extract end by L' give

$$\frac{(1+R')fA}{(1+gR')L'} = c_3 S_m + c_4 \quad (A-8b)$$

An expression for x_2 can be obtained from a balance around stage 1.

$$x_2 = \frac{V_1 y_1}{L_2} + \frac{L_1 x_1}{L_2} - \frac{V_0 y_0}{L_2}$$

Since

$$V_0 y_0 = F' y_{F'} + R'B x_B = q_{F'}A + \frac{R'(1-g)L_1 x_1}{1+R'}$$

and

$$y_1 = K_1 x_1 = \frac{K_1(1+R')fA}{(1+gR')L_1}$$

the following expression for x_2 is obtained after dropping the stage subscripts:

$$x_2 = \frac{fA}{(L')^2(1+gR')}\left[K'V'(1+R') + L'(1+R') - \frac{L'(1+gR')q_{F'}}{f} - L'R'(1-g) \right]$$

Substituting for x_2 in Eq. (A-8a) gives

$$\frac{fA}{(L')^2(1+gR')}\left[K'V'(1+R)+L'(1+R')-L'R'(1-g)\right.$$
$$\left.-\frac{L'(1+gR')q_{F'}}{f}\right]=c_3(S_m)^2+c_4 \quad \text{(A-8c)}$$

Subtracting (A-8b) from (A-8c) and solving for c_3 give

$$c_3=\frac{fA[K'V'(1+R')-L'R'(1-g)-L'(1+gR')q_{F'}/f]}{(L')^2(1+gR')(S_m{}^2-S_m)}$$

Equation (A-8b) can be rearranged to give an expression for c_4 in terms of c_3:

$$c_4=\frac{(1+R')fA}{(1+gR')L'}-c_3S_m$$

Substituting for c_4 in Eq. (A-8a) gives

$$x_m=c_3(S_m{}^m-S_m)+\frac{(1+R')fA}{(1+gR')L'}$$

Substituting for c_3 and canceling an S_m in the first term give the following equation for the lower (or extract) section:

$$x_m=\frac{fA[K'V'(1+R')-L'R'(1-g)-L'(1+gR')q_{F'}/f]}{(L')^2(1+gR')(S_m-1)}(S_m{}^{m-1}-1)$$
$$+\frac{(1+R')fA}{(1+gR')L'} \quad \text{(A-10)}$$

The equation for the entire column is obtained by combining Eqs. (A-9) and (A-10) with a component balance around the feed stage.

$$V_M y_M+L_{M+2}x_{M+2}+Fy_F=V_{M+1}y_{M+1}+L_{M+1}x_{M+1}$$

Since the rates are assumed constant within each section, and since $x_{M+1}=y_{M+1}/K_{M+1}$,

$$V'y_M+Lx_{M+2}+Fy_F=Vy_{M+1}+\frac{L'y_{M+1}}{K_{M+1}}$$

The subscript on the feed stage K will be retained until the decision is made to regard it as K or K'.

The y_{M+1} can be eliminated by a balance cutting V_N, L_{N+1}, V_{M+1}, and L_{M+2} and remembering that

$$L_{N+1}x_{N+1}=q_sA+\frac{R}{1+R}V_N y_N$$

and

$$y_N=\frac{(1+R)(1-f)A}{V_N}$$

Then

$$y_{M+1}=\frac{(1-f)A}{V}+\frac{L}{V}x_{M+2}-\frac{q_sA}{V}$$

Substituting for y_{M+1} in the feed-stage balance and replacing $F y_F$ with $q_F A$ give

$$V' y_M + q_F A = (1 - f)A - q_s A + \frac{L'(1 - f)A}{K_{M+1} V} + \frac{LL' x_{M+2}}{K_{M+1} V} - \frac{L' q_s A}{K_{M+1} V}$$

Using $y_M = K' x_M$ and Eq. (A-10) to substitute for y_M and Eq. (A-9) to substitute for x_{M+2} gives the following equation for the entire column:

$$\frac{K'V'fA[K'V'(1 + R') - L'R'(1 - g) - q_{F'}L'(1 + gR')/f]}{(L')^2(1 + gR')(S_m - 1)}(S_m{}^{M-1} - 1)$$

$$+ \frac{K'V'(1 + R')fA}{L'(1 + gR')} + q_F A = (1 - f)A - q_s A + \frac{L'(1 - f)A}{K_{M+1} V} - \frac{L' q_s A}{K_{M+1} V}$$

$$+ \frac{LL'(1 - f)A[(L/K)(1 + R) - RV - Vq_s/(1 - f)]}{K_{M+1} K V^3 (S_n{}^N - S_n{}^{N-1})}(S_n{}^N - S_n{}^{M+2})$$

$$+ \frac{LL'(1 + R)(1 - f)A}{K_{M+1} K V^2}$$

This equation after considerable algebraic manipulation was simplified to the following final form which is Eq. (8-3):

$$f = \frac{(1 - S_n{}^{N-M}) + q_s(S_n{}^{N-M} - S_n) + R(1 - S_n) + h q_{F'} S_n{}^{N-M}(1 - S_m{}^M)}{(1 - S_n{}^{N-M}) + h S_n{}^{N-M}(1 - S_m{}^M) + R(1 - S_n)}$$
$$+ h[(1 + R')/(1 + gR')]S_m{}^M S_n{}^{N-M}(1 - S_m)$$

(8-3)

If $K_{M+1} = K$: $\quad h = \dfrac{L - KV}{L' - K'V'} = \dfrac{L}{L'} \dfrac{1 - S_n}{1 - S_m}$

If $K_{M+1} = K'$: $\quad h = \dfrac{L/K - V}{L'/K' - V'} = \dfrac{K'}{K} \dfrac{L}{L'} \dfrac{1 - S_n}{1 - S_m}$

The expression used for h depends upon the condition of the intermediate feed. For extraction and for distillation with a vapor feed, the feed is more similar to the light or raffinate phase and $K_{M+1} = K$. For a bubble-point feed, $K_{M+1} = K'$ should be used.

REFERENCES

1. Hougen, O. A., and K. M. Watson: "Chemical Process Principles," Part III, Kinetics and Catalysis, 1st ed., pp. 938–940, John Wiley & Sons, Inc., New York, 1947.
2. Mickley, H. S., T. K. Sherwood, and C. E. Reed: "Applied Mathematics in Chemical Engineering," pp. 95–99, 2d ed., McGraw-Hill Book Company, Inc., New York, 1957.
3. Wylie, Jr., C. R.: "Advanced Engineering Mathematics," 1st ed., pp. 502–503, McGraw-Hill Book Company, Inc., New York, 1951.

APPENDIX B

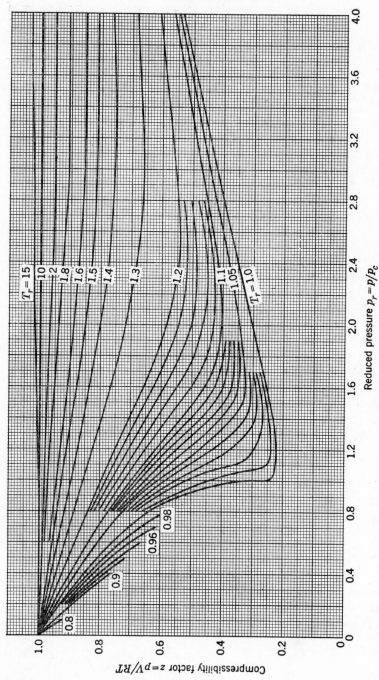

Reduced pressure $p_r = p/p_c$

Compressibility factor $z = pV/RT$

$T_r = 15$
10
2
1.8
1.6
1.5
1.4
1.3
1.2
1.1
1.05
$T_r = 1.0$

0.8
0.9
0.96
0.98

FIG. B-1a. Compressibility-factor chart for critical compressibility factor ($z_c = p_cV_c/RT_c$) of 0.27. For compounds with $z_c = 0.23$, 0.25, and 0.29 see the original tables of Lyderson, Greenkorn, and Hougen, "Generalized Thermodynamic Properties of Pure Fluids," Report 4, Engineering Experiment Station, University of Wisconsin, Madison, Wis., 1955.

623

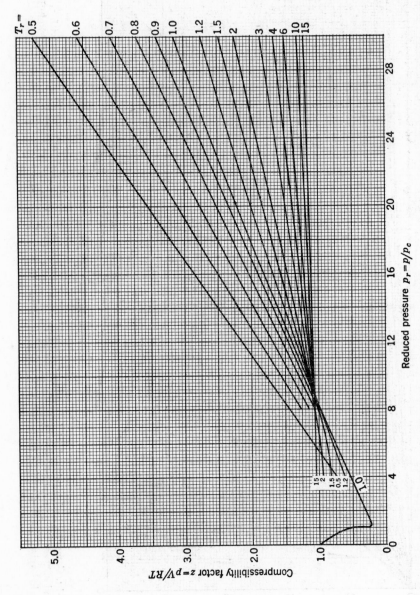

$T_r =$
0.5
0.6
0.7
0.8
0.9
1.0
1.2
1.5
2
3
4
6
10
15

Compressibility factor $z = pV/RT$

Reduced pressure $p_r = p/p_c$

15
2
1.5
0.5
1.2
1.0

FIG. B-1b. Compressibility-factor chart for critical compressibility factor ($z_c = p_c V_c / R T_c$) of 0.27. For compounds with $z_c = 0.23$, 0.25, and 0.29 see the original tables of Lyderson, Greenkorn, and Hougen, "Generalized Thermodynamic Properties of Pure Fluids," Report 4, Engineering Experiment Station, University of Wisconsin, Madison, Wis., 1955.

624

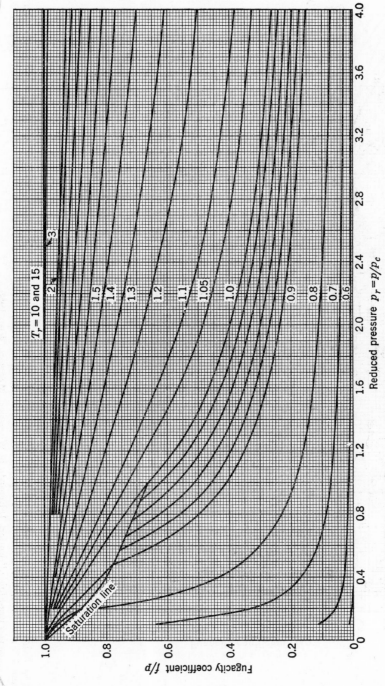

FIG. B-2a. Fugacity coefficient f/p chart for critical compressibility factor ($z_c = p_c V_c / R T_c$) of 0.27. For compounds with $z_c = 0.23$, 0.25, and of 0.29 see the original tables of Lyderson, Greenkorn, and Hougen, "Generalized Thermodynamic Properties of Pure Fluids," Report 4, Engineering Experiment Station, University of Wisconsin, Madison, Wis., 1955.

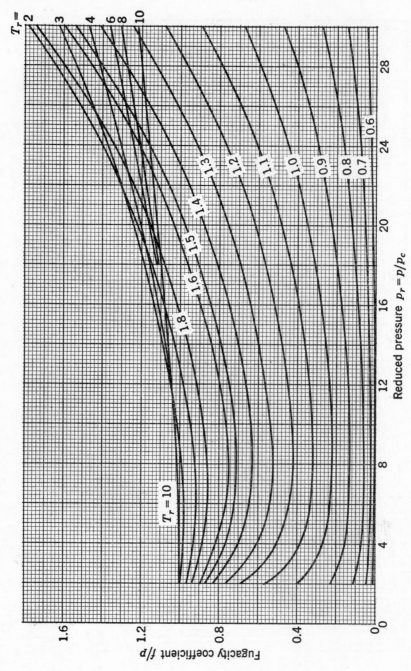

FIG. B-2b. Fugacity-coefficient (f/p) chart for critical compressibility factor $(z_c = p_c V_c / R T_c)$ of 0.27. For compounds with $z_c = 0.23, 0.25,$ and 0.29 see the original tables of Lyderson, Greenkorn, and Hougen, "Generalized Thermodynamic Properties of Pure Fluids," Report 4, Engineering Experiment Station, University of Wisconsin, Madison, Wis., 1955.

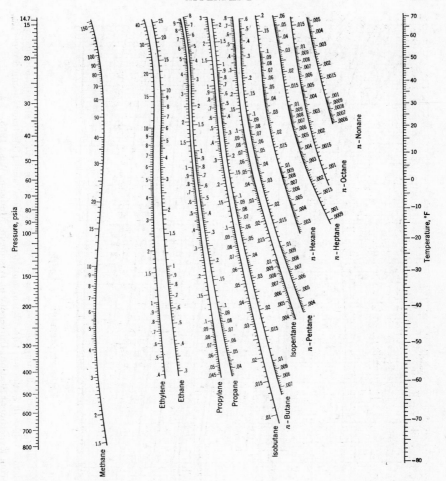

F IG. B-3a. Distribution coefficients ($K = y/x$) in light hydrocarbon systems, low-temperature range. [*C. L. Depriester, Chemical Eng. Progr. Symposium Ser.*, **49**(7), 1 (1953).]

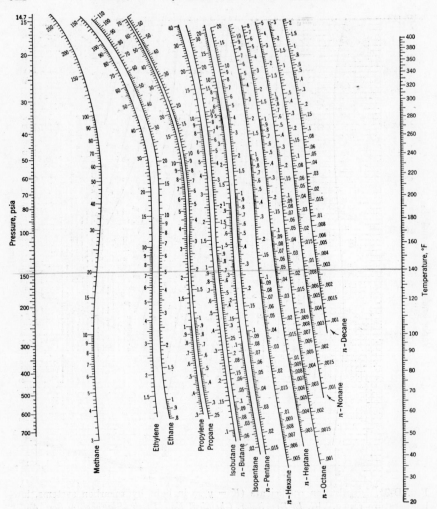

FIG. B-3b. Distribution coefficients $(K = y/x)$ in light hydrocarbon systems, high-temperature range. [*C. L. Depriester, Chem. Eng. Progr. Symposium Ser.*, **49**(7), 1 (1953).]

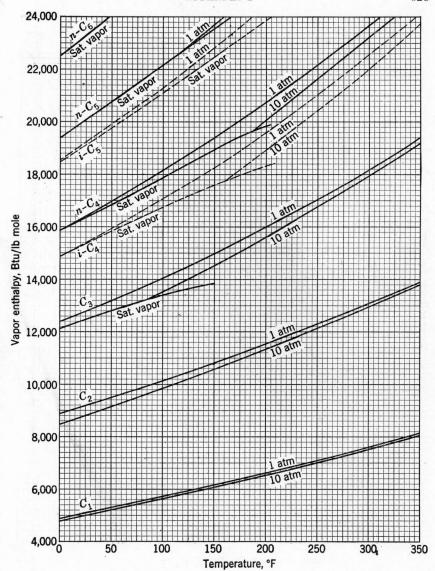

FIG. B-4a. Enthalpies of hydrocarbon gases, C_1 to C_6. Reference state is saturated liquid at $-200°F$. Dashed lines are used for isoparaffins. (*Partial representation of data from Maxwell, "Data Book on Hydrocarbons," pp. 98–107, D. Van Nostrand Company, Inc., Princeton, N.J., 1950.*)

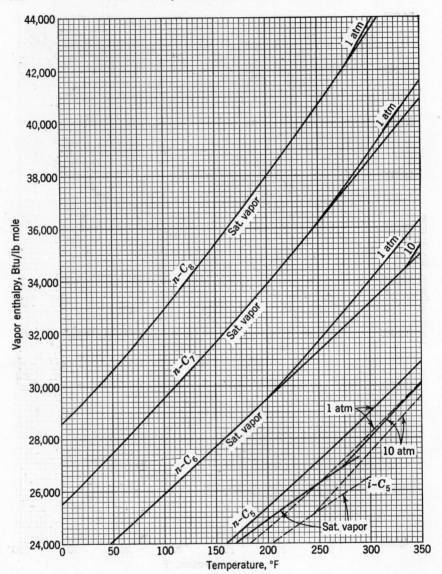

Fig. B-4b. Enthalpies of hydrocarbon gases, C₅ to C₈. Reference state is saturated liquid at −200°F. Dashed lines are used for isoparaffins. (*Partial representation of data from Maxwell, "Data Book on Hydrocarbons," pp. 98–107, D. Van Nostrand Company, Inc., Princeton, N.J., 1950.*)

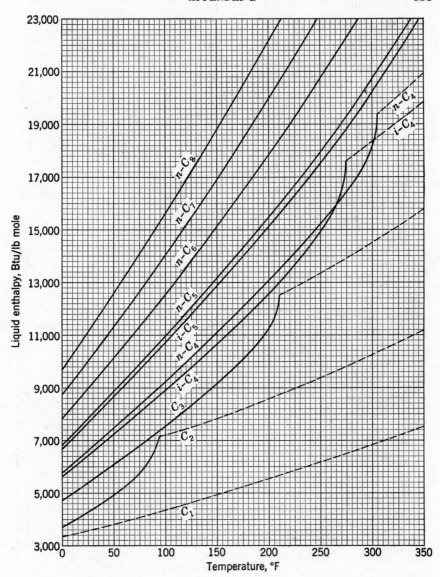

FIG. B-5. Enthalpies of hydrocarbon liquids, C_1 to C_8. Reference state is saturated liquid at $-200°F$. Solid lines are for saturated liquid; dashed lines are for gas in solution above critical point. (*Partial representation of data from Maxwell, "Data Book on Hydrocarbons," pp. 98–107, D. Van Nostrand Company, Inc., Princeton, N.J., 1950.*)

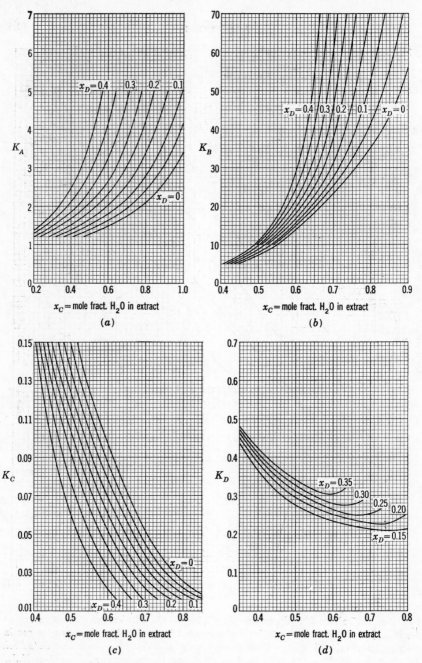

Fig. B-6. (a) Acetone-distribution coefficients ($K = y/x$ where y refers to raffinate phase) vs extract composition in the acetone (A)–chloroform (B)–water (C)–acetic acid (D) system; (b) chloroform-distribution coefficients vs extract composition; (c) water-distribution coefficients vs extract composition; (d) acetic acid distribution coefficients vs extract composition. [Data of Brancker, Hunter, and Nash, J. Phys. Chem., **44**, 683 (1940), with some tie lines calculated by J. C. Smith, Ind. Eng. Chem., **36**, 68 (1944).]

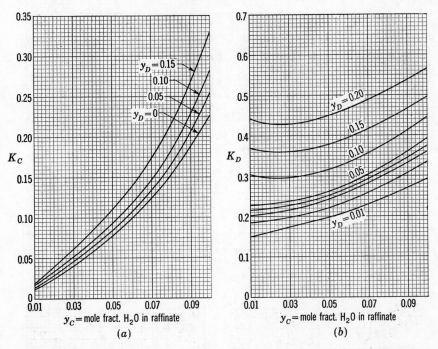

FIG. B-7. (a) Water-distribution coefficients vs raffinate composition; (b) acetic acid distribution coefficients vs raffinate composition. (See Fig. B-6 for system and literature references.)

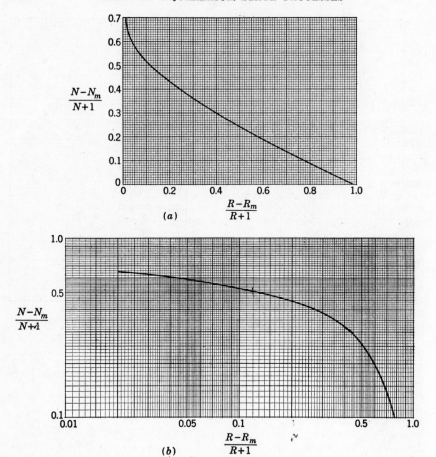

FIG. B-8. (a) Gilliland chart; (b) Gilliland chart on log-log scales. [*Gilliland, Ind. Eng. Chem.*, **32**, 1220 (1940).]

NAME INDEX

SUBJECT INDEX

639

DAy
Ro

AUG 9 - '93